STUDENT'S
SOLUTIONS MANUAL
JUDITH A. PENNA

ELEMENTARY ALGEBRA
CONCEPTS AND APPLICATIONS
SIXTH EDITION

Marvin L. Bittinger
Indiana University—Purdue University at Indianapolis

David J. Ellenbogen
Community College of Vermont

Addison
Wesley

Boston San Francisco New York
London Toronto Sydney Tokyo Singapore Madrid
Mexico City Munich Paris Cape Town Hong Kong Montreal

ISBN   0-201-65871-2

2 3 4 5 6 7 8 9 10 VG 04 03 02 01

Contents

Chapter 1

Introduction to Algebraic Expressions

Exercise Set 1.1

1. Substitute 9 for a and multiply.
   $3a = 3 \cdot 9 = 27$

2. 56

3. Substitute 2 for t and add.
   $t + 6 = 2 + 6 = 8$

4. 4

5. $\dfrac{x+y}{4} = \dfrac{2+14}{4} = \dfrac{16}{4} = 4$

6. 5

7. $\dfrac{m-n}{2} = \dfrac{20-6}{2} = \dfrac{14}{2} = 7$

8. 3

9. $\dfrac{a}{b} = \dfrac{45}{9} = 5$

10. 6

11. $\dfrac{9m}{q} = \dfrac{9 \cdot 6}{18} = \dfrac{54}{18} = 3$

12. 3

13. $bh = (6 \text{ ft})(4 \text{ ft})$
    $= (6)(4)(\text{ft})(\text{ft})$
    $= 24 \text{ ft}^2$, or 24 square feet

14. 24 hr

15. $A = \dfrac{1}{2}bh$

    $= \dfrac{1}{2}(5 \text{ cm})(6 \text{ cm})$

    $= \dfrac{1}{2}(5)(6)(\text{cm})(\text{cm})$

    $= \dfrac{5}{2} \cdot 6 \text{ cm}^2$

    $= 15 \text{ cm}^2$, or 15 square centimeters

16. (a) 150 sec;
    (b) 450 sec;
    (c) 10 min

17. $\dfrac{h}{a} = \dfrac{10}{37}$, or about 0.270

18. 26 cm^2

19. Let a represent Jan's age. Then we have $j + 8$, or $8 + j$.

20. $4a$

21. $b + 6$, or $6 + b$

22. Let w represent Lou's weight; $w + 7$, or $7 + w$

23. $c - 9$

24. $d - 4$

25. $q + 6$, or $6 + q$

26. $z + 11$, or $11 + z$

27. Let s represent Phil's speed. Then we have $9s$, or $s9$.

28. $d + c$, or $c + d$

29. $y - x$

30. Let a represent Lorrie's age; $a - 2$

31. $x \div w$, or $\dfrac{x}{w}$

32. Let s and t represent the numbers; $s \div t$, or $\dfrac{s}{t}$

33. $n - m$

34. $q - p$

35. Let l and h represent the box's length and height, respectively. Then we have $l + h$, or $h + l$.

36. $d + f$, or $f + d$

37. $9 \cdot 2m$

38. Let p represent Paula's speed and w represent the wind speed; $p - 2w$

39. Let y represent "some number." Then we have $\dfrac{1}{4}y$, or $\dfrac{y}{4}$, or $y/4$, or $y \div 4$.

40. Let m and n represent the numbers; $\frac{1}{3}(m+n)$, or $\frac{m+n}{3}$

41. Let x represent the number of women attending. Then we have 64% of x, or $0.64x$.

42. Let y represent "a number;" 38% of y, or $0.38y$

43. $\$50 - x$

44. $65t$ mi

45.
$$\underline{x + 17 = 32} \quad \text{Writing the equation}$$
$$15 + 17 \;?\; 32 \quad \text{Substituting 15 for } x$$
$$32 \mid 32 \quad 32 = 32 \text{ is TRUE.}$$

Since the left-hand and right-hand sides are the same, 15 is a solution.

46. No

47.
$$\underline{a - 28 = 75} \quad \text{Writing the equation}$$
$$93 - 28 \;?\; 75 \quad \text{Substituting 93 for } a$$
$$65 \mid 75 \quad 65 = 75 \text{ is FALSE.}$$

Since the left-hand and right-hand sides are not the same, 93 is not a solution.

48. Yes

49.
$$\underline{\frac{t}{7} = 9}$$
$$\frac{63}{7} \;?\; 9$$
$$9 \mid 9 \quad 9 = 9 \text{ is TRUE.}$$

Since the left-hand and right-hand sides are the same, 63 is a solution.

50. No

51.
$$\underline{\frac{108}{x} = 36}$$
$$\frac{108}{3} \;?\; 36$$
$$36 \mid 36 \quad 36 = 36 \text{ is TRUE.}$$

Since the left-hand and right-hand sides are the same, 3 is a solution.

52. No

53. Let x represent the number.

What number added to 73 is 201?

Translating: x $+$ 73 $=$ 201

$x + 73 = 201$

54. Let w represent the number; $7w = 2303$

55. Let y represent the number.

Rewording: 42 times what number is 2352?

Translating: 42 $\cdot$ y $=$ 2352

$42y = 2352$

56. Let x represent the number; $x + 345 = 987$

57. Let s represent the number of squares your opponent controls.

Rewording: The number of squares your opponent controls   added to   35 is 64.

Translating: s $+$ 35 $= 64$

$s + 35 = 64$

58. Let y represent the number of hours the carpenter worked; $25y = 53,400$

59. Let x represent the total amount of waste generated, in millions of tons.

Rewording: 27% of   the total amount of waste   is   56 million tons.

Translating: 27% $\cdot$ x $=$ 56

$27\% \cdot x = 56$, or $0.27x = 56$

60. Let m represent the length of the average commute in the West, in minutes; $m = 24.5 - 1.8$

61. ◈

62. ◈

63. ◈

64. ◈

65. Area of sign: $A = \frac{1}{2}(3 \text{ ft})(2.5 \text{ ft}) = 3.75 \text{ ft}^2$

Cost of sign: $\$90(3.75) = \$337.50$

66. 158.75 cm^2

67. When x is twice y, then y is one-half x, so
$$y = \frac{12}{2} = 6.$$
$$\frac{x-y}{3} = \frac{12-6}{3} = \frac{6}{3} = 2$$

68. 9

69. When a is twice b, then b is one-half a, so $b = \frac{16}{2} = 8.$
$$\frac{a+b}{4} = \frac{16+8}{4} = \frac{24}{4} = 6$$

70. 4

71. The next whole number is one more than $w + 3$:
$$w + 3 + 1 = w + 4$$

72. d

73. Let a and b represent the numbers. Then we have
$$\frac{1}{3} \cdot \frac{1}{2} \cdot ab.$$

74. $l + w + l + w$, or $2l + 2w$

75. $s + s + s + s$, or $4s$

76. $a + 9$

77. ◈

Exercise Set 1.2

1. $x + 7$ Changing the order

2. $2 + a$

3. $c + ab$

4. $3y + x$

5. $3y + 9x$

6. $7b + 3a$

7. $5(1 + a)$

8. $9(5 + x)$

9. $a \cdot 2$ Changing the order

10. yx

11. ts

12. $x4$

13. $5 + ba$

14. $x + y3$

15. $(a + 1)5$

16. $(x + 5)9$

17. $a + (5 + b)$

18. $5 + (m + r)$

19. $(r + t) + 7$

20. $(x + 2) + y$

21. $ab + (c + d)$

22. $m + (np + r)$

23. $8(xy)$

24. $9(ab)$

25. $(2a)b$

26. $(9r)p$

27. $(3 \cdot 2)(a + b)$

28. $(5x)(2 + y)$

29. a) $r + (t + 6) = (t + 6) + r$ Using the commutative law
$$= (6 + t) + r \quad \text{Using the commutative law again}$$

   b) $r + (t + 6) = (t + 6) + r$ Using the commutative law
$$= t + (6 + r) \quad \text{Using the associative law}$$

Answers may vary.

30. Answers may vary; $v + (w + 5)$; $(v + 5) + w$

31. a) $(17a)b = b(17a)$ Using the commutative law
$$= b(a17) \quad \text{Using the commutative law again}$$

   b) $(17a)b = (a17)b$ Using the commutative law
$$= a(17b) \quad \text{Using the associative law}$$

Answers may vary.

32. Answers may vary; $3(yx)$; $(3x)y$

33. $(5 + x) + 2$
  $= (x + 5) + 2$ Commutative law
  $= x + (5 + 2)$ Associative law
  $= x + 7$ Simplifying

34. $(2a)4 = 4(2a)$ Commutative law
    $= (4 \cdot 2)a$ Associative law
    $= 8a$ Simplifying

35. $(m3)7 = m(3 \cdot 7)$ Associative law
      $= (3 \cdot 7)m$ Commutative law
      $= 21m$ Simplifying

36. $4 + (9 + x)$
  $= (4 + 9) + x$ Associative law
  $= x + (4 + 9)$ Commutative law
  $= x + 13$ Simplifying

37. $4(a + 3) = 4 \cdot a + 4 \cdot 3 = 4a + 12$

38. $3x + 15$

39. $6(1 + x) = 6 \cdot 1 + 6 \cdot x = 6 + 6x$

40. $6v + 24$

41. $3(x + 1) = 3 \cdot x + 3 \cdot 1 = 3x + 3$

42. $9x + 27$

43. $8(3 + y) = 8 \cdot 3 + 8 \cdot y = 24 + 8y$

44. $7s + 35$

45. $9(2x + 6) = 9 \cdot 2x + 9 \cdot 6 = 18x + 54$

46. $54m + 63$

47. $5(r + 2 + 3t) = 5 \cdot r + 5 \cdot 2 + 5 \cdot 3t = 5r + 10 + 15t$

48. $20x + 32 + 12p$

49. $(a + b)2 = a(2) + b(2) = 2a + 2b$

50. $7x + 14$

51. $(x + y + 2)5 = x(5) + y(5) + 2(5) = 5x + 5y + 10$

52. $12 + 6a + 6b$

53. $x + xyz + 19$
  The terms are separated by plus signs. They are x, xyz, and 19.

54. $9, 17a, abc$

55. $2a + \dfrac{a}{b} + 5b$
  The terms are separated by plus signs. They are $2a$, $\dfrac{a}{b}$, and $5b$.

56. $3xy, 20, \dfrac{4a}{b}$

57. $2a + 2b = 2(a + b)$ The common factor is 2.
  Check: $2(a + b) = 2 \cdot a + 2 \cdot b = 2a + 2b$

58. $5(y + z)$

59. $7 + 7y = 7 \cdot 1 + 7 \cdot y$ The common factor is 7.
      $= 7(1 + y)$ Using the distributive law
  Check: $7(1 + y) = 7 \cdot 1 + 7 \cdot y = 7 + 7y$

60. $113(1 + x)$

61. $18x + 3 = 3 \cdot 6x + 3 \cdot 1 = 3(6x + 1)$
  Check: $3(6x + 1) = 3 \cdot 6x + 3 \cdot 1 = 18x + 3$

62. $5(4a + 1)$

63. $5x + 10 + 15y = 5 \cdot x + 5 \cdot 2 + 5 \cdot 3y = 5(x + 2 + 3y)$
  Check: $5(x + 2 + 3y) = 5 \cdot x + 5 \cdot 2 + 5 \cdot 3y = 5x + 10 + 15y$

64. $3(1 + 9b + 2c)$

65. $12x + 9 = 3 \cdot 4x + 3 \cdot 3 = 3(4x + 3)$
  Check: $3(4x + 3) = 3 \cdot 4x + 3 \cdot 3 = 12x + 9$

66. $6(x + 1)$

67. $3a + 9b = 3 \cdot a + 3 \cdot 3b = 3(a + 3b)$
  Check: $3(a + 3b) = 3 \cdot a + 3 \cdot 3b = 3a + 9b$

68. $5(a + 3b)$

69. $44x + 11y + 22z = 11 \cdot 4x + 11 \cdot y + 11 \cdot 2z = 11(4x + y + 2z)$
  Check: $11(4x + y + 2z) = 11 \cdot 4x + 11 \cdot y + 11 \cdot 2z = 44x + 11y + 22z$

70. $7(2a + 8b + 1)$

71. ◈

72. ◈

73. Let k represent Kara's salary. Then we have $2k$.

74. $\dfrac{1}{2}m$, or $\dfrac{m}{2}$, or $m/2$, or $m \div 2$

75.

76.

77. The expressions are equivalent.

$$8 + 4(a + b) = 8 + 4a + 4b = 4(2 + a + b)$$

78. The expressions are not equivalent.

Let $m = 1$. Then we have:

$$7 \div 3 \cdot 1 = \frac{7}{3} \cdot 1 = \frac{7}{3}, \text{ but}$$

$$1 \cdot 3 \div 7 = 3 \div 7 = \frac{3}{7}.$$

79. The expressions are equivalent.

$$(rt + st)5 = 5(rt + st) = 5 \cdot t(r + s) = 5t(r + s)$$

80. The expression are equivalent.

$$yax + ax = (y + 1)ax = (1 + y)ax = ax(1 + y) = xa(1 + y)$$

81. The expressions are not equivalent.

Let $x = 1$ and $y = 0$. Then we have:

$$30 \cdot 0 + 1 \cdot 15 = 0 + 15 = 15, \text{ but}$$

$$5[2(1 + 3 \cdot 0)] = 5[2(1)] = 5 \cdot 2 = 10.$$

82. The expressions are equivalent.

$$[c(2 + 3b)]5 = 5[c(2 + 3b)] = 5c(2 + 3b) = 10c + 15bc$$

83.

84.

Exercise Set 1.3

1. We write two factorizations of 50. There are other factorizations as well.

$$2 \cdot 25, \ 5 \cdot 10$$

List all of the factors of 50:

1, 2, 5, 10, 25, 50

2. $2 \cdot 35, \ 5 \cdot 14$; 1, 2, 5, 7, 10, 14, 35, 70

3. We write two factorizations of 42. There are other factorizations as well.

$$2 \cdot 21, \ 6 \cdot 7$$

List all of the factors of 42:

1, 2, 3, 6, 7, 14, 21, 42

4. $2 \cdot 30, \ 5 \cdot 12$; 1, 2, 3, 4, 5, 6, 10, 12, 15, 20, 30, 60

5. $26 = 2 \cdot 13$

6. $3 \cdot 5$

7. We begin factoring 30 in any way that we can and continue factoring until each factor is prime.

$$30 = 2 \cdot 15 = 2 \cdot 3 \cdot 5$$

8. $5 \cdot 11$

9. We begin by factoring 20 in any way that we can and continue factoring until each factor is prime.

$$20 = 4 \cdot 5 = 2 \cdot 2 \cdot 5$$

10. $2 \cdot 5 \cdot 5$

11. We begin by factoring 27 in any way that we can and continue factoring until each factor is prime.

$$27 = 3 \cdot 9 = 3 \cdot 3 \cdot 3$$

12. $2 \cdot 7 \cdot 7$

13. We begin by factoring 18 in any way that we can and continue factoring until each factor is prime.

$$18 = 2 \cdot 9 = 2 \cdot 3 \cdot 3$$

14. $2 \cdot 3 \cdot 3 \cdot 3$

15. We begin by factoring 40 in any way that we can and continue factoring until each factor is prime.

$$40 = 4 \cdot 10 = 2 \cdot 2 \cdot 2 \cdot 5$$

16. $2 \cdot 2 \cdot 2 \cdot 7$

17. 43 has exactly two different factors, 43 and 1. Thus, 43 is prime.

18. $2 \cdot 2 \cdot 2 \cdot 3 \cdot 5$

19. $210 = 2 \cdot 105 = 2 \cdot 3 \cdot 35 = 2 \cdot 3 \cdot 5 \cdot 7$

20. Prime

21. $115 = 5 \cdot 23$

22. $11 \cdot 13$

23. $\dfrac{10}{14} = \dfrac{2 \cdot 5}{2 \cdot 7}$ Factoring numerator and denominator

$$= \frac{2}{2} \cdot \frac{5}{7}$$ Rewriting as a product of two fractions

$$= 1 \cdot \frac{5}{7}$$ $\dfrac{2}{2} = 1$

$$= \frac{5}{7}$$ Using the identity property of 1

24. $\dfrac{2}{3}$

25. $\dfrac{16}{56} = \dfrac{2 \cdot 8}{7 \cdot 8} = \dfrac{2}{7} \cdot \dfrac{8}{8} = \dfrac{2}{7} \cdot 1 = \dfrac{2}{7}$

26. $\dfrac{8}{3}$

27. $\dfrac{6}{48} = \dfrac{1 \cdot 6}{8 \cdot 6}$ Factoring and using the identity property of 1 to write 6 as $1 \cdot 6$

     $= \dfrac{1}{8} \cdot \dfrac{6}{6}$

     $= \dfrac{1}{8} \cdot 1 = \dfrac{1}{8}$

28. $\dfrac{6}{35}$

29. $\dfrac{49}{7} = \dfrac{7 \cdot 7}{1 \cdot 7} = \dfrac{7}{1} \cdot \dfrac{7}{7} = \dfrac{7}{1} \cdot 1 = 7$

30. 12

31. $\dfrac{19}{76} = \dfrac{1 \cdot 19}{4 \cdot 19}$ Factoring and using the identity property of 1 to write 19 as $1 \cdot 19$

     $= \dfrac{1 \cdot \cancel{19}}{4 \cdot \cancel{19}}$ Removing a factor equal to 1: $\dfrac{19}{19} = 1$

     $= \dfrac{1}{4}$

32. $\dfrac{1}{3}$

33. $\dfrac{150}{25} = \dfrac{6 \cdot 25}{1 \cdot 25}$ Factoring and using the identity property of 1 to write 25 as $1 \cdot 25$

     $= \dfrac{6 \cdot \cancel{25}}{1 \cdot \cancel{25}}$ Removing a factor equal to 1: $\dfrac{25}{25} = 1$

     $= \dfrac{6}{1}$

     $= 6$ Simplifying

34. 5

35. $\dfrac{75}{80} = \dfrac{5 \cdot 15}{5 \cdot 16}$ Factoring the numerator and the denominator

     $= \dfrac{\cancel{5} \cdot 15}{\cancel{5} \cdot 16}$ Removing a factor equal to 1: $\dfrac{5}{5} = 1$

     $= \dfrac{15}{16}$

36. $\dfrac{21}{25}$

37. $\dfrac{120}{82} = \dfrac{2 \cdot 60}{2 \cdot 41}$ Factoring

     $= \dfrac{\cancel{2} \cdot 60}{\cancel{2} \cdot 41}$ Removing a factor equal to 1: $\dfrac{2}{2} = 1$

     $= \dfrac{60}{41}$

38. $\dfrac{5}{3}$

39. $\dfrac{210}{98} = \dfrac{2 \cdot 7 \cdot 15}{2 \cdot 7 \cdot 7}$ Factoring

     $= \dfrac{\cancel{2} \cdot \cancel{7} \cdot 15}{\cancel{2} \cdot \cancel{7} \cdot 7}$ Removing a factor equal to 1: $\dfrac{2 \cdot 7}{2 \cdot 7} = 1$

     $= \dfrac{15}{7}$

40. $\dfrac{2}{5}$

41. $\dfrac{1}{2} \cdot \dfrac{3}{7} = \dfrac{1 \cdot 3}{2 \cdot 7}$ Multiplying numerators and denominators

     $= \dfrac{3}{14}$

42. $\dfrac{44}{25}$

43. $\dfrac{9}{2} \cdot \dfrac{3}{4} = \dfrac{9 \cdot 3}{2 \cdot 4} = \dfrac{27}{8}$

44. 1

45. $\dfrac{1}{8} + \dfrac{3}{8} = \dfrac{1 + 3}{8}$ Adding numerators; keeping the common denominator

     $= \dfrac{4}{8}$

     $= \dfrac{1 \cdot \cancel{4}}{2 \cdot \cancel{4}} = \dfrac{1}{2}$ Simplifying

46. $\dfrac{5}{8}$

47. $\dfrac{4}{9} + \dfrac{13}{18} = \dfrac{4}{9} \cdot \dfrac{2}{2} + \dfrac{13}{18}$ Using 18 as the common denominator

     $= \dfrac{8}{18} + \dfrac{13}{18}$

     $= \dfrac{21}{18}$

     $= \dfrac{7 \cdot \cancel{3}}{6 \cdot \cancel{3}} = \dfrac{7}{6}$ Simplifying

48. $\dfrac{4}{3}$

49. $\frac{3}{a} \cdot \frac{b}{7} = \frac{3b}{7a}$ Multiplying numerators and denominators

50. $\frac{xy}{5z}$

51. $\frac{4}{a} + \frac{3}{a} = \frac{7}{a}$ Adding numerators; keeping the common denominator

52. $\frac{2}{a}$

53. $\frac{3}{10} + \frac{8}{15} = \frac{3}{10} \cdot \frac{3}{3} + \frac{8}{15} \cdot \frac{2}{2}$ Using 30 as the common denominator

$= \frac{9}{30} + \frac{16}{30}$

$= \frac{25}{30}$

$= \frac{5 \cdot 5}{6 \cdot 5} = \frac{5}{6}$ Simplifying

54. $\frac{31}{24}$

55. $\frac{9}{7} - \frac{2}{7} = \frac{7}{7} = 1$

56. 2

57. $\frac{13}{18} - \frac{4}{9} = \frac{13}{18} - \frac{4}{9} \cdot \frac{2}{2}$ Using 18 as the common denominator

$= \frac{13}{18} - \frac{8}{18}$

$= \frac{5}{18}$

58. $\frac{31}{45}$

59. $\frac{5}{7} - \frac{5}{21} = \frac{5}{7} \cdot \frac{3}{3} - \frac{5}{21}$ Using 21 as the common denominator

$= \frac{15}{21} - \frac{5}{21}$

$= \frac{10}{21}$

60. 0

61. $\frac{7}{6} \div \frac{3}{5} = \frac{7}{6} \cdot \frac{5}{3}$ Multiplying by the reciprocal of the divisor

$= \frac{35}{18}$

62. $\frac{28}{15}$

63. $\frac{8}{9} \div \frac{4}{15} = \frac{8}{9} \cdot \frac{15}{4} = \frac{2 \cdot 4 \cdot 3 \cdot 5}{3 \cdot 3 \cdot 4} = \frac{10}{3}$

64. $\frac{1}{4}$

65. $12 \div \frac{3}{7} = \frac{12}{1} \cdot \frac{7}{3} = \frac{4 \cdot 3 \cdot 7}{1 \cdot 3} = 28$

66. $\frac{1}{2}$

67. Note that we have a number divided by itself. Thus, the result is 1. We can also do this exercise as follows:

$\frac{7}{13} \div \frac{7}{13} = \frac{7}{13} \cdot \frac{13}{7} = \frac{7 \cdot 13}{7 \cdot 13} = 1$

68. $\frac{51}{20}$

69. $\frac{\frac{2}{7}}{\frac{5}{3}} = \frac{2}{7} \div \frac{5}{3} = \frac{2}{7} \cdot \frac{3}{5} = \frac{2 \cdot 3}{7 \cdot 5} = \frac{6}{35}$

70. $\frac{15}{8}$

71. $\frac{9}{\frac{1}{2}} = 9 \div \frac{1}{2} = \frac{9}{1} \cdot \frac{2}{1} = \frac{9 \cdot 2}{1 \cdot 1} = 18$

72. $\frac{7}{15}$

73. ◈

74. ◈

75. $5(x + 3) = 5(3 + x)$ Commutative law of addition
Answers may vary.

76. Answers may vary; $(a + b) + 7$, or $7 + (b + a)$

77. ◈

78. ◈

79. We need to find the smallest number that has both 6 and 8 as factors. Starting with 6 we list some numbers with a factor of 6, and starting with 8 we also list some numbers with a factor of 8. Then we find the first number that is on both lists.

6, 12, 18, 24, 30, 36, ...

8, 16, 24, 32, 40, 48, ...

Since 24 is the smallest number that is on both lists, the carton should be 24 in. long.

80.

Product	56	63	36	72	140	96	168
Factor	7	7	2	36	14	8	8
Factor	8	9	18	2	10	12	21
Sum	15	16	20	38	24	20	29

81. $\dfrac{16 \cdot 9 \cdot 4}{15 \cdot 8 \cdot 12} = \dfrac{\cancel{4} \cdot 4 \cdot \cancel{3} \cdot \cancel{3} \cdot \cancel{2} \cdot 2}{\cancel{3} \cdot 5 \cdot \cancel{2} \cdot \cancel{4} \cdot \cancel{3} \cdot \cancel{4}} = \dfrac{2}{5}$

82. 1

83. $\dfrac{27pqrs}{9prst} = \dfrac{3 \cdot \cancel{9} \cdot \cancel{p} \cdot q \cdot \cancel{r} \cdot \cancel{s}}{\cancel{9} \cdot \cancel{p} \cdot \cancel{r} \cdot \cancel{s} \cdot t} = \dfrac{3q}{t}$

84. $\dfrac{8}{3}$

85. $\dfrac{15 \cdot 4xy \cdot 9}{6 \cdot 25x \cdot 15y} = \dfrac{\cancel{15} \cdot \cancel{2} \cdot 2 \cdot \cancel{x} \cdot \cancel{y} \cdot \cancel{3} \cdot 3}{\cancel{2} \cdot \cancel{3} \cdot 25 \cdot \cancel{x} \cdot \cancel{15} \cdot \cancel{y}} = \dfrac{6}{25}$

86. $\dfrac{5}{2}$

87. $\dfrac{\dfrac{27ab}{15mn}}{\dfrac{18bc}{25np}} = \dfrac{27ab}{15mn} \div \dfrac{18bc}{25np} = \dfrac{27ab}{15mn} \cdot \dfrac{25np}{18bc} = $

$\dfrac{27ab \cdot 25np}{15mn \cdot 18bc} = \dfrac{\cancel{3} \cdot \cancel{9} \cdot a \cdot \cancel{b} \cdot \cancel{5} \cdot 5 \cdot \cancel{n} \cdot p}{\cancel{3} \cdot \cancel{5} \cdot m \cdot \cancel{n} \cdot 2 \cdot \cancel{9} \cdot \cancel{b} \cdot c} = \dfrac{5ap}{2mc}$

88. $\dfrac{2yc}{b}$

89. $A = lw = \left(\dfrac{4}{5}\text{ m}\right)\left(\dfrac{7}{9}\text{ m}\right)$

$= \left(\dfrac{4}{5}\right)\left(\dfrac{7}{9}\right)(\text{m})(\text{m})$

$= \dfrac{28}{45}\text{ m}^2, \text{ or } \dfrac{28}{45}\text{ square meters}$

90. $\dfrac{25}{28}\text{ m}^2$

91. $P = 4s = 4\left(3\dfrac{5}{9}\text{ m}\right) = 4 \cdot \dfrac{32}{9}\text{ m} = \dfrac{128}{9}\text{ m, or}$

$14\dfrac{2}{9}\text{ m}$

92. $\dfrac{142}{45}\text{ m, or } 3\dfrac{7}{45}\text{ m}$

93. ◈

Exercise Set 1.4

1. The real number -19 corresponds to 19°F below zero, and the real number 59 corresponds to 59°F above zero.

2. $-2, 5$

3. The real number -150 corresponds to burning 150 calories, and the real number 65 corresponds to consuming 65 calories.

4. 1200, -800

5. The real number -1286 corresponds to 1286 ft below sea level. The real number 29,029 corresponds to 29,029 ft above sea level.

6. Jets: -34, Strikers: 34

7. The real number 750 corresponds to a \$750 deposit, and the real number -125 corresponds to a \$125 withdrawal.

8. 22, -9

9. The real numbers 20, -150, and 300 correspond to the interception of the missile, the loss of the starship, and the capture of the base, respectively.

10. -10, 235

11. Since $\dfrac{10}{3} = 3\dfrac{1}{3}$, its graph is $\dfrac{1}{3}$ of a unit to the right of 3.

12.

13. The graph of -4.3 is $\dfrac{3}{10}$ of a unit to the left of -4.

14.

15.

16.

17. $\dfrac{7}{8}$ means $7 \div 8$, so we divide.

```
      0.8 7 5
  8 ) 7.0 0 0
      6 4
      ─────
        6 0
        5 6
      ─────
          4 0
          4 0
      ─────
             0
```

We have $\dfrac{7}{8} = 0.875$.

18. -0.125

19. We first find decimal notation for $\frac{3}{4}$. Since $\frac{3}{4}$ means $3 \div 4$, we divide.

$$
\begin{array}{r}
0.7\,5 \\
4\,\overline{)\,3.0\,0} \\
2\,8 \\
\hline
2\,0 \\
2\,0 \\
\hline
0
\end{array}
$$

Thus, $\frac{3}{4} = 0.75$, so $-\frac{3}{4} = -0.75$.

20. $0.8\overline{3}$

21. $\frac{7}{6}$ means $7 \div 6$, so we divide.

$$
\begin{array}{r}
1.1\,6\,6 \\
6\,\overline{)\,7.0\,0\,0} \\
6 \\
\hline
1\,0 \\
6 \\
\hline
4\,0 \\
3\,6 \\
\hline
4\,0 \\
3\,6 \\
\hline
4
\end{array}
$$

We have $\frac{7}{6} = 1.1\overline{6}$.

22. $0.41\overline{6}$

23. $\frac{2}{3}$ means $2 \div 3$, so we divide.

$$
\begin{array}{r}
0.6\,6\,6\,... \\
3\,\overline{)\,2.0\,0\,0} \\
1\,8 \\
\hline
2\,0 \\
1\,8 \\
\hline
2\,0 \\
1\,8 \\
\hline
2
\end{array}
$$

We have $\frac{2}{3} = 0.\overline{6}$.

24. 0.25

25. We first find decimal notation for $\frac{1}{2}$. Since $\frac{1}{2}$ means $1 \div 2$, we divide.

$$
\begin{array}{r}
0.5 \\
2\,\overline{)\,1.0} \\
1\,0 \\
\hline
0
\end{array}
$$

Thus, $\frac{1}{2} = 0.5$, so $-\frac{1}{2} = -0.5$.

26. -0.375

27. Since the denominator is 100, we know that $\frac{13}{100} = 0.13$. We could also divide 13 by 100 to find this result.

28. -0.35

29. Since -8 is to the left of 2, we have $-8 < 2$.

30. $9 > 0$

31. Since 7 is to the right of 0, we have $7 > 0$.

32. $8 > -8$

33. Since -6 is to the left of 6, we have $-6 < 6$.

34. $0 > -7$

35. Since -8 is to the left of -5, we have $-8 < -5$.

36. $-4 < -3$

37. Since -5 is to the right of -11, we have $-5 > -11$.

38. $-3 > -4$

39. Since -12.5 is to the left of -9.4, we have $-12.5 < -9.4$.

40. $-10.3 > -14.5$

41. We convert to decimal notation. $\frac{5}{12} = 0.41\overline{6}$ and $\frac{11}{25} = 0.44$. Thus, $\frac{5}{12} < \frac{11}{25}$.

42. $-\frac{14}{17} < -\frac{27}{35}$

43. $-7 > x$ has the same meaning as $x < -7$.

44. $9 < a$

45. $-10 \leq y$ has the same meaning as $y \geq -10$.

46. $t \leq 12$

47. $-3 \geq -11$ is true, since $-3 > -11$ is true.

48. False

49. $0 \geq 8$ is false, since neither $0 > 8$ nor $0 = 8$ is true.

50. True

51. $-8 \leq -8$ is true because $-8 = -8$ is true.

52. True

53. $|-23| = 23$ since -23 is 23 units from 0.

54. 47

55. $|17| = 17$ since 17 is 17 units from 0.

56. 3.1

57. $|5.6| = 5.6$ since 5.6 is 5.6 units from 0.

58. $\dfrac{2}{5}$

59. $|329| = 329$ since 329 is 329 units from 0.

60. 456

61. $\left|-\dfrac{9}{7}\right| = \dfrac{9}{7}$ since $-\dfrac{9}{7}$ is $\dfrac{9}{7}$ units from 0.

62. 8.02

63. $|0| = 0$ since 0 is 0 units from itself.

64. 1.07

65. $|x| = |-8| = 8$

66. 5

67. $-83, -4.7, 0, \dfrac{5}{9}, 8.31, 62$

68. 62

69. $-83, 0, 62$

70. $\pi, \sqrt{17}$

71. All are real numbers.

72. 0, 62

73. ◈

74. ◈

75. $3xy = 3 \cdot 2 \cdot 7 = 42$

76. $5 + ab$; $ba + 5$; $5 + ba$

77. ◈

78. ◈

79. ◈

80. $-17, -12, 5, 13$

81. List the numbers as they occur on the number line, from left to right: $-23, -17, 0, 4$

82. $-\dfrac{4}{3}, \dfrac{4}{9}, \dfrac{4}{8}, \dfrac{4}{6}, \dfrac{4}{5}, \dfrac{4}{3}, \dfrac{4}{2}$

83. $-\dfrac{2}{3}, \dfrac{1}{2}, -\dfrac{3}{4}, -\dfrac{5}{6}, \dfrac{3}{8}, \dfrac{1}{6}$ can be written in decimal notation as $-0.66\overline{6}, 0.5, -0.75, -0.83\overline{3}, 0.375, 0.16\overline{6}$, respectively. Listing from least to greatest (in fractional form), we have
$$-\dfrac{5}{6}, -\dfrac{3}{4}, -\dfrac{2}{3}, \dfrac{1}{6}, \dfrac{3}{8}, \dfrac{1}{2}.$$

84. $|-5| > |-2|$

85. $|4| = 4$ and $|-7| = 7$, so $|4| < |-7|$.

86. $|-8| = |8|$

87. $|23| = 23$ and $|-23| = 23$, so $|23| = |-23|$.

88. $|-3| < |5|$

89. $|-19| = 19$ and $|-27| = 27$, so $|-19| < |-27|$.

90. $-7, 7$

91. x represents an integer whose distance from 0 is less than 3 units. Thus, $x = -2, -1, 0, 1, 2$.

92. $-4, -3, 3, 4$

93. $0.1\overline{1} = \dfrac{0.3\overline{3}}{3} = \dfrac{\frac{1}{3}}{3} = \dfrac{1}{3} \cdot \dfrac{1}{3} = \dfrac{1}{9}$

94. $\dfrac{3}{3}$

95. $5.5\overline{5} = 50(0.1\overline{1}) = 50 \cdot \dfrac{1}{9} = \dfrac{50}{9}$
(See Exercise 93.)

96. $\dfrac{70}{9}$

97. ◈

Exercise Set 1.5

1. Start at 4. Move 7 units to the left.

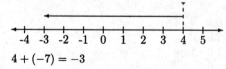

$4 + (-7) = -3$

2. -3

3. Start at −5. Move 9 units to the right.

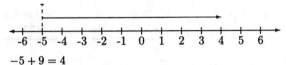

$-5 + 9 = 4$

4. 5

5. Start at 8. Move 8 units to the left.

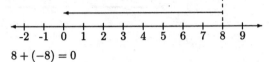

$8 + (-8) = 0$

6. 0

7. Start at −3. Move 5 units to the left.

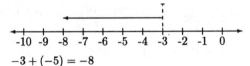

$-3 + (-5) = -8$

8. −10

9. $-15 + 0$ One number is 0. The answer is the other number. $-15 + 0 = -15$

10. −6

11. $0 + (-8)$ One number is 0. The answer is the other number. $0 + (-8) = -8$

12. −2

13. $12 + (-12)$ The numbers have the same absolute value. The sum is 0. $12 + (-12) = 0$

14. 0

15. $-24 + (-17)$ Two negatives. Add the absolute values, getting 41. Make the answer negative.
$-24 + (-17) = -41$

16. −42

17. $-15 + 15$ The numbers have the same absolute value. The sum is 0. $-15 + 15 = 0$

18. 0

19. $18 + (-11)$ The absolute values are 18 and 11. The difference is $18 - 11$, or 7. The positive number has the larger absolute value, so the answer is positive.
$18 + (-11) = 7$

20. 3

21. $10 + (-12)$ The absolute values are 10 and 12. The difference is $12 - 10$, or 2. The negative number has the larger absolute value, so the answer is negative.
$10 + (-12) = -2$

22. −4

23. $-3 + 14$ The absolute values are 3 and 14. The difference is $14 - 3$, or 11. The positive number has the larger absolute value, so the answer is positive.
$-3 + 14 = 11$

24. 7

25. $-14 + (-19)$ Two negatives. Add the absolute values, getting 33. Make the answer negative.
$-14 + (-19) = -33$

26. 2

27. $19 + (-19)$ The numbers has the same absolute value. The sum is 0. $19 + (-19) = 0$

28. −26

29. $23 + (-5)$ The absolute values are 23 and 5. The difference is $23 - 5$ or 18. The positive number has the larger absolute value, so the answer is positive.
$23 + (-5) = 18$

30. −22

31. $-23 + (-9)$ Two negatives. Add the absolute values, getting 32. Make the answer negative.
$-23 + (-9) = -32$

32. 32

33. $40 + (-40)$ The numbers have the same absolute value. The sum is 0. $40 + (-40) = 0$

34. 0

35. $85 + (-65)$ The absolute values are 85 and 65. The difference is $85 - 65$, or 20. The positive number has the larger absolute value, so the answer is positive.
$85 + (-65) = 20$

36. 45

37. $-3.6 + 1.9$ The absolute values are 3.6 and 1.9. The difference is $3.6 - 1.9$, or 1.7. The negative number has the larger absolute value, so the answer is negative. $-3.6 + 1.9 = -1.7$

38. −1.8

39. $-5.4 + (-3.7)$ Two negatives. Add the absolute values, getting 9.1. Make the answer negative.
$-5.4 + (-3.7) = -9.1$

40. -13.2

41. $\dfrac{-3}{5} + \dfrac{4}{5}$ The absolute values are $\dfrac{3}{5}$ and $\dfrac{4}{5}$. The difference is $\dfrac{4}{5} - \dfrac{3}{5}$, or $\dfrac{1}{5}$. The positive number has the larger absolute value, so the answer is positive.
$$\dfrac{-3}{5} + \dfrac{4}{5} = \dfrac{1}{5}$$

42. $\dfrac{1}{7}$

43. $\dfrac{-4}{7} + \dfrac{-2}{7}$ Two negatives. Add the absolute values, getting $\dfrac{6}{7}$.   Make the answer negative.
$$\dfrac{-4}{7} + \dfrac{-2}{7} = \dfrac{-6}{7}$$

44. $\dfrac{-7}{9}$

45. $-\dfrac{2}{5} + \dfrac{1}{3}$ The absolute values are $\dfrac{2}{5}$ and $\dfrac{1}{3}$. The difference is $\dfrac{6}{15} - \dfrac{5}{15}$, or $\dfrac{1}{15}$. The negative number has the larger absolute value, so the answer is negative.
$$-\dfrac{2}{5} + \dfrac{1}{3} = -\dfrac{1}{15}$$

46. $\dfrac{5}{26}$

47. $\dfrac{-4}{9} + \dfrac{2}{3}$ The absolute values are $\dfrac{4}{9}$ and $\dfrac{2}{3}$. The difference is $\dfrac{6}{9} - \dfrac{4}{9}$, or $\dfrac{2}{9}$. The positive number has the larger absolute value, so the answer is positive.
$$\dfrac{-4}{9} + \dfrac{2}{3} = \dfrac{2}{9}$$

48. $\dfrac{1}{6}$

49. $35 + (-14) + (-19) + (-5)$
$= 35 + [(-14) + (-19) + (-5)]$ Using the associative law of addition
$= 35 + (-38)$ Adding the negatives
$= -3$ Adding a positive and a negative

50. -62

51. $-4.9 + 8.5 + 4.9 + (-8.5)$
Note that we have two pairs of numbers with different signs and the same absolute value: -4.9 and 4.9, 8.5 and -8.5. The sum of each pair is 0, so the result is $0 + 0$, or 0.

52. 37.9

53. Rewording:

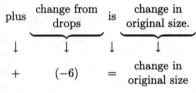

Since $-5 + 8 + (-6)$
$= 3 + (-6)$
$= -3$,
the class has 3 students less than the original class size.

54. Maya owed \$69 at the end of August.

55. Rewording:

	1998 loss	plus	1999 loss	plus
Translating:	$-26,500$	$+$	$(-10,200)$	$+$

2000 profit	is	total profit or loss.
$32,400$	$=$	total profit or loss.

Since $-26,500 + (-10,200) + 32,400$
$= -36,700 + 32,400$
$= -4300$,
the loss was \$4300.

56. The total gain was 22 yd.

57. Rewording:

	Original balance	plus	change from writing first check	plus
Translating:	350	$+$	(-530)	$+$

deposit	plus	change from writing second check	is	new balance.
75	$+$	(-90)	$=$	new balance

Since $350 + (-530) + (75) + (-90)$

$$= (350 + 75) + [-530 + (-90)]$$
$$= 425 + (-620)$$
$$= -195,$$

Leah's account is $195 overdrawn.

58. Lyle owes $85 on his credit card.

59. Rewording:

$$\underbrace{\text{First change}} \quad \text{plus} \quad \underbrace{\text{second change}} \quad \text{plus}$$
$$\downarrow \qquad \downarrow \qquad \downarrow \qquad \downarrow$$

Translating: $\quad \dfrac{3}{16} \quad + \quad \left(-\dfrac{1}{2}\right) \quad +$

$$\underbrace{\text{third change}} \quad \text{is} \quad \underbrace{\text{total change.}}$$
$$\downarrow \qquad \downarrow \qquad \downarrow$$
$$\dfrac{1}{4} \quad = \text{total change.}$$

Since $\dfrac{3}{16} + \left(-\dfrac{1}{2}\right) + \dfrac{1}{4}$

$$= \left(\dfrac{3}{16} + \dfrac{4}{16}\right) + \left(-\dfrac{8}{16}\right)$$
$$= \dfrac{7}{16} + \left(-\dfrac{8}{16}\right)$$
$$= \left(-\dfrac{1}{16}\right),$$

the value of the stock had fallen $\$\dfrac{1}{16}$ at the end of the day.

60. The elevation of the peak is 13,796 ft above sea level.

61. $7a + 5a = (7 + 5)a \quad$ Using the distributive law
$$= 12a$$

62. $11x$

63. $-3x + 12x = (-3 + 12)x \quad$ Using the distributive law
$$= 9x$$

64. $-5m$

65. $5t + 8t = (5 + 8)t = 13t$

66. $14a$

67. $7m + (-9m) = [7 + (-9)]m = -2m$

68. 0

69. $-5a + (-2a) = [-5 + (-2)]a = -7a$

70. $-7n$

71. $\quad -3 + 8x + 4 + (-10x)$

$$= -3 + 4 + 8x + (-10x) \quad \text{Using the commutative law of addition}$$
$$= (-3 + 4) + [8 + (-10)]x \quad \text{Using the distributive law}$$
$$= 1 - 2x \quad \text{Adding}$$

72. $7a + 2$

73. Perimeter $= 8 + 5x + 9 + 7x$
$$= 8 + 9 + 5x + 7x$$
$$= (8 + 9) + (5 + 7)x$$
$$= 17 + 12x$$

74. $10a + 13$

75. Perimeter $= 9 + 6n + 7 + 8n + 4n$
$$= 9 + 7 + 6n + 8n + 4n$$
$$= (9 + 7) + (6 + 8 + 4)n$$
$$= 16 + 18n$$

76. $19n + 11$

77. ◈

78. ◈

79. $7(3z + y + 2) = 7 \cdot 3z + 7 \cdot y + 7 \cdot 2 = 21z + 7y + 14$

80. $\dfrac{28}{3}$

81. ◈

82. ◈

83. Starting with the final value, we "undo" the rise and drop in value by adding their opposites. The result is the original value.

Rewording: $\quad \underbrace{\text{Final value}} \quad \text{plus} \quad \underbrace{\text{opposite of rise}} \quad \text{plus}$
$$\downarrow \qquad \downarrow \qquad \downarrow \qquad \downarrow$$

Translating: $\quad 64\dfrac{3}{8} \quad + \quad \left(-2\dfrac{3}{8}\right) \quad +$

$$\underbrace{\text{opposite of drop}} \quad \text{is original value.}$$
$$\downarrow \qquad \downarrow \qquad \downarrow$$
$$3\dfrac{1}{4} \quad = \text{original value.}$$

Since $64\frac{3}{8} + \left(-2\frac{3}{8}\right) + 3\frac{1}{4} = 62 + 3\frac{1}{4}$

$$= 65\frac{1}{4},$$

the stock's original value was $\$65\frac{1}{4}$.

84. $\$55.50$

85.
$$4x + \underline{} + (-9x) + (-2y)$$
$$= 4x + (-9x) + \underline{} + (-2y)$$
$$= [4 + (-9)]x + \underline{} + (-2y)$$
$$= -5x + \underline{} + (-2y)$$

This expression is equivalent to $-5x - 7y$, so the missing term is the term which yields $-7y$ when added to $-2y$. Since $-5y + (-2y) = -7y$, the missing term is $-5y$.

86. $-15b$

87.
$$3m + 2n + \underline{} + (-2m)$$
$$= 2n + \underline{} + (-2m) + 3m$$
$$= 2n + \underline{} + (-2 + 3)m$$
$$= 2n + \underline{} + m$$

This expression is equivalent to $2n + (-6m)$, so the missing term is the term which yields $-6m$ when added to m. Since $-7m + m = -6m$, the missing term is $-7m$.

88. $-3y$

89. Note that, in order for the sum to be 0, the two missing terms must be the opposites of the given terms. Thus, the missing terms are $-7t$ and -23.

90. $\frac{7}{2}x$

91. $-3 + (-3) + 2 + (-2) + 1 = -5$

Since the total is 5 under par after the five rounds and $-5 = -1 + (-1) + (-1) + (-1) + (-1)$, the golfer was 1 under par on average.

Exercise Set 1.6

1. The opposite of 39 is -39 because $39 + (-39) = 0$.

2. 17

3. The opposite of -9 is 9 because $-9 + 9 = 0$.

4. $-\frac{7}{2}$

5. The opposite of -3.14 is 3.14 because $-3.14 + 3.14 = 0$.

6. -48.2

7. If $x = 23$, then $-x = -(23) = -23$. (The opposite of 23 is -23.)

8. 26

9. If $x = -\frac{14}{3}$, then $-x = -\left(-\frac{14}{3}\right) = \frac{14}{3}$.
$\left(\text{The opposite of } -\frac{14}{3} \text{ is } \frac{14}{3}.\right)$

10. $-\frac{1}{328}$

11. If $x = 0.101$, then $-x = -(0.101) = -0.101$.
(The opposite of 0.101 is -0.101.)

12. 0

13. If $x = -72$, then $-(-x) = -(-72) = 72$
(The opposite of the opposite of 72 is 72.)

14. 29

15. If $x = -\frac{2}{5}$, then $-(-x) = -\left[-\left(-\frac{2}{5}\right)\right] = -\frac{2}{5}$.
$\left(\text{The opposite of the opposite of } -\frac{2}{5} \text{ is } -\frac{2}{5}.\right)$

16. -9.1

17. When we change the sign of -1 we obtain 1.

18. 7

19. When we change the sign of 7 we obtain -7.

20. -10

21. $-3 - 5$ is read "negative three minus five."
$$-3 - 5 = -3 + (-5) = -8$$

22. Negative four minus seven; $= -11$

23. $2 - (-9)$ is read "two minus negative nine."
$$2 - (-9) = 2 + 9 = 11$$

24. Five minus negative eight; 13

25. $4 - 6$ is read "four minus six."
$$4 - 6 = 4 + (-6) = -2$$

26. Nine minus twelve; -3

27. $-5 - (-7)$ is read "negative five minus negative seven."
$$-5 - (-7) = -5 + 7 = 2$$

28. Negative two minus negative five; 3

29. $6 - 8 = 6 + (-8) = -2$

30. -9

31. $0 - 5 = 0 + (-5) = -5$

32. -8

33. $3 - 9 = 3 + (-9) = -6$

34. -10

35. $0 - 10 = 0 + (-10) = -10$

36. -7

37. $-9 - (-3) = -9 + 3 = -6$

38. -4

39. Note that we are subtracting a number from itself. The result is 0. We could also do this exercise as follows:

$$-8 - (-8) = -8 + 8 = 0$$

40. 0

41. $14 - 19 = 14 + (-19) = -5$

42. -4

43. $30 - 40 = 30 + (-40) = -10$

44. -7

45. $-7 - (-9) = -7 + 9 = 2$

46. -5

47. $-9 - (-9) = -9 + 9 = 0$
(See Exercise 39.)

48. 0

49. $5 - 5 = 5 + (-5) = 0$
(See Exercise 39.)

50. 0

51. $4 - (-4) = 4 + 4 = 8$

52. 12

53. $-7 - 4 = -7 + (-4) = -11$

54. -14

55. $6 - (-10) = 6 + 10 = 16$

56. 15

57. $-14 - 2 = -14 + (-2) = -16$

58. -19

59. $-4 - (-3) = -4 + 3 = -1$

60. -1

61. $5 - (-6) = 5 + 6 = 11$

62. 17

63. $0 - 6 = 0 + (-6) = -6$

64. -5

65. $-3 - (-1) = -3 + 1 = -2$

66. -3

67. $-9 - 16 = -9 + (-16) = -25$

68. -21

69. $0 - (-1) = 0 + 1 = 1$

70. 5

71. $-9 - 0 = -9 + 0 = -9$

72. -8

73. $12 - (-5) = 12 + 5 = 17$

74. 10

75. $18 - 63 = 18 + (-63) = -45$

76. -23

77. $-18 - 63 = -18 + (-63) = -81$

78. -68

79. $-45 - 4 = -45 + (-4) = -49$

80. -58

81. $1.5 - 9.4 = 1.5 + (-9.4) = -7.9$

82. -5.5

83. $0.825 - 1 = 0.825 + (-1) = -0.175$

84. -0.928

85. $\dfrac{3}{7} - \dfrac{5}{7} = \dfrac{3}{7} + \left(-\dfrac{5}{7}\right) = -\dfrac{2}{7}$

86. $-\dfrac{7}{11}$

87. $\dfrac{-2}{9} - \dfrac{5}{9} = \dfrac{-2}{9} + \left(\dfrac{-5}{9}\right) = \dfrac{-7}{9}$, or $-\dfrac{7}{9}$

88. $-\dfrac{4}{5}$

89. $-\dfrac{2}{13} - \left(-\dfrac{5}{13}\right) = -\dfrac{2}{13} + \dfrac{5}{13} = \dfrac{3}{13}$

90. $\dfrac{5}{17}$

91. We subtract the smaller number from the larger.
Translate: $3.8 - (-5.2)$
Simplify: $3.8 - (-5.2) = 3.8 + 5.2 = 9$

92. $-2.1 - (-5.9)$; 3.8

93. We subtract the smaller number from the larger.
Translate: $114 - (-79)$
Simplify: $114 - (-79) = 114 + 79 = 193$

94. $23 - (-17)$; 40

95. $-21 - 37 = -21 + (-37) = -58$

96. -26

97. $9 - (-25) = 9 + 25 = 34$

98. 26

99. $25 - (-12) - 7 - (-2) + 9 = 25 + 12 + (-7) + 2 + 9 = 41$

100. -22

101. $-31 + (-28) - (-14) - 17 = (-31) + (-28) + 14 + (-17) = -62$

102. 22

103. $-34 - 28 + (-33) - 44 = (-34) + (-28) + (-33) + (-44) = -139$

104. 5

105. $-93 + (-84) - (-93) - (-84)$
Note that we are subtracting -93 from -93 and -84 from -84. Thus, the result will be 0. We could also do this exercise as follows:
$-93 + (-84) - (-93) - (-84) = -93 + (-84) + 93 + 84 = 0$

106. 4

107. $-7x - 4y = -7x + (-4y)$, so the terms are $-7x$ and $-4y$.

108. $7a, -9b$

109. $9 - 5t - 3st = 9 + (-5t) + (-3st)$, so the terms are 9, $-5t$, and $-3st$.

110. $-4, -3x, 2xy$

111. $4x - 7x$
$= 4x + (-7x)$ Adding the opposite
$= (4 + (-7))x$ Using the distributive law
$= -3x$

112. $-11a$

113. $7a - 12a + 4$
$= 7a + (-12a) + 4$ Adding the opposite
$= (7 + (-12))a + 4$ Using the distributive law
$= -5a + 4$

114. $-22x + 7$

115. $-8n - 9 + n$
$= -8n + (-9) + n$ Adding the opposite
$= -8n + n + (-9)$ Using the commutative law of addition
$= -7n - 9$ Adding like terms

116. $9n - 15$

117. $3x + 5 - 9x$
$= 3x + 5 + (-9x)$
$= 3x + (-9x) + 5$
$= -6x + 5$

118. $3a - 5$

119. $2 - 6t - 9 - 2t$
$= 2 + (-6t) + (-9) + (-2t)$
$= 2 + (-9) + (-6t) + (-2t)$
$= -7 - 8t$

120. $-2b - 12$

121. $5y + (-3x) - 9x + 1 - 2y + 8$
$= 5y + (-3x) + (-9x) + 1 + (-2y) + 8$
$= 5y + (-2y) + (-3x) + (-9x) + 1 + 8$
$= 3y - 12x + 9$

122. $46 + 3x + 6z$

123. $13x - (-2x) + 45 - (-21) - 7x$
$= 13x + 2x + 45 + 21 + (-7x)$
$= 13x + 2x + (-7x) + 45 + 21$
$= 8x + 66$

124. $6x + 39$

125. We subtract the lower temperature from the higher temperature:
$44 - (-56) = 44 + 56 = 100$
The temperature dropped 100°F.

126. \$165

127. We subtract the lower elevation from the higher elevation:
$29,028 - (-1312) = 30,340$
The difference in elevation is 30,340 ft.

128. 14,494 ft

129. We subtract the lower elevation from the higher elevation:
$-40 - (-156) = -40 + 156 = 116$
Lake Assal is 116 m lower than the Valdes Peninsula.

130. 1767 m

131. ◈

132. ◈

133. Area $= lw = (36 \text{ ft})(12 \text{ ft}) = 432 \text{ ft}^2$

134. $2 \cdot 2 \cdot 2 \cdot 2 \cdot 2 \cdot 3 \cdot 3 \cdot 3$

135. ◈

136. ◈

137. True. For example, for $m = 5$ and $n = 3$, $5 > 3$ and $5 - 3 > 0$, or $2 > 0$. For $m = -4$ and $n = -9$, $-4 > -9$ and $-4 - (-9) > 0$, or $5 > 0$.

138. False. For example, let $m = -3$ and $n = -5$. Then $-3 > -5$, but $-3 + (-5) = -8 \not> 0$.

139. False. For example, let $m = 2$ and $n = -2$. Then 2 and -2 are opposites, but $2 - (-2) = 4 \neq 0$.

140. True. For example, for $m = 4$ and $n = -4$, $4 = -(-4)$ and $4 + (-4) = 0$; for $m = -3$ and $n = 3$, $-3 = -3$ and $-3 + 3 = 0$.

141. ◈

142. ◈

Exercise Set 1.7

1. $-4 \cdot 9 = -36$ Think: $4 \cdot 9 = 36$, make the answer negative.

2. -21

3. $-8 \cdot 7 = -56$ Think: $8 \cdot 7 = 56$, make the answer negative.

4. -18

5. $8 \cdot (-3) = -24$

6. -45

7. $-9 \cdot 8 = -72$

8. -30

9. $-6 \cdot (-7) = 42$ Multiplying absolute values; the answer is positive.

10. 10

11. $-5 \cdot (-9) = 45$ Multiplying absolute values; the answer is positive.

12. 18

13. $17 \cdot (-10) = -170$

14. 120

15. $-12 \cdot 12 = -144$

16. 195

17. $-25 \cdot (-48) = 1200$

18. -1677

19. $-3.5 \cdot (-28) = 98$

20. -203.7

21. $6 \cdot (-13) = -78$

22. -63

23. $-7 \cdot (-3.1) = 21.7$

24. 12.8

25. $\frac{2}{3} \cdot \left(-\frac{3}{5}\right) = -\left(\frac{2 \cdot 3}{3 \cdot 5}\right) = -\left(\frac{2}{5} \cdot \frac{3}{3}\right) = -\frac{2}{5}$

26. $-\frac{10}{21}$

27. $-\dfrac{3}{8}\cdot\left(-\dfrac{2}{9}\right)=\dfrac{\cancel{3}\cdot\cancel{2}\cdot 1}{4\cdot\cancel{2}\cdot\cancel{3}\cdot 3}=\dfrac{1}{12}$

28. $\dfrac{1}{4}$

29. $(-5.3)(2.1)=-11.13$

30. -40.85

31. $-\dfrac{5}{9}\cdot\dfrac{3}{4}=-\dfrac{5\cdot\cancel{3}}{\cancel{3}\cdot 3\cdot 4}=-\dfrac{5}{12}$

32. -6

33. $\quad 3\cdot(-7)\cdot(-2)\cdot 6$

$\quad=-21\cdot(-12)\qquad$ Multiplying the first two

$\qquad\qquad\qquad\qquad$ numbers and the last two

$\qquad\qquad\qquad\qquad$ numbers

$\quad=252$

34. 756

35. 0, The product of 0 and any real number is 0.

36. 0

37. $-\dfrac{1}{3}\cdot\dfrac{1}{4}\cdot\left(-\dfrac{3}{7}\right)=-\dfrac{1}{12}\cdot\left(-\dfrac{3}{7}\right)=\dfrac{3}{12\cdot 7}=$

$\dfrac{\cancel{3}\cdot 1}{\cancel{3}\cdot 4\cdot 7}=\dfrac{1}{28}$

38. $\dfrac{3}{35}$

39. $-2\cdot(-5)\cdot(-3)\cdot(-5)=10\cdot 15=150$

40. 30

41. 0, The product of 0 and any real number is 0.

42. 0

43. $(-8)(-9)(-10)=72(-10)=-720$

44. 5040

45. $(-6)(-7)(-8)(-9)(-10)=42\cdot 72\cdot(-10)=$
$3024\cdot(-10)=-30,240$

46. $151,200$

47. $28\div(-7)=-4\qquad$ Check: $-4\cdot(-7)=28$

48. -8

49. $\dfrac{36}{-9}=-4\qquad -4\cdot(-9)=36$

50. -2

51. $\dfrac{-16}{8}=-2\qquad$ Check: $-2\cdot 8=-16$

52. 8

53. $\dfrac{-48}{-12}=4\qquad$ Check: $4(-12)=-48$

54. 7

55. $\dfrac{-72}{9}=-8\qquad$ Check: $-8\cdot 9=-72$

56. -2

57. $-100\div(-50)=2\qquad$ Check: $2(-50)=-100$

58. -25

59. $-108\div 9=-12\qquad$ Check: $-12\cdot 9=-108$

60. $\dfrac{64}{7}$

61. $\dfrac{400}{-50}=-8\qquad$ Check: $-8\cdot(-50)=400$

62. $\dfrac{300}{13}$

63. Undefined

64. 0

65. $-4.8\div 1.2=-4\qquad$ Check: $-4(1.2)=-4.8$

66. -3

67. $\dfrac{0}{-9}=0$

68. Undefined

69. Undefined

70. 0

71. $\dfrac{-8}{3}=\dfrac{8}{-3}\ $ and $\ \dfrac{-8}{3}=-\dfrac{8}{3}$

72. $\dfrac{12}{-7},\ -\dfrac{12}{7}$

73. $\dfrac{29}{-35}=\dfrac{-29}{35}\ $ and $\ \dfrac{29}{-35}=-\dfrac{29}{35}$

74. $\dfrac{-9}{14},\ -\dfrac{9}{14}$

75. $-\dfrac{7}{3}=\dfrac{-7}{3}\ $ and $\ -\dfrac{7}{3}=\dfrac{7}{-3}$

76. $\dfrac{-4}{15},\ \dfrac{4}{-15}$

77. $\frac{-x}{2} = \frac{x}{-2}$ and $\frac{-x}{2} = -\frac{x}{2}$

78. $\frac{-9}{a}$, $-\frac{9}{a}$

79. The reciprocal of $\frac{4}{-5}$ is $\frac{-5}{4}$ $\left(\text{or equivalently, } -\frac{5}{4}\right)$ because $\frac{4}{-5} \cdot \frac{-5}{4} = 1$.

80. $-\frac{9}{2}$

81. The reciprocal of $-\frac{47}{13}$ is $-\frac{13}{47}$ because $-\frac{47}{13} \cdot \left(-\frac{13}{47}\right) = 1$.

82. $-\frac{12}{31}$

83. The reciprocal of -10 is $\frac{1}{-10}$ $\left(\text{or equivalently, } -\frac{1}{10}\right)$ because $-10\left(\frac{1}{-10}\right) = 1$.

84. $\frac{1}{13}$

85. The reciprocal of 4.3 is $\frac{1}{4.3}$ because $4.3\left(\frac{1}{4.3}\right) = 1$. Since $\frac{1}{4.3} = \frac{1}{4.3} \cdot \frac{10}{10} = \frac{10}{43}$, the reciprocal can also be expressed as $\frac{10}{43}$.

86. $\frac{1}{-8.5}$, or $-\frac{1}{8.5}$

87. The reciprocal of $\frac{-9}{4}$ is $\frac{4}{-9}$ $\left(\text{or equivalently, } -\frac{4}{9}\right)$ because $\frac{-9}{4} \cdot \frac{4}{-9} = 1$.

88. $\frac{11}{-6}$, or $-\frac{11}{6}$

89. The reciprocal of -1 is $\frac{1}{-1}$, or -1 because $(-1)(-1) = 1$.

90. $\frac{5}{3}$

91. $\left(\frac{-7}{4}\right)\left(-\frac{3}{5}\right)$
$= \left(-\frac{7}{4}\right)\left(-\frac{3}{5}\right)$ Rewriting $\frac{-7}{4}$ as $-\frac{7}{4}$
$= \frac{21}{20}$

92. $\frac{5}{18}$

93. $\left(\frac{-6}{5}\right)\left(\frac{2}{-11}\right)$
$= \left(\frac{-6}{5}\right)\left(\frac{-2}{11}\right)$ Rewriting $\frac{2}{-11}$ as $\frac{-2}{11}$
$= \frac{12}{55}$

94. $\frac{35}{12}$

95. $\frac{-3}{8} + \frac{-5}{8} = \frac{-8}{8} = -1$

96. $\frac{3}{5}$

97. $\left(\frac{-9}{5}\right)\left(\frac{5}{-9}\right)$
Note that this is the product of reciprocals. Thus, the result is 1.

98. $\frac{5}{28}$

99. $\left(-\frac{3}{11}\right) + \left(-\frac{6}{11}\right) = -\frac{9}{11}$

100. $-\frac{6}{7}$

101. $\frac{7}{8} \div \left(-\frac{1}{2}\right) = \frac{7}{8} \cdot \left(-\frac{2}{1}\right) = -\frac{14}{8} = -\frac{7 \cdot 2}{2 \cdot 4 \cdot 1} = -\frac{7}{4}$

102. $-\frac{9}{8}$

103. $\frac{9}{5} \cdot \frac{-20}{3} = \frac{9}{5}\left(-\frac{20}{3}\right) = -\frac{180}{15} = -\frac{3 \cdot 3 \cdot 4 \cdot 5}{5 \cdot 3 \cdot 1} = -12$

104. $-\frac{7}{36}$

105. $\left(-\frac{18}{7}\right) + \left(-\frac{3}{7}\right) = -\frac{21}{7} = -3$

106. -3

107. $-\frac{5}{9} \div \left(-\frac{5}{9}\right)$
Note that we have a number divided by itself. Thus, the result is 1.

108. $\frac{5}{3}$

109. $-44.1 \div (-6.3) = 7$ Do the long division. The answer is positive.

110. -2

111. $\frac{1}{9} - \frac{2}{9} = -\frac{1}{9}$

112. $-\frac{4}{7}$

113. $\frac{-3}{10} + \frac{2}{5} = \frac{-3}{10} + \frac{2}{5} \cdot \frac{2}{2} = \frac{-3}{10} + \frac{4}{10} = \frac{1}{10}$

114. $\frac{1}{9}$

115. $\frac{7}{10} \div \left(\frac{-3}{5}\right) = \frac{7}{10} \div \left(-\frac{3}{5}\right) = \frac{7}{10} \cdot \left(-\frac{5}{3}\right) = -\frac{35}{30} =$
$-\frac{7 \cdot \cancel{5}}{2 \cdot \cancel{5} \cdot 3} = -\frac{7}{6}$

116. $-\frac{3}{2}$

117. $\frac{5}{7} - \frac{1}{-7} = \frac{5}{7} - \left(-\frac{1}{7}\right) = \frac{5}{7} + \frac{1}{7} = \frac{6}{7}$

118. $\frac{5}{9}$

119. $\frac{-4}{15} + \frac{2}{-3} = \frac{-4}{15} + \frac{-2}{3} = \frac{-4}{15} + \frac{-2}{3} \cdot \frac{5}{5} = \frac{-4}{15} + \frac{-10}{15} =$
$\frac{-14}{15}$, or $-\frac{14}{15}$

120. $-\frac{1}{2}$

121.

122. ◈

123. $\frac{264}{468} = \frac{\cancel{2} \cdot \cancel{2} \cdot 2 \cdot \cancel{3} \cdot 11}{\cancel{2} \cdot \cancel{2} \cdot \cancel{3} \cdot 3 \cdot 13} = \frac{22}{39}$

124. $12x - 2y - 9$

125. ◈

126. ◈

127. Consider the sum $2 + 3$. Its reciprocal is $\frac{1}{2+3}$, or
$\frac{1}{5}$, but $\frac{1}{2} + \frac{1}{3} = \frac{5}{6}$.

128. $-1, 1$

129. When n is negative, $-n$ is positive, so $\frac{m}{-n}$ is the
quotient of a negative and a positive number and,
thus, is negative.

130. Positive

131. When n is negative, $-n$ is positive, so $\frac{-n}{m}$ is the
quotient of a positive and a negative number and,
thus, is negative. When m is negative, $-m$ is posi-
tive, so $-m \cdot \left(\frac{-n}{m}\right)$ is the product of a positive and
a negative number and, thus, is negative.

132. Positive

133. $m + n$ is the sum of two negative numbers, so it is
negative; $\frac{m}{n}$ is the quotient of two negative numbers,
so it is positive. Then $(m + n) \cdot \frac{m}{n}$ is the product
of a negative and a positive number and, thus, is
negative.

134. Positive

135. a) m and n have different signs;
   b) either m or n is zero;
   c) m and n have the same sign

136. $a(-b) + ab = a[-b + b]$ Distributive law
                    $= a(0)$ Law of opposites
                    $= 0$ Multiplicative property of 0
Therefore, $a(-b)$ is the opposite of ab by the law of
opposites.

137. ◈

Exercise Set 1.8

1. $\underbrace{4 \cdot 4 \cdot 4}_{3\ \text{factors}} = 4^3$

2. 6^4

3. $\underbrace{x \cdot x \cdot x \cdot x \cdot x \cdot x \cdot x}_{7\ \text{factors}} = x^7$

4. y^6

5. $3t \cdot 3t \cdot 3t \cdot 3t \cdot 3t = (3t)^5$

6. $(5m)^5$

7. $2^4 = 2 \cdot 2 \cdot 2 \cdot 2 = 4 \cdot 4 = 16$

8. 125

9. $(-3)^2 = (-3)(-3) = 9$

10. 49

11. $-3^2 = -(3 \cdot 3) = -9$

12. -49

13. $4^3 = 4 \cdot 4 \cdot 4 = 16 \cdot 4 = 64$

14. 9

15. $(-5)^4 = (-5)(-5)(-5)(-5) = 25 \cdot 25 = 625$

16. 625

17. $7^1 = 7$ (1 factor)

18. -1

19. $(3t)^4 = (3t)(3t)(3t)(3t) =$
$3 \cdot 3 \cdot 3 \cdot 3 \cdot t \cdot t \cdot t \cdot t = 81t^4$

20. $25t^2$

21. $(-7x)^3 = (-7x)(-7x)(-7x) =$
$(-7)(-7)(-7)(x)(x)(x) = -343x^3$

22. $625x^4$

23. $\begin{aligned} 5 + 3 \cdot 7 &= 5 + 21 && \text{Multiplying} \\ &= 26 && \text{Adding} \end{aligned}$

24. -5

25. $\begin{aligned} 8 \cdot 7 + 6 \cdot 5 &= 56 + 30 && \text{Multiplying} \\ &= 86 && \text{Adding} \end{aligned}$

26. 51

27. $\begin{aligned} 19 - 5 \cdot 3 + 3 &= 19 - 15 + 3 && \text{Multiplying} \\ &= 4 + 3 && \text{Subtracting and add-} \\ &= 7 && \text{ing from left to right} \end{aligned}$

28. 9

29. $\begin{aligned} 9 \div 3 + 16 \div 8 &= 3 + 2 && \text{Dividing} \\ &= 5 && \text{Adding} \end{aligned}$

30. 28

31. $84 \div 28 - 84 \div 28$

Note that we are subtracting a number, $84 \div 28$, from itself. Thus, the result is 0.

32. 21

33. $\begin{aligned} & 4 - 8 \div 2 + 3^2 \\ &= 4 - 8 \div 2 + 9 && \text{Simplifying the exponential} \\ & && \text{expression} \\ &= 4 - 4 + 9 && \text{Dividing} \\ &= 0 + 9 && \text{Subtracting and} \\ &= 9 && \text{adding from left to right} \end{aligned}$

34. 298

35. $\begin{aligned} & 9 - 3^2 \div 9(-1) \\ &= 9 - 9 \div 9(-1) && \text{Simplifying the exponential} \\ & && \text{expression} \\ &= 9 - 1(-1) && \text{Dividing and} \\ &= 9 + 1 && \text{multiplying from left to right} \\ &= 10 && \text{Adding} \end{aligned}$

36. 11

37. $\begin{aligned} (8 - 2 \cdot 3) - 9 &= (8 - 6) - 9 && \text{Multiplying} \\ & && \text{inside the parentheses} \\ &= 2 - 9 && \text{Subtracting} \\ & && \text{inside the parentheses} \\ &= -7 \end{aligned}$

38. -36

39. $(-24) \div (-3) \cdot \left(-\dfrac{1}{2}\right) = 8 \cdot \left(-\dfrac{1}{2}\right) = -\dfrac{8}{2} = -4$

40. 32

41. $\begin{aligned} & 13(-10)^2 + 45 \div (-5) \\ &= 13(100) + 45 \div (-5) && \text{Simplifying the} \\ & && \text{exponential expression} \\ &= 1300 + 45 \div (-5) && \text{Multiplying and} \\ &= 1300 - 9 && \text{dividing from left to right} \\ &= 1291 && \text{Subtracting} \end{aligned}$

42. 13

43. $2^4 + 2^3 - 10 \div (-1)^4 = 16 + 8 - 10 \div 1 =$
$16 + 8 - 10 = 24 - 10 = 14$

44. 33

45. $5 + 3(2-9)^2 = 5 + 3(-7)^2 = 5 + 3 \cdot 49 = 5 + 147 = 152$

46. 13

47. $[2 \cdot (5 - 8)]^2 = [2 \cdot (-3)]^2 = (-6)^2 = 36$

48. 12

49. $\dfrac{7+2}{5^2 - 4^2} = \dfrac{9}{25 - 16} = \dfrac{9}{9} = 1$

50. 2

51. $8(-7) + |6(-5)| = -56 + |-30| = -56 + 30 = -26$

52. 49

53. $\dfrac{(-2)^3 + 4^2}{3 - 5^2 + 3 \cdot 6} = \dfrac{-8 + 16}{3 - 25 + 3 \cdot 6} = \dfrac{8}{3 - 25 + 18} =$
$\dfrac{8}{-22 + 18} = \dfrac{8}{-4} = -2$

54. -5

55. $\dfrac{27 - 2 \cdot 3^2}{8 \div 2^2 - (-2)^2} = \dfrac{27 - 2 \cdot 9}{8 \div 4 - 4} = \dfrac{27 - 18}{2 - 4} = \dfrac{9}{-2} = -\dfrac{9}{2}$

56. 5

57. $7 - 5x = 7 - 5 \cdot 3$ Substituting 3 for x
$ = 7 - 15$ Multiplying
$ = -8$ Subtracting

58. -7

59. $ 24 \div t^3$
$= 24 \div (-2)^3$ Substituting -2 for t
$= 24 \div (-8)$ Simplifying the exponential
 expression
$= -3$ Dividing

60. 16

61. $ 45 \div 3 \cdot a = 45 \div 3 \cdot (-1)$ Substituting -1 for a
$ = 15 \cdot (-1)$ Dividing
$ = -15$ Multiplying

62. -125

63. $ 5x \div 15x^2$
$= 5 \cdot 3 \div 15(3)^2$ Substituting 3 for x
$= 5 \cdot 3 \div 15 \cdot 9$ Simplifying the exponential
 expression
$= 15 \div 15 \cdot 9$ Multiplying and dividing
$= 1 \cdot 9$ in order from
$= 9$ left to right

64. 8

65. $(12 \cdot 17) \div (17 \cdot 12)$

Since $12 \cdot 17$ and $17 \cdot 12$ are equivalent expressions, we have a number divided by itself so the result is 1.

66. 20

67. $-x^2 - 5x = -(-3)^2 - 5(-3) = -9 - 5(-3) = -9 + 15 = 6$

68. 24

69. $\dfrac{3a - 4a^2}{a^2 - 20} = \dfrac{3 \cdot 5 - 4(5)^2}{(5)^2 - 20} = \dfrac{3 \cdot 5 - 4 \cdot 25}{25 - 20} = $
$\dfrac{15 - 100}{5} = \dfrac{-85}{5} = -17$

70. 0

71. $-(9x + 1) = -9x - 1$ Removing parentheses and changing the sign of each term

72. $-3x - 5$

73. $-(7 - 2x) = -7 + 2x$ Removing parentheses and changing the sign of each term

74. $-6x + 7$

75. $-(4a - 3b + 7c) = -4a + 3b - 7c$

76. $-5x + 2y + 3z$

77. $-(3x^2 + 5x - 1) = -3x^2 - 5x + 1$

78. $-8x^3 + 6x - 5$

79. $ 5x - (2x + 7)$
$= 5x - 2x - 7$ Removing parentheses and changing the sign of each term
$= 3x - 7$ Collecting like terms

80. $5y - 9$

81. $2a - (5a - 9) = 2a - 5a + 9 = -3a + 9$

82. $8n + 7$

83. $2x + 7x - (4x + 6) = 2x + 7x - 4x - 6 = 5x - 6$

84. $a - 7$

85. $9t - 5r - 2(3r + 6t) = 9t - 5r - 6r - 12t = -3t - 11r$

86. $-2m - 6n$

87. $ 15x - y - 5(3x - 2y + 5z)$
$= 15x - y - 15x + 10y - 25z$ Multiplying each term in parentheses by -5
$= 9y - 25z$

88. $-16a + 27b - 32c$

89. $3x^2 + 7 - (2x^2 + 5) = 3x^2 + 7 - 2x^2 - 5$
$ = x^2 + 2$

90. 0

91. $5t^3 + t - 3(t + 2t^3) = 5t^3 + t - 3t - 6t^3$
$ = -t^3 - 2t$

92. $2n^2 - n$

93. $ 12a^2 - 3ab + 5b^2 - 5(-5a^2 + 4ab - 6b^2)$
$= 12a^2 - 3ab + 5b^2 + 25a^2 - 20ab + 30b^2$
$= 37a^2 - 23ab + 35b^2$

94. $-20a^2 + 29ab + 48b^2$

95. $\quad -7t^3 - t^2 - 3(5t^3 - 3t)$
$\quad = -7t^3 - t^2 - 15t^3 + 9t$
$\quad = -22t^3 - t^2 + 9t$

96. $9t^4 - 45t^3 + 17t$

97. $\quad 5(2x - 7) - [4(2x - 3) + 2]$
$\quad = 5(2x - 7) - [8x - 12 + 2]$
$\quad = 5(2x - 7) - [8x - 10]$
$\quad = 10x - 35 - 8x + 10$
$\quad = 2x - 25$

98. $42x - 23$

99. ◈

100. ◈

101. Let x represent "a number." Then we have $2x + 9$.

102. Let x and y represent the numbers; $\frac{1}{2}(x + y)$.

103. ◈

104. ◈

105. $\quad 5t - \{7t - [4r - 3(t - 7)] + 6r\} - 4r$
$\quad = 5t - \{7t - [4r - 3t + 21] + 6r\} - 4r$
$\quad = 5t - \{7t - 4r + 3t - 21 + 6r\} - 4r$
$\quad = 5t - \{10t + 2r - 21\} - 4r$
$\quad = 5t - 10t - 2r + 21 - 4r$
$\quad = -5t - 6r + 21$

106. $-4z$

107. $\quad \{x - [f - (f - x)] + [x - f]\} - 3x$
$\quad = \{x - [f - f + x] + [x - f]\} - 3x$
$\quad = \{x - [x] + [x - f]\} - 3x$
$\quad = \{x - x + x - f\} - 3x$
$\quad = x - f - 3x$
$\quad = -2x - f$

108. ◈

109. ◈

110. True

111. False; let $m = 1$ and $n = 2$. Then $-2 + 1 = -(2 - 1) = -1$, but $-(2 + 1) = -3$.

112. True

113. False; let $m = 2$ and $n = 3$. Then $3(-3 - 2) = 3(-5) = -15$, but $-3^2 + 3 \cdot 2 = -9 + 6 = -3$.

114. False

115. True; $-m(-n + m) = mn - m^2 = m(n - m)$

116. True

117. $[x + 3(2 - 5x) \div 7 + x](x - 3)$
When $x = 3$, the factor $x - 3$ is 0, so the product is 0.

118. 1

119. $\quad 4 \cdot 20^3 + 17 \cdot 20^2 + 10 \cdot 20 + 0 \cdot 2$
$\quad = 4 \cdot 8000 + 17 \cdot 400 + 10 \cdot 20 + 0 \cdot 2$
$\quad = 32,000 + 6800 + 200 + 0$
$\quad = 39,000$

120. 1, 5, 0

Chapter 2

Equations, Inequalities, and Problem Solving

Exercise Set 2.1

1. $x + 8 = 23$

$x + 8 - 8 = 23 - 8$ Subtracting 8 from both sides

$x = 15$ Simplifying

Check: $x + 8 = 23$

$15 + 8 \; ? \; 23$

$23 \mid 23$ TRUE

The solution is 15.

2. 3

3. $t + 9 = -4$

$t + 9 - 9 = -4 - 9$ Subtracting 9 from both sides

$t = -13$

Check: $t + 9 = -4$

$-13 + 9 \; ? \; -4$

$-4 \mid -4$ TRUE

The solution is -13.

4. 34

5. $y + 7 = -3$

$y + 7 - 7 = -3 - 7$

$y = -10$

Check: $y + 7 = -3$

$-10 + 7 \; ? \; -3$

$-3 \mid -3$ TRUE

The solution is -10.

6. -21

7. $-5 = x + 8$

$-5 - 8 = x + 8 - 8$

$-13 = x$

Check: $-5 = x + 8$

$-5 \; ? \; -13 + 8$

$-5 \mid -5$ TRUE

The solution is -13.

8. -31

9. $x - 9 = 6$

$x - 9 + 9 = 6 + 9$

$x = 15$

Check: $x - 9 = 6$

$15 - 9 \; ? \; 6$

$6 \mid 6$ TRUE

The solution is 15.

10. 13

11. $y - 6 = -14$

$y - 6 + 6 = -14 + 6$

$y = -8$

Check: $y - 6 = -14$

$-8 - 6 \; ? \; -14$

$-14 \mid -14$ TRUE

The solution is -8.

12. -15

13. $9 + t = 3$

$-9 + 9 + t = -9 + 3$

$t = -6$

Check: $9 + t = 3$

$9 - 6 \; ? \; 3$

$3 \mid 3$ TRUE

The solution is -6.

14. 18

15. $12 = -7 + y$

$7 + 12 = 7 + (-7) + y$

$19 = y$

Check: $12 = -7 + y$

$12 \; ? \; -7 + 19$

$12 \mid 12$ TRUE

The solution is 19.

16. 24

17. $-5 + t = -9$

$5 + (-5) + t = 5 + (-9)$

$t = -4$

Check: $-5 + t = -9$

$-5 + (-4) \; ? \; -9$

$-9 \mid -9$ TRUE

The solution is -4.

18. -15

19.
$$r + \frac{1}{3} = \frac{8}{3}$$
$$r + \frac{1}{3} - \frac{1}{3} = \frac{8}{3} - \frac{1}{3}$$
$$r = \frac{7}{3}$$

Check:
$$\begin{array}{c} r + \frac{1}{3} = \frac{8}{3} \\ \hline \frac{7}{3} + \frac{1}{3} \; ? \; \frac{8}{3} \\ \frac{8}{3} \; \bigg| \; \frac{8}{3} \quad \text{TRUE} \end{array}$$

The solution is $\frac{7}{3}$.

20. $\frac{1}{4}$

21.
$$x + \frac{3}{5} = -\frac{7}{10}$$
$$x + \frac{3}{5} - \frac{3}{5} = -\frac{7}{10} - \frac{3}{5}$$
$$x = -\frac{7}{10} - \frac{3}{5} \cdot \frac{2}{2}$$
$$x = -\frac{7}{10} - \frac{6}{10}$$
$$x = -\frac{13}{10}$$

Check:
$$\begin{array}{c} x + \frac{3}{5} = -\frac{7}{10} \\ \hline -\frac{13}{10} + \frac{3}{5} \; ? \; -\frac{7}{10} \\ -\frac{13}{10} + \frac{6}{10} \; \bigg| \\ -\frac{7}{10} \; \bigg| \; -\frac{7}{10} \quad \text{TRUE} \end{array}$$

The solution is $-\frac{13}{10}$.

22. $-\frac{3}{2}$

23.
$$x - \frac{5}{6} = \frac{7}{8}$$
$$x - \frac{5}{6} + \frac{5}{6} = \frac{7}{8} + \frac{5}{6}$$
$$x = \frac{7}{8} \cdot \frac{3}{3} + \frac{5}{6} \cdot \frac{4}{4}$$
$$x = \frac{21}{24} + \frac{20}{24}$$
$$x = \frac{41}{24}$$

Check:
$$\begin{array}{c} x - \frac{5}{6} = \frac{7}{8} \\ \hline \frac{41}{24} - \frac{5}{6} \; ? \; \frac{7}{8} \\ \frac{41}{24} - \frac{20}{24} \; \bigg| \; \frac{21}{24} \\ \frac{21}{24} \; \bigg| \; \frac{21}{24} \quad \text{TRUE} \end{array}$$

The solution is $\frac{41}{24}$.

24. $\frac{19}{12}$

25.
$$-\frac{1}{5} + z = -\frac{1}{4}$$
$$\frac{1}{5} - \frac{1}{5} + z = \frac{1}{5} - \frac{1}{4}$$
$$z = \frac{1}{5} \cdot \frac{4}{4} - \frac{1}{4} \cdot \frac{5}{5}$$
$$z = \frac{4}{20} - \frac{5}{20}$$
$$z = -\frac{1}{20}$$

Check:
$$\begin{array}{c} -\frac{1}{5} + z = -\frac{1}{4} \\ \hline -\frac{1}{5} + \left(-\frac{1}{20}\right) \; ? \; -\frac{1}{4} \\ -\frac{4}{20} + \left(-\frac{1}{20}\right) \; \bigg| \; -\frac{5}{20} \\ -\frac{5}{20} \; \bigg| \; -\frac{5}{20} \quad \text{TRUE} \end{array}$$

The solution is $-\frac{1}{20}$.

26. $-\frac{5}{8}$

27.
$$m + 3.9 = 5.4$$
$$m + 3.9 - 3.9 = 5.4 - 3.9$$
$$m = 1.5$$

Check:
$$\begin{array}{c} m + 3.9 = 5.4 \\ \hline 1.5 + 3.9 \; ? \; 5.4 \\ 5.4 \; \bigg| \; 5.4 \quad \text{TRUE} \end{array}$$

The solution is 1.5.

28. 3.4

29.
$$-9.7 = -4.7 + y$$
$$4.7 + (-9.7) = 4.7 + (-4.7) + y$$
$$-5 = y$$

Check: $\dfrac{-9.7 = -4.7 + y}{}$

$-9.7 \ ? \ -4.7 + (-5)$
$-9.7 \ | \ -9.7$ TRUE

The solution is -5.

30. -10.6

31. $5x = 80$

$\dfrac{5x}{5} = \dfrac{80}{5}$ Dividing both sides by 5

$1 \cdot x = 16$ Simplifying

$x = 16$ Identity property of 1

Check: $\dfrac{5x = 80}{}$

$5 \cdot 16 \ ? \ 80$
$80 \ | \ 80$ TRUE

The solution is 16.

32. 13

33. $9t = 36$

$\dfrac{9t}{9} = \dfrac{36}{9}$ Dividing both sides by 9

$1 \cdot t = 4$ Simplifying

$t = 4$ Identity property of 1

Check: $\dfrac{9t = 36}{}$

$9 \cdot 4 \ ? \ 36$
$36 \ | \ 36$ TRUE

The solution is 4.

34. 12

35. $84 = 7x$

$\dfrac{84}{7} = \dfrac{7x}{7}$ Dividing both sides by 7

$12 = 1 \cdot x$
$12 = x$

Check: $\dfrac{84 = 7x}{}$

$84 \ ? \ 7 \cdot 12$
$84 \ | \ 84$ TRUE

The solution is 12.

36. 8

37. $-x = 23$

$-1 \cdot x = 23$
$-1 \cdot (-1 \cdot x) = -1 \cdot 23$
$1 \cdot x = -23$
$x = -23$

Check: $\dfrac{-x = 23}{}$

$-(-23) \ ? \ 23$
$23 \ | \ 23$ TRUE

The solution is -23.

38. -100

39. $-t = -8$

The equation states that the opposite of t is the opposite of 8. Thus, $t = 8$. We could also do this exercise as follows.

$-t = -8$

$-1(-t) = -1(-8)$ Multiplying both sides by -1

$t = 8$

Check: $\dfrac{-t = -8}{}$

$-(8) \ ? \ -8$
$-8 \ | \ -8$ TRUE

The solution is 8.

40. 68

41. $7x = -49$

$\dfrac{7x}{7} = \dfrac{-49}{7}$

$1 \cdot x = -7$
$x = -7$

Check: $\dfrac{7x = -49}{}$

$7(-7) \ ? \ -49$
$-49 \ | \ -49$ TRUE

The solution is -7.

42. -4

43. $-12x = 72$

$\dfrac{-12x}{-12} = \dfrac{72}{-12}$

$1 \cdot x = -6$
$x = -6$

Check: $\dfrac{-12x = 72}{}$

$-12(-6) \ ? \ 72$
$72 \ | \ 72$ TRUE

The solution is -6.

44. -7

45. $-3.4t = -20.4$

$\dfrac{-3.4t}{-3.4} = \dfrac{-20.4}{-3.4}$

$1 \cdot t = 6$
$t = 6$

Check: $\dfrac{-3.4t = -20.4}{}$

$-3.4(6) \ ? \ -20.4$
$-20.4 \ | \ -20.4$ TRUE

The solution is 6.

46. 8

47.
$$\frac{a}{4} = 13$$
$$\frac{1}{4} \cdot a = 13$$
$$4 \cdot \frac{1}{4} \cdot a = 4 \cdot 13$$
$$a = 52$$

Check: $\dfrac{a}{4} = 13$

$$\frac{\dfrac{52}{4} \; ? \; 13}{13 \;\Big|\; 13} \quad \text{TRUE}$$

The solution is 52.

48. −88

49.
$$\frac{3}{4}x = 27$$
$$\frac{4}{3} \cdot \frac{3}{4}x = \frac{4}{3} \cdot 27$$
$$1 \cdot x = \frac{4 \cdot \cancel{3} \cdot 3 \cdot 3}{\cancel{3} \cdot 1}$$
$$x = 36$$

Check: $\dfrac{3}{4}x = 27$

$$\frac{\dfrac{3}{4} \cdot 36 \; ? \; 27}{27 \;\Big|\; 27} \quad \text{TRUE}$$

The solution is 36.

50. 20

51.
$$\frac{-t}{5} = 9$$
$$5 \cdot \frac{1}{5} \cdot (-t) = 5 \cdot 9$$
$$-t = 45$$
$$-1(-t) = -1 \cdot 45$$
$$t = -45$$

Check: $\dfrac{-t}{5} = 9$

$$\frac{\dfrac{-(-45)}{5} \; ? \; 9}{\dfrac{45}{5}}$$
$$9 \;\Big|\; 9 \quad \text{TRUE}$$

The solution is −45.

52. −54

53.
$$\frac{2}{7} = \frac{x}{3}$$
$$\frac{2}{7} = \frac{1}{3} \cdot x$$
$$3 \cdot \frac{2}{7} = 3 \cdot \frac{1}{3} \cdot x$$
$$\frac{6}{7} = x$$

Check: $\dfrac{2}{7} = \dfrac{x}{3}$

$$\frac{2}{7} \; ? \; \frac{6/7}{3}$$
$$\frac{6}{7} \cdot \frac{1}{3}$$
$$\frac{6}{21}$$
$$\frac{2}{7} \;\Big|\; \frac{2}{7} \quad \text{TRUE}$$

The solution is $\dfrac{6}{7}$.

54. $\dfrac{5}{9}$

55.
$$-\frac{3}{5}r = -\frac{3}{5}$$

The solution of the equation is the number that is multiplied by $-\dfrac{3}{5}$ to get $-\dfrac{3}{5}$. That number is 1. We could also do this exercise as follows:

$$-\frac{3}{5}r = -\frac{3}{5}$$
$$-\frac{5}{3} \cdot \left(-\frac{3}{5}r\right) = -\frac{5}{3}\left(-\frac{3}{5}\right)$$
$$r = 1$$

Check: $-\dfrac{3}{5}r = -\dfrac{3}{5}$

$$-\frac{3}{5} \cdot 1 \; ? \; -\frac{3}{5}$$
$$-\frac{3}{5} \;\Big|\; -\frac{3}{5} \quad \text{TRUE}$$

The solution is 1.

56. $\dfrac{2}{3}$

57.
$$\frac{-3r}{2} = -\frac{27}{4}$$
$$-\frac{3}{2}r = -\frac{27}{4}$$
$$-\frac{2}{3}\cdot\left(-\frac{3}{2}r\right) = -\frac{2}{3}\cdot\left(-\frac{27}{4}\right)$$
$$r = \frac{2\cdot3\cdot3\cdot3}{3\cdot2\cdot2}$$
$$r = \frac{9}{2}$$

Check:
$$\frac{-3r}{2} = -\frac{27}{4}$$
$$\frac{-\frac{3}{2}\cdot\frac{9}{2}}{} \;?\; -\frac{27}{4}$$
$$-\frac{27}{4} \;\bigg|\; -\frac{27}{4} \qquad \text{TRUE}$$

The solution is $\frac{9}{2}$.

58. -1

59.
$$4.5 + t = -3.1$$
$$4.5 + t - 4.5 = -3.1 - 4.5$$
$$t = -7.6$$
The solution is -7.6.

60. 24

61.
$$-8.2x = 20.5$$
$$\frac{-8.2x}{-8.2} = \frac{20.5}{-8.2}$$
$$x = -2.5$$
The solution is -2.5.

62. -5.5

63.
$$12 = y + 29$$
$$12 - 29 = y + 29 - 29$$
$$-17 = y$$
The solution is -17.

64. -128

65.
$$a - \frac{1}{6} = -\frac{2}{3}$$
$$a - \frac{1}{6} + \frac{1}{6} = -\frac{2}{3} + \frac{1}{6}$$
$$a = -\frac{4}{6} + \frac{1}{6}$$
$$a = -\frac{3}{6}$$
$$a = -\frac{1}{2}$$
The solution is $-\frac{1}{2}$.

66. $-\frac{14}{9}$

67.
$$-24 = \frac{8x}{5}$$
$$-24 = \frac{8}{5}x$$
$$\frac{5}{8}(-24) = \frac{5}{8}\cdot\frac{8}{5}x$$
$$-\frac{5\cdot8\cdot3}{8\cdot1} = x$$
$$-15 = x$$
The solution is -15.

68. $-\frac{1}{2}$

69.
$$-\frac{4}{3}t = -16$$
$$-\frac{3}{4}\left(-\frac{4}{3}t\right) = -\frac{3}{4}(-16)$$
$$t = \frac{3\cdot4\cdot4}{4}$$
$$t = 12$$
The solution is 12.

70. $-\frac{17}{35}$

71.
$$-483.297 = -794.053 + t$$
$$-483.297 + 794.053 = -794.053 + t + 794.053$$
$$310.756 = t \qquad \text{Using a calculator}$$
The solution is 310.756.

72. -8655

73. ◈

74. ◈

75.
$$9 - 2\cdot5^2 + 7$$
$$= 9 - 2\cdot25 + 7 \quad \text{Simplifying the exponential expression}$$
$$= 9 - 50 + 7 \quad \text{Multiplying}$$
$$= -41 + 7 \quad \text{Subtracting and}$$
$$= -34 \quad \text{adding from left to right}$$

76. 41

77.
$$16 \div (2 - 3\cdot2) + 5$$
$$= 16 \div (2 - 6) + 5 \quad \text{Simplifying inside}$$
$$= 16 \div (-4) + 5 \quad \text{the parentheses}$$
$$= -4 + 5 \quad \text{Dividing}$$
$$= 1 \quad \text{Adding}$$

78. -16

79. ◈

80. ◈

81. $2x = x + x$

$2x = 2x$ Adding on the right side

This is an identity.

82. Contradiction

83. $5x = 0$

$$\frac{5x}{5} = \frac{0}{5}$$

$x = 0$

The solution is 0.

84. Identity

85. $x + 8 = 3 + x + 7$

$x + 8 = 10 + x$ Adding on the right side

$x + 8 - x = 10 + x - x$

$8 = 10$

This is a contradiction.

86. 0

87. $2|x| = -14$

$$\frac{2|x|}{2} = -\frac{14}{2}$$

$|x| = -7$

Since the absolute value of a number is always non-negative, this is a contradiction.

88. $-2, 2$

89. $mx = 9.4m$

$$\frac{mx}{m} = \frac{9.4m}{m}$$

$x = 9.4$

The solution is 9.4.

90. 4

91.

$$\frac{7cx}{2a} = \frac{21}{a} \cdot c$$

$$\frac{7c}{2a} \cdot x = \frac{21}{a} \cdot c$$

$$\frac{2a}{7c} \cdot \frac{7c}{2a} \cdot x = \frac{2a}{7c} \cdot \frac{21}{a} \cdot \frac{c}{1}$$

$$x = \frac{2 \cdot \cancel{a} \cdot 3 \cdot 7 \cdot \cancel{c}}{7 \cdot \cancel{c} \cdot \cancel{a} \cdot 1}$$

$$x = 6$$

The solution is 6.

92. 2

93.

$5a = ax - 3a$

$5a + 3a = ax - 3a + 3a$

$8a = ax$

$$\frac{8a}{a} = \frac{ax}{a}$$

$8 = x$

The solution is 8.

94. $-13, 13$

95.

$x - 4720 = 1634$

$x - 4720 + 4720 = 1634 + 4720$

$x = 6354$

$x + 4720 = 6354 + 4720$

$x + 4720 = 11,074$

96. 250

97. ◈

Exercise Set 2.2

1.

$5x + 3 = 38$

$5x + 3 - 3 = 38 - 3$ Subtracting 3 from both sides

$5x = 35$ Simplifying

$$\frac{5x}{5} = \frac{35}{5}$$ Dividing both sides by 4

$x = 7$ Simplifying

Check: $\dfrac{5x + 3 = 38}{}$

$5 \cdot 7 + 3 \; ? \; 38$

$35 + 3 \;\big|$

$38 \;\big|\; 38$ TRUE

The solution is 7.

2. 8

3.

$8x + 4 = 68$

$8x + 4 - 4 = 68 - 4$ Subtracting 4 from both sides

$8x = 64$ Simplifying

$$\frac{8x}{8} = \frac{64}{8}$$ Dividing both sides by 8

$x = 8$ Simplifying

Check: $\dfrac{8x + 4 = 68}{}$

$8 \cdot 8 + 4 \; ? \; 68$

$64 + 4 \;\big|$

$68 \;\big|\; 68$ TRUE

The solution is 8.

4. 9

5.
$$7t - 8 = 27$$
$$7t - 8 + 8 = 27 + 8 \qquad \text{Adding 8 to both sides}$$
$$7t = 35$$
$$\frac{7t}{7} = \frac{35}{7} \qquad \text{Dividing both sides by 7}$$
$$t = 5$$

Check:
$$\begin{array}{c|c} \multicolumn{2}{c}{7t - 8 = 27} \\ \hline 7 \cdot 5 - 8 \ ? \ 27 & \\ 35 - 8 & \\ 27 & 27 \qquad \text{TRUE} \end{array}$$

The solution is 5.

6. 3

7.
$$3x - 9 = 33$$
$$3x - 9 + 9 = 33 + 9$$
$$3x = 42$$
$$\frac{3x}{3} = \frac{42}{3}$$
$$x = 14$$

Check:
$$\begin{array}{c|c} \multicolumn{2}{c}{3x - 9 = 33} \\ \hline 3 \cdot 14 - 9 \ ? \ 33 & \\ 42 - 9 & \\ 33 & 33 \qquad \text{TRUE} \end{array}$$

The solution is 14.

8. 10

9.
$$8z + 2 = -54$$
$$8z + 2 - 2 = -54 - 2$$
$$8z = -56$$
$$\frac{8z}{8} = \frac{-56}{8}$$
$$z = -7$$

Check:
$$\begin{array}{c|c} \multicolumn{2}{c}{8z + 2 = -54} \\ \hline 8(-7) + 2 \ ? \ -54 & \\ -56 + 2 & \\ -54 & -54 \qquad \text{TRUE} \end{array}$$

The solution is −7.

10. −6

11.
$$-39 = 1 + 8x$$
$$-39 - 1 = 1 + 8x - 1$$
$$-40 = 8x$$
$$\frac{-40}{8} = \frac{8x}{8}$$
$$-5 = x$$

Check:
$$\begin{array}{c|c} \multicolumn{2}{c}{-39 = 1 + 8x} \\ \hline -39 \ ? \ 1 + 8(-5) & \\ & 1 - 40 \\ -39 & -39 \qquad \text{TRUE} \end{array}$$

The solution is −5.

12. −11

13.
$$9 - 4x = 37$$
$$9 - 4x - 9 = 37 - 9$$
$$-4x = 28$$
$$\frac{-4x}{-4} = \frac{28}{-4}$$
$$x = -7$$

Check:
$$\begin{array}{c|c} \multicolumn{2}{c}{9 - 4x = 37} \\ \hline 9 - 4(-7) \ ? \ 37 & \\ 9 + 28 & \\ 37 & 37 \qquad \text{TRUE} \end{array}$$

The solution is −7.

14. −24

15.
$$-7x - 24 = -129$$
$$-7x - 24 + 24 = -129 + 24$$
$$-7x = -105$$
$$\frac{-7x}{-7} = \frac{-105}{-7}$$
$$x = 15$$

Check:
$$\begin{array}{c|c} \multicolumn{2}{c}{-7x - 24 = -129} \\ \hline -7 \cdot 15 - 24 \ ? \ -129 & \\ -105 - 24 & \\ -129 & -129 \qquad \text{TRUE} \end{array}$$

The solution is 15.

16. 19

17.
$$48 = 5x + 7x$$
$$48 = 12x \qquad \text{Combining like terms}$$
$$\frac{48}{12} = \frac{12x}{12} \qquad \text{Dividing both sides by 12}$$
$$4 = x$$

Check:
$$\begin{array}{c|c} \multicolumn{2}{c}{48 = 5x + 7x} \\ \hline 48 \ ? \ 5 \cdot 4 + 7 \cdot 4 & \\ & 20 + 28 \\ 48 & 48 \qquad \text{TRUE} \end{array}$$

The solution is 4.

18. 5

19.
$$27 - 6x = 99$$
$$27 - 6x - 27 = 99 - 27$$
$$-6x = 72$$
$$\frac{-6x}{-6} = \frac{72}{-6}$$
$$x = -12$$

Check: $\quad\dfrac{27 - 6x = 99}{}$

$$27 - 6(-12) \ ? \ 99$$
$$27 + 72 \ \Big|$$
$$\qquad 99 \ \Big| \ 99 \qquad \text{TRUE}$$

The solution is -12.

20. 3

21. $\quad 4x + 3x = 42 \qquad$ Combining like terms
$$7x = 42$$
$$\frac{7x}{7} = \frac{42}{7}$$
$$x = 6$$

Check: $\quad\dfrac{4x + 3x = 42}{}$

$$4 \cdot 6 + 3 \cdot 6 \ ? \ 42$$
$$24 + 18 \ \Big|$$
$$\qquad 42 \ \Big| \ 42 \qquad \text{TRUE}$$

The solution is 6.

22. 4

23.
$$-2a + 5a = 24$$
$$3a = 24$$
$$\frac{3a}{3} = \frac{24}{3}$$
$$a = 8$$

Check: $\quad\dfrac{-2a + 5a = 24}{}$

$$-2 \cdot 8 + 5 \cdot 8 \ ? \ 24$$
$$-16 + 40 \ \Big|$$
$$\qquad 24 \ \Big| \ 24 \qquad \text{TRUE}$$

The solution is 8.

24. -3

25.
$$-7y - 8y = -15$$
$$-15y = -15$$
$$\frac{-15y}{-15} = \frac{-15}{-15}$$
$$y = 1$$

Check: $\quad\dfrac{-7y - 8y = -15}{}$

$$-7 \cdot 1 - 8 \cdot 1 \ ? \ -15$$
$$-7 - 8 \ \Big|$$
$$\qquad -15 \ \Big| \ -15 \qquad \text{TRUE}$$

The solution is 1.

26. 4

27.
$$10.2y - 7.3y = -58$$
$$2.9y = -58$$
$$\frac{2.9y}{2.9} = \frac{-58}{2.9}$$
$$y = -\frac{58}{2.9}$$
$$y = -20$$

Check:

$$\dfrac{10.2y - 7.3y = -58}{}$$
$$10.2(-20) - 7.3(-20) \ ? \ -58$$
$$-204 + 146 \ \Big|$$
$$\qquad -58 \ \Big| \ -58 \qquad \text{TRUE}$$

The solution is -20.

28. -20

29.
$$x + \frac{1}{3}x = 8$$
$$\left(1 + \frac{1}{3}\right)x = 8$$
$$\frac{4}{3}x = 8$$
$$\frac{3}{4} \cdot \frac{4}{3}x = \frac{3}{4} \cdot 8$$
$$x = 6$$

Check: $\quad\dfrac{x + \dfrac{1}{3}x = 8}{}$

$$6 + \frac{1}{3} \cdot 6 \ ? \ 8$$
$$6 + 2 \ \Big|$$
$$\qquad 8 \ \Big| \ 8 \qquad \text{TRUE}$$

The solution is 6.

30. 8

31.
$$9y - 35 = 4y$$
$$9y = 4y + 35 \qquad \text{Adding 35 and simplifying}$$
$$9y - 4y = 35 \qquad \text{Subtracting } 4y \text{ and}$$
$$\qquad\qquad\qquad \text{simplifying}$$
$$5y = 35 \qquad \text{Collecting like terms}$$
$$\frac{5y}{5} = \frac{35}{5} \qquad \text{Dividing both sides by 5}$$
$$y = 7$$

Check: $\quad\dfrac{9y - 35 = 4y}{}$

$$9 \cdot 7 - 35 \ ? \ 4 \cdot 7$$
$$63 - 35 \ \Big| \ 28$$
$$\qquad 28 \ \Big| \ 28 \qquad \text{TRUE}$$

The solution is 7.

32. -3

33.
$$6x - 5 = 7 + 2x$$
$$6x - 5 - 2x = 7 + 2x - 2x \quad \text{Subtracting } 2x \text{ on both sides}$$
$$4x - 5 = 7 \quad \text{Simplifying}$$
$$4x - 5 + 5 = 7 + 5 \quad \text{Adding 5 on both sides}$$
$$4x = 12 \quad \text{Simplifying}$$
$$\frac{4x}{4} = \frac{12}{4} \quad \text{Dividing by 4 on both sides}$$
$$x = 3$$

Check:
$$6x - 5 = 7 + 2x$$
$$\begin{array}{c|c} 6 \cdot 3 - 5 \ ? \ 7 + 2 \cdot 3 \\ 18 - 5 & 7 + 6 \\ 13 & 13 \end{array} \quad \text{TRUE}$$

The solution is 3.

34. 5

35.
$$6x + 3 = 2x + 3$$
$$6x - 2x = 3 - 3$$
$$4x = 0$$
$$\frac{4x}{4} = \frac{0}{4}$$
$$x = 0$$

Check:
$$6x + 3 = 2x + 3$$
$$\begin{array}{c|c} 6 \cdot 0 + 3 \ ? \ 2 \cdot 0 + 3 \\ 0 + 3 & 0 + 3 \\ 3 & 3 \end{array} \quad \text{TRUE}$$

The solution is 0.

36. 4

37.
$$5 - 2x = 3x - 7x + 25$$
$$5 - 2x = -4x + 25$$
$$4x - 2x = 25 - 5$$
$$2x = 20$$
$$\frac{2x}{2} = \frac{20}{2}$$
$$x = 10$$

Check:
$$5 - 2x = 3x - 7x + 25$$
$$\begin{array}{c|c} 5 - 2 \cdot 10 \ ? \ 3 \cdot 10 - 7 \cdot 10 + 25 \\ 5 - 20 & 30 - 70 + 25 \\ -15 & -40 + 25 \\ -15 & -15 \end{array} \quad \text{TRUE}$$

The solution is 10.

38. 10

39.
$$7 + 3x - 6 = 3x + 5 - x$$
$$3x + 1 = 2x + 5 \quad \text{Combining like terms on each side}$$
$$3x - 2x = 5 - 1$$
$$x = 4$$

Check:
$$7 + 3x - 6 = 3x + 5 - x$$
$$\begin{array}{c|c} 7 + 3 \cdot 4 - 6 \ ? \ 3 \cdot 4 + 5 - 4 \\ 7 + 12 - 6 & 12 + 5 - 4 \\ 19 - 6 & 17 - 4 \\ 13 & 13 \end{array} \quad \text{TRUE}$$

The solution is 4.

40. 0

41.
$$4y - 4 + y + 24 = 6y + 20 - 4y$$
$$5y + 20 = 2y + 20$$
$$5y - 2y = 20 - 20$$
$$3y = 0$$
$$y = 0$$

Check:
$$4y - 4 + y + 24 = 6y + 20 - 4y$$
$$\begin{array}{c|c} 4 \cdot 0 - 4 + 0 + 24 \ ? \ 6 \cdot 0 + 20 - 4 \cdot 0 \\ 0 - 4 + 0 + 24 & 0 + 20 - 0 \\ 20 & 20 \end{array} \quad \text{TRUE}$$

The solution is 0.

42. 7

43.
$$\frac{5}{4}x + \frac{1}{4}x = 2x + \frac{1}{2} + \frac{3}{4}x$$

The number 4 is the least common denominator, so we multiply by 4 on both sides.
$$4\left(\frac{5}{4}x + \frac{1}{4}x\right) = 4\left(2x + \frac{1}{2} + \frac{3}{4}x\right)$$
$$4 \cdot \frac{5}{4}x + 4 \cdot \frac{1}{4}x = 4 \cdot 2x + 4 \cdot \frac{1}{2} + 4 \cdot \frac{3}{4}x$$
$$5x + x = 8x + 2 + 3x$$
$$6x = 11x + 2$$
$$6x - 11x = 2$$
$$-5x = 2$$
$$\frac{-5x}{-5} = \frac{2}{-5}$$
$$x = -\frac{2}{5}$$

Check:
$$\frac{5}{4}x + \frac{1}{4}x = 2x + \frac{1}{2} + \frac{3}{4}x$$
$$\begin{array}{c|c} \frac{5}{4}\left(-\frac{2}{5}\right) + \frac{1}{4}\left(-\frac{2}{5}\right) \ ? \ 2\left(-\frac{2}{5}\right) + \frac{1}{2} + \frac{3}{4}\left(-\frac{2}{5}\right) \\ -\frac{1}{2} - \frac{1}{10} & -\frac{4}{5} + \frac{1}{2} - \frac{3}{10} \\ -\frac{5}{10} - \frac{1}{10} & -\frac{8}{10} + \frac{5}{10} - \frac{3}{10} \\ -\frac{6}{10} & -\frac{6}{10} \end{array} \quad \text{TRUE}$$

The solution is $-\frac{2}{5}$.

44. $\dfrac{1}{2}$

45. $\dfrac{2}{3}+\dfrac{1}{4}t=6$

The number 12 is the least common denominator, so we multiply by 12 on both sides.

$$12\left(\dfrac{2}{3}+\dfrac{1}{4}t\right)=12\cdot 6$$
$$12\cdot\dfrac{2}{3}+12\cdot\dfrac{1}{4}t=72$$
$$8+3t=72$$
$$3t=72-8$$
$$3t=64$$
$$t=\dfrac{64}{3}$$

Check: $\dfrac{2}{3}+\dfrac{1}{4}t=6$

$$\begin{array}{c|c}\dfrac{2}{3}+\dfrac{1}{4}\left(\dfrac{64}{3}\right) \ ? \ 6 & \\[4pt] \dfrac{2}{3}+\dfrac{16}{3} & \\[4pt] \dfrac{18}{3} & \\[4pt] 6 & 6 \quad \text{TRUE}\end{array}$$

The solution is $\dfrac{64}{3}$.

46. $-\dfrac{2}{3}$

47. $\dfrac{2}{3}+4t=6t-\dfrac{2}{15}$

The number 15 is the least common denominator, so we multiply by 15 on both sides.

$$15\left(\dfrac{2}{3}+4t\right)=15\left(6t-\dfrac{2}{15}\right)$$
$$15\cdot\dfrac{2}{3}+15\cdot4t=15\cdot6t-15\cdot\dfrac{2}{15}$$
$$10+60t=90t-2$$
$$10+2=90t-60t$$
$$12=30t$$
$$\dfrac{12}{30}=t$$
$$\dfrac{2}{5}=t$$

Check: $\dfrac{2}{3}+4t=6t-\dfrac{2}{15}$

$$\begin{array}{c|c}\dfrac{2}{3}+4\cdot\dfrac{2}{5} \ ? \ 6\cdot\dfrac{2}{5}-\dfrac{2}{15} & \\[4pt] \dfrac{2}{3}+\dfrac{8}{5} & \dfrac{12}{5}-\dfrac{2}{15} \\[4pt] \dfrac{10}{15}+\dfrac{24}{15} & \dfrac{36}{15}-\dfrac{2}{15} \\[4pt] \dfrac{34}{15} & \dfrac{34}{15} \quad \text{TRUE}\end{array}$$

The solution is $\dfrac{2}{5}$.

48. -3

49. $\dfrac{1}{3}x+\dfrac{2}{5}=\dfrac{4}{15}+\dfrac{3}{5}x-\dfrac{2}{3}$

The number 15 is the least common denominator, so we multiply by 15 on both sides.

$$15\left(\dfrac{1}{3}x+\dfrac{2}{5}\right)=15\left(\dfrac{4}{15}+\dfrac{3}{5}x-\dfrac{2}{3}\right)$$
$$15\cdot\dfrac{1}{3}x+15\cdot\dfrac{2}{5}=15\cdot\dfrac{4}{15}+15\cdot\dfrac{3}{5}x-15\cdot\dfrac{2}{3}$$
$$5x+6=4+9x-10$$
$$5x+6=-6+9x$$
$$5x-9x=-6-6$$
$$-4x=-12$$
$$\dfrac{-4x}{-4}=\dfrac{-12}{-4}$$
$$x=3$$

Check: $\dfrac{1}{3}x+\dfrac{2}{5}=\dfrac{4}{15}+\dfrac{3}{5}x-\dfrac{2}{3}$

$$\begin{array}{c|c}\dfrac{1}{3}\cdot3+\dfrac{2}{5} \ ? \ \dfrac{4}{15}+\dfrac{3}{5}\cdot3-\dfrac{2}{3} & \\[4pt] 1+\dfrac{2}{5} & \dfrac{4}{15}+\dfrac{9}{5}-\dfrac{2}{3} \\[4pt] \dfrac{5}{5}+\dfrac{2}{5} & \dfrac{4}{15}+\dfrac{27}{15}-\dfrac{10}{15} \\[4pt] \dfrac{7}{5} & \dfrac{21}{15} \\[4pt] \dfrac{7}{5} & \dfrac{7}{5} \quad \text{TRUE}\end{array}$$

The solution is 3.

50. -3

51.
$$2.1x + 45.2 = 3.2 - 8.4x$$
Greatest number of decimal places is 1
$$10(2.1x + 45.2) = 10(3.2 - 8.4x)$$
Multiplying by 10 to clear decimals
$$10(2.1x) + 10(45.2) = 10(3.2) - 10(8.4x)$$
$$21x + 452 = 32 - 84x$$
$$21x + 84x = 32 - 452$$
$$105x = -420$$
$$x = \frac{-420}{105}$$
$$x = -4$$

Check:
$$\begin{array}{c|c} \multicolumn{2}{c}{2.1x + 45.2 = 3.2 - 8.4x} \\ \hline 2.1(-4) + 45.2 \ ? \ 3.2 - 8.4(-4) & \\ -8.4 + 45.2 & 3.2 + 33.6 \\ 36.8 & 36.8 \quad \text{TRUE} \end{array}$$

The solution is -4.

52. $\dfrac{4}{5}$, or 0.8

53.
$$0.76 + 0.21t = 0.96t - 0.49$$
Greatest number of decimal places is 2
$$100(0.76 + 0.21t) = 100(0.96t - 0.49)$$
Multiplying by 100 to clear decimals
$$100(0.76) + 100(0.21t) = 100(0.96t) - 100(0.49)$$
$$76 + 21t = 96t - 49$$
$$76 + 49 = 96t - 21t$$
$$125 = 75t$$
$$\frac{125}{75} = t$$
$$\frac{5}{3} = t, \text{ or}$$
$$1.\overline{6} = t$$

The answer checks. The solution is $\dfrac{5}{3}$, or $1.\overline{6}$.

54. 1

55.
$$\frac{2}{5}x - \frac{3}{2}x = \frac{3}{4}x + 2$$
The least common denominator is 20.
$$20\left(\frac{2}{5}x - \frac{3}{2}x\right) = 20\left(\frac{3}{4}x + 2\right)$$
$$20 \cdot \frac{2}{5}x - 20 \cdot \frac{3}{2}x = 20 \cdot \frac{3}{4}x + 20 \cdot 2$$
$$8x - 30x = 15x + 40$$
$$-22x = 15x + 40$$
$$-22x - 15x = 40$$
$$-37x = 40$$
$$\frac{-37x}{-37} = \frac{40}{-37}$$
$$x = -\frac{40}{37}$$

Check:
$$\begin{array}{c|c} \multicolumn{2}{c}{\frac{2}{5}x - \frac{3}{2}x = \frac{3}{4}x + 2} \\ \hline \frac{2}{5}\left(-\frac{40}{37}\right) - \frac{3}{2}\left(-\frac{40}{37}\right) \ ? \ \frac{3}{4}\left(-\frac{40}{37}\right) + 2 & \\ -\frac{16}{37} + \frac{60}{37} & -\frac{30}{37} + \frac{74}{37} \\ \frac{44}{37} & \frac{44}{37} \quad \text{TRUE} \end{array}$$

The solution is $-\dfrac{40}{37}$.

56. $\dfrac{32}{7}$

57.
$$7(2a - 1) = 21$$
$$14a - 7 = 21 \qquad \text{Using the distributive law}$$
$$14a = 21 + 7 \quad \text{Adding 7}$$
$$14a = 28$$
$$a = 2 \qquad \text{Dividing by 14}$$

Check:
$$\begin{array}{c|c} \multicolumn{2}{c}{7(2a - 1) = 21} \\ \hline 7(2 \cdot 2 - 1) \ ? \ 21 & \\ 7(4 - 1) & \\ 7 \cdot 3 & \\ 21 & 21 \quad \text{TRUE} \end{array}$$

The solution is 2.

58. $\dfrac{9}{2}$

59.
$$35 = 5(3x + 1)$$
$$35 = 15x + 5 \qquad \text{Using the distributive law}$$
$$35 - 5 = 15x$$
$$30 = 15x$$
$$2 = x$$

Check:
$$\begin{array}{c|c} \multicolumn{2}{c}{35 = 5(3x + 1)} \\ \hline 35 \ ? \ 5(3 \cdot 2 + 1) & \\ & 5(6 + 1) \\ & 5 \cdot 7 \\ 35 & 35 \quad \text{TRUE} \end{array}$$

The solution is 2.

60. 1

61.
$$2(3 + 4m) - 6 = 48$$
$$6 + 8m - 6 = 48$$
$$8m = 48 \quad \text{Combining like terms}$$
$$m = 6$$

Check:

$$\frac{2(3+4m)-6=48}{\begin{array}{c|c} 2(3+4\cdot 6)-6 \; ? \; 48 \\ 2(3+24)-6 \\ 2\cdot 27-6 \\ 54-6 \\ 48 & 48 \end{array}}$$ TRUE

The solution is 6.

62. 9

63. $7r - (2r + 8) = 32$
$7r - 2r - 8 = 32$
$5r - 8 = 32$ Combining like terms
$5r = 32 + 8$
$5r = 40$
$r = 8$

Check:

$$\frac{7r-(2r+8)=32}{\begin{array}{c|c} 7\cdot 8 - (2\cdot 8 + 8) \; ? \; 32 \\ 56 - (16 + 8) \\ 56 - 24 \\ 32 & 32 \end{array}}$$ TRUE

The solution is 8.

64. 8

65. $13 - 3(2x - 1) = 4$
$13 - 6x + 3 = 4$
$16 - 6x = 4$
$-6x = 4 - 16$
$-6x = -12$
$x = 2$

Check:

$$\frac{13-3(2x-1)=4}{\begin{array}{c|c} 13 - 3(2\cdot 2 - 1) \; ? \; 4 \\ 13 - 3(4 - 1) \\ 13 - 3\cdot 3 \\ 13 - 9 \\ 4 & 4 \end{array}}$$ TRUE

The solution is 2.

66. 17

67. $3(t - 2) = 9(t + 2)$
$3t - 6 = 9t + 18$
$-6 - 18 = 9t - 3t$
$-24 = 6t$
$-4 = t$

Check:

$$\frac{3(t-2)=9(t+2)}{\begin{array}{c|c} 3(-4-2) \; ? \; 9(-4+2) \\ 3(-6) & 9(-2) \\ -18 & -18 \end{array}}$$ TRUE

The solution is -4.

68. $-\dfrac{5}{3}$

69. $7(5x - 2) = 6(6x - 1)$
$35x - 14 = 36x - 6$
$-14 + 6 = 36x - 35x$
$-8 = x$

Check:

$$\frac{7(5x-2)=6(6x-1)}{\begin{array}{c|c} 7(5(-8)-2) \; ? \; 6(6(-8)-1) \\ 7(-40-2) & 6(-48-1) \\ 7(-42) & 6(-49) \\ -294 & -294 \end{array}}$$ TRUE

The solution is -8.

70. -12

71. $19 - (2x + 3) = 2(x + 3) + x$
$19 - 2x - 3 = 2x + 6 + x$
$16 - 2x = 3x + 6$
$16 - 6 = 3x + 2x$
$10 = 5x$
$2 = x$

Check:

$$\frac{19-(2x+3)=2(x+3)+x}{\begin{array}{c|c} 19 - (2\cdot 2 + 3) \; ? \; 2(2+3)+2 \\ 19 - (4 + 3) & 2\cdot 5 + 2 \\ 19 - 7 & 10 + 2 \\ 12 & 12 \end{array}}$$ TRUE

The solution is 2.

72. 1

73. $\dfrac{1}{4}(3t - 4) = 5$

$4\cdot \dfrac{1}{4}(3t - 4) = 4\cdot 5$

$3t - 4 = 20$

$3t = 24$ Adding 4 to both sides

$t = 8$ Dividing both sides by 3

Check:

$$\frac{\dfrac{1}{4}(3t-4)=5}{\begin{array}{c|c} \dfrac{1}{4}(3\cdot 8 - 4) \; ? \; 5 \\ \dfrac{1}{4}(24 - 4) \; ? \\ \dfrac{1}{4}\cdot 20 \\ 5 & 5 \end{array}}$$ TRUE

The solution is 8.

74. 11

75. $\dfrac{4}{3}(5x+1) = 8$

$\dfrac{3}{4} \cdot \dfrac{4}{3}(5x+1) = \dfrac{3}{4} \cdot 8$

$5x + 1 = 6$

$5x = 5$

$x = 1$

Check: $\dfrac{4}{3}(5x+1) = 8$

$\dfrac{4}{3}(5 \cdot 1 + 1) \; ? \; 8$

$\dfrac{4}{3}(6) \; \Big| \; 8$

$8 \; \Big| \; 8$ TRUE

The solution is 1.

76. 6

77. $\dfrac{3}{2}(2x+5) = -\dfrac{15}{2}$

$\dfrac{2}{3} \cdot \dfrac{3}{2}(2x+5) = \dfrac{2}{3}\left(-\dfrac{15}{2}\right)$

$2x + 5 = -5$

$2x = -10$

$x = -5$

Check: $\dfrac{3}{2}(2x+5) = -\dfrac{15}{2}$

$\dfrac{3}{2}(2(-5)+5) \; ? \; -\dfrac{15}{2}$

$\dfrac{3}{2}(-10+5) \; \Big|$

$\dfrac{3}{2}(-5) \; \Big|$

$-\dfrac{15}{2} \; \Big| \; -\dfrac{15}{2}$ TRUE

The solution is -5.

78. $\dfrac{16}{15}$

79. $\dfrac{3}{4}\left(3x - \dfrac{1}{2}\right) - \dfrac{2}{3} = \dfrac{1}{3}$

$\dfrac{9}{4}x - \dfrac{3}{8} - \dfrac{2}{3} = \dfrac{1}{3}$

Multiplying by the number 24 will clear all the fractions, so we multiply by 24 on both sides.

$24\left(\dfrac{9}{4}x - \dfrac{3}{8} - \dfrac{2}{3}\right) = 24 \cdot \dfrac{1}{3}$

$24 \cdot \dfrac{9}{4}x - 24 \cdot \dfrac{3}{8} - 24 \cdot \dfrac{2}{3} = 8$

$54x - 9 - 16 = 8$

$54x - 25 = 8$

$54x = 8 + 25$

$54x = 33$

$x = \dfrac{33}{54}$

$x = \dfrac{11}{18}$

The check is left to the student. The solution is $\dfrac{11}{18}$.

80. $-\dfrac{5}{32}$

81. $0.7(3x+6) = 1.1 - (x+2)$

$2.1x + 4.2 = 1.1 - x - 2$

$10(2.1x + 4.2) = 10(1.1 - x - 2)$ Clearing decimals

$21x + 42 = 11 - 10x - 20$

$21x + 42 = -10x - 9$

$21x + 10x = -9 - 42$

$31x = -51$

$x = -\dfrac{51}{31}$

The check is left to the student. The solution is $-\dfrac{51}{31}$.

82. $\dfrac{39}{14}$

83. $a + (a-3) = (a+2) - (a+1)$

$a + a - 3 = a + 2 - a - 1$

$2a - 3 = 1$

$2a = 1 + 3$

$2a = 4$

$a = 2$

Check: $a + (a-3) = (a+2) - (a+1)$

$2 + (2-3) \; ? \; (2+2) - (2+1)$

$2 - 1 \; \Big| \; 4 - 3$

$1 \; \Big| \; 1$ TRUE

The solution is 2.

84. -7.4

85. ◈

86.

87. $3 - 5a = 3 - 5 \cdot 2 = 3 - 10 = -7$

88. 15

89. $7x - 2x = 7(-3) - 2(-3) = -21 + 6 = -15$

90. -28

91. ◈

92. ◈

93.
$$8.43x - 2.5(3.2 - 0.7x) = -3.455x + 9.04$$
$$8.43x - 8 + 1.75x = -3.455x + 9.04$$
$$10.18x - 8 = -3.455x + 9.04$$
$$10.18x + 3.455x = 9.04 + 8$$
$$13.635x = 17.04$$
$$x = 1.\overline{2497}$$

The solution is $1.\overline{2497}$.

94. 4.423346424

95.
$$-2[3(x - 2) + 4] = 4(5 - x) - 2x$$
$$-2[3x - 6 + 4] = 20 - 4x - 2x$$
$$-2[3x - 2] = 20 - 6x$$
$$-6x + 4 = 20 - 6x$$
$$4 = 20 \qquad \text{Adding } 6x \text{ to both sides}$$

This is contradiction.

96. $-\dfrac{7}{2}$

97.
$$3(x + 4) = 3(4 + x)$$
$$3x + 12 = 12 + 3x$$
$$3x + 12 - 12 = 12 + 3x - 12$$
$$3x = 3x$$

This is an identity.

98. Contradiction

99.
$$2x(x + 5) - 3(x^2 + 2x - 1) = 9 - 5x - x^2$$
$$2x^2 + 10x - 3x^2 - 6x + 3 = 9 - 5x - x^2$$
$$-x^2 + 4x + 3 = 9 - 5x - x^2$$
$$4x + 3 = 9 - 5x \qquad \text{Adding } x^2$$
$$4x + 5x = 9 - 3$$
$$9x = 6$$
$$x = \frac{2}{3}$$

The solution is $\dfrac{2}{3}$.

100. -2

101.
$$9 - 3x = 2(5 - 2x) - (1 - 5x)$$
$$9 - 3x = 10 - 4x - 1 + 5x$$
$$9 - 3x = 9 + x$$
$$9 - 9 = x + 3x$$
$$0 = 4x$$
$$0 = x$$

The solution is 0.

102. Identity

103. $[7 - 2(8 \div (-2))]x = 0$

Since $7 - 2(8 \div (-2)) \neq 0$ and the product on the left side of the equation is 0, then x must be 0.

104. $\dfrac{52}{45}$

105.
$$\frac{5x + 3}{4} + \frac{25}{12} = \frac{5 + 2x}{3}$$
$$12\left(\frac{5x + 3}{4} + \frac{25}{12}\right) = 12\left(\frac{5 + 2x}{3}\right)$$
$$12\left(\frac{5x + 3}{4}\right) + 12 \cdot \frac{25}{12} = 4(5 + 2x)$$
$$3(5x + 3) + 25 = 4(5 + 2x)$$
$$15x + 9 + 25 = 20 + 8x$$
$$15x + 34 = 20 + 8x$$
$$7x = -14$$
$$x = -2$$

The solution is -2.

Exercise Set 2.3

1. We substitute 10 for t and calculate M.
$$M = \frac{1}{5} \cdot 10 = 2$$
The storm is 2 miles away.

2. 3450 watts

3. We substitute 21,345 for n and calculate f.
$$f = \frac{21,345}{15} = 1423$$
There are 1423 full-time equivalent students.

4. 54 in^2

5. We substitute 84 for c and 8 for w and calculate D.
$$D = \frac{c}{w} = \frac{84}{8} = 10.5$$
The calorie density is 10.5 calories/oz.

6. $\dfrac{43}{3}$ m/cycle, or $14.\overline{3}$ m/cycle

7. Substitute 1 for t and calculate n.

$$n = 0.5t^4 + 3.45t^3 - 96.65t^2 + 347.7t$$
$$= 0.5(1)^4 + 3.45(1)^3 - 96.65(1)^2 + 347.7(1)$$
$$= 0.5 + 3.45 - 96.65 + 347.7$$
$$= 255$$

255 mg of ibuprofen remains in the bloodstream.

8. 42

9. $A = bh$

$\dfrac{A}{h} = \dfrac{bh}{h}$ Dividing both sides by h

$\dfrac{A}{h} = b$

10. $h = \dfrac{A}{b}$

11. $d = rt$

$\dfrac{d}{t} = \dfrac{rt}{t}$ Dividing both sides by t

$\dfrac{d}{t} = r$

12. $t = \dfrac{d}{r}$

13. $I = Prt$

$\dfrac{I}{rt} = \dfrac{Prt}{rt}$ Dividing both sides by rt

$\dfrac{I}{rt} = P$

14. $t = \dfrac{I}{Pr}$

15. $H = 65 - m$

$H + m = 65$ Adding m to both sides

$m = 65 - H$ Subtracting H from both sides

16. $h = d + 64$

17. $P = 2l + 2w$

$P - 2w = 2l + 2w - 2w$ Subtracting $2w$ from both sides

$P - 2w = 2l$

$\dfrac{P - 2w}{2} = \dfrac{2l}{2}$ Dividing both sides by 2

$\dfrac{P - 2w}{2} = l$, or

$\dfrac{P}{2} - w = l$

18. $w = \dfrac{P - 2l}{2}$, or $w = \dfrac{P}{2} - l$

19. $A = \pi r^2$

$\dfrac{A}{r^2} = \dfrac{\pi r^2}{r^2}$

$\dfrac{A}{r^2} = \pi$

20. $r^2 = \dfrac{A}{\pi}$

21. $A = \dfrac{1}{2}bh$

$2A = 2 \cdot \dfrac{1}{2}bh$ Multiplying both sides by 2

$2A = bh$

$\dfrac{2A}{b} = \dfrac{bh}{b}$ Dividing both sides by h

$\dfrac{2A}{b} = h$

22. $b = \dfrac{2A}{h}$

23. $E = mc^2$

$\dfrac{E}{c^2} = \dfrac{mc^2}{c^2}$ Dividing both sides by c^2

$\dfrac{E}{c^2} = m$

24. $c^2 = \dfrac{E}{m}$

25. $Q = \dfrac{c + d}{2}$

$2Q = 2 \cdot \dfrac{c + d}{2}$ Multiplying both sides by 2

$2Q = c + d$

$2Q - c = c + d - c$ Subtracting c from both sides

$2Q - c = d$

26. $p = 2Q + q$

27. $A = \dfrac{a + b + c}{3}$

$3A = 3 \cdot \dfrac{a + b + c}{3}$ Multiplying both sides by 3

$3A = a + b + c$

$3A - a - c = a + b + c - a - c$ Subtracting a and c from both sides

$3A - a - c = b$

28. $c = 3A - a - b$

29. $M = \dfrac{A}{s}$

$s \cdot M = s \cdot \dfrac{A}{s}$ Multiplying both sides by s

$sM = A$

30. $b = \dfrac{Pc}{a}$

31. $A = at + bt$

$A = t(a + b)$ Factoring

$\dfrac{A}{a + b} = t$ Dividing both sides by $a + b$

32. $x = \dfrac{S}{r + s}$

33. $A = \dfrac{1}{2}ah + \dfrac{1}{2}bh$

$2A = 2\left(\dfrac{1}{2}ah + \dfrac{1}{2}bh\right)$

$2A = ah + bh$

$2A = h(a + b)$

$\dfrac{2A}{a + b} = h$

34. $P = \dfrac{A}{1 + rt}$

35. $R = r + \dfrac{400(W - L)}{N}$

$N \cdot R = N\left(r + \dfrac{400(W - L)}{N}\right)$

Multiplying both sides by N

$NR = Nr + 400(W - L)$

$NR = Nr + 400W - 400L$

$NR + 400L = Nr + 400W$ Adding $400L$ to both sides

$400L = Nr + 400W - NR$ Adding $-NR$ to both sides

$L = \dfrac{Nr + 400W - NR}{400}$

36. $r^2 = \dfrac{360A}{\pi S}$

37. ◈

38. ◈

39. $0.79(38.4)0$

One factor is 0, so the product is 0.

40. 9.18

41. $20 \div (-4) \cdot 2 - 3$

$= -5 \cdot 2 - 3$ Dividing and

$= -10 - 3$ multiplying from left to right

$= -13$ Subtracting

42. 65

43. ◈

44. ◈

45. $K = 19.18w + 7h - 9.52a + 92.4$

$2627 = 19.18(82) + 7(185) - 9.52a + 92.4$

$2627 = 1572.76 + 1295 - 9.52a + 92.4$

$2627 = 2960.16 - 9.52a$

$-333.16 = -9.52a$

$35 \approx a$

The man is about 35 years old.

46. $T = t - \dfrac{h}{100}, \ 0 \le h \le 12{,}000$

47. $c = \dfrac{w}{a} \cdot d$

$ac = a \cdot \dfrac{w}{a} \cdot d$

$ac = wd$

$a = \dfrac{wd}{c}$

48. About 76.4 in.

49. $\dfrac{y}{z} \div \dfrac{z}{t} = 1$

$\dfrac{y}{z} \cdot \dfrac{t}{z} = 1$

$\dfrac{yt}{z^2} = 1$

$\dfrac{z^2}{t} \cdot \dfrac{yt}{z^2} = \dfrac{z^2}{t} \cdot 1$

$y = \dfrac{z^2}{t}$

50. $c = \dfrac{d}{a - b}$

51. $qt = r(s + t)$

$qt = rs + rt$

$qt - rt = rs$

$t(q - r) = rs$

$t = \dfrac{rs}{q - r}$

52. $a = \dfrac{c}{3 + b + d}$

53. We subtract the minimum output for a well-insulated house with a square feet from the minimum output for a poorly-insulated house with a square feet. Let S represent the number of BTU's saved.

$$S = 50a - 30a$$
$$S = 20a$$

54. $K = 917 + 13.2276w + 2.3622h - 6a$

55. $K = 19.18\left(\dfrac{w}{2.2046}\right) + 7\left(\dfrac{h}{0.3937}\right) - 9.52a + 92.4$

$\qquad K = 8.70w + 17.78h - 9.52a + 92.4$

Exercise Set 2.4

1. $82\% = 82 \times 0.01$ Replacing % by $\times\ 0.01$
$\qquad\quad = 0.82$

2. 0.49

3. $9\% = 9 \times 0.01$ Replacing % by $\times\ 0.01$
$\qquad\ = 0.09$

4. 0.913

5. $43.7\% = 43.7 \times 0.01 = 0.437$

6. 0.02

7. $0.46\% = 0.46 \times 0.01 = 0.0046$

8. 0.048

9. 0.29

First move the decimal point	0.29.
two places to the right;	└─↑
then write a % symbol:	29%

10. 78%

11. 0.998

First move the decimal point	0.99.8
two places to the right;	└─↑
then write a % symbol:	99.8%

12. 35.8%

13. 1.92

First move the decimal point	1.92.
two places to the right;	└─↑
then write a % symbol:	192%

14. 139%

15. 2.1

First move the decimal point	2.10.
two places to the right;	└─↑
then write a % symbol:	210%

16. 920%

17. 0.0068

First move the decimal point	0.00.68
two places to the right;	└─↑
then write a % symbol:	0.68%

18. 0.95%

19. $\dfrac{3}{8}$ $\left(\text{Note: } \dfrac{3}{8} = 0.375\right)$

First move the decimal point	0.37.5
two places to the right;	└─↑
then write a % symbol:	37.5%

20. 75%

21. $\dfrac{7}{25}$ $\left(\text{Note: } \dfrac{7}{25} = 0.28\right)$

First move the decimal point	0.28.
two places to the right;	└─↑
then write a % symbol:	28%

22. 80%

23. $\dfrac{2}{3}$ $\left(\text{Note: } \dfrac{2}{3} = 0.66\overline{6}\right)$

First move the decimal point	0.66.$\overline{6}$
two places to the right;	└─↑
then write a % symbol:	66.$\overline{6}$%

Since $0.\overline{6} = \dfrac{2}{3}$, this can also be expressed as $66\dfrac{2}{3}\%$.

24. $83.\overline{3}\%$, or $83\dfrac{1}{3}\%$

25. *Translate*.

$\underbrace{\text{What percent}}_{y}$ of $\underset{\downarrow}{68}$ $\underset{=}{\text{is}}$ $\underset{17}{17}$?

We solve the equation and then convert to percent notation.

$$y \cdot 68 = 17$$
$$y = \dfrac{17}{68}$$
$$y = 0.25 = 25\%$$

The answer is 25%.

26. 26%

27. *Translate*.

$$\underbrace{\text{What percent}}_{\downarrow}_{y} \text{ of } 125 \overset{\downarrow}{\cdot} \overset{\downarrow}{125} \overset{\downarrow}{=} \overset{\downarrow}{30}\text{?}$$

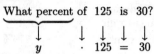

We solve the equation and then convert to percent notation.

$$y \cdot 125 = 30$$
$$y = \frac{30}{125}$$
$$y = 0.24 = 24\%$$

The answer is 24%.

28. 19%

29. *Translate*.

$$14 \text{ is } 30\% \text{ of } \underbrace{\text{what number?}}$$
$$\overset{\downarrow}{14} \overset{\downarrow}{=} \overset{\downarrow}{30\%} \overset{\downarrow}{\cdot} \overset{\downarrow}{y}$$

We solve the equation.

$$14 = 0.3y \qquad (30\% = 0.3)$$
$$\frac{14}{0.3} = y$$
$$46.\overline{6} = y$$

The answer is $46.\overline{6}$, or $46\frac{2}{3}$, or $\frac{140}{3}$.

30. 225

31. *Translate*.

$$0.3 \text{ is } 12\% \text{ of } \underbrace{\text{what number?}}$$
$$\overset{\downarrow}{0.3} \overset{\downarrow}{=} \overset{\downarrow}{12\%} \overset{\downarrow}{\cdot} \overset{\downarrow}{y}$$

We solve the equation.

$$0.3 = 0.12y \qquad (12\% = 0.12)$$
$$\frac{0.3}{0.12} = y$$
$$2.5 = y$$

The answer is 2.5.

32. 4

33. *Translate*.

$$\underbrace{\text{What number}}_{y} \text{ is } 35\% \text{ of } 240\text{?}$$
$$\overset{\downarrow}{y} \overset{\downarrow}{=} \overset{\downarrow}{35\%} \overset{\downarrow}{\cdot} \overset{\downarrow}{240}$$

We solve the equation.

$$y = 0.35 \cdot 240 \qquad (35\% = 0.35)$$
$$y = 84 \qquad \text{Multiplying}$$

The answer is 84.

34. 10,000

35. *Translate*.

$$\underbrace{\text{What percent}}_{y} \text{ of } 60 \overset{\downarrow}{\cdot} \overset{\downarrow}{60} \overset{\downarrow}{=} \overset{\downarrow}{75}\text{?}$$

We solve the equation and then convert to percent notation.

$$y \cdot 60 = 75$$
$$y = \frac{75}{60}$$
$$y = 1.25 = 125\%$$

The answer is 125%.

36. 100%

37. *Translate*.

$$\text{What is } 2\% \text{ of } 40\text{?}$$
$$\overset{\downarrow}{x} \overset{\downarrow}{=} \overset{\downarrow}{2\%} \overset{\downarrow}{\cdot} \overset{\downarrow}{40}$$

We solve the equation.

$$x = 0.02 \cdot 40 \qquad (2\% = 0.02)$$
$$x = 0.8 \qquad \text{Multiplying}$$

The answer is 0.8.

38. 0.8

39. Observe that 25 is half of 50. Thus, the answer is 0.5, or 50%. We could also do this exercise by translating to an equation.

Translate.

$$25 \text{ is } \underbrace{\text{what percent}}_{y} \text{ of } 50\text{?}$$
$$\overset{\downarrow}{25} \overset{\downarrow}{=} \overset{\downarrow}{y} \overset{\downarrow}{\cdot} \overset{\downarrow}{50}$$

We solve the equation and convert to percent notation.

$$25 = y \cdot 50$$
$$\frac{25}{50} = y$$
$$0.5 = y, \text{ or } 50\% = y$$

The answer is 50%.

40. 400

41. Let $I =$ the amount of interest Sarah will pay. Then we have:

$$I \text{ is } 8\% \text{ of } \$3500.$$
$$\overset{\downarrow}{I} \overset{\downarrow}{=} \overset{\downarrow}{0.08} \overset{\downarrow}{\cdot} \overset{\downarrow}{\$3500}$$
$$I = \$280$$

Sarah will pay $280 interest.

42. $168

43. Let p = the number of people who voted in the 2000 presidential election, in millions. Then we have:

48.62 is 48.36% of p.
$$\downarrow \quad \downarrow \quad \downarrow \quad \downarrow\downarrow$$
$$48.62 = 0.4836 \cdot p$$
$$\frac{48.62}{0.4836} = p$$
$$100.54 \approx p$$

About 100.54 million, or 100,540,000, people voted in the 2000 presidential election.

44. About $11.9 billion

45. If n = the number of women who had babies in good or excellent health, we have:

n is 8% of 300.
$$\downarrow\downarrow \quad \downarrow \quad \downarrow \quad \downarrow$$
$$n = 0.08 \cdot 300$$
$$n = 24$$

24 women had babies in good or excellent health.

46. 285

47. Let a = the number of pounds of almonds the average American consumes each year. Then we have:

a is 25% of 2.25.
$$\downarrow\downarrow \quad \downarrow \quad \downarrow \quad \downarrow$$
$$a = 0.25 \cdot 2.25$$
$$a = 0.5625$$

The average American consumes 0.5625 lb of almonds each year.

48. 7410

49. Let b = the number of bowlers you would expect to be left-handed. Then we have:

b is 17% of 160.
$$\downarrow\downarrow \quad \downarrow \quad \downarrow \quad \downarrow$$
$$b = 0.17 \cdot 160$$
$$b \approx 27$$

You would expect 27 bowlers to be left-handed.

50. 7%

51. Let p = the percent that were correct. Then we have:

76 is what percent of 88?
$$\downarrow\downarrow \qquad \downarrow \qquad \downarrow\downarrow$$
$$76 = \quad p \quad \cdot 88$$
$$\frac{76}{88} = p$$
$$0.86\overline{4} = p, \text{ or}$$
$$86.\overline{4}\% = p$$

86.$\overline{4}$% of the items were correct.

52. 52%

53. When the sales tax is 5%, the total amount paid is 105% of the cost of the merchandise. Let c = the cost of the merchandise. Then we have:

$37.80 is 105% of c.
$$\downarrow \quad \downarrow \quad \downarrow \quad \downarrow\downarrow$$
$$37.80 = 1.05 \cdot c$$
$$\frac{37.80}{1.05} = c$$
$$36 = c$$

The price of the merchandise was $36.

54. $940

55. When the sales tax is 5%, the total amount paid is 105% of the cost of the merchandise. Let c = the amount the school group owes, or the cost of the software without tax. Then we have:

$157.41 is 106% of c.
$$\downarrow \quad \downarrow \quad \downarrow \quad \downarrow\downarrow$$
$$157.41 = 1.06 \cdot c$$
$$\frac{157.41}{1.06} = c$$
$$148.5 = c$$

The school group owes $148.50.

56. $138.95

57. A self-employed person must earn 120% as much as a non-self-employed person. Let a = the amount Roy would need to earn, in dollars per hour, on his own for a comparable income. Then we have:

a is 120% of $15.
$$\downarrow\downarrow \quad \downarrow \quad \downarrow \quad \downarrow$$
$$a = 1.2 \cdot 15$$
$$a = 18$$

Roy would need to earn $18 per hour on his own.

58. $14.40 per hour

59. The number of calories in a serving of Light Style Bread is 85% of the number of calories in a serving of regular bread. Let c = the number of calories in a serving of regular bread. Then we have:

140 calories is 85% of c.
$$\downarrow \qquad \downarrow \quad \downarrow \quad \downarrow\downarrow$$
$$140 \quad = 0.85 \cdot c$$
$$\frac{140}{0.85} = c$$
$$165 \approx c$$

There are about 165 calories in a serving of regular bread.

60. 58 calories

61.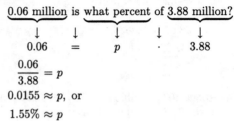

62. ◈

63. Let n represent "some number." Then we have $n+5$, or $5 + n$.

64. Let w represent Tino's weight; $w - 4$

65. $8 \cdot 2a$, or $2a \cdot 8$.

66. Let m and n represent the numbers; $mn+1$, or $1+mn$

67. ◈

68. ◈

69. Let $p =$ the population of Bardville. Then we have:

1332 is 15% of 48% of the population.

$$\begin{array}{cccccc} \downarrow & \downarrow & \downarrow & \downarrow & \downarrow & \downarrow \\ 1332 & = & 0.15 & \cdot & 0.48 & \cdot & p \end{array}$$

$$\frac{1332}{0.15(0.48)} = p$$

$$18,500 = p$$

The population of Bardville is 18,500.

70. Rollie's: $12.83; Sound Warp: $12.97

71. The new price is 125% of the old price. Let $p =$ the new price. Then we have:

p is 125% of $20,800.

$$\begin{array}{ccccc} \downarrow\downarrow & \downarrow & \downarrow & \downarrow \\ p = & 1.25 & \cdot & 20,800 \end{array}$$

$$p = 26,000$$

Now let $x =$ the percent of the new price represented by the old price. We have:

$20,800 is what percent of $26,000.

$$\begin{array}{ccccc} \downarrow & \downarrow & \downarrow & \downarrow & \downarrow \\ 20,800 & = & x & \cdot & 26,000 \end{array}$$

$$\frac{20,800}{26,000} = x$$

$$0.8 = x, \text{ or}$$

$$80\% = x$$

The old price is $100\% - 80\%$, or 20% lower than the new price.

72. $35.\overline{135}\%$, or $35\frac{5}{37}\%$

73. The number of births increased by $3.94 - 3.88$, or 0.06 million. Let $p =$ the percent of increase. Then we have:

0.06 million is what percent of 3.88 million?

$$\begin{array}{ccccc} \downarrow & \downarrow & \downarrow & \downarrow & \downarrow \\ 0.06 & = & p & \cdot & 3.88 \end{array}$$

$$\frac{0.06}{3.88} = p$$

$$0.0155 \approx p, \text{ or}$$

$$1.55\% \approx p$$

The number of births increased by about 1.55%.

74. ◈

75. ◈

Exercise Set 2.5

1. *Familiarize*. Let $x =$ the number. Then "three less than twice a number" translates to $2x - 3$.

***Translate*.**

Three less than twice a number is 19.

$$\begin{array}{cc} \downarrow & \downarrow \ \downarrow \\ 2x - 3 & = \ 19 \end{array}$$

***Carry out*.** We solve the equation.

$$\begin{aligned} 2x - 3 &= 19 \\ 2x &= 22 \quad \text{Adding 3} \\ x &= 11 \quad \text{Dividing by 2} \end{aligned}$$

***Check*.** Twice, or two times, 11 is 22. Three less than 22 is 19. The answer checks.

***State*.** The number is 11.

2. 8

3. *Familiarize*. Let $a =$ the number. Then "five times the sum of 3 and some number" translates to $5(a+3)$.

***Translate*.**

Five times the sum of 3 and some number is 70.

$$\begin{array}{cc} \downarrow & \downarrow \ \downarrow \\ 5(a + 3) & = \ 70 \end{array}$$

***Carry out*.** We solve the equation.

$$\begin{aligned} 5(a + 3) &= 70 \\ 5a + 15 &= 70 \quad \text{Using the distributive law} \\ 5a &= 55 \quad \text{Subtracting 15} \\ a &= 11 \quad \text{Dividing by 5} \end{aligned}$$

***Check*.** The sum of 3 and 11 is 14, and $5 \cdot 14 = 70$. The answer checks.

***State*.** The number is 11.

4. 13

5. *Familiarize*. Let $p =$ the regular price of the shoes. At 15% off, Amy paid 85% of the regular price.

Translate.

$63.75 is 85% of the regular price.

$$63.75 = 0.85 \cdot p$$

Carry out. We solve the equation.

$$63.75 = 0.85p$$
$$\frac{63.75}{0.08} = p \qquad \text{Dividing both sides by 0.85}$$
$$75 = p$$

Check. 85% of $75, or 0.85($75), is $63.75. The answer checks.

State. The regular price was $75.

6. $90

7. *Familiarize*. Let $b =$ the price of the book itself. When the sales tax rate is 5%, the tax paid on the book is 5% of b, or $0.05b$.

Translate.

Price of book plus sales tax is $89.25.

$$b + 0.05b = 89.25$$

Carry out. We solve the equation.

$$b + 0.05b = 89.25$$
$$1.05b = 89.25$$
$$b = \frac{89.25}{1.05}$$
$$b = 85$$

Check. 5% of $85, or 0.05($85), is $4.25 and $85 + $4.25 is $89.25, the total cost. The answer checks.

State. The book itself cost $85.

8. $95

9. *Familiarize*. Let $d =$ Kouros' distance, in miles, from the start after 8 hr. Then the distance from the finish line is $2d$.

Translate.

Distance from start plus distance from finish is 188 mi.

$$d + 2d = 188$$

Carry out. We solve the equation.

$$d + 2d = 188$$
$$3d = 188$$
$$d = \frac{188}{3}, \text{ or } 62\frac{2}{3}$$

Check. If Kouros is $\frac{188}{3}$ mi from the start, then he is $2 \cdot \frac{188}{3}$, or $\frac{376}{3}$ mi from the finish. Since $\frac{188}{3} + \frac{376}{3} = \frac{564}{3} = 188$, the total distance run, the answer checks.

State. Kouros had run $62\frac{2}{3}$ mi.

10. $699\frac{1}{3}$ mi

11. *Familiarize*. Let $x =$ the first page number. Then $x + 1 =$ the second page number, and $x + 2 =$ the third page number.

Translate.

The sum of three consecutive page numbers is 60.

$$x + (x+1) + (x+2) = 60$$

Carry out. We solve the equation.

$$x + (x+1) + (x+2) = 60$$
$$3x + 3 = 60 \quad \text{Combining like terms}$$
$$3x = 57 \quad \text{Subtracting 3 from both sides}$$
$$x = 19 \quad \text{Dividing both sides by 3}$$

If x is 19, then $x + 1$ is 20 and $x + 2 = 21$.

Check. 19, 20, and 21 are consecutive integers, and $19 + 20 + 21 = 60$. The result checks.

State. The page numbers are 19, 20, and 21.

12. 32, 33, 34

13. *Familiarize*. Let $x =$ the smaller odd number. Then $x + 2 =$ the next odd number.

Translate. We reword the problem.

Smaller odd number + next odd number is 60.

$$x + (x+2) = 60$$

Carry out. We solve the equation.

$$x + (x+2) = 60$$
$$2x + 2 = 60 \quad \text{Combining like terms}$$
$$2x = 58 \quad \text{Subtracting 2 from both sides}$$
$$x = 29 \quad \text{Dividing both sides by 2}$$

If x is 29, then $x + 2$ is 31.

Check. 29 and 31 are consecutive odd integers, and their sum is 60. The answer checks.

State. The integers are 29 and 31.

14. 53, 55

15. *Familiarize*. Let x = the first even integer. Then $x + 2$ = the next even integer.

***Translate*.**

$$\underbrace{\text{The sum of two}}_{x + (x + 2)}\text{ consecutive even integers} \quad \underset{=}{\text{is }} \underset{126}{126.}$$

***Carry out*.** We solve the equation.

$$
\begin{aligned}
x + (x + 2) &= 126 \\
2x + 2 &= 126 \quad \text{Combining like terms} \\
2x &= 124 \quad \text{Subtracting 2 from both} \\
&\qquad\qquad\quad\text{sides} \\
x &= 62 \quad \text{Dividing both sides by 2}
\end{aligned}
$$

If x is 62, then $x + 2$ is 64.

***Check*.** 62 and 64 are consecutive even integers, and $62 + 64 = 126$. The result checks.

***State*.** The numbers are 62 and 64.

16. 24, 26

17. *Familiarize*. Let b = the bride's age. Then $b + 19$ = the groom's age.

***Translate*.**

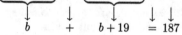

$$\underbrace{\text{Bride's age}}_{b} \text{ plus } \underbrace{\text{groom's age}}_{b + 19} \underset{=}{\text{is }} \underset{187}{187.}$$

***Carry out*.** We solve the equation.

$$
\begin{aligned}
b + (b + 19) &= 187 \\
2b + 19 &= 187 \\
2b &= 168 \\
b &= 84
\end{aligned}
$$

If b is 84, then $b + 19$ is 103.

***Check*.** 103 is 19 more than 84, and $84 + 103 = 187$. The answer checks.

***State*.** The bride was 84 yr old, and the groom was 103 yr old.

18. Man: 97 yr; woman: 91 yr

19. *Familiarize*. We draw a picture. We let x = the measure of the first angle. Then $3x$ = the measure of the second angle, and $x + 30$ = the measure of the third angle.

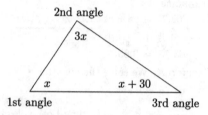

Recall that the measures of the angles of any triangle add up to 180°.

***Translate*.**

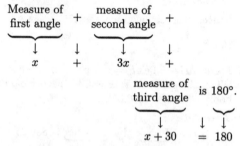

$$\underbrace{\text{Measure of}}_{x}\text{ first angle} + \underbrace{\text{measure of}}_{3x}\text{ second angle} +$$

$$\underbrace{\text{measure of}}_{x + 30}\text{ third angle} \quad \underset{=}{\text{is }} \underset{180}{180°.}$$

***Carry out*.** We solve the equation.

$$
\begin{aligned}
x + 3x + (x + 30) &= 180 \\
5x + 30 &= 180 \\
5x &= 150 \\
x &= 30
\end{aligned}
$$

Possible answers for the angle measures are as follows:

First angle: $x = 30°$

Second angle: $3x = 3(30)° = 90°$

Third angle: $x + 30° = 30° + 30° = 60°$

***Check*.** Consider 30°, 90°, and 60°. The second angle is three times the first, and the third is 30° more than the first. The sum of the measures of the angles is 180°. These numbers check.

***State*.** The measure of the first angle is 30°, the measure of the second angle is 90°, and the measure of the third angle is 60°.

20. 22.5°, 90°, 67.5°

21. *Familiarize*. Let x = the measure of the first angle. Then $3x$ = the measure of the second angle, and $x + 3x + 10 = 4x + 10$ = the measure of the third angle. Recall that the sum of the measures of the angles of a triangle is 180°.

Translate.

Measure of first angle $+$ measure of second angle $+$

$\downarrow$ $\qquad$ $\downarrow$ $\qquad$ $\downarrow$ $\qquad$ $\downarrow$

$x \qquad + \qquad 3x \qquad +$

measure of third angle $\quad$ is 180°.

$\downarrow \qquad \downarrow \quad \downarrow$

$(4x + 10) \quad = \quad 180$

Carry out. We solve the equation.

$$x + 3x + (4x + 10) = 180$$
$$8x + 10 = 180$$
$$8x = 170$$
$$x = 21.25$$

If x is 21.25, then $3x$ is 63.75, and $4x + 10$ is 95.

Check. Consider 21.25°, 63.75°, and 95°. The second is three times the first, and the third is 10° more than the sum of the other two. The sum of the measures of the angles is 180°. These numbers check.

State. The measure of the third angle is 95°.

22. 70°

23. *Familiarize*. The page numbers are consecutive integers. If we let p = the smaller number, then $p+1$ = the larger number.

Translate. We reword the problem.

First integer $+$ Second integer $= 385$

$\downarrow \qquad \downarrow \qquad \downarrow \qquad \downarrow \; \downarrow$

$x \qquad + \quad (x + 1) \quad = 385$

Carry out. We solve the equation.

$$x + (x + 1) = 385$$
$$2x + 1 = 385 \quad \text{Combining like terms}$$
$$2x = 384 \quad \text{Adding } -1 \text{ on both sides}$$
$$x = 192 \quad \text{Dividing on both sides by 2}$$

Check. If $x = 192$, then $x + 1 = 193$. These are consecutive integers, and $192 + 193 = 385$. The answer checks.

State. The page numbers are 192 and 193.

24. 140, 141

25. *Familiarize*. Let s = the length of the shortest side, in mm. Then $s + 2$ and $s + 4$ represent the lengths of the other two sides. The perimeter is the sum of the lengths of the sides.

Translate.

Length of first side $\;$ plus $\;$ length of second side $\;$ plus

$\downarrow$ $\qquad\qquad$ $\downarrow$ $\qquad\qquad$ $\downarrow$ $\qquad\qquad$ $\downarrow$

$s \qquad + \qquad (s + 2) \qquad +$

length of third side $\quad = \; \underline{195 \text{ mm}}$

$\downarrow \qquad\qquad \downarrow$

$(s + 4) \qquad = \qquad 195$

Carry out. We solve the equation.

$$s + (s + 2) + (s + 4) = 195$$
$$3s + 6 = 195$$
$$3s = 189$$
$$s = 63$$

If s is 63, then $s + 2$ is 65 and $s + 4$ is 67.

Check. The numbers 63, 65, and 67 are consecutive odd integers. Their sum is 195. These numbers check.

State. The lengths of the sides of the triangle are 63 mm, 65 mm, and 67 mm.

26. Width: 100 ft; length: 160 ft; area: 16,000 ft^2

27. *Familiarize*. We draw a picture. Let l = the length of the state, in miles. Then $l - 90$ = the width.

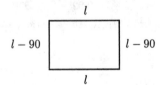

The perimeter is the sum of the lengths of the sides.

Translate. We use the definition of perimeter to write an equation.

Width $+$ Width $+$ Length $+$ Length is 1280.

$\downarrow \quad \downarrow \quad \downarrow \quad \downarrow \quad \downarrow \quad \downarrow \quad \downarrow \quad \downarrow \; \downarrow$

$(l - 90) + (l - 90) + \quad l \quad + \quad l \quad = 1280$

Carry out. We solve the equation.

$$(l - 90) + (l - 90) + l + l = 1280$$
$$4l - 180 = 1280$$
$$4l = 1460$$
$$l = 365$$

Then $l - 90 = 275$.

Check. The width, 275 mi, is 90 mi less than the length, 365 mi. The perimeter is 275 mi + 275 mi + 365 mi + 365 mi, or 1280 mi. This checks.

State. The length is 365 mi, and the width is 275 mi.

28. Length: 27.9 cm; width: 21.6 cm

29. *Familiarize*. Let a = the amount Sarah invested. The investment grew by 28% of a, or $0.28a$.

Translate.

$$\underbrace{\text{Amount invested}}_{a} \underset{+}{\text{plus}} \underbrace{\text{amount of growth}}_{0.28a} \underset{=}{\text{is \$448.}} \underset{448}{}$$

Carry out. We solve the equation.

$$a + 0.28a = 448$$
$$1.28a = 448$$
$$a = 350$$

Check. 28% of \$350 is 0.28(\$350), or \$98, and \$350 + \$98 = \$448. The answer checks.

State. Sarah invested \$350.

30. \$6600

31. *Familiarize*. Let b = the balance in the account at the beginning of the month. The balance grew by 2% of b, or $0.02b$.

Translate.

$$\underbrace{\text{Original balance}}_{b} \underset{+}{\text{plus}} \underbrace{\text{amount of growth}}_{0.02b} \underset{=}{\text{is \$870.}} \underset{870}{}$$

Carry out. We solve the equation.

$$b + 0.02b = 870$$
$$1.02b = 870$$
$$b \approx \$852.94$$

Check. 2% of \$852.94 is 0.02(\$852.94), or \$17.06, and \$852.94 + \$17.06 = \$870. The answer checks.

State. The balance at the beginning of the month was \$852.94.

32. \$6540

33. *Familiarize*. The total cost is the initial charge plus the mileage charge. Let d = the distance, in miles, that Courtney can travel for \$12. The mileage charge is the cost per mile times the number of miles traveled or $0.75d$.

Translate.

$$\underbrace{\text{Initial charge}}_{3} \underset{+}{\text{plus}} \underbrace{\text{mileage charge}}_{0.75d} \underset{=}{\text{is \$12.}} \underset{12}{}$$

Carry out. We solve the equation.

$$3 + 0.75d = 12$$
$$0.75d = 9$$
$$d = 12$$

Check. A 12-mi taxi ride from the airport would cost \$3 + 12(\$0.75), or \$3 + \$9, or \$12. The answer checks.

State. Courtney can travel 12 mi from the airport for \$12.

34. 15 mi

35. *Familiarize*. The total cost is the daily charge plus the mileage charge. Let d = the distance that can be traveled, in miles, in one day for \$100. The mileage charge is the cost per mile times the number of miles traveled, or $0.39d$.

Translate.

$$\underbrace{\text{Daily rate}}_{49.95} \underset{+}{\text{plus}} \underbrace{\text{mileage charge}}_{0.39d} \underset{=}{\text{is \$100.}} \underset{100}{}$$

Carry out. We solve the equation.

$$49.95 + 0.39d = 100$$
$$0.39d = 50.05$$
$$d = 128.\overline{3}, \text{or} 128\frac{1}{3}$$

Check. For a trip of $128\frac{1}{3}$ mi, the mileage charge is $\$0.39\left(128\frac{1}{3}\right)$, or \$50.05, and \$49.95 + \$50.05 = \$100. The answer checks.

State. They can travel $128\frac{1}{3}$ mi in one day and stay within their budget.

36. 80 mi

37. *Familiarize*. Let x = the measure of one angle. Then $90 - x$ = the measure of its complement.

Translate.

$$\underbrace{\text{Measure of one angle}}_{x} \underset{=}{\text{is 15°}} \underset{15}{} \underset{+}{\text{more than}} \underbrace{\text{twice the measure of its complement.}}_{2(90 - x)}$$

Carry out. We solve the equation.

$$x = 15 + 2(90 - x)$$
$$x = 15 + 180 - 2x$$
$$x = 195 - 2x$$
$$3x = 195$$
$$x = 65$$

If x is 65, then $90 - x$ is 25.

Check. The sum of the angle measures is 90°. Also, 65° is 15° more than twice its complement, 25°. The answer checks.

State. The angle measures are 65° and 25°.

38. 105°, 75°

39. *Familiarize.* We will use the equation
$$T = \frac{1}{4}N + 40.$$

Translate. We substitute 80 for T.
$$80 = \frac{1}{4}N + 40$$

Carry out. We solve the equation.
$$80 = \frac{1}{4}N + 40$$
$$40 = \frac{1}{4}N$$
$$160 = N \qquad \text{Multiplying by 4 on both sides}$$

Check. When $N = 160$, we have $T = \frac{1}{4} \cdot 160 + 40 = 40 + 40 = 80$. The answer checks.

State. A cricket chirps 160 times per minute when the temperature is 80°F.

40. 2020

41. ◈

42. ◈

43. Since -9 is to the left of 5 on the number line, we have $-9 < 5$.

44. $1 < 3$

45. Since -4 is to the left of 7 on the number line, we have $-4 < 7$.

46. $-9 > -12$

47. ◈

48. ◈

49. *Familiarize.* Let c = the amount the meal originally cost. The 15% tip is calculated on the original cost of the meal, so the tip is 0.15c.

Translate.

Original cost plus tip less $10 is $32.55.

$$c + 0.15c - 10 = 32.55$$

Carry out. We solve the equation.
$$c + 0.15c - 10 = 32.55$$
$$1.15c - 10 = 32.55$$
$$1.15c = 42.55$$
$$c = 37$$

Check. If the meal originally cost $37, the tip was 15% of $37, or 0.15($37), or $5.55. Since $37+$5.55−$10 = $32.55, the answer checks.

State. The meal originally cost $37.

50. 19

51. *Familiarize.* Let s = one score. Then four score = $4s$ and four score and seven = $4s + 7$.

Translate. We reword .

1776 plus four score and seven is 1863

$$1776 + (4s + 7) = 1863$$

Carry out. We solve the equation.
$$1776 + (4s + 7) = 1863$$
$$4s + 1783 = 1863$$
$$4s = 80$$
$$s = 20$$

Check. If a score is 20 years, then four score and seven represents 87 years. Adding 87 to 1776 we get 1863. This checks.

State. A score is 20.

52. 4, 16

53. *Familiarize.* We let x = the length of the original rectangle. Then $\frac{3}{4}x$ = the width. We draw a picture of the enlarged rectangle. Each dimension is increased by 2 cm, so $x + 2$ = the length of the enlarged rectangle and $\frac{3}{4}x + 2$ = the width.

$$\frac{3}{4}x + 2 \quad \boxed{ x+2 } \quad \frac{3}{4}x + 2$$

$$x + 2$$

Translate. We use the perimeter of the enlarged rectangle to write an equation.

Width + Width + Length + Length is

$$\left(\frac{3}{4}x + 2\right) + \left(\frac{3}{4}x + 2\right) + (x + 2) + (x + 2) =$$

Perimeter.

$$50$$

Carry out.

$$\left(\frac{3}{4}x+2\right)+\left(\frac{3}{4}x+2\right)+(x+2)+(x+2)=50$$

$$\frac{7}{2}x+8=50$$

$$2\left(\frac{7}{2}x+8\right)=2\cdot50$$

$$7x+16=100$$

$$7x=84$$

$$x=12$$

Then $\frac{3}{4}x=\frac{3}{4}(12)=9$.

Check. If the dimensions of the original rectangle are 12 cm and 9 cm, then the dimensions of the enlarged rectangle are 14 cm and 11 cm. The perimeter of the enlarged rectangle is $11+11+14+14=50$ cm. Also, 9 is $\frac{3}{4}$ of 12. These values check.

State. The length is 12 cm, and the width is 9 cm.

54. 87°, 89°, 91°, 93°

55. Familiarize. Let $x=$ the first even number. Then the next four even numbers are $x+2$, $x+4$, $x+6$, and $x+8$. The sum of the measures of the angles of an n-sided polygon is given by the formula $(n-2)\cdot180°$. Thus, the sum of the measures of the angles of a pentagon is $(5-2)\cdot180°$, or $3\cdot180°$, or 540°.

Translate.

The sum of the measures of the angles is 540°.

$$x+(x+2)+(x+4)+(x+6)+(x+8)=540$$

Carry out. We solve the equation.

$$x+(x+2)+(x+4)+(x+6)+(x+8)=540$$

$$5x+20=540$$

$$5x=520$$

$$x=104$$

If x is 104, then the other numbers are 106, 108, 110, and 112.

Check. The numbers 104, 106, 108, 110, and 112 are consecutive odd numbers. Their sum is 540. The answer checks.

State. The measures of the angles are 104°, 106°, 108°, 110°, and 112°.

56. 120

57. Familiarize. Let $p=$ the price before the two discounts. With the first 10% discount, the price becomes 90% of p, or $0.9p$. With the second 10% discount, the final price is 90% of $0.9p$, or $0.9(0.9p)$.

Translate.

The final price is $77.75.

$$0.9(0.9p)=77.75$$

Carry out. We solve the equation.

$$0.9(0.9p)=77.75$$

$$p=\frac{77.75}{0.9(0.9)}$$

$$p\approx95.99$$

Check. 90% of $95.99 is $86.39 and 90% of $86.39 is $77.75. The answer checks.

State. The price before the two discounts was $95.99.

58. 30

59. Familiarize. Let $n=$ the number of CD's purchased. Assume that two or more CD's were purchased. Then the first CD costs $8.49 and the total cost of the remaining $n-1$ CD's is $3.99(n-1)$. The shipping and handling costs are $2.47 for the first CD, $2.28 for the second, and a total of $1.99(n-2)$ for the remaining $n-2$ CD's. Then the total cost of the shipment is $8.49+3.99(n-1)+2.47+2.28+1.99(n-2)$.

Translate.

Total cost of shipment was $65.07.

$$8.49+3.99(n-1)+2.47+2.28+1.99(n-2)=65.07$$

Carry out. We solve the equation.

$$8.49+3.99(n-1)+2.47+2.28+1.99(n-2)=65.07$$

$$8.49+3.99n-3.99+2.47+2.28+1.99n-3.98=65.07$$

$$5.27+5.98n=65.07$$

$$5.98n=59.80$$

$$n=10$$

Check. If 10 CD's are purchased, the total cost of the CD's is $8.49+$3.99(9)=44.40. The total shipping and handling costs are $2.47+$2.28+$1.99(8)=$20.67. Then the total cost of the order is $44.40+$20.67=$65.07.

State. There were 10 CD's in the shipment.

60. 76

61. Familiarize. At $0.30 per $\frac{1}{5}$ mile, the mileage charge can also be given as 5($0.30), or $1.50 per mile. Since it took 20 min to complete what is usually a 10-min drive, the taxi was stopped in traffic for 20 − 10, or 10, min. Let d = the distance, in miles, that Glenda traveled.

Translate.

Initial charge	plus	mileage charge	plus	charge for being stopped in traffic	is $13.
↓	↓	↓	↓	↓	↓ ↓
2	+	1.5d	+	0.2(10)	= 13

Carry out. We solve the equation.

$$2 + 1.5d + 0.2(10) = 13$$
$$2 + 1.5d + 2 = 13$$
$$1.5d + 4 = 13$$
$$1.5d = 9$$
$$d = 6$$

Check. The mileage charge for traveling 6 mi is $1.50(6) = $9. The charge for being stopped in traffic is $0.20(10) = $2. Since $2 + $9 + $2 = $13, the answer checks.

State. Glenda traveled 6 mi.

62. ◈

63. ◈

64. Width: 23.31 cm; length: 27.56 cm

65. **Familiarize**. Let s = the length of the first side, in cm. Then $s + 3.25$ = the length of the second side, and $(s + 3.25) + 4.35$, or $s + 7.6$ = the length of the third side.

Translate.

The perimeter	is	26.87 cm.
↓	↓	↓
$s + (s + 3.25) + (s + 7.6)$ =		26.87

Carry out. We solve the equation.

$$s + (s + 3.25) + (s + 7.6) = 26.87$$
$$3s + 10.85 = 26.87$$
$$3s = 16.02$$
$$s = 5.34$$

If $s = 5.34$, then $s + 3.25 = 8.59$, and $s + 7.6 = 12.94$.

Check. Consider sides of 5.34 cm, 8.59 cm, and 12.94 cm. The second side is 3.25 cm longer than the first side, and the third side is 4.35 cm longer than the second side. The sum of the lengths of the sides is 26.87. The answer checks.

State. The lengths of the sides are 5.34 cm, 8.59 cm, and 12.94 cm.

Exercise Set 2.6

1. $x > -2$

   a) Since $5 > -2$ is true, 5 is a solution.

   b) Since $0 > -2$ is true, 0 is a solution.

   c) Since $-1.9 > -2$ is true, -1.9 is a solution.

   d) Since $-7.3 > -2$ is false, -7.3 is not a solution.

   e) Since $1.6 > -2$ is true, 1.6 is a solution.

2. a) Yes, b) No, c) Yes, d) Yes, e) No

3. $x \geq 6$

   a) Since $-6 \geq 6$ is false, -6 is not a solution.

   b) Since $0 \geq 6$ is false, 0 is not a solution.

   c) Since $6 \geq 6$ is true, 6 is a solution.

   d) Since $6.01 \geq 6$ is true, 6.01 is a solution.

   e) Since $-3\frac{1}{2} \geq 6$ is false, $-3\frac{1}{2}$ is not a solution.

4. a) Yes, b) Yes, c) Yes, d) No, e) Yes

5. The solutions of $x \leq 7$ are shown by shading the point 7 and all points to the left of 7. The closed circle at 7 indicates that 7 is part of the graph.

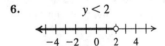

6. $y < 2$

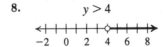

7. The solutions of $t > -2$ are those numbers greater than -2. They are shown on the graph by shading all points to the right of -2. The open circle at -2 indicates that -2 is not part of the graph.

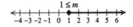

8. $y > 4$

$$\overset{\longleftarrow |\ |\ |\ |\ |\ |\ \circ\ |\ |\ |\ |\ \longrightarrow}{\underset{-2\quad 0\quad 2\quad 4\quad 6\quad 8}{}}$$

9. The solutions of $1 \leq m$, or $m \geq 1$, are those numbers greater than or equal to 1. They are shown by shading the point 1 and all points to the right of 1. The closed circle at 1 indicates that 1 is part of the graph.

$$\overset{1 \leq m}{\overset{\longleftarrow |\ |\ |\ |\ |\ \bullet\ |\ |\ |\ |\ |\ \longrightarrow}{\underset{-4\ -3\ -2\ -1\ 0\ 1\ 2\ 3\ 4\ 5\ 6}{}}}$$

10.
$$t \geq 0$$

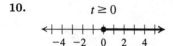

11. In order to be a solution of the inequality
$-3 < x \leq 5$, a number must be a solution of both
$-3 < x$ and $x \leq 5$. The solution set is graphed as
follows:

$$-3 < x \leq 5$$

The open circle at -3 means that -3 is not part of
the graph. The closed circle at 5 means that 5 is part
of the graph.

12.
$$-5 \leq x < 2$$

13. In order to be a solution of the inequality $0 < x < 3$,
a number must be a solution of both $0 < x$ and $x < 3$.
The solution set is graphed as follows:

$$0 < x < 3$$

The open circles at 0 and at 3 mean that 0 and 3 are
not part of the graph.

14.
$$-5 \leq x \leq 0$$

15. All points to the right of -4 are shaded. The
open circle at -4 indicates that -4 is not part
of the graph. Using set-builder notation we have
$\{x | x > -4\}$.

16. $\{x | x < 3\}$

17. The point 2 and all points to the left of 2 are shaded.
Using set-builder notation we have $\{x | x \leq 2\}$.

18. $\{x | x \geq -2\}$

19. All points to the left of -1 are shaded. The open circle at -1 indicates that -1 is not part of the graph.
Using set-builder notation we have $\{x | x < -1\}$.

20. $\{x | x > 1\}$

21. The point 0 and all points to the right of 0 are
shaded. Using set-builder notation we have
$\{x | x \geq 0\}$.

22. $\{x | x \leq 0\}$

23.
$$\begin{aligned} y + 2 &> 9 \\ y + 2 - 2 &> 9 - 2 \qquad \text{Adding } -2 \text{ to both sides} \\ y &> 7 \qquad \text{Simplifying} \end{aligned}$$

The solution set is $\{y | y > 7\}$. The graph is as follows:

24. $\{y | y > 3\}$

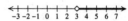

25.
$$\begin{aligned} x + 8 &\leq -10 \\ x + 8 - 8 &\leq -10 - 8 \qquad \text{Subtracting 8 from} \\ & \qquad\qquad\qquad\quad \text{both sides} \\ x &\leq -18 \qquad \text{Simplifying} \end{aligned}$$

The solution set is $\{x | x \leq -18\}$. The graph is as
follows:

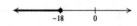

26. $\{x | x \leq -21\}$

27.
$$\begin{aligned} x - 3 &< 7 \\ x - 3 + 3 &< 7 + 3 \\ x &< 10 \end{aligned}$$

The solution set is $\{x | x < 10\}$. The graph is as
follows:

28. $\{x | x < 17\}$

29.
$$\begin{aligned} 5 &\leq t + 8 \\ 5 - 8 &\leq t + 8 - 8 \\ -3 &\leq t \end{aligned}$$

The solution set is $\{t | -3 \leq t\}$, or $\{t | t \geq -3\}$. The
graph is as follows:

30. $\{t | t \geq -5\}$

31.
$$\begin{aligned} y - 7 &> -12 \\ y - 7 + 7 &> -12 + 7 \\ y &> -5 \end{aligned}$$

The solution set is $\{y | y > -5\}$. The graph is as
follows:

32. $\{y|y > -6\}$

33.
$$2x + 4 \leq x + 9$$
$$2x + 4 - 4 \leq x + 9 - 4 \quad \text{Adding } -4$$
$$2x \leq x + 5 \quad \text{Simplifying}$$
$$2x - x \leq x + 5 - x \quad \text{Adding } -x$$
$$x \leq 5 \quad \text{Simplifying}$$

The solution set is $\{x|x \leq 5\}$. The graph is as follows:

34. $\{x|x \leq -3\}$

35.
$$5x - 6 \geq 4x - 1$$
$$5x - 6 + 6 \geq 4x - 1 + 6 \quad \text{Adding 6 to both sides}$$
$$5x \geq 4x + 5$$
$$5x - 4x \geq 4x + 5 - 4x \quad \text{Adding } -4x \text{ to both sides}$$
$$x \geq 5$$

The solution set is $\{x|x \geq 5\}$.

36. $\{x|x \geq 20\}$

37.
$$y + \frac{1}{3} \leq \frac{5}{6}$$
$$y + \frac{1}{3} - \frac{1}{3} \leq \frac{5}{6} - \frac{1}{3}$$
$$y \leq \frac{5}{6} - \frac{2}{6}$$
$$y \leq \frac{3}{6}$$
$$y \leq \frac{1}{2}$$

The solution set is $\left\{y \middle| y \leq \frac{1}{2}\right\}$.

38. $\left\{x \middle| x \leq \frac{1}{4}\right\}$

39.
$$t - \frac{1}{8} > \frac{1}{2}$$
$$t - \frac{1}{8} + \frac{1}{8} > \frac{1}{2} + \frac{1}{8}$$
$$t > \frac{4}{8} + \frac{1}{8}$$
$$t > \frac{5}{8}$$

The solution set is $\left\{t \middle| t > \frac{5}{8}\right\}$.

40. $\left\{y \middle| y > \frac{7}{12}\right\}$

41.
$$-9x + 17 > 17 - 8x$$
$$-9x + 17 - 17 > 17 - 8x - 17 \quad \text{Adding } -17$$
$$-9x > -8x$$
$$-9x + 9x > -8x + 9x \quad \text{Adding } 9x$$
$$0 > x$$

The solution set is $\{x|x < 0\}$.

42. $\{n|n < 0\}$

43. $-23 < -t$

The inequality states that the opposite of 23 is less than the opposite of t. Thus, t must be less than 23, so the solution set is $\{t|t < 23\}$. To solve this inequality using the addition principle, we would proceed as follows:
$$-23 < -t$$
$$t - 23 < 0 \quad \text{Adding } t \text{ to both sides}$$
$$t < 23 \quad \text{Adding 23 to both sides}$$

The solution set is $\{t|t < 23\}$.

44. $\{x|x < -19\}$

45.
$$5x < 35$$
$$\frac{1}{5} \cdot 5x < \frac{1}{5} \cdot 35 \quad \text{Multiplying by } \frac{1}{5}$$
$$x < 7$$

The solution set is $\{x|x < 7\}$. The graph is as follows:

46. $\{x|x \geq 4\}$

47. $9y \le 81$

$\frac{1}{9} \cdot 9y \le \frac{1}{9} \cdot 81$ Multiplying by $\frac{1}{9}$

$y \le 9$

The solution set is $\{y|y \le 9\}$. The graph is as follows:

48. $\{t|t < 35\}$

49. $-7x < 13$

$-\frac{1}{7} \cdot (-7x) > -\frac{1}{7} \cdot 13$ Multiplying by $-\frac{1}{7}$

$\quad\quad$ The symbol has to be reversed.

$x > -\frac{13}{7}$ Simplifying

The solution set is $\left\{x \middle| x > -\frac{13}{7}\right\}$.

50. $\left\{y \middle| y < \frac{17}{8}\right\}$

51. $-24 > 8t$

$-3 > t$

The solution set is $\{t|t < -3\}$.

52. $\{x|x > 4\}$

53. $7y \ge -2$

$\frac{1}{7} \cdot 7y \ge \frac{1}{7}(-2)$ Multiplying by $\frac{1}{7}$

$y \ge -\frac{2}{7}$

The solution set is $\left\{y \middle| y \ge -\frac{2}{7}\right\}$.

54. $\left\{x \middle| x > -\frac{3}{5}\right\}$

55. $-2y \le \frac{1}{5}$

$-\frac{1}{2} \cdot (-2y) \ge -\frac{1}{2} \cdot \frac{1}{5}$

$\quad\quad$ The symbol has to be reversed.

$y \ge -\frac{1}{10}$

The solution set is $\left\{y \middle| y \ge -\frac{1}{10}\right\}$.

56. $\left\{x \middle| x \le -\frac{1}{10}\right\}$

57. $-\frac{8}{5} > -2x$

$-\frac{1}{2} \cdot \left(-\frac{8}{5}\right) < -\frac{1}{2} \cdot (-2x)$

$\frac{8}{10} < x$

$\frac{4}{5} < x$, or $x > \frac{4}{5}$

The solution set is $\left\{x \middle| \frac{4}{5} < x\right\}$, or $\left\{x \middle| x > \frac{4}{5}\right\}$.

58. $\left\{y \middle| y < \frac{1}{16}\right\}$

59. $7 + 3x < 34$

$7 + 3x - 7 < 34 - 7$ Adding -7 to both sides

$3x < 27$ Simplifying

$x < 9$ Multiplying both sides by $\frac{1}{3}$

The solution set is $\{x|x < 9\}$.

60. $\{y|y < 8\}$

61. $6 + 5y \ge 26$

$6 + 5y - 6 \ge 26 - 6$ Adding -6

$5y \ge 20$

$y \ge 4$ Multiplying by $\frac{1}{5}$

The solution set is $\{y|y \ge 4\}$.

62. $\{x|x \ge 8\}$

63. $4t - 5 \le 23$

$4t - 5 + 5 \le 23 + 5$ Adding 5 to both sides

$4t \le 28$

$\frac{1}{4} \cdot 4t \le \frac{1}{4} \cdot 28$ Multiplying both sides by $\frac{1}{4}$

$x \le 7$

The solution set is $\{x|x \le 7\}$.

64. $\{y|y \le 6\}$

65.
$$13x - 7 < -46$$
$$13x - 7 + 7 < -46 + 7$$
$$13x < -39$$
$$\frac{1}{13} \cdot 13x < \frac{1}{13} \cdot (-39)$$
$$x < -3$$

The solution set is $\{x | x < -3\}$.

66. $\{y | y < -6\}$

67.
$$16 < 4 - 3y$$
$$16 - 4 < 4 - 3y - 4 \quad \text{Adding } -4 \text{ to both sides}$$
$$12 < -3y$$
$$-\frac{1}{3} \cdot 12 > -\frac{1}{3} \cdot (-3y) \quad \text{Multiplying by } -\frac{1}{3}$$
$$\quad \boxed{} \text{ The symbol has to be reversed.}$$
$$-4 > y$$

The solution set is $\{y | -4 > y\}$, or $\{y | y < -4\}$.

68. $\{x | -2 > x\}$, or $\{x | x < -2\}$

69.
$$39 > 3 - 9x$$
$$39 - 3 > 3 - 9x - 3 \quad \text{Adding } -3$$
$$36 > -9x$$
$$-\frac{1}{9} \cdot 36 < -\frac{1}{9} \cdot (-9x) \quad \text{Multiplying by } -\frac{1}{9}$$
$$\quad \boxed{} \text{ The symbol has to be reversed.}$$
$$-4 < x$$

The solution set is $\{x | -4 < x\}$, or $\{x | x > -4\}$.

70. $\{y | -5 < y\}$, or $\{y | y > -5\}$

71.
$$5 - 6y > 25$$
$$-5 + 5 - 6y > -5 + 25$$
$$-6y > 20$$
$$-\frac{1}{6} \cdot (-6y) < -\frac{1}{6} \cdot 20$$
$$\quad \boxed{} \text{ The symbol has to be reversed.}$$
$$y < -\frac{20}{6}$$
$$y < -\frac{10}{3}$$

The solution set is $\left\{ y \middle| y < -\frac{10}{3} \right\}$.

72. $\{y | y < -3\}$

73.
$$-3 < 8x + 7 - 7x$$
$$-3 < x + 7 \quad \text{Collecting like terms}$$
$$-3 - 7 < x + 7 - 7$$
$$-10 < x$$

The solution set is $\{x | -10 < x\}$, or $\{x | x > -10\}$.

74. $\{x | -13 < x\}$, or $\{x | x > -13\}$

75.
$$6 - 4y > 4 - 3y$$
$$6 - 4y + 4y > 4 - 3y + 4y \quad \text{Adding } 4y$$
$$6 > 4 + y$$
$$-4 + 6 > -4 + 4 + y \quad \text{Adding } -4$$
$$2 > y, \text{ or } y < 2$$

The solution set is $\{y | 2 > y\}$, or $\{y | y < 2\}$.

76. $\{y | 2 > y\}$, or $\{y | y < 2\}$

77.
$$7 - 9y \le 4 - 8y$$
$$7 - 9y + 9y \le 4 - 8y + 9y$$
$$7 \le 4 + y$$
$$-4 + 7 \le -4 + 4 + y$$
$$3 \le y, \text{ or } y \ge 3$$

The solution set is $\{y | 3 \le y\}$, or $\{y | y \ge 3\}$.

78. $\{y | 2 \le y\}$, or $\{y | y \ge 2\}$

79.
$$33 - 12x < 4x + 97$$
$$33 - 12x - 97 < 4x + 97 - 97$$
$$-64 - 12x < 4x$$
$$-64 - 12x + 12x < 4x + 12x$$
$$-64 < 16x$$
$$-4 < x$$

The solution set is $\{x | -4 < x\}$, or $\{x | x > -4\}$.

80. $\left\{ x \middle| x < \frac{9}{5} \right\}$

81.
$$2.1x + 43.2 > 1.2 - 8.4x$$
$$10(2.1x + 43.2) > 10(1.2 - 8.4x) \quad \text{Multiplying by } 10 \text{ to clear decimals}$$
$$21x + 432 > 12 - 84x$$
$$21x + 84x > 12 - 432 \quad \text{Adding } 84x \text{ and } -432$$
$$105x > -420$$
$$x > -4 \quad \text{Multiplying by } \frac{1}{105}$$

The solution set is $\{x | x > -4\}$.

82. $\left\{ y \middle| y \le \frac{5}{3} \right\}$

83.
$$0.7n - 15 + n \ge 2n - 8 - 0.4n$$
$$1.7n - 15 \ge 1.6n - 8 \quad \text{Collecting like terms}$$
$$10(1.7n - 15) \ge 10(1.6n - 8) \quad \text{Multiplying by } 10$$
$$17n - 150 \ge 16n - 80$$
$$17n - 16n \ge -80 + 150 \quad \text{Adding } -16n \text{ and } 150$$
$$n \ge 70$$

The solution set is $\{n | n \ge 70\}$

84. $\{t | t > 1\}$

85.
$$\frac{x}{3} - 4 \le 1$$
$$3\left(\frac{x}{3} - 4\right) \le 3 \cdot 1 \qquad \text{Multiplying by 3 to}$$
$$\qquad\qquad\qquad\qquad \text{to clear the fraction}$$
$$x - 12 \le 3 \qquad \text{Simplifying}$$
$$x \le 15 \qquad \text{Adding 12}$$
The solution set is $\{x | x \le 15\}$.

86. $\{x | x > 2\}$

87.
$$3 < 5 - \frac{t}{7}$$
$$-2 < -\frac{t}{7}$$
$$-7(-2) > -7\left(-\frac{t}{7}\right)$$
$$14 > t$$
The solution set is $\{t | t < 14\}$.

88. $\{x | x > 35\}$

89.
$$4(2y - 3) < 36$$
$$8y - 12 < 36 \qquad \text{Removing parentheses}$$
$$8y < 48 \qquad \text{Adding 12}$$
$$y < 6 \qquad \text{Multiplying by } \frac{1}{8}$$
The solution set is $\{y | y < 6\}$.

90. $\{y | y > 5\}$

91.
$$3(t - 2) \ge 9(t + 2)$$
$$3t - 6 \ge 9t + 18$$
$$3t - 9t > 18 + 6$$
$$-6t \ge 24$$
$$t \le -4 \qquad \text{Multiplying by } -\frac{1}{6} \text{ and}$$
$$\qquad\qquad\qquad \text{reversing the symbol}$$
The solution set is $\{t | t \le -4\}$.

92. $\left\{t \middle| t < -\frac{5}{3}\right\}$

93.
$$3(r - 6) + 2 < 4(r + 2) - 21$$
$$3r - 18 + 2 < 4r + 8 - 21$$
$$3r - 16 < 4r - 13$$
$$-16 + 13 < 4r - 3r$$
$$-3 < r, \text{ or } r > -3$$
The solution set is $\{r | r > -3\}$.

94. $\{t | t > -12\}$

95.
$$\frac{2}{3}(2x - 1) \ge 10$$
$$\frac{3}{2} \cdot \frac{2}{3}(2x - 1) \ge \frac{3}{2} \cdot 10 \qquad \text{Multiplying by } \frac{3}{2}$$
$$2x - 1 \ge 15$$
$$2x \ge 16$$
$$x \ge 8$$
The solution set is $\{x | x \ge 8\}$.

96. $\{x | x \le 7\}$

97.
$$\frac{3}{4}\left(3x - \frac{1}{2}\right) - \frac{2}{3} < \frac{1}{3}$$
$$\frac{3}{4}\left(3x - \frac{1}{2}\right) < 1 \qquad \text{Adding } \frac{2}{3}$$
$$\frac{9}{4}x - \frac{3}{8} < 1 \qquad \text{Removing parentheses}$$
$$8 \cdot \left(\frac{9}{4}x - \frac{3}{8}\right) < 8 \cdot 1 \qquad \text{Clearing fractions}$$
$$18x - 3 < 8$$
$$18x < 11$$
$$x < \frac{11}{18}$$
The solution set is $\left\{x \middle| x < \frac{11}{18}\right\}$.

98. $\left\{x \middle| x > -\frac{5}{32}\right\}$

99. ◈

100. ◈

101. Let n represent "some number." Then we have $n + 3$, or $3 + n$.

102. Let x and y represent the numbers; $2(x + y)$

103. Let x represent "a number." Then we have $2x - 3$.

104. Let y represent "a number;" $2y + 5$, or $5 + 2y$

105. ◈

106. ◈

107.
$$6[4 - 2(6 + 3t)] > 5[3(7 - t) - 4(8 + 2t)] - 20$$
$$6[4 - 12 - 6t] > 5[21 - 3t - 32 - 8t] - 20$$
$$6[-8 - 6t] > 5[-11 - 11t] - 20$$
$$-48 - 36t > -55 - 55t - 20$$
$$-48 - 36t > -75 - 55t$$
$$-36t + 55t > -75 + 48$$
$$19t > -27$$
$$t > -\frac{27}{19}$$
The solution set is $\left\{t \middle| t > -\frac{27}{19}\right\}$.

108. $\left\{x \mid x \le \dfrac{5}{6}\right\}$

109. $-(x+5) \ge 4a - 5$
$-x - 5 \ge 4a - 5$
$-x \ge 4a - 5 + 5$
$-x \ge 4a$
$-1(-x) \le -1 \cdot 4a$
$x \le -4a$

The solution set is $\{x \mid x \le -4a\}$.

110. $\{x \mid x > 7\}$

111. $\quad y < ax + b$ Assume $a > 0$.
$y - b < ax$
$\dfrac{y - b}{a} < x$ Since $a > 0$, the inequality symbol stays the same.

The solution set is $\left\{x \mid x > \dfrac{y-b}{a}\right\}$.

112. $\left\{x \mid x < \dfrac{y-b}{a}\right\}$

113. $|x| < 3$

a) Since $|3.2| = 3.2$, and $3.2 < 3$ is false, 3.2 is not a solution.

b) Since $|-2| = 2$ and $2 < 3$ is true, -2 is a solution.

c) Since $|-3| = 3$ and $3 < 3$ is false, -3 is not a solution.

d) Since $|-2.9| = 2.9$ and $2.9 < 3$ is true, -2.9 is a solution.

e) Since $|3| = 3$ and $3 < 3$ is false, 3 is not a solution.

f) Since $|1.7| = 1.7$ and $1.7 < 3$ is true, 1.7 is a solution.

114.

115. $|x| > -3$

Since absolute value is always nonnegative, the absolute value of any real number will be greater than -3. Thus, the solution set is $\{x \mid x \text{ is a real number}\}$, or $(-\infty, \infty)$.

116. $\emptyset$

1. Let n represent the number. Then we have
$n \ge 7$.

2. Let n represent the number; $n \ge 5$.

3. Let b represent the weight of the baby, in kilograms. Then we have
$b > 2$.

4. Let p represent the number of people who attended the concert; $75 < p < 100$.

5. Let s represent the average speed, in mph. Then we have
$90 < s < 110$.

6. Let n represent the number of people who attended the Million Man March; $n \ge 400,000$.

7. Let a represent the number of people who attended the Million Man March. Then we have
$a \le 1,200,000$.

8. Let a represent the amount of acid, in liters; $a \le 40$.

9. Let c represent the cost, per gallon, of gasoline. Then we have
$c \ge \$1.50$.

10. Let t represent the temperature; $t \le -2$.

11. *Familiarize.* Let c = the number of copies Myra has made. The total cost of the copies is the setup fee of \$5 plus \$4 times the number of copies, or $\$4 \cdot c$.

Translate.

Setup fee	plus	copying cost	cannot exceed	\$65.
$\downarrow$	$\downarrow$	$\downarrow$	$\downarrow$	$\downarrow$
5	+	$4c$	$\le$	65

Carry out. We solve the inequality.
$5 + 4c \le 65$
$4c \le 60$
$c \le 15$

Check. As a partial check, we show that Myra can have 15 copies made and not exceed her \$65 budget.
$\$5 + \$4 \cdot 15 = 5 + 60 = \$65$

State. Myra can have 15 or fewer copies made and stay within her budget.

12. 25 people

13. *Familiarize*. Let m represent the number of miles per day. Then the cost per day for those miles is $0.46m$. The total cost is the daily rate plus the daily mileage cost. The total cost cannot exceed $200. In other words the total cost must be less than or equal to $200, the daily budget.

Translate.

$$\underbrace{\text{Daily rate}}\ +\ \underbrace{\text{Mileage cost}}\ \underbrace{\leq}\ \underbrace{\text{Budget.}}$$
$$42.95\quad +\quad 0.46m\quad\ \leq\quad\ 200$$

Carry out.

$$42.95 + 0.46m \leq 200$$
$$4295 + 46m \leq 20,000 \quad \text{Clearing decimals}$$
$$46m \leq 15,705$$
$$m \leq \frac{15,705}{46}$$
$$m \leq 341.4 \quad \text{Rounding to the nearest tenth}$$

Check. We can check to see if the solution set seems reasonable.

When $m = 342$, the total cost is

$42.95 + 0.46(342)$, or $200.27.

When $m = 341.4$, the total cost is

$42.95 + 0.46(341.4)$, or $199.99.

When $m = 341$, the total cost is

$42.95 + 0.46(341)$, or $199.81.

From these calculations it would appear that $m \leq 341.4$ is the correct solution.

State. To stay within the budget, the number of miles the Letsons drive must not exceed 341.4.

14. 5 minutes or more

15. *Familiarize*. Let t = the number of hours the car is parked. Then $2t$ = the number of half-hours it is parked. The total parking cost is the initial $0.45 charge plus $0.25 per half hour, or $0.25 \cdot 2t$.

Translate.

$$\underbrace{\text{Initial charge}}\ \text{plus}\ \underbrace{\text{charge for time parked}}\ \underbrace{\text{is at least}}\ \$2.20.$$
$$0.45\quad +\quad 0.25 \cdot 2t\quad\ \geq\quad\ 2.20$$

Carry out. We solve the inequality.

$$0.45 + 0.25 \cdot 2t \geq 2.20$$
$$0.45 + 0.5t \geq 2.2$$
$$0.5t \geq 1.75$$
$$t \geq 3.5$$

Check. As a partial check, we can show that the parking charge for 3.5 hr is $2.20. Note that in 3.5 hr there are 2(3.5), or 7, half-hours.

$$\$0.45 + \$0.25(7) = \$0.45 + \$1.75 = \$2.20.$$

State. Laura's car is generally parked for 3.5 hr or more.

16. More than 2.5 hr

17. *Familiarize*. Let c = the number of courses for which Angelica registers. Her total tuition is the $35 registration fee plus $375 times the number of courses for which she registers, or $375 \cdot c$.

Translate.

$$\underbrace{\text{Registration fee}}\ \text{plus}\ \underbrace{\text{fee for courses}}\ \underbrace{\text{cannot exceed}}\ \$1000.$$
$$35\quad +\quad 375 \cdot c\quad\ \leq\quad\ 1000$$

Carry out. We solve the inequality.

$$35 + 375c \leq 1000$$
$$375c \leq 965$$
$$c \leq 2.57\overline{3}$$

Check. Although the solution set of the inequality is all numbers less than or equal to $2.57\overline{3}$, since c represents the number of courses for which Angelica registers, we round down to 2. If she registers for 2 courses, her tuition is $35 + $375 \cdot 2$, or $785 which does not exceed $1000. If she registers for 3 courses, her tuition is $35 + $375 \cdot 3$, or $1160 which exceeds $1000.

State. Angelica can register for at most 2 courses.

18. Mileages less than or equal to 525.8 mi

19. *Familiarize*. The average of the four scores is their sum divided by the number of tests, 4. We let s represent Nadia's score on the last test.

Translate. The average of the four scores is given by

$$\frac{82 + 76 + 78 + s}{4}.$$

Since this average must be at least 80, this means that it must be greater than or equal to 80. Thus, we can translate the problem to the inequality

$$\frac{82 + 76 + 78 + s}{4} \geq 80.$$

Carry out. We first multiply by 4 to clear the fraction.

$$4\left(\frac{82 + 76 + 78 + s}{4}\right) \geq 4 \cdot 80$$
$$82 + 76 + 78 + s \geq 320$$
$$236 + s \geq 320$$
$$s \geq 84$$

Check. As a partial check, we show that Nadia can get a score of 84 on the fourth test and have an average of at least 80:

$$\frac{82 + 76 + 78 + 84}{4} = \frac{320}{4} = 80.$$

State. Scores of 84 and higher will earn Nadia at least a B.

20. Scores greater than or equal to 97

21. **Familiarize.** Let s = the number of servings of fruits or vegetables Dale eats on Saturday.

Translate.

$$\underbrace{\text{Average number of fruit or vegetable servings}}_{\displaystyle\frac{4+6+7+4+6+4+s}{7}} \quad \underbrace{\text{is at least}}_{\displaystyle \geq} \quad \underbrace{5.}_{\displaystyle 5}$$

Carry out. We first multiply by 7 to clear the fraction.

$$7\left(\frac{4+6+7+4+6+4+s}{7}\right) \geq 7 \cdot 5$$
$$4+6+7+4+6+4+s \geq 35$$
$$31 + s \geq 35$$
$$s \geq 4$$

Check. As a partial check, we show that Dale can eat 4 servings of fruits or vegetables on Saturday and average at least 5 servings per day for the week:

$$\frac{4+6+7+4+6+4+4}{7} = \frac{35}{7} = 5$$

State. Dale should eat at least 4 servings of fruits or vegetables on Saturday.

22. 8 credits or more

23. **Familiarize.** Let m represent the number of minutes Monroe practices on the seventh day.

Translate.

$$\underbrace{\text{Average practice time}}_{\displaystyle\frac{15+28+30+0+15+25+m}{7}} \quad \underbrace{\begin{array}{c}\text{is at}\\\text{least}\end{array}}_{\displaystyle \geq} \quad \underbrace{20 \text{ min.}}_{\displaystyle 20}$$

Carry out. We solve the inequality.

$$\frac{15+28+30+0+15+25+m}{7} \geq 20$$
$$7\left(\frac{15+28+30+0+15+25+m}{7}\right) \geq 7 \cdot 20$$
$$15+28+30+0+15+25+m \geq 140$$
$$113 + m \geq 140$$
$$m \geq 27$$

Check. As a partial check, we show that if Monroe practices 27 min on the seventh day he meets expectations.

$$\frac{15+28+30+0+15+25+27}{7} = 20$$

State. Monroe must practice 27 min or more on the seventh day in order to meet expectations.

24. 21 calls or more

25. **Familiarize.** We first make a drawing. We let l represent the length, in feet.

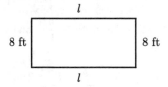

The perimeter is $P = 2l + 2w$, or $2l + 2 \cdot 8$, or $2l + 16$.

Translate. We translate to 2 inequalities.

$$\underbrace{\text{The perimeter}}_{\displaystyle 2l + 16} \quad \underbrace{\text{is at least}}_{\displaystyle \geq} \quad \underbrace{200 \text{ ft.}}_{\displaystyle 200}$$

$$\underbrace{\text{The perimeter}}_{\displaystyle 2l + 16} \quad \underbrace{\text{is at most}}_{\displaystyle \leq} \quad \underbrace{200 \text{ ft.}}_{\displaystyle 200}$$

Carry out. We solve each inequality.

$$2l + 16 \geq 200 \qquad 2l + 16 \leq 200$$
$$2l \geq 184 \qquad\qquad 2l \leq 184$$
$$l \geq 92 \qquad\qquad\quad l \leq 92$$

Check. We check to see if the solutions seem reasonable.

When $l = 91$ ft, $P = 2 \cdot 91 + 16$, or 198 ft.

When $l = 92$ ft, $P = 2 \cdot 92 + 16$, or 200 ft.

When $l = 93$ ft, $P = 2 \cdot 93 + 16$, or 202 ft.

From these calculations, it appears that the solutions are correct.

State. Lengths greater than or equal to 92 ft will make the perimeter at least 200 ft. Lengths less than or equal to 92 ft will make the perimeter at most 200 ft.

26. Lengths greater than 6 cm

27. **Familiarize.** We first make a drawing. Let w = the width, in feet. Then $2w$ = the length.

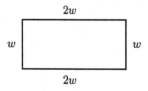

The perimeter is $P = 2l + 2w = 2 \cdot 2w + 2w = 4w + 2w = 6w$.

Translate.

$$
\underbrace{\text{The perimeter}}_{\downarrow \atop 6w} \quad \underbrace{\text{cannot exceed}}_{\downarrow \atop \leq} \quad \underbrace{\text{70 ft.}}_{\downarrow \atop 70}
$$

Carry out. We solve the inequality.

$$6w \leq 70$$
$$w \leq \frac{35}{3}, \text{ or } 11\frac{2}{3}$$

Check. As a partial check we show that the perimeter is 70 ft when the width is $\frac{35}{3}$ ft and the length is $2 \cdot \frac{35}{3}$, or $\frac{70}{3}$ ft.

$$P = 2 \cdot \frac{70}{3} + 2 \cdot \frac{35}{3} = \frac{140}{3} + \frac{70}{3} = \frac{210}{3} = 70$$

State. Widths less than or equal to $11\frac{2}{3}$ ft will meet the given conditions.

28. George worked more than 12 hr; Joan worked more than 15 hr

29. Familiarize. Let t = the number of 15-min units of time for a road call. Rick's Automotive charges $\$50 + \$15 \cdot t$ for a road call, and Twin City Repair charges $\$70 + \$10 \cdot t$.

Translate.

$$
\underbrace{\text{Rick's charge}}_{\downarrow \atop 50 + 15t} \quad \underbrace{\text{is less than}}_{\downarrow \atop <} \quad \underbrace{\text{Twin City's charge.}}_{\downarrow \atop 70 + 10t}
$$

Carry out. We solve the inequality.

$$50 + 15t < 70 + 10t$$
$$15t < 20 + 10t$$
$$5t < 20$$
$$t < 4$$

Check. We check to see if the solution seems reasonable. When $t = 3$, Rick's charges $\$50 + \$15 \cdot 3$, or $\$95$, and Twin City charges $\$70 + \$10 \cdot 3$, or $\$100$. When $t = 4$, Rick's charges $\$50 + \$15 \cdot 4$, or $\$110$, and Twin City charges $\$70 + \$10 \cdot 4$, or $\$110$. When $t = 5$, Rick's charges $\$50 + \$15 \cdot 5$, or $\$125$, and Twin City charges $\$70 + \$10 \cdot 5$, or $\$120$. From these calculations, it appears that the solution is correct.

State. It would be more economical to call Rick's for a service call of less than 4 15-min time units, or of less than 1 hr.

30. At most $\$49.02$

31. Familiarize. We first make a drawing. We let l represent the length.

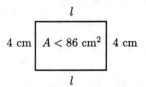

The area is the length times the width, or $4l$.

Translate.

$$
\underbrace{\text{Area}}_{\downarrow \atop 4l} \quad \underbrace{\text{is less than}}_{\downarrow \atop <} \quad \underbrace{\text{86 cm}^2.}_{\downarrow \atop 86}
$$

Carry out.

$$4l < 86$$
$$l < 21.5$$

Check. We check to see if the solution seems reasonable.

When $l = 22$, the area is $22 \cdot 4$, or 88 cm^2.

When $l = 21.5$, the area is $21.5(4)$, or 86 cm^2.

When $l = 21$, the area is $21 \cdot 4$, or 84 cm^2.

From these calculations, it would appear that the solution is correct.

State. The area will be less than 86 cm^2 for lengths less than 21.5 cm.

32. Lengths greater than or equal to 16.5 yd

33. Familiarize. Let v = the blue book value of the car. Since the car was repaired, we know that $\$8500$ does not exceed $0.8v$ or, in other words, $0.8v$ is at least $\$8500$.

Translate.

$$
\underbrace{\substack{\text{80\% of the} \\ \text{blue book value}}}_{\downarrow \atop 0.8v} \quad \underbrace{\text{is at least}}_{\downarrow \atop \geq} \quad \underbrace{\$8500.}_{\downarrow \atop 8500}
$$

Carry out.

$$0.8v \geq 8500$$
$$v \geq \frac{8500}{0.8}$$
$$v \geq 10,625$$

Check. As a partial check, we show that 80% of $\$10,625$ is at least $\$8500$:

$$0.8(\$10,625) = \$8500$$

State. The blue book value of the car was at least $\$10,625$.

34. More than $\$16,800$

35. Familiarize. We will use the formula $F = \frac{9}{5}C + 32$.

Translate.

$$\underbrace{\text{Fahrenheit temperature}}_{\downarrow} \quad \underbrace{\text{is above}}_{\downarrow} \quad 98.6°.$$
$$F \qquad\qquad\qquad > \qquad 98.6$$

Substituting $\frac{9}{5}C + 32$ for F, we have

$$\frac{9}{5}C + 32 > 98.6.$$

Carry out. We solve the inequality.

$$\frac{9}{5}C + 32 > 98.6$$
$$\frac{9}{5}C > 66.6$$
$$C > \frac{333}{9}$$
$$C > 37$$

Check. We check to see if the solution seems reasonable.

When $C = 36$, $\frac{9}{5} \cdot 36 + 32 = 96.8$.

When $C = 37$, $\frac{9}{5} \cdot 37 + 32 = 98.6$.

When $C = 38$, $\frac{9}{5} \cdot 38 + 32 = 100.4$.

It would appear that the solution is correct, considering that rounding occurred.

State. The human body is feverish for Celsius temperatures greater than 37°.

36. Temperatures less than 31.3°C

37. Familiarize. Let $r =$ the amount of fat in a serving of the regular peanut butter, in grams. If reduced fat peanut butter has at least 25% less fat than regular peanut butter, then it has at most 75% as much fat as the regular peanut butter.

Translate.

$$\underbrace{\text{12 g of fat}}_{\downarrow} \quad \underbrace{\text{is at most}}_{\downarrow} \quad \underbrace{75\%}_{\downarrow} \quad \underbrace{\text{of}}_{\downarrow} \quad \underbrace{\begin{array}{c}\text{the amount of}\\\text{fat in regular}\\\text{peanut butter.}\end{array}}_{\downarrow}$$
$$12 \qquad\quad \le \qquad 0.75 \quad \cdot \qquad\qquad r$$

Carry out.

$$12 \le 0.75r$$
$$16 \le r$$

Check. As a partial check, we show that 12 g of fat does not exceed 75% of 16 g of fat:

$$0.75(16) = 12$$

State. Regular peanut butter contains at least 16 g of fat per serving.

38. They contain at least $6\frac{2}{3}$ g of fat.

39. Familiarize. Let $d =$ the depth of the well, in feet. Then the cost on the pay-as-you-go plan is $500 + $8d$. The cost of the guaranteed-water plan is $4000. We want to find the values of d for which the pay-as-you-go plan costs less than the guaranteed-water plan.

Translate.

$$\underbrace{\begin{array}{c}\text{Cost of pay-as-}\\\text{you-go plan}\end{array}}_{\downarrow} \quad \underbrace{\text{is less than}}_{\downarrow} \quad \underbrace{\begin{array}{c}\text{cost of}\\\text{guaranteed-}\\\text{water plan}\end{array}}_{\downarrow}$$
$$500 + 8d \qquad\qquad < \qquad\qquad 4000$$

Carry out.

$$500 + 8d < 4000$$
$$8d < 3500$$
$$d < 437.5$$

Check. We check to see that the solution is reasonable.

When $d = 437$, $\$500 + \$8 \cdot 437 = \$3996 < \$4000$

When $d = 437.5$, $\$500 + \$8(437.5) = \$4000$

When $d = 438$, $\$500 + \$8(438) = \$4004 > \$4000$

From these calculations, it appears that the solution is correct.

State. It would save a customer money to use the pay-as-you-go plan for a well of less than 437.5 ft.

40. 8 mi or more

41. Familiarize. $R = -0.012t + 20.8$

In the formula R represents the world record in the 200-m dash and t represents the years since 1920. When $t = 0(1920)$, the record was $-0.012(0) + 20.8$, or 20.8 sec. When $t = 2(1922)$, the record was $-0.012(2) + 20.8$, or 20.776 sec. For what values of t will $-0.012t + 20.8$ be less than 19.8?

Translate. The record is to be less than 19.8. We have the inequality

$$R < 19.8.$$

To find the t values which satisfy this condition we substitute $-0.012t + 20.8$ for R.

$$-0.012t + 20.8 < 19.8$$

Carry out.

$$-0.012t + 20.8 < 19.8$$
$$-0.012t < -1$$
$$t > \frac{-1}{-0.012}$$
$$t > 83.\overline{3}, \text{ or } 83\frac{1}{3}$$

Check. We check to see if the solution set we obtained seems reasonable.

When $t = 83\frac{1}{4}$, $R = -0.012(83.25) + 20.8 = 19.801$.

When $t = 83\frac{1}{3}$, $R = -0.012\left(\frac{250}{3}\right) + 20.8 = 19.8$.

When $t = 83\frac{1}{2}$, $R = -0.012(83.5) + 20.8 = 19.798$.

Since $r = 19.8$ when $t = 83\frac{1}{3}$ and R decreases as t increases, R will be less than 19.8 when t is greater than $83\frac{1}{3}$.

State. The world record will be less than 19.8 seconds when t is greater than $83\frac{1}{3}$ years $\Big($more than $83\frac{1}{3}$ years after 1920$\Big)$. This occurs in years after 2003.

42. Years after 2005

43. Familiarize. Let $w =$ the number of weeks after July 1. After w weeks the water level has dropped $\frac{2}{3}w$ ft.

Translate.

Original depth	minus	drop in water level	does not exceed	21 ft.
↓	↓	↓	↓	↓
25	−	$\frac{2}{3}w$	≤	21

Carry out. We solve the inequality.

$$25 - \frac{2}{3}w \leq 21$$

$$-\frac{2}{3}w \leq -4$$

$$w \geq -\frac{3}{2}(-4)$$

$$w \geq 6$$

Check. As a partial check we show that the water level is 21 ft 6 weeks after July 1.

$$25 - \frac{2}{3} \cdot 6 = 25 - 4 = 21 \text{ ft}$$

Since the water level continues to drop during the weeks after July 1, the answer seems reasonable.

State. The water level will not exceed 21 ft for dates at least 6 weeks after July 1.

44. When the puppy is more than 18 weeks old

45. Familiarize. Let $h =$ the height of the triangle, in ft. Recall that the formula for the area of a triangle with base b and height h is $A = \frac{1}{2}bh$.

Translate.

Area	is at least	3 ft².
↓	↓	↓
$\frac{1}{2}\left(1\frac{1}{2}\right)h$	≥	3

Carry out. We solve the inequality.

$$\frac{1}{2}\left(1\frac{1}{2}\right)h \geq 3$$

$$\frac{1}{2} \cdot \frac{3}{2} \cdot h \geq 3$$

$$\frac{3}{4}h \geq 3$$

$$h \geq \frac{4}{3} \cdot 3$$

$$h \geq 3$$

Check. As a partial check, we show that the area of the triangle is 3 ft² when the height is 4 ft.

$$\frac{1}{2}\left(1\frac{1}{2}\right)(4) = \frac{1}{2} \cdot \frac{3}{2} \cdot \frac{4}{1} = 3$$

State. The height should be at least 4 ft.

46. Heights less than or equal to 3 ft

47. Familiarize. We will use the equation $y = 0.027x + 0.19$.

Translate.

The cost	is at most	\$6.
↓	↓	↓
$0.027x + 0.19$	≤	6

Carry out. We solve the inequality.

$$0.027x + 0.19 \leq 6$$

$$0.027x \leq 5.81$$

$$x \leq 215.2 \quad \text{Rounding to the nearest tenth}$$

Check. As a partial check, we show that the cost for driving 215.2 mi is \$6.

$$0.027(215.2) + 0.19 \approx \$6$$

State. The cost will be at most \$6 for mileages less than or equal to 215.2 mi.

48. 2001 and beyond

49. ◈

50. ◈

51. $\dfrac{9-5}{6-4} = \dfrac{4}{2} = 2$

52. $\dfrac{1}{2}$

53. $\dfrac{8-(-2)}{1-4} = \dfrac{10}{-3}$, or $-\dfrac{10}{3}$

54. $-\dfrac{1}{5}$

55.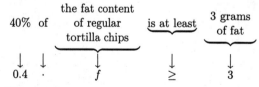

56. ◈

57. *Familiarize*. Let $h =$ the number of hours the car has been parked. Then $h-1 =$ the number of hours after the first hour.

Translate.

Charge for first hour	plus	charge for additional hours	exceeds	\$16.50.
↓	↓	↓	↓	↓
4.00	+	2.50(h − 1)	>	16.50

Carry out. We solve the inequality.

$$4.00 + 2.50(h-1) > 16.50$$
$$40 + 25(h-1) > 165 \quad \text{Multiplying by 10 to clear decimals}$$
$$40 + 25h - 25 > 165$$
$$25h + 15 > 165$$
$$25h > 150$$
$$h > 6$$

Check. We check to see if this solution seems reasonable.

When $h = 5$, $4.00 + 2.50(5-1) = 14.00$.

When $h = 6$, $4.00 + 2.50(6-1) = 16.50$.

When $h = 7$, $4.00 + 2.50(7-1) = 19.00$.

It appears that the solution is correct.

State. The charge exceeds \$16.50 when the car has been parked for more than 6 hr.

58. Temperatures between $-15°C$ and $-9\frac{4}{9}°C$

59. Since $8^2 = 64$, the length of a side must be less than or equal to 8 cm (and greater than 0 cm, of course). We can also use the five-step problem-solving procedure.

Familiarize. Let s represent the length of a side of the square. The area s is the square of the length of a side, or s^2.

Translate.

The area	is no more than	64 cm².
↓	↓	↓
s^2	≤	64

Carry out.

$$s^2 \le 64$$
$$s^2 - 64 \le 0$$
$$(s+8)(s-8) \le 0$$

We know that $(s+8)(s-8) = 0$ for $s = -8$ or $s = 8$. Now $(s+8)(s-8) < 0$ when the two factors have opposite signs. That is:

$$s+8>0 \quad and \ s-8<0 \ or \ s+8<0 \quad and \ s-8>0$$
$$s>-8 \ and \quad s<8 \ or \quad s<-8 \ and \quad s>8$$

This can be expressed   This is not possible.

as $-8 < s < 8$.

Then $(s+8)(s-8) \le 0$ for $-8 \le s \le 8$.

Check. Since the length of a side cannot be negative we only consider positive values of s, or $0 < s \le 8$. We check to see if this solution seems reasonable.

When $s = 7$, the area is 7^2, or 49 cm².

When $s = 8$, the area is 8^2, or 64 cm².

When $s = 9$, the area is 9^2, or 81 cm².

From these calculations, it appears that the solution is correct.

State. Sides of length 8 cm or less will allow an area of no more than 64 cm². (Of course, the length of a side must be greater than 0 also.)

60. 47 and 49

61. *Familiarize*. Let $f =$ the fat content of a serving of regular tortilla chips, in grams. A product that contains 60% less fat than another product has 40% of the fat content of that product. If Reduced Fat Tortilla Pops cannot be labeled lowfat, then they contain at least 3 g of fat.

Translate.

40%	of	the fat content of regular tortilla chips	is at least	3 grams of fat
↓	↓	↓	↓	↓
0.4	·	f	≥	3

Carry out.

$$0.4f \ge 3$$
$$f \ge 7.5$$

Check. As a partial check, we show that 40% of 7.5 g is not less than 3 g.

$$0.4(7.5) = 3$$

State. A serving of regular tortilla chips contains at least 7.5 g of fat.

62. Between 5 and 9 hr

63. **Familiarize**. Let p = the price of Neoma's tenth book. If the average price of each of the first 9 books is \$12, then the total price of the 9 books is $9 \cdot \$12$, or \$108. The average price of the first 10 books will be $\dfrac{\$108 + p}{10}$.

Translate.

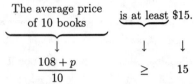

$$\frac{108 + p}{10} \qquad \geq \qquad 15$$

Carry out. We solve the inequality.

$$\frac{108 + p}{10} \geq 15$$
$$108 + p \geq 150$$
$$p \geq 42$$

Check. As a partial check, we show that the average price of the 10 books is \$15 when the price of the tenth book is \$42.

$$\frac{\$108 + \$42}{10} = \frac{\$150}{10} = \$15$$

State. Neoma's tenth book should cost at least \$42 if she wants to select a \$15 book for her free book.

64.

65. Let p = the total purchases for the year. Solving $10\%p > 25$, we get $p > 250$. Thus, when a customer's purchases are more than \$250 for the year, the customer saves money by purchasing a card.

Chapter 3

Introduction to Graphing

1. We go to the top of the bar that is above the body weight 100 lb. Then we move horizontally from the top of the bar to the vertical scale listing numbers of drinks. It appears that consuming approximately 2 drinks in one hour will give a 100 lb person a blood-alcohol level of 0.08%.

2. Approximately 3.5 drinks

3. From 4 on the vertical scale we move horizontally until we reach a bar whose top is above the horizontal line on which we are moving. The first such bar corresponds to a body weight of 220 lb. This means that for body weights represented by bars to the left of this one, consuming 4 drinks will yield a blood-alcohol level of 0.08%. The bar immediately to the left of the 220-pound bar represents 200 pounds. Thus, we can conclude an individual weighs more than 220 lb if 4 drinks are consumed in one hour without reaching a blood-alcohol level of 0.08%.

4. The individual weighs more than 240 lb.

5. *Familiarize*. Since there are 272 million Americans and about one-third of them live in the South, there are about $\frac{1}{3} \cdot 272$, or $\frac{272}{3}$ million Southerners. The pie chart indicates that 3% of Americans choose brown as their favorite color. Let b = the number of Southerners, in millions, who choose brown as their favorite color.

   Translate. We reword and translate the problem.

   What is 3% of $\frac{272}{3}$ million?
   $$b = 3\% \cdot \frac{272}{3}$$

   Carry out. We solve the equation.
   $$b = 0.03 \cdot \frac{272}{3} = 2.72$$

   Check. We repeat the calculations. The answer checks.

   State. About 2.72 million, or 2,720,000 Southerners choose brown as their favorite color.

6. About 5,440,000

7. *Familiarize*. Since there are 272 million Americans and about one-eighth are senior citizens, there are about $\frac{1}{8} \cdot 272$ million, or 34 million senior citizens. The pie chart indicates that 4% of Americans choose black as their favorite color. Let b = the number of senior citizens who choose black as their favorite color.

   Translate. We reword and translate the problem.

   What is 4% of 34 million?
   $$b = 4\% \cdot 34,000,000$$

   Carry out. We solve the equation.
   $$b = 0.04 \cdot 34,000,000 = 1,360,000$$

   Check. We repeat the calculations. The answer checks.

   State. About 1,360,000 senior citizens choose black as their favorite color.

8. About 1,360,000

9. *Familiarize*. From the pie chart we see that 9.9% of solid waste is plastic. We let x = the amount of plastic, in millions of tons, in the waste generated in 1998.

   Translate. We reword the problem.

   What is 9.9% of 210?
   $$x = 9.9\% \cdot 210$$

   Carry out.
   $$x = 0.099 \cdot 210 = 20.79$$

   Check. We can repeat the calculation. The result checks.

   State. In 1998, about 20.79 million tons of waste was plastic.

10. About 1.7 lb

11. *Familiarize*. From the pie chart we see that 5.5% of solid waste is glass. From Exercise 9 we know that Americans generated 210 million tons of waste in 1998. Then the amount of this that is glass is

    0.055(210), or 11.55 million tons

    We let x = the amount of glass, in millions of tons, that Americans recycled in 1998.

Translate. We reword the problem.

What is 26% of 11.55 million tons?

$$x = 26\% \cdot 11.55$$

Carry out.

$$x = 0.26(11.55) \approx 3.0$$

Check. We go over the calculations again. The result checks.

State. Americans recycled about 3.0 million tons of glass in 1998.

12. About 0.02 lb

13. Locate 1997 on the horizontal scale and then move up to the line that represents CD sales. Now move to the vertical axis and read that about 70% of recordings sold in 1997 were CDs.

14. About 20%

15. Locate 25% on the vertical scale, midway between 20% and 30%. Move right to the line representing cassette sales and then move down to the horizontal axis. We see that in 1995 approximately 25% of the recordings sold were cassettes.

16. 1998

17. The line slants upward most steeply from 1994 to 1995, so sales of CDs increased the most from 1994 to 1995.

18. 1996 to 1997

19. Starting at the origin:

$(1,2)$ is 1 unit right and 2 units up;

$(-2,3)$ is 2 units left and 3 units up;

$(4,-1)$ is 4 units right and 1 unit down;

$(-5,-3)$ is 5 units left and 3 units down;

$(4,0)$ is 4 units right and 0 units up or down;

$(0,-2)$ is 0 units right or left and 2 units down.

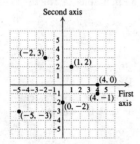

20.

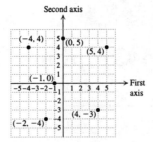

21. Starting at the origin:

$(4,4)$ is 4 units right and 4 units up;

$(-2,4)$ is 2 units left and 4 units up;

$(5,-3)$ is 5 units right and 3 units down;

$(-5,-5)$ is 5 units left and 5 units down;

$(0,4)$ is 0 units right or left and 4 units up;

$(0,-4)$ is 0 units right or left and 4 units down;

$(3,0)$ is 3 units right and 0 units up or down;

$(-4,0)$ is 4 units left and 0 units up or down.

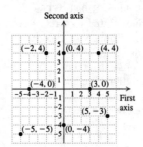

22.

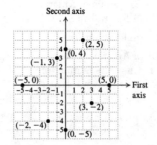

23.

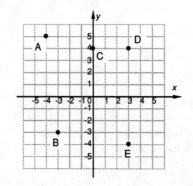

Point A is 4 units left and 5 units up. The coordinates of A are $(-4, 5)$.

Point B is 3 units left and 3 units down. The coordinates of B are $(-3, -3)$.

Point C is 0 units right or left and 4 units up. The coordinates of C are $(0,4)$.

Point D is 3 units right and 4 units up. The coordinates of D are $(3,4)$.

Point E is 3 units right and 4 units down. The coordinates of E are $(3, -4)$.

24. $A: (3,3)$, $B: (0,-4)$, $C: (-5,0)$, $D: (-1,-1)$, $E: (2,0)$

25.

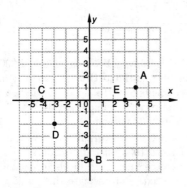

Point A is 4 units right and 1 unit up. The coordinates of A are $(4,1)$.

Point B is 0 units right or left and 5 units down. The coordinates of B are $(0, -5)$.

Point C is 4 units left and 0 units up or down. The coordinates of C are $(-4, 0)$.

Point D is 3 units left and 2 units down. The coordinates of D are $(-3, -2)$.

Point E is 3 units right and 0 units up or down. The coordinates of E are $(3,0)$.

26. $A: (-5,1)$, $B: (0,5)$, $C: (5,3)$, $D: (0,-1)$, $E: (2,-4)$

27. Since the first coordinate is positive and the second coordinate negative, the point $(7, -2)$ is located in quadrant IV.

28. III

29. Since both coordinates are negative, the point $(-4, -3)$ is in quadrant III.

30. IV

31. Since both coordinates are positive, the point $(2, 1)$ is in quadrant I.

32. II

33. Since the first coordinate is negative and the second coordinate is positive, the point $(-4.9, 8.3)$ is in quadrant II.

34. I

35. First coordinates are positive in the quadrants that lie to the right of the origin, or in quadrants I and IV.

36. III and IV

37. Points for which both coordinates are positive lie in quadrant I, and points for which both coordinates are negative life in quadrant III. Thus, both coordinates have the same sign in quadrants I and III.

38. II and IV

39. a) Draw a line segment connecting the points $(1991, 62.1)$ and $(1999, 49.6)$. Then locate 1995 on the horizontal axis and move up to the line. Now move horizontally to the vertical axis and estimate that the birth rate among teenagers in 1995 was approximately 56 births per 1000 females.

   b) Extend the line segment on the graph until it is above 2003 on the horizontal axis. Then, from 2003, move up to the line and across to the vertical axis. We predict that the birth rate among teenagers in 2003 will be approximately 43 per 1000 females.

40. a) Approximately 23 million participants

   b) Approximately 15 million participants

41. Draw a horizontal axis for the year and a vertical axis for the percentage of people age 26 to 34 who smoke. Number the axes with a scale that will permit us to view both the given data and the desired data. Then plot the points $(1985, 45.7)$ and $(1998, 32.5)$ and draw a line segment connecting them.

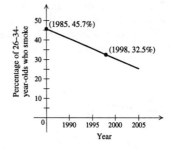

a) Locate 1990 on the horizontal scale and move up to the line. Then move horizontally to the vertical axis and estimate that approximately 40% of people age 26 to 34 smoked in 1990.

b) Extend the line segment on the graph until it is above 2003 on the horizontal axis. Then, from 2003, move up to the line and across to the vertical axis. We predict that in 2003 about 27.5% of people age 26 to 34 will smoke.

42.

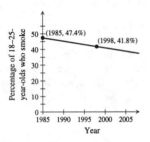

a) About 45%

b) About 40%

43. Draw a horizontal axis for the year and a vertical axis for U.S. college enrollment, in millions. Number the axes with a scale that will permit us to view both the given data and the desired data. Then plot the points $(1990, 60.3)$ and $(2000, 68.3)$ and draw a line segment connecting them.

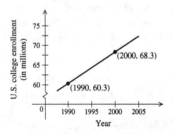

a) Locate 1996 on the horizontal axis and move up to the line. Then move horizontally to the vertical axis and estimate that U.S. college enrollment in 1996 was about 65 million students.

b) Extend the line segment on the graph until it is above 2005 on the horizontal axis. Then, from 2005, move up to the line and across to the vertical axis. We predict that in 2005 U.S. college enrollment will be about 72 million students.

44.

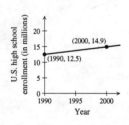

a) About 14 million students

b) About 16 million students

45. Draw a horizontal axis for the year and a vertical axis for the number of U.S. residents over the age of 65, in millions. Number the axes with a scale that will permit us to view both the given data and the desired data. Then plot the points $(1990, 31)$ and $(2000, 34.4)$ and draw a line segment connecting them.

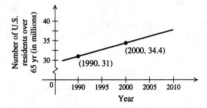

a) Since 1995 is midway between 1990 and 2000, it is reasonable to estimate that in 1995 the number of U.S. residents over the age of 65 will be about midway between 31 million and 34.4 million or approximately 33 million.

We could also use the graph to make this estimate. Locate 1995 on the horizontal axis and move up to the line. Then move horizontally to the vertical axis and estimate that in 1995 the number of U.S. residents over the age of 65 was about 33 million.

b) Extend the line segment on the graph until it is above 2010 on the horizontal axis. Then, from 2010, move up to the line and across to the vertical axis. We predict that in 2010 the number of U.S. residents over the age of 65 will be about 38 million.

46.

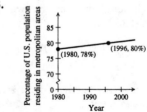

a) About 79.5%

b) About 81.5%

47. ◈

48. ◈

49. $4 \cdot 3 - 6 \cdot 5 = 12 - 30 = -18$

50. -31

51. $-\frac{1}{2}(-6) + 3 = 3 + 3 = 6$

52. 1

53.
$$3x - 2y = 6$$
$$-2y = -3x + 6 \quad \text{Adding } -3x \text{ to both sides}$$
$$-\frac{1}{2}(-2y) = -\frac{1}{2}(-3x + 6)$$
$$y = -\frac{1}{2}(-3x) - \frac{1}{2}(6)$$
$$y = \frac{3}{2}x - 3$$

54. $y = \frac{7}{4}x - \frac{7}{2}$

55. ◈

56. ◈

57. If the coordinates of a point are reciprocals of each other, they have the same sign. Thus, the point could be in quadrant I or quadrant III.

58. II or IV

59.

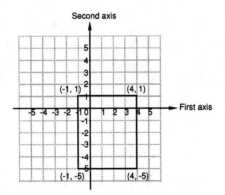

The coordinates of the fourth vertex are $(-1, -5)$.

60. $(5,2)$, $(-7,2)$, or $(3,-8)$

61. Answers may vary.

We select eight points such that the sum of the coordinates for each point is 7.

$(0,7)$	$0 + 7 = 7$
$(1,6)$	$1 + 6 = 7$
$(2,5)$	$2 + 5 = 7$
$(3,4)$	$3 + 4 = 7$
$(4,3)$	$4 + 3 = 7$
$(5,2)$	$5 + 2 = 7$
$(6,1)$	$6 + 1 = 7$
$(7,0)$	$7 + 0 = 7$

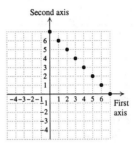

62. Answers may vary.

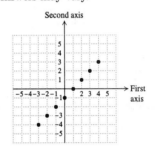

63. Plot the three given points and observe that the coordinates of the fourth vertex are $(5,3)$

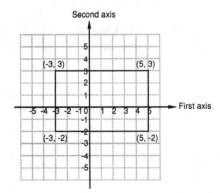

The length of the rectangle is 8 units, and the width is 5 units.

$$P = 2l + 2w$$
$$P = 2 \cdot 8 + 2 \cdot 5 = 16 + 10 = 26 \text{ units}$$

64. $\frac{65}{2}$ sq units

65. Latitude 32.5° North,

Longitude 64.5° West

66. Latitude 27° North,

Longitude 81° West

67. ◈

68. ◈

Exercise Set 3.2

---

1. We substitute 0 for x and 2 for y (alphabetical order of variables).

$$
\begin{array}{c|c}
\multicolumn{2}{c}{y = 7x + 1} \\
\hline
2 \; ? \; 7 \cdot 0 + 1 & \\
& 0 + 1 \\
2 & 1 \qquad \text{FALSE}
\end{array}
$$

Since $2 = 1$ is false, the pair $(0, 2)$ is not a solution.

2. Yes

3. We substitute 4 for x and 2 for y.

$$
\begin{array}{c|c}
\multicolumn{2}{c}{3y + 2x = 12} \\
\hline
3 \cdot 2 + 2 \cdot 4 \; ? \; 12 & \\
6 + 8 & \\
14 & 12 \quad \text{FALSE}
\end{array}
$$

Since $14 = 12$ is false, the pair $(4,2)$ is not a solution.

4. No

5. We substitute 2 for a and -1 for b.

$$
\begin{array}{c|c}
\multicolumn{2}{c}{4a - 3b = 11} \\
\hline
4 \cdot 2 - 3(-1) \; ? \; 11 & \\
8 + 3 & \\
11 & 11 \quad \text{TRUE}
\end{array}
$$

Since $11 = 11$ is true, the pair $(2, -1)$ is a solution.

6. Yes

7. To show that a pair is a solution, we substitute, replacing x with the first coordinate and y with the second coordinate in each pair.

$$
\begin{array}{c|c} \qquad \begin{array}{c|c}
\multicolumn{2}{c}{y = x - 2} \\
\hline
1 \; ? \; 3 - 2 & \\
1 & 1 \quad \text{TRUE}
\end{array} \qquad
\begin{array}{c|c}
\multicolumn{2}{c}{y = x - 2} \\
\hline
-4 \; ? \; -2 - 2 & \\
-4 & -4 \quad \text{TRUE}
\end{array}
\end{array}
$$

In each case the substitution results in a true equation. Thus, $(3, 1)$ and $(-2, -4)$ are both solutions of $y = x - 2$. We graph these points and sketch the line passing through them.

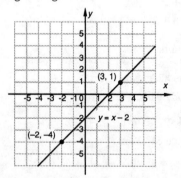

The line appears to pass through $(5, 3)$ also. We check to determine if $(5, 3)$ is a solution of $y = x - 2$.

$$
\begin{array}{c|c}
\multicolumn{2}{c}{y = x - 2} \\
\hline
3 \; ? \; 5 - 2 & \\
3 & 3 \qquad \text{TRUE}
\end{array}
$$

Thus, $(5, 3)$ is another solution. There are other correct answers, including $(-3, -5)$, $(-1, -3)$, $(0, -2)$, $(1, -1)$, $(2, 0)$, and $(4, 2)$.

8.
$$
\begin{array}{c|c}
\multicolumn{2}{c}{y = x + 3} \\
\hline
2 \; ? \; -1 + 3 & \\
2 & 2 \qquad \text{TRUE}
\end{array} \qquad
\begin{array}{c|c}
\multicolumn{2}{c}{y = x + 3} \\
\hline
7 \; ? \; 4 + 3 & \\
7 & 7 \qquad \text{TRUE}
\end{array}
$$

$(0, 3)$; answers may vary

9. To show that a pair is a solution, we substitute, replacing x with the first coordinate and y with the second coordinate in each pair.

$$
\begin{array}{c|c}
\multicolumn{2}{c}{y = \dfrac{1}{2}x + 3} \\
\hline
5 \; ? \; \dfrac{1}{2} \cdot 4 + 3 & \\
& 2 + 3 \\
5 & 5 \qquad \text{TRUE}
\end{array} \qquad
\begin{array}{c|c}
\multicolumn{2}{c}{y = \dfrac{1}{2}x + 3} \\
\hline
2 \; ? \; \dfrac{1}{2}(-2) + 3 & \\
& -1 + 3 \\
2 & 2 \qquad \text{TRUE}
\end{array}
$$

In each case the substitution results in a true equation. Thus, $(4, 5)$ and $(-2, 2)$ are both solutions of $y = \dfrac{1}{2}x + 3$. We graph these points and sketch the line passing through them.

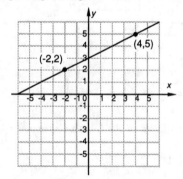

The line appears to pass through $(0, 3)$ also. We check to determine if $(0, 3)$ is a solution of $y = \dfrac{1}{2}x + 3$.

$$
\begin{array}{c|c}
\multicolumn{2}{c}{y = \dfrac{1}{2}x + 3} \\
\hline
3 \; ? \; \dfrac{1}{2} \cdot 0 + 3 & \\
3 & 3 \qquad \text{TRUE}
\end{array}
$$

Thus, $(0, 3)$ is another solution. There are other correct answers, including $(-6, 0)$, $(-4, 1)$, $(2, 4)$, and $(6, 6)$.

10.

$$y = \frac{1}{2}x - 1$$

$$\begin{array}{c|c} 2 \ ? \ \frac{1}{2} \cdot 6 - 1 & \\ 3 - 1 & \\ 2 & 2 \qquad \text{TRUE} \end{array}$$

$$y = \frac{1}{2}x - 1$$

$$\begin{array}{c|c} -1 \ ? \ \frac{1}{2} \cdot 0 - 1 & \\ -1 & -1 \qquad \text{TRUE} \end{array}$$

$(2, 0)$; answers may vary

11. To show that a pair is a solution, we substitute, replacing x with the first coordinate and y with the second coordinate in each pair.

$$y + 3x = 7$$

$$\begin{array}{c|c} 1 + 3 \cdot 2 \ ? \ 7 & \\ 1 + 6 & \\ 7 & 7 \ \text{TRUE} \end{array}$$

$$y + 3x = 7$$

$$\begin{array}{c|c} -5 + 3 \cdot 4 \ ? \ 7 & \\ -5 + 12 & \\ 7 & 7 \ \text{TRUE} \end{array}$$

In each case the substitution results in a true equation. Thus, $(2, 1)$ and $(4, -5)$ are both solutions of $y + 3x = 7$. We graph these points and sketch the line passing through them.

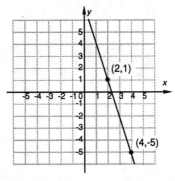

The line appears to pass through $(1, 4)$ also. We check to determine if $(1, 4)$ is a solution of $y + 3x = 7$.

$$y + 3x = 7$$

$$\begin{array}{c|c} 4 + 3 \cdot 1 \ ? \ 7 & \\ 4 + 3 & \\ 7 & 7 \ \text{TRUE} \end{array}$$

Thus, $(1, 4)$ is another solution. There are other correct answers, including $(3, -2)$.

12.

$$2y + x = 5$$

$$\begin{array}{c|c} 2 \cdot 3 - 1 \ ? \ 5 & \\ 6 - 1 & \\ 5 & 5 \ \text{TRUE} \end{array}$$

$$2y + x = 5$$

$$\begin{array}{c|c} 2(-1) + 7 \ ? \ 5 & \\ -2 + 7 & \\ 5 & 5 \ \text{TRUE} \end{array}$$

$(1, 2)$; answers may vary

13. To show that a pair is a solution, we substitute, replacing x with the first coordinate and y with the second coordinate in each pair.

$$4x - 2y = 10$$

$$\begin{array}{c|c} 4 \cdot 0 - 2(-5) \ ? \ 10 & \\ 10 & 10 \ \text{TRUE} \end{array}$$

$$4x - 2y = 10$$

$$\begin{array}{c|c} 4 \cdot 4 - 2 \cdot 3 \ ? \ 10 & \\ 16 - 6 & \\ 10 & 10 \ \text{TRUE} \end{array}$$

In each case the substitution results in a true equation. Thus, $(0, -5)$ and $(4, 3)$ are both solutions of $4x - 2y = 10$. We graph these points and sketch the line passing through them.

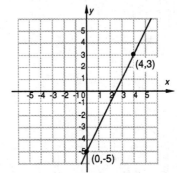

The line appears to pass through $(2, -1)$ also. We check to determine if $(2, -1)$ is a solution of $4x - 2y = 10$.

$$4x - 2y = 10$$

$$\begin{array}{c|c} 4 \cdot 2 - 2(-1) \ ? \ 10 & \\ 8 + 2 & \\ 10 & 10 \ \text{TRUE} \end{array}$$

Thus, $(2, -1)$ is another solution. There are other correct answers, including $(1, -3)$, $(2, -1)$, $(3, 1)$, and $(5, 5)$.

14.

$$6x - 3y = 3$$

$$\begin{array}{c|c} 6 \cdot 1 - 3 \cdot 1 \ ? \ 3 & \\ 6 - 3 & \\ 3 & 3 \ \text{TRUE} \end{array}$$

$$6x - 3y = 3$$

$$\begin{array}{c|c} 6(-1) - 3(-3) \ ? \ 3 & \\ -6 + 9 & \\ 3 & 3 \ \text{TRUE} \end{array}$$

$(0, -1)$; answers may vary

15. $y = x - 1$

The equation is equivalent to $y = x + (-1)$. The y-intercept is $(0, -1)$. We find two other pairs.

When $x = 3$, $y = 3 - 1 = 2$.

When $x = -5$, $y = -5 - 1 = -6$.

x	y
0	-1
3	2
-5	-6

Plot these points, draw the line they determine, and label the graph $y = x - 1$.

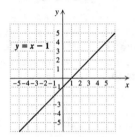

16.

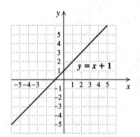

17. $y = x$

The equation is equivalent to $y = x + 0$. The y-intercept is $(0, 0)$. We find two other points.

When $x = -2$, $y = -2$.
When $x = 3$, $y = 3$.

x	y
0	0
-2	-2
3	3

Plot these points, draw the line they determine, and label the graph $y = x$.

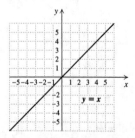

18.

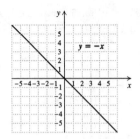

19. $y = \frac{1}{2}x$

The equation is equivalent to $y = \frac{1}{2}x + 0$. The y-intercept is $(0, 0)$. We find two other points.

When $x = -4$, $y = \frac{1}{2}(-4) = -2$.

When $x = 4$, $y = \frac{1}{2} \cdot 4 = 2$.

x	y
0	0
-4	-2
4	2

Plot these points, draw the line they determine, and label the graph $y = \frac{1}{2}x$.

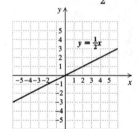

20.

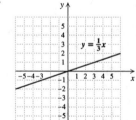

21. $y = x + 2$

The equation is in the form $y = mx + b$. The y-intercept is $(0, 2)$. We find two other points.

When $x = -2$, $y = -2 + 2 = 0$.

When $x = 3$, $y = 3 + 2 = 5$.

x	y
0	2
-2	0
3	5

Plot these points, draw the line they determine, and label the graph $y = x + 2$.

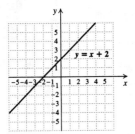

22.

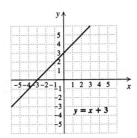

23. $y = 3x - 2 = 3x + (-2)$

The y-intercept is $(0, -2)$. We find two other points.

When $x = -2$, $y = 3(-2) + 2 = -6 + 2 = -4$.

When $x = 1$, $y = 3 \cdot 1 + 2 = 3 + 2 = 5$.

x	y
0	-2
-2	-4
1	5

Plot these points, draw the line they determine, and label the graph $y = 3x + 2$.

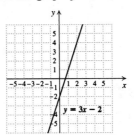

24.

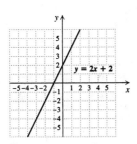

25. $y = \dfrac{1}{2}x + 1$

The y-intercept is $(0, 1)$. We find two other points using multiples of 2 for x to avoid fractions.

When $x = -4$, $y = \dfrac{1}{2}(-4) + 1 = -2 + 1 = -1$.

When $x = 4$, $y = \dfrac{1}{2} \cdot 4 + 1 = 2 + 1 = 3$.

x	y
0	1
-4	-1
4	3

Plot these points, draw the line they determine, and label the graph $y = \dfrac{1}{2}x + 1$.

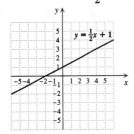

26.

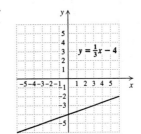

27. $x + y = -5$

$\qquad y = -x - 5$

$\qquad y = -x + (-5)$

The y-intercept is $(0, -5)$. We find two other points.

When $x = -4$, $y = -(-4) - 5 = 4 - 5 = -1$.

When $x = -1$, $y = -(-1) - 5 = 1 - 5 = -4$.

x	y
0	-5
-4	-1
-1	-4

Plot these points, draw the line they determine, and label the graph $x + y = -5$.

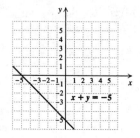

28.

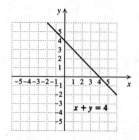

29. $y = \dfrac{5}{3}x - 2 = \dfrac{5}{3}x + (-2)$

The y-intercept is $(0, -2)$. We find two other points using multiples of 3 for x to avoid fractions.

When $x = -3$, $y = \dfrac{5}{3}(-3) - 2 = -5 - 2 = -7.$

When $x = 3$, $y = \dfrac{5}{3} \cdot 3 - 2 = 5 - 2 = 3.$

x	y
0	-2
-3	-7
3	3

Plot these points, draw the line they determine, and label the graph $y = \dfrac{5}{3}x - 2$.

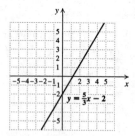

30.

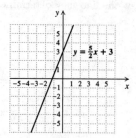

31. $x + 2y = 8$

$2y = -x + 8$

$y = -\dfrac{1}{2}x + 4$

The y-intercept is $(0, 4)$. We find two other points using multiples of 2 for x to avoid fractions.

When $x = -2$, $y = -\dfrac{1}{2}(-2) + 4 = 1 + 4 = 5.$

When $x = 4$, $y = -\dfrac{1}{2} \cdot 4 + 4 = -2 + 4 = 2.$

x	y
0	4
-2	5
4	2

Plot these points, draw the line they determine, and label the graph $x + 2y = 8$.

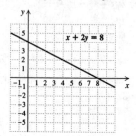

32.

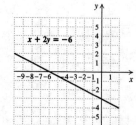

33. $y = \dfrac{3}{2}x + 1$

The y-intercept is $(0, 1)$. We find two other points using multiples of 2 for x to avoid fractions.

When $x = -4$, $y = \dfrac{3}{2}(-4) + 1 = -6 + 1 = -5.$

When $x = 2$, $y = \dfrac{3}{2} \cdot 2 + 1 = 3 + 1 = 4.$

x	y
0	1
-4	-5
2	4

Plot these points, draw the line they determine, and label the graph $y = \dfrac{3}{2}x + 1$.

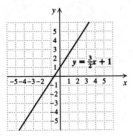

34.

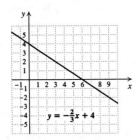

35. $6x - 3y = 9$
$$-3y = -6x + 9$$
$$y = 2x - 3$$
$$y = 2x + (-3)$$

The y-intercept is $(0, -3)$. We find two other points.

When $x = -1$, $y = 2(-1) - 3 = -2 - 3 = -5$.

When $x = 3$, $y = 2 \cdot 3 - 3 = 6 - 3 = 3$.

x	y
0	-3
-1	-5
3	3

Plot these points, draw the line they determine, and label the graph $6x - 3y = 9$.

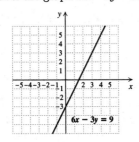

36.

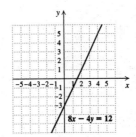

37. $8y + 2x = -4$
$$8y = -2x - 4$$
$$y = -\frac{1}{4}x - \frac{1}{2}$$
$$y = -\frac{1}{4}x + \left(-\frac{1}{2}\right)$$

The y-intercept is $\left(0, -\dfrac{1}{2}\right)$. We find two other points.

When $x = -2$, $y = -\dfrac{1}{4}(-2) - \dfrac{1}{2} = \dfrac{1}{2} - \dfrac{1}{2} = 0$.

When $x = 2$, $y = -\dfrac{1}{4} \cdot 2 - \dfrac{1}{2} = -\dfrac{1}{2} - \dfrac{1}{2} = -1$.

x	y
0	$-\dfrac{1}{2}$
-2	0
2	-1

Plot these points, draw the line they determine, and label the graph $8y + 2x = -4$.

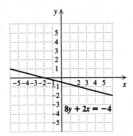

38.

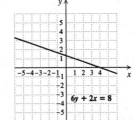

39. We graph $w = \dfrac{1}{2}t + 5$. Since the number of gallons of bottled water consumed cannot be negative in this application, we select only nonnegative values for t.

If $t = 0$, $w = \dfrac{1}{2} \cdot 0 + 5 = 5$.

If $t = 4$, $w = \dfrac{1}{2} \cdot 4 + 5 = 2 + 5 = 7$.

If $t = 10$, $w = \dfrac{1}{2} \cdot 10 + 5 = 5 + 5 = 10$.

t	w
0	5
4	7
10	10

We plot the points and draw the graph.

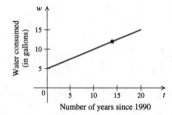

To predict the number of gallons consumed per person in 2004 we find the second coordinate associated with 14. (2004 is 14 years after 1990.) Locate the point on the line that is above 14 and then find the value on the vertical axis that corresponds to that point. That value is 12, so we predict that 12 gallons of bottled water will be consumed per person in 2004.

40.

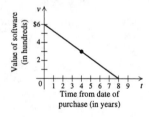

$300

41. We graph $t + w = 15$, or $w = -t + 15$. Since time cannot be negative in this application, we select only nonnegative values for t.

If $t = 0$, $w = -0 + 15 = 15$.

If $t = 2$, $w = -2 + 15 = 13$.

If $t = 5$, $w = -5 + 15 = 10$.

t	w
0	15
2	13
5	10

We plot the points and draw the graph. Since the likelihood of death cannot be negative, the graph stops at the horizontal axis.

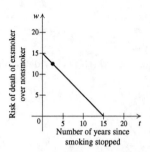

To estimate how much more likely it is for Sandy to die from lung cancer than Polly, we find the second coordinate associated with $2\dfrac{1}{2}$. Locate the point on the line that is above $2\dfrac{1}{2}$ and then find the value on the vertical axis that corresponds to that point. That value is about $12\dfrac{1}{2}$, so it is $12\dfrac{1}{2}$ times more likely for Sandy to die from lung cancer than Polly.

42.

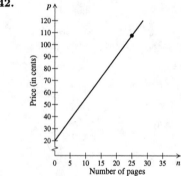

110¢, or $1.10

43. We graph $T = \dfrac{6}{5}c + 1$. Since the number of credits cannot be negative, we select only nonnegative values for c.

If $c = 5$, $T = \dfrac{6}{5} \cdot 5 + 1 = 6 + 1 = 7$.

If $c = 10$, $T = \dfrac{6}{5} \cdot 10 + 1 = 12 + 1 = 13$.

If $c = 15$, $T = \dfrac{6}{5} \cdot 15 + 1 = 18 + 1 = 19$.

c	T
5	7
10	13
15	19

We plot the points and draw the graph.

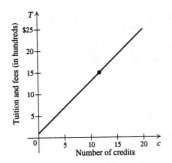

Four three-credit courses total 4 · 3, or 12, credits. To estimate the cost of tuition and fees for a student who is registered for 12 credits, we find the second coordinate associated with 12. Locate the point on the line that is above 12 and then find the value on the vertical axis that corresponds to that point. That value is about 15, so tuition and fees will cost about $1500.

44.

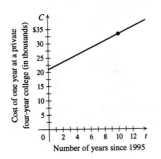

$33,500

45. We graph $n = \frac{5}{2}d + 20$.

When $d = 0$, $n = \frac{5}{2} \cdot 0 + 20 = 20$.

When $d = 2$, $n = \frac{5}{2} \cdot 2 + 20 = 25$.

When $d = 8$, $n = \frac{5}{2} \cdot 8 + 20 = 40$.

d	n
0	20
2	25
8	40

We plot the points and draw the graph.

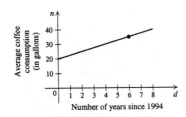

To estimate coffee consumption in 2000, we first note that 2000 is 6 years after 1994. Then we find the second coordinate associated with 6. Locate the point on the line that is above 6 and find the value on the vertical axis that corresponds to that point. That value is about 35, so 35 gal of coffee were consumed by the average U.S. consumer in 2000.

46.

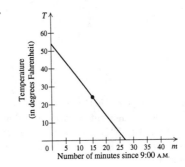

24°F

47. ◈

48. ◈

49. $5x + 3 \cdot 0 = 12$

$5x + 0 = 12$

$5x = 12$

$x = \frac{12}{5}$

Check: $\dfrac{5x + 3 \cdot 0 = 12}{}$

$5 \cdot \dfrac{12}{5} + 3 \cdot 0 \; ? \; 12$

$\dfrac{12 + 0}{} \Big|$

$12 \;\Big|\; 12$ TRUE

The solution is $\dfrac{12}{5}$.

50. $\dfrac{9}{2}$

51. $7 \cdot 0 - 4y = 10$

$0 - 4y = 10$

$y = -\dfrac{5}{2}$

Check: $\dfrac{7 \cdot 0 - 4y = 10}{}$

$7 \cdot 0 - 4\left(-\dfrac{5}{2}\right) \; ? \; 10$

$\dfrac{0 + 10}{} \Big|$

$10 \;\Big|\; 10$ TRUE

The solution is $-\dfrac{5}{2}$.

52. $p = \dfrac{w}{q+1}$

53. $\quad Ax + By = C$
$\qquad\qquad By = C - Ax \qquad$ Subtracting Ax
$\qquad\qquad y = \dfrac{C - Ax}{B} \qquad$ Dividing by B

54. $Q = 2A - T$

55. ◈

56. ◈

57. Let s represent the gear that Lauren uses on the southbound portion of her ride and n represent the gear she uses on the northbound portion. Then we have $s + n = 18$. We graph this equation, using only positive integer values for s and n.

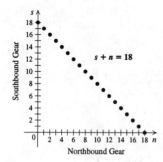

58. $x + y = 5$, or $y = -x + 5$

59. Note that the sum of the coordinates of each point on the graph is 2. Thus, we have $x + y = 2$, or $y = -x + 2$.

60. $y = x + 2$

61. Note that when $x = 0$, $y = -5$ and when $y = 0$, $x = 3$. An equation that fits this situation is $5x - 3y = 15$, or $y = \dfrac{5}{3}x - 5$.

62.

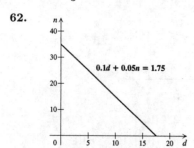

5 dimes, 25 nickels; 10 dimes, 15 nickels; 12 dimes, 11 nickels

63. The equation is $25d + 5l = 225$.

Since the number of dinners cannot be negative, we choose only nonnegative values of d when graphing the equation. The graph stops at the horizontal axis since the number of lunches cannot be negative.

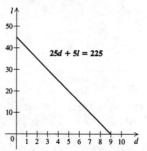

We see that three points on the graph are $(1, 40)$, $(5, 20)$, and $(8, 5)$. Thus, three combinations of dinners and lunches that total $225 are

$\qquad$ 1 dinner, 40 lunches,

$\qquad$ 5 dinners, 20 lunches,

$\qquad$ 8 dinners, 5 lunches.

64.

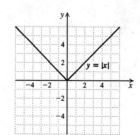

65. $y = -|x|$

x	y
-3	-3
-2	-2
-1	-1
0	0
1	-1
2	-2
3	-3

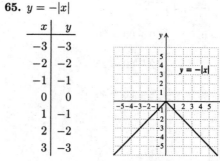

66.

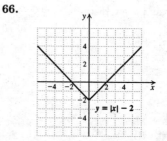

67. $y = -|x| + 2$

x	y
-3	-1
-2	0
-1	1
0	2
1	1
2	0
3	-1

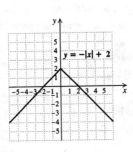

68.

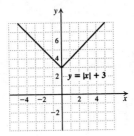

69.

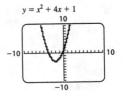

y = −2.8x + 3.5

70.

y = 4.5x + 2.1

71.

y = 2.8x − 3.5

72.

y = −4.5x − 2.1

73.

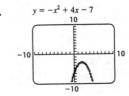

$y = x^2 + 4x + 1$

74.

$y = -x^2 + 4x - 7$

75.

Exercise Set 3.3

1. (a) The graph crosses the y-axis at $(0, 5)$, so the y-intercept is $(0, 5)$.

    (b) The graph crosses the x-axis at $(2, 0)$, so the x-intercept is $(2, 0)$.

2. (a) $(0, 3)$; (b) $(4, 0)$

3. (a) The graph crosses the y-axis at $(0, -4)$, so the y-intercept is $(0, -4)$.

    (b) The graph crosses the x-axis at $(3, 0)$, so the x-intercept is $(3, 0)$.

4. (a) $(0, 5)$; (b) $(-3, 0)$

5. (a) The graph crosses the y-axis at $(0, -2)$, so the y-intercept is $(0, -2)$.

    (b) The graph crosses the x-axis at $(-3, 0)$ and also at $(3, 0)$, so the x-intercepts are $(-3, 0)$ and $(3, 0)$.

6. (a) $(0, 1)$; (b) $(-3, 0)$

7. (a) The graph crosses the y-axis at $(0, 4)$, so the y-intercept is $(0, 4)$.

    (b) The graph crosses the x-axis at $(-3, 0)$, $(3, 0)$, and $(5, 0)$. Each of these points is an x-intercept.

8. (a) $(0, -3)$; (b) $(-2, 0)$, $(2, 0)$, $(5, 0)$

9. $5x + 3y = 15$

    (a) To find the y-intercept, let $x = 0$. This is the same as ignoring the x-term and then solving.

$$3y = 15$$
$$y = 5$$

The y-intercept is $(0, 5)$.

(b) To find the x-intercept, let $y = 0$. This is the same as ignoring the y-term and then solving.

$$5x = 15$$
$$x = 3$$

The x-intercept is $(3, 0)$.

10. (a) $(0, 10)$

(b) $(4, 0)$

11. $7x - 2y = 28$

(a) To find the y-intercept, let $x = 0$. This is the same as ignoring the x-term and then solving.

$$-2y = 28$$
$$y = -14$$

The y-intercept is $(0, -14)$.

(b) To find the x-intercept, let $y = 0$. This is the same as ignoring the y-term and then solving.

$$7x = 28$$
$$x = 4$$

The x-intercept is $(4, 0)$.

12. (a) $(0, -8)$

(b) $(6, 0)$.

13. $-4x + 3y = 10$

(a) To find the y-intercept, let $x = 0$. This is the same as ignoring the x-term and then solving.

$$3y = 10$$
$$y = \frac{10}{3}$$

The y-intercept is $\left(0, \frac{10}{3}\right)$.

(b) To find the x-intercept, let $y = 0$. This is the same as ignoring the y-term and then solving.

$$-4x = 10$$
$$x = -\frac{5}{2}$$

The x-intercept is $\left(-\frac{5}{2}, 0\right)$.

14. (a) $\left(0, \frac{7}{3}\right)$

(b) $\left(-\frac{7}{2}, 0\right)$

15. $y = 9$

Observe that this is the equation of a horizontal line 9 units above the x-axis. Thus, (a) the y-intercept is $(0, 9)$ and (b) there is no x-intercept.

16. (a) None

(b) $(8, 0)$

17. $x + 2y = 6$

Find the y-intercept:

$$2y = 6 \quad \text{Ignoring the } x\text{-term}$$
$$y = 3$$

The y-intercept is $(0, 3)$.

Find the x-intercept:

$$x = 6 \quad \text{Ignoring the } y\text{-term}$$

The x-intercept is $(6, 0)$.

To find a third point we replace x with 2 and solve for y.

$$2 + 2y = 6$$
$$2y = 4$$
$$y = 2$$

The point $(2, 2)$ appears to line up with the intercepts, so we draw the graph.

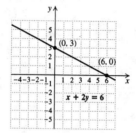

18.

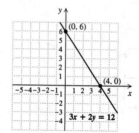

19. $6x + 9y = 36$

Find the y-intercept:

$$9y = 36 \quad \text{Ignoring the } x\text{-term}$$
$$y = 4$$

The y-intercept is $(0, 4)$.

Find the x-intercept:

$$6x = 36 \quad \text{Ignoring the } y\text{-term}$$
$$x = 6$$

The x-intercept is $(6, 0)$.

To find a third point we replace x with -3 and solve for y.

$$6(-3) + 9y = 36$$
$$-18 + 9y = 36$$
$$9y = 54$$
$$y = 6$$

The point $(-3, 6)$ appears to line up with the intercepts, so we draw the graph.

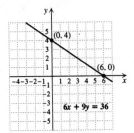

20.

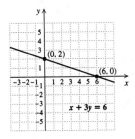

21. $-x + 3y = 9$

Find the y-intercept:

$$3y = 9 \quad \text{Ignoring the } x\text{-term}$$
$$y = 3$$

The y-intercept is $(0, 3)$.

Find the x-intercept:

$$-x = 9 \quad \text{Ignoring the } y\text{-term}$$
$$x = -9$$

The x-intercept is $(-9, 0)$.

To find a third point we replace x with 3 and solve for y.

$$-3 + 3y = 9$$
$$3y = 12$$
$$y = 4$$

The point $(3, 4)$ appears to line up with the intercepts, so we draw the graph.

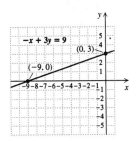

22.

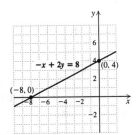

23. $2x - y = 8$

Find the y-intercept:

$$-y = 8 \quad \text{Ignoring the } x\text{-term}$$
$$y = -8$$

The y-intercept is $(0, -8)$.

Find the x-intercept:

$$2x = 8 \quad \text{Ignoring the } y\text{-term}$$
$$x = 4$$

The x-intercept is $(4, 0)$.

To find a third point we replace x with 2 and solve for y.

$$2 \cdot 2 - y = 8$$
$$4 - y = 8$$
$$-y = 4$$
$$y = -4$$

The point $(2, -4)$ appears to line up with the intercepts, so we draw the graph.

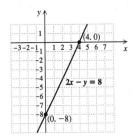

24.

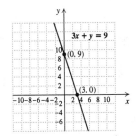

25. $y = -3x + 6$

Find the y-intercept:

$$y = 6 \quad \text{Ignoring the } x\text{-term}$$

The y-intercept is $(0, 6)$.

Find the x−intercept:

$$0 = -3x + 6 \quad \text{Replacing } y \text{ with } 0$$
$$3x = 6$$
$$x = 2$$

The x-intercept is $(2, 0)$.

To find a third point we replace x with 3 and find y.

$$y = -3 \cdot 3 + 6 = -9 + 6 = -3$$

The point $(3, -3)$ appears to line up with the intercepts, so we draw the graph.

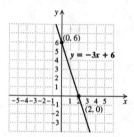

26.

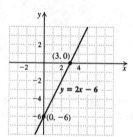

27. $5x - 10 = 5y$

We can leave the equation in the given form or rewrite it in the form $Ax + By = C$. We will use the given form.

Find the y-intercept:

$$-10 = 5y \quad \text{Ignoring the } x\text{-term}$$
$$-2 = y$$

The y-intercept is $(0, -2)$.

To find the x-intercept, let $y = 0$.

$$5x - 10 = 5 \cdot 0$$
$$5x - 10 = 0$$
$$5x = 10$$
$$x = 2$$

The x-intercept is $(2, 0)$.

To find a third point we replace x with 5 and solve for y.

$$5 \cdot 5 - 10 = 5y$$
$$25 - 10 = 5y$$
$$15 = 5y$$
$$3 = y$$

The point $(5, 3)$ appears to line up with the intercepts, so we draw the graph.

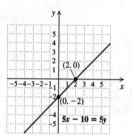

28.

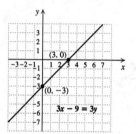

29. $2x - 5y = 10$

Find the y-intercept:

$$-5y = 10 \quad \text{Ignoring the } x\text{-term}$$
$$y = -2$$

The y-intercept is $(0, -2)$.

Find the x-intercept:

$$2x = 10 \quad \text{Ignoring the } y\text{-term}$$
$$x = 5$$

The x-intercept is $(5, 0)$.

To find a third point we replace x with -5 and solve for y.

$$2(-5) - 5y = 10$$
$$-10 - 5y = 10$$
$$-5y = 20$$
$$y = -4$$

The point $(-5, -4)$ appears to line up with the intercepts, so we draw the graph.

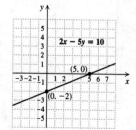

30.

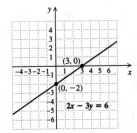

31. $6x + 2y = 12$

Find the y-intercept:

$$2y = 12 \quad \text{Ignoring the } x\text{-term}$$
$$y = 6$$

The y-intercept is $(0, 6)$.

Find the x-intercept:

$$6x = 12 \quad \text{Ignoring the } y\text{-term}$$
$$x = 2$$

The x-intercept is $(2, 0)$.

To find a third point we replace x with 3 and solve for y.

$$6 \cdot 3 + 2y = 12$$
$$18 + 2y = 12$$
$$2y = -6$$
$$y = -3$$

The point $(3, -3)$ appears to line up with the intercepts, so we draw the graph.

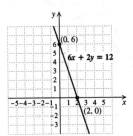

32.

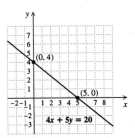

33. $x - 1 = y$

We can leave the equation in the given form or rewrite it in the form $Ax + By = C$. We will use the given form.

Find the y-intercept:

$$-1 = y \quad \text{Ignoring the } x\text{-term}$$

The y-intercept is $(0, -1)$.

To find the x-intercept, let $y = 0$.

$$x - 1 = 0$$
$$x = 1$$

The x-intercept is $(1, 0)$.

To find a third point we replace x with -3 and solve for y.

$$-3 - 1 = y$$
$$-4 = y$$

The point $(-3, -4)$ appears to line up with the intercepts, so we draw the graph.

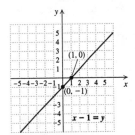

34.

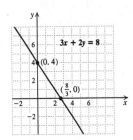

35. $2x - 6y = 18$

Find the y-intercept:

$$-6y = 18 \quad \text{Ignoring the } x\text{-term}$$
$$y = -3$$

The y-intercept is $(0, -3)$.

Find the x-intercept:

$$2x = 18 \quad \text{Ignoring the } y\text{-term}$$
$$x = 9$$

The x-intercept is $(9, 0)$.

To find a third point we replace x with 3 and solve for y.

$$2 \cdot 3 - 6y = 18$$
$$6 - 6y = 18$$
$$-6y = 12$$
$$y = -2$$

The point $(3, -2)$ appears to line up with the intercepts, so we draw the graph.

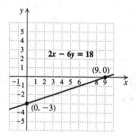

36.

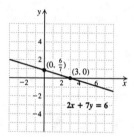

37. $4x - 3y = 12$

Find the y-intercept:

$$-3y = 12 \quad \text{Ignoring the } x\text{-term}$$
$$y = -4$$

The y-intercept is $(0, -4)$.

Find the x-intercept:

$$4x = 12 \quad \text{Ignoring the } y\text{-term}$$
$$x = 3$$

The x-intercept is $(3, 0)$.

To find a third point we replace x with 6 and solve for y.

$$4 \cdot 6 - 3y = 12$$
$$24 - 3y = 12$$
$$-3y = -12$$
$$y = 4$$

The point $(6, 4)$ appears to line up with the intercepts, so we draw the graph.

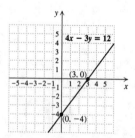

38.

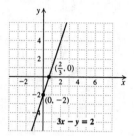

39. $-3x = 6y - 2$

We can leave the equation in the given form or rewrite it in the form $Ax + By = C$. We will use the given form.

To find the y-intercept, let $x = 0$.

$$-3 \cdot 0 = 6y - 2$$
$$0 = 6y - 2$$
$$2 = 6y$$
$$\frac{1}{3} = y$$

The y-intercept is $\left(0, \dfrac{1}{3}\right)$.

Find the x-intercept:

$$-3x = -2 \quad \text{Ignoring the } y\text{-term}$$
$$x = \frac{2}{3}$$

The x-intercept is $\left(\dfrac{2}{3}, 0\right)$.

To find a third point we replace x with -4 and solve for y.

$$-3(-4) = 6y - 2$$
$$12 = 6y - 2$$
$$14 = 6y$$
$$\frac{7}{3} = y$$

The point $\left(-4, \dfrac{7}{3}\right)$ appears to line up with the intercepts, so we draw the graph.

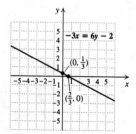

40.

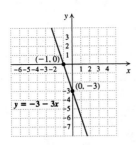

41. $3 = 2x - 5y$

Find the y-intercept:

$$3 = -5y \quad \text{Ignoring the } x\text{-term}$$
$$-\frac{3}{5} = y$$

The y-intercept is $\left(0, -\frac{3}{5}\right)$.

Find the x-intercept:

$$3 = 2x \quad \text{Ignoring the } y\text{-term}$$
$$\frac{3}{2} = x$$

The x-intercept is $\left(\frac{3}{2}, 0\right)$.

To find a third point we replace x with -1 and solve for y.

$$3 = 2(-1) - 5y$$
$$3 = -2 - 5y$$
$$5 = -5y$$
$$-1 = y$$

The point $(-1, -1)$ appears to line up with the intercepts, so we draw the graph.

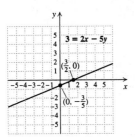

42.

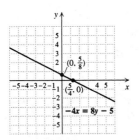

43. $x + 2y = 0$

Find the y-intercept:

$$2y = 0 \quad \text{Ignoring the } x\text{-term}$$
$$y = 0$$

The y-intercept is $(0, 0)$. Note that this is also the x-intercept.

In order to graph the line, we will find a second point.

When $x = 4$, $4 + 2y = 0$
$$2y = -4$$
$$y = -2.$$

Thus, a second point is $(4, -2)$.

To find a third point we replace x with -2 and solve for y.

$$-2 + 2y = 0$$
$$2y = 2$$
$$y = 1$$

The point $(-2, 1)$ appears to line up with the other two points, so we draw the graph.

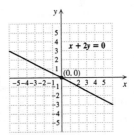

44.

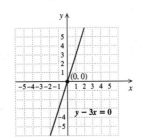

45. $y = 5$

Any ordered pair $(x, 5)$ is a solution. The variable y must be 5, but the x variable can be any number we choose. A few solutions are listed below. Plot these points and draw the line.

x	y
-3	5
0	5
2	5

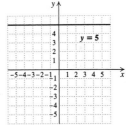

46.

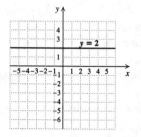

47. $x = 4$

Any ordered pair $(4, y)$ is a solution. The variable x must be 4, but the y variable can be any number we choose. A few solutions are listed below. Plot these points and draw the line.

x	y
4	-2
4	0
4	4

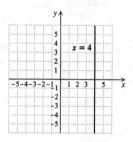

48.

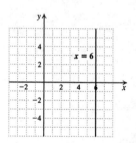

49. $y = -2$

Any ordered pair $(x, -2)$ is a solution. The variable y must be -2, but the x variable can be any number we choose. A few solutions are listed below. Plot these points and draw the line.

x	y
-3	-2
0	-2
4	-2

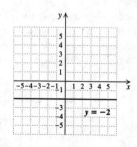

50.

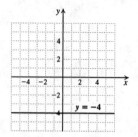

51. $x = -1$

Any ordered pair $(-1, y)$ is a solution. The variable x must be -1, but the y variable can be any number we choose. A few solutions are listed below. Plot these points and draw the line.

x	y
-1	-3
-1	0
-1	2

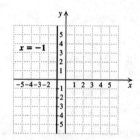

52.

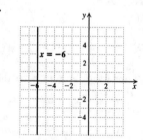

53. $y = 7$

Any ordered pair $(x, 7)$ is a solution. A few solutions are listed below. Plot these points and draw the line.

x	y
-3	7
0	7
4	7

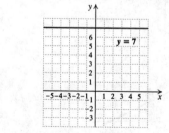

54.

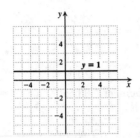

55. $x = 1$

Any ordered pair $(1, y)$ is a solution. A few solutions are listed below. Plot these points and draw the line.

x	y
1	-1
1	4
1	5

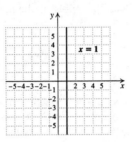

56.

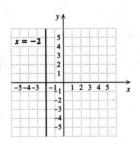

57. $y = 0$

Any ordered pair $(x, 0)$ is a solution. A few solutions are listed below. Plot these points and draw the line.

x	y
-4	0
0	0
2	0

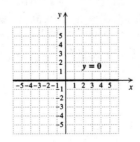

58.

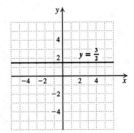

59. $x = -\dfrac{5}{2}$

Any ordered pair $\left(-\dfrac{5}{2}, y \right)$ is a solution. A few solutions are listed below. Plot these points and draw the line.

x	y
$-\dfrac{5}{2}$	-3
$-\dfrac{5}{2}$	0
$-\dfrac{5}{2}$	5

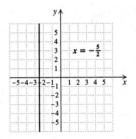

60.

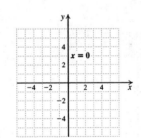

61. $-5y = 15$

Observe that $-5y = 15$ is equivalent to $y = -3$. Thus, the graph is a horizontal line 3 units below the x-axis.

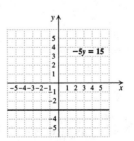

62.

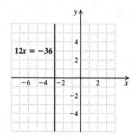

63. $35 + 7y = 0$

$$7y = -35$$

$$y = -5$$

The graph is a horizontal line 5 units below the x-axis.

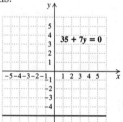

64.

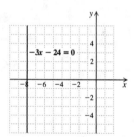

65. Note that every point on the horizontal line passing through $(0, -1)$ has -1 as the y-coordinate. Thus, the equation of the line is $y = -1$.

66. $x = -1$

67. Note that every point on the vertical line passing through $(4, 0)$ has 4 as the x-coordinate. Thus, the equation of the line is $x = 4$.

68. $y = -5$

69. Note that every point on the horizontal line passing through $(0, 0)$ has 0 as the y-coordinate. Thus, the equation of the line is $y = 0$.

70. $x = 0$

71. ◈

72. ◈

73. $d - 7$

74. $w + 5$, or $5 + w$

75. Let x represent "a number." Then we have $2 + x$, or $x + 2$.

76. Let y represent "a number;" $3y$

77. Let x and y represent the numbers. Then we have $2(x + y)$.

78. Let m and n represent the numbers; $\frac{1}{2}(m + n)$

79. ◈

80. ◈

81. The x-axis is a horizontal line, so it is of the form $y = b$. All points on the x-axis are of the form $(x, 0)$, so b must be 0 and the equation is $y = 0$.

82. $y = 5$

83. A line parallel to the y-axis has an equation of the form $x = a$. Since the x-coordinate of one point on the line is -2, then $a = -2$ and the equation is $x = -2$.

84. $(6, 6)$

85. Since the x-coordinate of the point of intersection must be -3 and y must equal x, the point of intersection is $(-3, -3)$.

86. $y = -\frac{5}{2}x + 5$, or $5x + 2y = 10$

87. The y-intercept is $(0, 5)$, so we have $y = mx + 5$. Another point on the line is $(-3, 0)$ so we have

$$0 = m(-3) + 5$$

$$-5 = -3m$$

$$\frac{5}{3} = m$$

The equation is $y = \frac{5}{3}x + 5$, or $5x - 3y = -15$.

88. 12

89. Substitute 0 for x and -8 for y.

$$4 \cdot 0 = C - 3(-8)$$

$$0 = C + 24$$

$$-24 = C$$

90.

91. Find the y-intercept:

$$2y = 50 \quad \text{Covering the } x\text{-term}$$
$$y = 25$$

The y-intercept is $(0, 25)$.

Find the x-intercept:

$$3x = 50 \qquad \text{Covering the } y\text{-term}$$
$$x = \frac{50}{3} = 16.\overline{6}$$

The x-intercept is $\left(\frac{50}{3}, 0\right)$, or $(16.\overline{6}, 0)$.

92. $\left(0, -\frac{80}{7}\right)$, or $(0, -11.\overline{428571})$; $(40, 0)$

93. From the equation we see that the y-intercept is $(0, -9)$.

To find the x-intercept, let $y = 0$.

$$0 = 0.2x - 9$$
$$9 = 0.2x$$
$$45 = x$$

The x-intercept is $(45, 0)$.

94. $(0, -15)$; $\left(\frac{150}{13}, 0\right)$, or $(11.\overline{538461}, 0)$

95. Find the y-intercept.

$$-20y = 1 \qquad \text{Covering the } x\text{-term}$$
$$y = -\frac{1}{20}, \text{ or } -0.05$$

The y-intercept is $\left(0, -\frac{1}{20}\right)$, or $(0, -0.05)$.

Find the x-intercept:

$$25x = 1 \qquad \text{Covering the } y\text{-term}$$
$$x = \frac{1}{25}, \text{ or } 0.04$$

The x-intercept is $\left(\frac{1}{25}, 0\right)$, or $(0.04, 0)$.

96. $\left(0, \frac{1}{25}\right)$, or $(0, 0.04)$; $\left(\frac{1}{50}, 0\right)$, or $(0.02, 0)$

Exercise Set 3.4

1. a) We divide the number of miles traveled by the number of gallons of gas used for that amount of driving.

Rate, in miles per gallon
$$= \frac{14,014 \text{ mi} - 13,741 \text{ mi}}{13 \text{ gal}}$$
$$= \frac{273 \text{ mi}}{13 \text{ gal}}$$
$$= 21 \text{ mi/gal}$$
$$= 21 \text{ miles per gallon}$$

b) We divide the cost of the rental by the number of days. From July 1 to July 4 is $4 - 1$, or 3 days.

Average cost, in dollars per day
$$= \frac{118 \text{ dollars}}{3 \text{ days}}$$
$$\approx 39.33 \text{ dollars/day}$$
$$\approx \$39.33 \text{ per day}$$

c) We divide the number of miles traveled by the number of days. In part (a) we found that the van was driven 273 mi, and in part (b) we found that it was rented for 3 days.

Rate, in miles per day
$$= \frac{273 \text{ mi}}{3 \text{ days}}$$
$$= 91 \text{ mi/day}$$
$$= 91 \text{ mi per day}$$

2. a) 16.5 mi per gal

b) \$46 per day

c) 115.5 mi per day

d) 40¢ per mi

3. a) From 2:00 to 5:00 is $5 - 2$, or 3 hr.

Average speed, in miles per hour
$$= \frac{18 \text{ mi}}{3 \text{ hr}}$$
$$= 6 \text{ mph}$$

b) From part (a) we know that the bike was rented for 3 hr.

Rate, in dollars per hour $= \dfrac{\$10.50}{3 \text{ hr}}$
$$= \$3.50 \text{ per hr}$$

c) Rate, in dollars per mile $= \dfrac{\$10.50}{18 \text{ mi}}$
$$\approx \$0.58 \text{ per mile}$$

4. a) 7 mph

    b) $6 per hr

    c) $0.86 per mi

5. a) It is 3 hr from 9:00 A.M. to noon and 5 more hours from noon to 5:00 P.M., so the typist worked $3 + 5$, or 8 hr.

    $$\text{Rate, in dollars per hour} = \frac{\$128}{8 \text{ hr}}$$
    $$= \$16 \text{ per hr}$$

    b) The number of pages typed is $48 - 12$, or 36.

    In part (a) we found that the typist worked 8 hr.

    $$\text{Rate, in pages per hour} = \frac{36 \text{ pages}}{8 \text{ hr}}$$
    $$= 4.5 \text{ pages per hr}$$

    c) In part (b) we found that 36 pages were typed.

    $$\text{Rate, in dollars per page} = \frac{\$128}{36 \text{ pages}}$$
    $$\approx \$3.56 \text{ per page}$$

6. a) $15 per hr

    b) 5.25 pages per hr

    c) $2.86 per page

7. The tuition increased $\$1318 - \$1239$, or \$79, in 1998–1996 or 2 yr.

    $$\text{Rate of increase} = \frac{\text{Change in tuition}}{\text{Change in time}}$$
    $$= \frac{\$79}{2 \text{ yr}}$$
    $$= \$39.50 \text{ per yr}$$

8. $170.67 per yr

9. a) The elevator traveled $34 - 5$, or 29 floors in $2{:}40 - 2{:}38$, or 2 min.

    $$\text{Average rate of travel} = \frac{29 \text{ floors}}{2 \text{ min}}$$
    $$= 14.5 \text{ floors per min}$$

    b) In part (a) we found that the elevator traveled 29 floors in 2 min. Note that $2 \text{ min} = 2 \times 1 \text{ min} = 2 \times 60 \text{ sec} = 120 \text{ sec}$.

    $$\text{Average rate of travel} = \frac{120 \text{ sec}}{29 \text{ floors}}$$
    $$\approx 4.14 \text{ sec per floor}$$

10. a) 1 driveway per hour

    b) 1 hr per driveway

11. a) Krakauer ascended $29,028 \text{ ft} - 27,600 \text{ ft}$, or 1428 ft. From 7:00 A.M. to noon it is $5 \text{ hr} = 5 \times 1 \text{ hr} = 5 \times 60 \text{ min} = 300 \text{ min}$. From noon to 1:25 P.M. is another 1 hr, 25 min, or $1 \text{ hr} + 25 \text{ min} = 60 \text{ min} + 25 \text{ min} = 85 \text{ min}$. The total time of the ascent is $300 \text{ min} + 85 \text{ min}$, or 385 min.

    $$\text{Rate, in feet per minute} = \frac{1428 \text{ ft}}{385 \text{ min}}$$
    $$\approx 3.71 \text{ ft per min}$$

    b) We use the information found in part (a).

    $$\text{Rate, in minutes per foot} = \frac{385 \text{ min}}{1428 \text{ ft}}$$
    $$\approx 0.27 \text{ min per ft}$$

12. a) 9.38 ft per min

    b) 0.11 min per ft

13. The rate is given in millions of crimes per year, so we list number of crimes, in millions, on the vertical axis and years on the horizontal axis. If we count by 10's of millions on the vertical axis we can easily reach 37 million without needed a terribly large graph. We plot the point (1996, 37 million). Then, to display the rate of growth, we move from that point to a point that represents 2.5 million fewer crimes 1 year later. The coordinates of this point are 1996+1, 37−2.5 million), or (1997, 34.5 million). Finally, we draw a line through the two points.

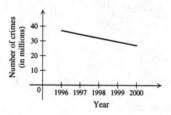

14.

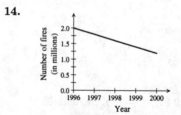

15. The rate is given in miles per hour, so we list the number of miles traveled on the vertical axis and the time of day on the horizontal axis. If we count by 100's of miles on the vertical axis we can easily reach 230 without needing a terribly large graph. We plot the point (3:00, 230). Then to display the rate of travel, we move from that point to a point that represents 90 more miles traveled 1 hour later. The

coordinates of this point are $(3:00 + 1 \text{ hr}, 230 + 90)$, or $(4:00, 320)$. Finally, we draw a line through the two points.

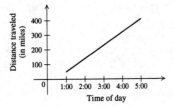

16.

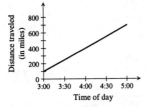

17. The rate is given in dollars per hour so we list money earned on the vertical axis and the time of day on the horizontal axis. We can count by $20 on the vertical axis and reach $50 without needing a terribly large graph. Next we plot the point $(2:00 \text{ P.M.}, \$50)$. To display the rate we move from that point to a point that represents $15 more 1 hour later. The coordinates of this point are $(2 + 1, \$50 + \$15)$, or $(3:00 \text{ P.M.}, \$65)$. Finally, we draw a line through the two points.

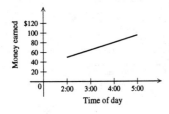

18.

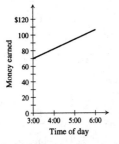

19. The rate is given in cost per minute so we list the amount of the telephone bill on the vertical axis and the number of additional minutes on the horizontal axis. We begin with $7.50 on the vertical axis and count by $0.50. A jagged line at the base of the axis indicates that we are not showing amounts smaller

than $7.50. We begin with 0 additional minutes on the horizontal axis and plot the point $(0, \$7.50)$. We move from there to a point that represents $0.10 more 1 minute later. The coordinates of this point are $(0+1 \text{ min}, \$7.50+\$0.10)$, or $(1 \text{ min}, \$7.60)$. Then we draw a line through the two points.

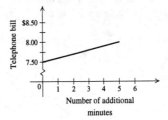

20.

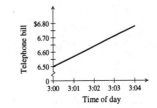

21. The points $(2:00, 7 \text{ haircuts})$ and $(5:00, 12 \text{ haircuts})$ are on the graph. This tells us that in the 3 hr between 2:00 and 5:00 there were $12 - 7 = 5$ haircuts completed. The rate is

$$\frac{5 \text{ haircuts}}{3 \text{ hr}} = \frac{5}{3}, \text{ or } 1\frac{2}{3} \text{ haircuts per hour.}$$

22. 4 manicures per hour

23. The points $(12:00, 100 \text{ mi})$ and $(2:00, 250 \text{ mi})$ are on the graph. This tells us that in the 2 hr between 12:00 and 2:00 the train traveled $250 - 100 = 150$ mi. The rate is

$$\frac{150 \text{ mi}}{2 \text{hr}} = 75 \text{ mi per hr.}$$

24. 87.5 mi per hr

25. The points $(15 \text{ min}, 150¢)$ and $(30 \text{ min}, 300¢)$ are on the graph. This tells us that in $30 - 15 = 15$ min the cost of the call increased $300¢ - 150¢ = 150¢$. The rate is

$$\frac{150¢}{15 \text{ min}} = 10¢ \text{ per min.}$$

26. 7¢ per min

27. The points $(2 \text{ yr}, \$2000)$ and $(4 \text{ yr}, \$1000)$ are on the graph. This tells us that in $4 - 2 = 2$ yr the value of the copier changes $\$1000 - \$2000 = -\$1000$. The rate is

$$\frac{-\$1000}{2 \text{ yr}} = -\$500 \text{ per yr.}$$

This means that the value of the copier is decreasing at a rate of $500 per yr.

28. −$0.15 billion per yr

29. The points (50 mi, 2 gal) and (200 mi, 8 gal) are on the graph. This tells us that when driven $200 - 50 = 150$ mi the vehicle consumed $8 - 2 = 6$ gal of gas. The rate is
$$\frac{6 \text{ gal}}{150 \text{ mi}} = 0.04 \text{ gal per mi}.$$

30. $0.08\overline{3}$ gal per mi

31. ◈

32. ◈

33. $-2 - (-7) = -2 + 7 = 5$

34. -6

35. $\dfrac{5 - (-4)}{-2 - 7} = \dfrac{9}{-9} = -1$

36. $-\dfrac{4}{3}$

37. $\dfrac{-4 - 8}{7 - (-2)} = \dfrac{-12}{9} = -\dfrac{4}{3}$

38. $-\dfrac{4}{5}$

39. ◈

40. ◈

41. Let $t =$ flight time and $a =$ altitude. While the plane is climbing at a rate of 6500 ft/min, the equation $a = 6500t$ describes the situation. Solving $34,000 = 6500t$, we find that the cruising altitude of 34,000 ft is reached after about 5.23 min. Thus we graph $a = 6500t$ for $0 \le t \le 5.23$.

The plane cruises at 34,000 ft for 3 min, so we graph $a = 34,000$ for $5.23 < t \le 8.23$. After 8.23 min the plane descends at a rate of 3500 ft/min and lands. The equation $a = 34,000 - 3500(t - 8.23)$, or $a = -3500t + 62,805$, describes this situation. Solving $0 = -3500t + 62,805$, we find that the plane lands after about 17.94 min. Thus we graph $a = -3500t + 62,805$ for $8.23 < t \le 17.94$. The entire graph is show below.

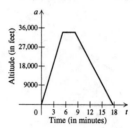

42.

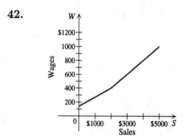

About $550

43. We begin with the graph in Exercise 29 showing the gas consumption of a Honda Odyssey. For each point (x, y) on this graph we can plot a point $(2x, y)$ on the graph that represents the gas consumption of the motorcycle.

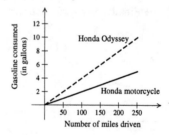

44.

45. Penny walks forward at a rate of $\dfrac{24 \text{ ft}}{3 \text{ sec}}$, or 8 ft per sec. In addition, the boat is traveling at a rate of 5 ft per sec. Thus, with respect to land, Penny is traveling at a rate of $8 + 5$, or 13 ft per sec.

46. 0.45 min per mi

47. First we find Annette's speed in minutes per kilometer.

$$\text{Speed} = \frac{15.5 \text{ min}}{7 \text{ km} - 4 \text{ km}} = \frac{15.5}{3} \frac{\text{min}}{\text{km}}$$

Now we convert min/km to min/mi.

$$\frac{15.5}{3} \frac{\text{min}}{\text{km}} \approx \frac{15.5}{3} \frac{\text{min}}{\text{km}} \cdot \frac{1 \text{ km}}{0.621 \text{ min}} \approx \frac{15.5}{1.863} \frac{\text{min}}{\text{mi}}$$

At a rate of $\dfrac{15.5}{1.863} \dfrac{\text{min}}{\text{mi}}$, to run a 5-mi race it would take $\dfrac{15.5}{1.863} \dfrac{\text{min}}{\text{mi}} \cdot 5 \text{ mi} \approx 41.6 \text{ min}.$

(Answers may vary slightly depending on the conversion factor used.)

48. 51.8 min

(Answers may vary slightly depending on the conversion factor used.)

49. First we find Ryan's rate. Then we double it to find Marcy's rate. Note that 50 minutes $= \dfrac{50}{60} \text{ hr} = \dfrac{5}{6} \text{ hr}.$

$$\text{Ryan's rate} = \frac{\text{change in number of bushels picked}}{\text{corresponding change in time}}$$

$$= \frac{5\frac{1}{2} - 4 \text{ bushels}}{\frac{5}{6} \text{ hr}}$$

$$= \frac{1\frac{1}{2} \text{ bushels}}{\frac{5}{6} \text{ hr}}$$

$$= \frac{3}{2} \cdot \frac{6}{5} \frac{\text{bushels}}{\text{hr}}$$

$$= \frac{9}{5} \text{ bushels per hour, or}$$

$$1.8 \text{ bushels per hour}$$

Then Marcy's rate is $2(1.8) = 3.6$ bushels per hour.

50. 27 candles per hour

Exercise Set 3.5

1. The rate can be found using the coordinates of any two points on the line. We use $(2, 30)$ and $(6, 90)$.

$$\text{Rate} = \frac{\text{change in number of calories burned}}{\text{corresponding change in time}}$$

$$= \frac{90 - 30 \text{ calories}}{6 - 2 \text{ min}}$$

$$= \frac{60 \text{ calories}}{4 \text{ min}}$$

$$= 15 \text{ calories per min}$$

2. 2.5 million people per year

3. The rate can be found using the coordinates of any two points on the line. We use $(35, 490)$ and $(45, 500)$, where 35 and 45 are in $1000's.

$$\text{Rate} = \frac{\text{change in score}}{\text{corresponding change in income}}$$

$$= \frac{500 - 490 \text{ points}}{45 - 35}$$

$$= \frac{10 \text{ points}}{10}$$

$$= 1 \text{ point per } \$1000 \text{ income}$$

4. $1\frac{1}{3}$ points per $1000 income

5. The rate can be found using the coordinates of any two points on the line. We use $(1993, 20)$ and $(1997, 17)$.

$$\text{Rate} = \frac{\text{change in percent}}{\text{corresponding change in time}}$$

$$= \frac{17\% - 20\%}{1997 - 1993}$$

$$= \frac{-3\%}{4 \text{ yr}}$$

$$= -\frac{3}{4}\% \text{ per yr, or } -0.75\% \text{ per yr}$$

6. -0.4% per yr, or $-\dfrac{2}{5}\%$ per yr

7. We can use any two points on the line, such as $(0, 1)$ and $(4, 4)$.

$$m = \frac{\text{change in } y}{\text{change in } x}$$

$$= \frac{4 - 1}{4 - 0} = \frac{3}{4}$$

8. $\dfrac{2}{3}$

9. We can use any two points on the line, such as $(1, 0)$ and $(3, 3)$.

$$m = \frac{\text{change in } y}{\text{change in } x}$$

$$= \frac{3 - 0}{3 - 1} = \frac{3}{2}$$

10. $\dfrac{1}{3}$

11. We can use any two points on the line, such as $(-3, -4)$ and $(0, -3)$.

$$m = \frac{\text{change in } y}{\text{change in } x}$$

$$= \frac{-3 - (-4)}{0 - (-3)} = \frac{1}{3}$$

12. 3

13. We can use any two points on the line, such as $(0,2)$ and $(2,0)$.
$$m = \frac{\text{change in } y}{\text{change in } x}$$
$$= \frac{2-0}{0-2} = \frac{2}{-2} = -1$$

14. $-\dfrac{1}{2}$

15. This is the graph of a horizontal line. Thus, the slope is 0.

16. $-\dfrac{3}{2}$

17. We can use any two points on the line, such as $(0,2)$ and $(3,1)$.
$$m = \frac{\text{change in } y}{\text{change in } x}$$
$$= \frac{1-2}{3-0} = -\frac{1}{3}$$

18. -2

19. This is the graph of a vertical line. Thus, the slope is undefined.

20. Undefined

21. We can use any two points on the line, such as $(-2,3)$ and $(2,2)$.
$$m = \frac{\text{change in } y}{\text{change in } x}$$
$$= \frac{2-3}{2-(-2)} = -\frac{1}{4}$$

22. 0

23. We can use any two points on the line, such as $(-2,-3)$ and $(2,3)$.
$$m = \frac{\text{change in } y}{\text{change in } x}$$
$$= \frac{3-(-3)}{2-(-2)} = \frac{6}{4} = \frac{3}{2}$$

24. $-\dfrac{2}{3}$

25. This is the graph of a horizontal line, so the slope is 0.

26. 5

27. We can use any two points on the line, such as $(-3,5)$ and $(0,-4)$.
$$m = \frac{\text{change in } y}{\text{change in } x}$$
$$= \frac{-4-5}{0-(-3)} = \frac{-9}{3} = -3$$

28. 0

29. $(1,2)$ and $(5,8)$
$$m = \frac{8-2}{5-1} = \frac{6}{4} = \frac{3}{2}$$

30. 2

31. $(-2,4)$ and $(3,0)$
$$m = \frac{4-0}{-2-3} = \frac{4}{-5} = -\frac{4}{5}$$

32. $-\dfrac{5}{6}$

33. $(-4,0)$ and $(5,7)$
$$m = \frac{7-0}{5-(-4)} = \frac{7}{9}$$

34. $\dfrac{2}{3}$

35. $(0,8)$ and $(-3,10)$
$$m = \frac{8-10}{0-(-3)} = \frac{8-10}{0+3} = \frac{-2}{3} = -\frac{2}{3}$$

36. $-\dfrac{1}{2}$

37. $(-2,3)$ and $(-6,5)$
$$m = \frac{5-3}{-6-(-2)} = \frac{2}{-6+2} = \frac{2}{-4} = -\frac{1}{2}$$

38. $-\dfrac{11}{8}$

39. $\left(-2,\dfrac{1}{2}\right)$ and $\left(-5,\dfrac{1}{2}\right)$

Observe that the points have the same y-coordinate. Thus, they lie on a horizontal line and its slope is 0. We could also compute the slope.
$$m = \frac{\frac{1}{2}-\frac{1}{2}}{-2-(-5)} = \frac{\frac{1}{2}-\frac{1}{2}}{-2+5} = \frac{0}{3} = 0$$

40. $\dfrac{4}{7}$

41. $(3,4)$ and $(9,-7)$
$$m = \frac{-7-4}{9-3} = \frac{-11}{6} = -\frac{11}{6}$$

42. Undefined

43. $(6, -4)$ and $(6, 5)$

Observe that the points have the same x-coordinate. Thus, they lie on a vertical line and its slope is undefined. We could also compute the slope.

$$m = \frac{-4 - 5}{6 - 6} = \frac{-9}{0}, \text{ undefined}$$

44. 0

45. The line $x = -3$ is a vertical line. The slope is undefined.

46. Undefined

47. The line $y = 4$ is a horizontal line. A horizontal line has slope 0.

48. 0

49. The line $x = 9$ is a vertical line. The slope is undefined.

50. Undefined

51. The line $y = -9$ is a horizontal line. A horizontal line has slope 0.

52. 0

53. The grade is expressed as a percent.

$$m = \frac{106}{1325} = 0.08 = 8\%$$

54. $0\frac{1}{20}$, or 0.05

55. The grade is expressed as a percent.

$$m = \frac{1}{12} = 0.08\overline{3} = 8.\overline{3}\%$$

56. 7%

57. $m = \dfrac{2.4}{8.2} = \dfrac{12}{41}$, or about 29%

58. 0.08, or 8%

59. Longs Peak rises $14,255 - 9600 = 4655$ ft.

$$m = \frac{4655}{15,840} \approx 0.29 \approx 29\%$$

60. About 64%

61. ◈

62. ◈

63. $ax + by = c$

$$by = c - ax \quad \text{Adding } -ax \text{ to both sides}$$
$$y = \frac{c - ax}{b} \quad \text{Dividing both sides by } b$$

64. $r = \dfrac{p + mn}{x}$

65. $ax - by = c$

$$-by = c - ax \quad \text{Adding } -ax \text{ to both sides}$$
$$y = \frac{c - ax}{-b} \quad \text{Dividing both sides by } -b$$

We could also express this result as $y = \dfrac{ax - c}{b}$.

66. $t = \dfrac{q - rs}{n}$

67.
$$\frac{2}{3}x - 5 = \frac{2}{3} \cdot 12 - 5 \quad \text{Substituting}$$
$$= 8 - 5$$
$$= 3$$

68. 2

69. ◈

70. ◈

71. If the line passes through $(4, -7)$ and never enters the first quadrant, then it slants down from left to right or is horizontal. This means that its slope is not positive ($m \le 0$). The line will slant most steeply if it passes through $(0, 0)$. In this case, $m = \dfrac{-7 - 0}{4 - 0} = -\dfrac{7}{4}$. Thus, the numbers the line could have for its slope are $\left\{ m \middle| -\dfrac{7}{4} \le m \le 0 \right\}$.

72. $\left\{ m \middle| m \ge \dfrac{5}{2} \right\}$

73. $x + y = 18$

$$y = 18 - x$$

The slope is $\dfrac{y}{x}$, or $\dfrac{18 - x}{x}$.

74. $\dfrac{1}{2}$

75. Let $t =$ the number of units each tick mark on the horizontal axis represents. Note that the graph drops 1 unit for every 6 tick marks of horizontal change. Then we have:

$$\frac{-1}{6t} = -\frac{2}{3}$$
$$-1 = -4t$$
$$\frac{1}{4} = t$$

Each tick mark on the horizontal axis represents $\frac{1}{4}$ unit.

Exercise Set 3.6

1. Slope $\frac{2}{5}$; y-intercept $(0,1)$

We plot $(0,1)$ and from there move up 2 units and right 5 units. This locates the point $(5,3)$. We plot $(5,3)$ and draw a line passing through $(0,1)$ and $(5,3)$.

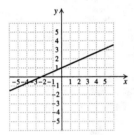

2.

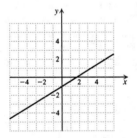

3. Slope $\frac{5}{3}$; y-intercept $(0,-2)$

We plot $(0,-2)$ and from there move up 5 units and right 3 units. This locates the point $(3,3)$. We plot $(3,3)$ and draw a line passing through $(0,-2)$ and $(3,3)$.

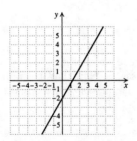

4.

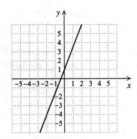

5. Slope $-\frac{3}{4}$; y-intercept $(0,5)$

We plot $(0,5)$. We can think of the slope as $\frac{-3}{4}$, so from $(0,5)$ we move down 3 units and right 4 units. This locates the point $(4,2)$. We plot $(4,2)$ and draw a line passing through $(0,5)$ and $(4,2)$.

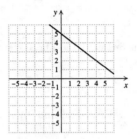

6.

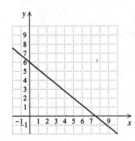

7. Slope 2; y-intercept $(0,-4)$

We plot $(0,-4)$. We can think of the slope as $\frac{2}{1}$, so from $(0,-4)$ we move up 2 units and right 1 unit. This locates the point $(1,-2)$. We plot $(1,-2)$ and draw a line passing through $(0,-4)$ and $(1,-2)$.

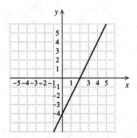

8.

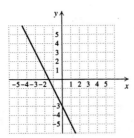

9. Slope -3; y-intercept $(0, 2)$

We plot $(0, 2)$. We can think of the slope as $\dfrac{-3}{1}$, so from $(0, 2)$ we move down 3 units and right 1 unit. This locates the point $(1, -1)$. We plot $(1, -1)$ and draw a line passing through $(0, 2)$ and $(1, -1)$.

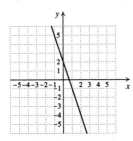

10.

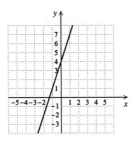

11. We read the slope and y-intercept from the equation.

$$y = \frac{3}{7}x + 5$$

The slope is $\dfrac{3}{7}$.   The y-intercept is $(0, 5)$.

12. $-\dfrac{3}{8}$, $(0, 6)$

13. We read the slope and y-intercept from the equation.

$$y = -\frac{5}{6}x + 2$$

The slope is $-\dfrac{5}{6}$.   The y-intercept is $(0, 2)$.

14. $\dfrac{7}{2}$, $(0, 4)$

15. $y = \dfrac{9}{4}x - 7$

$$y = \frac{9}{4}x + (-7)$$

The slope is $\dfrac{9}{4}$, and the y-intercept is $(0, -7)$.

16. $\dfrac{2}{9}$; $(0, -1)$

17. $y = -\dfrac{2}{5}x$

$$y = -\frac{2}{5}x + 0$$

The slope is $-\dfrac{2}{5}$, and the y-intercept is $(0, 0)$.

18. $\dfrac{4}{3}$; $(0, 0)$

19. We solve for y to rewrite the equation in the form $y = mx + b$.

$$-2x + y = 4$$
$$y = 2x + 4$$

The slope is 2, and the y-intercept is $(0, 4)$.

20. 5; $(0, 5)$

21. $3x - 4y = 12$

$$-4y = -3x + 12$$
$$y = -\frac{1}{4}(-3x + 12)$$
$$y = \frac{3}{4}x - 3$$

The slope is $\dfrac{3}{4}$, and the y-intercept is $(0, -3)$.

22. $\dfrac{3}{2}$; $(0, -9)$

23. $x - 5y = -8$

$$-5y = -x - 8$$
$$y = -\frac{1}{5}(-x - 8)$$
$$y = \frac{1}{5}x + \frac{8}{5}$$

The slope is $\dfrac{1}{5}$, and the y-intercept is $\left(0, \dfrac{8}{5}\right)$.

24. $\dfrac{1}{6}$; $\left(0, -\dfrac{3}{2}\right)$

25. Observe that this is the equation of a horizontal line that lies 4 units above the x-axis. Thus, the slope is 0, and the y-intercept is $(0, 4)$. We could also write the equation in slope-intercept form.

$$y = 4$$
$$y = 0x + 4$$

The slope is 0, and the y-intercept is $(0, 4)$.

26. $0;\ (0, 8)$

27. We use the slope-intercept equation, substituting 3 for m and 7 for b:
$$y = mx + b$$
$$y = 3x + 7$$

28. $y = -4x - 2$

29. We use the slope-intercept equation, substituting $\frac{7}{8}$ for m and -1 for b:
$$y = mx + b$$
$$y = \frac{7}{8}x - 1$$

30. $y = \frac{5}{7}x + 4$

31. We use the slope-intercept equation, substituting $-\frac{5}{3}$ for m and -8 for b:
$$y = mx + b$$
$$y = -\frac{5}{3}x - 8$$

32. $y = \frac{3}{4}x + 23$

33. Since the slope is 0, we know that the line is horizontal. Its y-intercept is $(0, 3)$, so the equation of the line must be $y = 3$.

We could also use the slope-intercept equation, substituting 0 for m and 3 for b.
$$y = mx + b$$
$$y = 0 \cdot x + 3$$
$$y = 3$$

34. $y = 7x$

35. $y = \frac{3}{5}x + 2$

First we plot the y-intercept $(0, 2)$. We can start at the y-intercept and use the slope, $\frac{3}{5}$, to find another point. We move up 3 units and right 5 units to get a new point $(5, 5)$. Thinking of the slope as $\frac{-3}{-5}$ we can start at $(0, 2)$ and move down 3 units and left 5 units to get another point $(-5, -1)$.

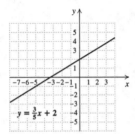

36.

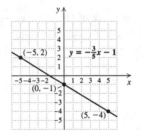

37. $y = -\frac{3}{5}x + 1$

First we plot the y-intercept $(0, 1)$. We can start at the y-intercept and, thinking of the slope as $\frac{-3}{5}$, find another point by moving down 3 units and right 5 units to the point $(5, -2)$. Thinking of the slope as $\frac{3}{-5}$ we can start at $(0, 1)$ and move up 3 units and left 5 units to get another point $(-5, 4)$.

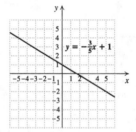

38.

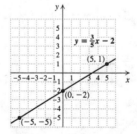

39. $y = \frac{5}{3}x + 3$

First we plot the y-intercept $(0, 3)$. We can start at the y-intercept and use the slope, $\frac{5}{3}$, to find another point. We move up 5 units and right 3 units to get a new point $(3, 8)$. Thinking of the slope as $\dfrac{-5}{-3}$ we can start at $(0, 3)$ and move down 5 units and left 3 units to get another point $(-3, -2)$.

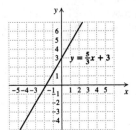

40.

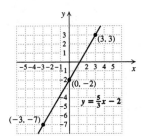

41. $y = -\dfrac{3}{2}x - 2$

First we plot the y-intercept $(0, -2)$. We can start at the y-intercept and, thinking of the slope as $\dfrac{-3}{2}$, find another point by moving down 3 units and right 2 units to the point $(2, -5)$. Thinking of the slope as $\dfrac{3}{-2}$ we can start at $(0, -2)$ and move up 3 units and left 2 units to get another point $(-2, 1)$.

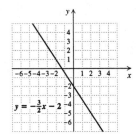

42.

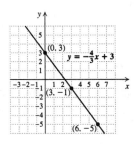

43. We first rewrite the equation in slope-intercept form.

$$2x + y = 1$$
$$y = -2x + 1$$

Now we plot the y-intercept $(0, 1)$. We can start at the y-intercept and, thinking of the slope as $\dfrac{-2}{1}$, find another point by moving down 2 units and right 1 unit to the point $(1, -1)$. In a similar manner, we can move from the point $(1, -1)$ to find a third point $(2, -3)$.

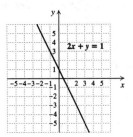

44.

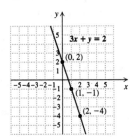

45. We first rewrite the equation in slope-intercept form.

$$3x - y = 4$$
$$-y = -3x + 4$$
$$y = 3x - 4 \quad \text{Multiplying by } -1$$

Now we plot the y-intercept $(0, -4)$. We can start at the y-intercept and, thinking of the slope as $\dfrac{3}{1}$, find another point by moving up 3 units and right 1 unit to the point $(1, -1)$. In a similar manner, we can move from the point $(1, -1)$ to find a third point $(2, 2)$.

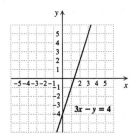

46.

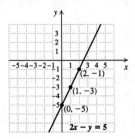

47. We first rewrite the equation in slope-intercept form.

$$2x + 3y = 9$$
$$3y = -2x + 9$$
$$y = \frac{1}{3}(-2x + 9)$$
$$y = -\frac{2}{3}x + 3$$

Now we plot the y-intercept $(0, 3)$. We can start at the y-intercept and, thinking of the slope as $\dfrac{-2}{3}$, find another point by moving down 2 units and right 3 units to the point $(3, 1)$. Thinking of the slope as $\dfrac{2}{-3}$ we can start at $(0, 3)$ and move up 2 units and left 3 units to get another point $(-3, 5)$.

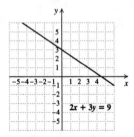

48.

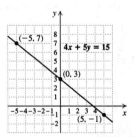

49. We first rewrite the equation in slope-intercept form.

$$x - 4y = 12$$
$$-4y = -x + 12$$
$$y = -\frac{1}{4}(-x + 12)$$
$$y = \frac{1}{4}x - 3$$

Now we plot the y-intercept $(0, -3)$. We can start at the y-intercept and use the slope, $\dfrac{1}{4}$, to find another point. We move up 1 unit and right 4 units to the point $(4, -2)$. Thinking of the slope as $\dfrac{-1}{-4}$ we can start at $(0, -3)$ and move down 1 unit and left 4 units to get another point $(-4, -4)$.

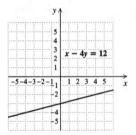

50.

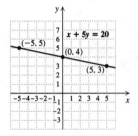

51. Two points on the graph are $(0, 9)$ and $(7, 16)$, so we see that the y-intercept will be $(0, 9)$. Now we find the slope:

$$m = \frac{16 - 9}{7 - 0} = \frac{7}{7} = 1$$

Then the equation is $y = x + 9$.

To graph the equation we first plot $(0, 9)$. We can think of the slope as $\dfrac{1}{1}$, so from the y-intercept we move up 1 unit and right 1 unit to the point $(1, 10)$. We plot $(1, 10)$ and draw the line passing through $(0, 9)$ and $(1, 10)$.

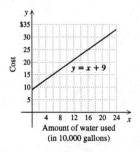

Since the slope is 1, the rate is $1 per 10,000 gallons.

52.

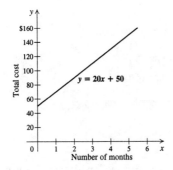

$20 per month

53. Two points on the graph are $(0, 16)$ and $(1, 16+1.5)$, or $(1, 17.5)$, so the y-intercept will be $(0, 16)$. Now we find the slope:
$$m = \frac{17.5 - 16}{1 - 0} = \frac{1.5}{1} = 1.5$$
Then the equation is $y = 1.5x + 16$.

54. $y = 0.07x + 4.95$

55. $y = \frac{2}{3}x + 7$: The slope is $\frac{2}{3}$, and the y-intercept is $(0, 7)$.

$y = \frac{2}{3}x - 5$: The slope is $\frac{2}{3}$, and the y-intercept is $(0, -5)$.

Since both lines have slope $\frac{2}{3}$ but different y-intercepts, their graphs are parallel.

56. No

57. The equation $y = 2x - 5$ represents a line with slope 2 and y-intercept $(0, -5)$. We rewrite the second equation in slope-intercept form.
$$4x + 2y = 9$$
$$2y = -4x + 9$$
$$y = \frac{1}{2}(-4x + 9)$$
$$y = -2x + \frac{9}{2}$$
The slope is -2 and the y-intercept is $\left(0, \frac{9}{2}\right)$. Since the lines have different slopes, their graphs are not parallel.

58. Yes

59. Rewrite each equation in slope-intercept form.
$$3x + 4y = 8$$
$$4y = -3x + 8$$
$$y = \frac{1}{4}(-3x + 8)$$
$$y = -\frac{3}{4}x + 2$$
The slope is $-\frac{3}{4}$, and the y-intercept is $(0, 2)$.
$$7 - 12y = 9x$$
$$-12y = 9x - 7$$
$$y = -\frac{1}{12}(9x - 7)$$
$$y = -\frac{3}{4}x + \frac{7}{12}$$
The slope is $-\frac{3}{4}$, and the y-intercept is $\left(0, \frac{7}{12}\right)$.

Since both lines have slope $-\frac{3}{4}$ but different y-intercepts, their graphs are parallel.

60. No

61. ◈

62. ◈

63. $y - k = m(x - h)$
$y = m(x - h) + k$ Adding k to both sides

64. $y = -2(x + 4) + 9$

65. $-5 - (-7) = -5 + 7 = 2$

66. 16

67. $-3 - 6 = -3 + (-6) = -9$

68. -10

69. ◈

70. ◈

71. Rewrite each equation in slope-intercept form.
$$3y = 5x - 3$$
$$y = \frac{1}{3}(5x - 3)$$
$$y = \frac{5}{3}x - 1$$
The slope is $\frac{5}{3}$.

$$3x + 5y = 10$$
$$5y = -3x + 10$$
$$y = \frac{1}{5}(-3x + 10)$$
$$y = -\frac{3}{5}x + 2$$

The slope is $-\frac{3}{5}$.

Since $\frac{5}{3}\left(-\frac{3}{5}\right) = -1$, the graphs of the equations are perpendicular.

72. Yes

73. Rewrite each equation in slope-intercept form.
$$3x + 5y = 10$$
$$5y = -3x + 10$$
$$y = \frac{1}{5}(-3x + 10)$$
$$y = -\frac{3}{5}x + 2$$

The slope is $-\frac{3}{5}$.
$$15x + 9y = 18$$
$$9y = -15x + 18$$
$$y = \frac{1}{9}(-15x + 18)$$
$$y = -\frac{5}{3}x + 2$$

The slope is $-\frac{5}{3}$.

Since $-\frac{3}{5}\left(-\frac{5}{3}\right) = 1 \neq -1$, the graphs of the equations are not perpendicular.

74. Yes

75. Since $x = 5$ represents a vertical line and $y = \frac{1}{2}$ represents a horizontal line, the graphs of the equations are perpendicular.

76. No

77. See the answer section in the text.

78. $y = -\frac{5}{2}x - \frac{10}{7}$

79. Rewrite each equation in slope-intercept form.
$$2x - 6y = 10$$
$$-6y = -2x + 10$$
$$y = \frac{1}{3}x - \frac{5}{3}$$

The slope of the line is $\frac{1}{3}$.

$$9x + 6y = 18$$
$$6y = -9x + 18$$
$$y = -\frac{3}{2}x + 3$$

The y-intercept of the line is $(0, 3)$.

The equation of the line is $y = \frac{1}{3}x + 3$.

80. $y = \frac{3}{2}x - 2$

81. Rewrite $2x + 5y = 6$ in slope-intercept form.
$$2x + 5y = 6$$
$$5y = -2x + 6$$
$$y = \frac{1}{5}(-2x + 6)$$
$$y = -\frac{2}{5}x + \frac{6}{5}$$

The slope is $-\frac{2}{5}$.

The slope of a line perpendicular to this line is a number m such that
$$-\frac{2}{5}m = -1, \text{ or}$$
$$m = \frac{5}{2}.$$

We graph the line whose equation we want to find. First we plot the given point $(2, 6)$. Now think of the slope as $\frac{-5}{-2}$. From the point $(2, 6)$ go down 5 units and left 2 units to the point $(0, 1)$. Plot this point and draw the graph.

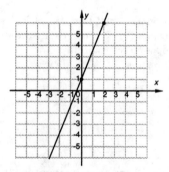

We see that the y-intercept is $(0, 1)$, so the desired equation is $y = \frac{5}{2}x + 1$.

82.

Exercise Set 3.7

1. $y - y_1 = m(x - x_1)$

We substitute 6 for m, 2 for x_1, and 7 for y_1.

$y - 7 = 6(x - 2)$

2. $y - 5 = 4(x - 3)$

3. $y - y_1 = m(x - x_1)$

We substitute $\frac{3}{5}$ for m, 9 for x_1, and 2 for y_1.

$y - 2 = \frac{3}{5}(x - 9)$

4. $y - 1 = \frac{2}{3}(x - 4)$

5. $y - y_1 = m(x - x_1)$

We substitute -4 for m, 3 for x_1, and 1 for y_1.

$y - 1 = -4(x - 3)$

6. $y - 2 = -5(x - 6)$

7. $y - y_1 = m(x - x_1)$

We substitute $\frac{3}{2}$ for m, 5 for x_1, and -4 for y_1.

$y - (-4) = \frac{3}{2}(x - 5)$

8. $y - (-1) = \frac{4}{3}(x - 7)$

9. $y - y_1 = m(x - x_1)$

We substitute $\frac{5}{4}$ for m, -2 for x_1, and 6 for y_1.

$y - 6 = \frac{5}{4}(x - (-2))$

10. $y - 4 = \frac{7}{2}(x - (-3))$

11. $y - y_1 = m(x - x_1)$

We substitute -2 for m, -4 for x_1, and -1 for y_1.

$y - (-1) = -2(x - (-4))$

12. $y - (-5) = -3(x - (-2))$

13. $y - y_1 = m(x - x_1)$

We substitute 1 for m, -2 for x_1, and 8 for y_1.

$y - 8 = 1(x - (-2))$

14. $y - 6 = -1(x - (-3))$

15. First we write the equation in point-slope form.

$y - y_1 = m(x - x_1)$

$y - 7 = 2(x - 5)$ Substituting

Next we find an equivalent equation of the form $y = mx + b$.

$y - 7 = 2(x - 5)$

$y - 7 = 2x - 10$

$y = 2x - 3$

16. $y = 3x - 16$

17. First we write the equation in point-slope form.

$y - y_1 = m(x - x_1)$

$y - (-2) = \frac{7}{4}(x - 4)$ Substituting

Next we find an equivalent equation of the form $y = mx + b$.

$y - (-2) = \frac{7}{4}(x - 4)$

$y + 2 = \frac{7}{4}x - 7$

$y = \frac{7}{4}x - 9$

18. $y = \frac{8}{3}x - 12$

19. First we write the equation in point-slope form.

$y - y_1 = m(x - x_1)$

$y - (-5) = -3(x - 1)$

Next we find an equivalent equation of the form $y = mx + b$.

$y - (-5) = -3(x - 1)$

$y + 5 = -3x + 3$

$y = -3x - 2$

20. $y = -2x + 5$

21. First we write the equation in point-slope form.

$y - y_1 = m(x - x_1)$

$y - (-1) = -4(x - (-2))$

Next we find an equivalent equation of the form $y = mx + b$.

$y - (-1) = -4(x - (-2))$

$y + 1 = -4(x + 2)$

$y + 1 = -4x - 8$

$y = -4x - 9$

22. $y = -5x - 9$

23. First we write the equation in point-slope form.

$$y - y_1 = m(x - x_1)$$
$$y - 5 = \frac{2}{3}(x - 6)$$

Next we find an equivalent equation of the form $y = mx + b$.

$$y - 5 = \frac{2}{3}(x - 6)$$
$$y - 5 = \frac{2}{3}x - 4$$
$$y = \frac{2}{3}x + 1$$

24. $y = \frac{3}{2}x + 1$

25. First we write the equation in point-slope form.

$$y - y_1 = m(x - x_1)$$
$$y - 2 = -\frac{5}{6}(x - 3)$$

Next we find an equivalent equation of the form $y = mx + b$.

$$y - 2 = -\frac{5}{6}(x - 3)$$
$$y - 2 = -\frac{5}{6}x + \frac{5}{2}$$
$$y = -\frac{5}{6}x + \frac{9}{2}$$

26. $y = -\frac{3}{4}x + \frac{13}{2}$

27. We plot $(1, 2)$, move up 4 and to the right 3 to $(4, 6)$ and draw the line.

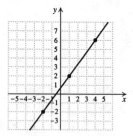

28.

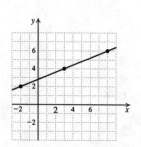

29. We plot $(2, 5)$, move down 3 and to the right 4 to $(6, 2)$ $\left(\text{since} -\frac{3}{4} = \frac{-3}{4} \right)$, and draw the line. We could also think of $-\frac{3}{4}$ and $\frac{3}{-4}$ and move up 3 and to the left 4 from the point $(2, 5)$ to $(-2, 8)$.

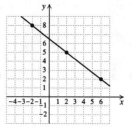

30.

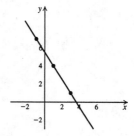

31. $y - 2 = \frac{1}{2}(x - 1)$ Point-slope form

The line has slope $\frac{1}{2}$ and passes through $(1, 2)$. We plot $(1, 2)$ and then find a second point by moving up 1 unit and right 2 units to $(3, 3)$. We draw the line through these points.

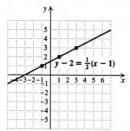

32.

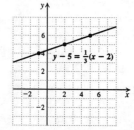

33. $y - 1 = -\dfrac{1}{2}(x - 3)$ Point-slope form

The line has slope $-\dfrac{1}{2}$, or $\dfrac{1}{-2}$ passes through $(3, 1)$. We plot $(3, 1)$ and then find a second point by moving up 1 unit and left 2 units to $(1, 2)$. We draw the line through these points.

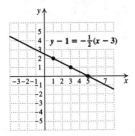

34.

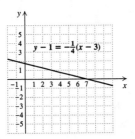

35. $y + 2 = \dfrac{1}{2}(x - 3)$, or $y - (-2) = \dfrac{1}{2}(x - 3)$

The line has slope $\dfrac{1}{2}$ and passes through $(3, -2)$. We plot $(3, -2)$ and then find a second point by moving up 1 unit and right 2 units to $(5, -1)$. We draw the line through these points.

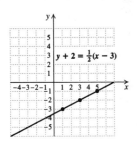

36.

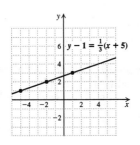

37. $y + 4 = 3(x + 1)$, or $y - (-4) = 3(x - (-1))$

The line has slope 3, or $\dfrac{3}{1}$, and passes through $(-1, -4)$. We plot $(-1, -4)$ and then find a second point by moving up 3 units and right 1 unit to $(0, -1)$. We draw the line through these points.

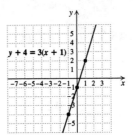

38.

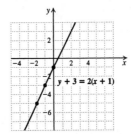

39. $y - 4 = -2(x + 1)$, or $y - 4 = -2(x - (-1))$

The line has slope -2, or $\dfrac{-2}{1}$, and passes through $(-1, 4)$. We plot $(-1, 4)$ and then find a second point by moving down 2 units and right 1 unit to $(0, 2)$. We draw the line through these points.

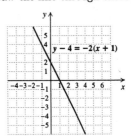

40.

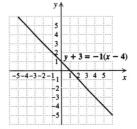

41. $y + 3 = -(x + 2)$, or $y - (-3) = -1(x - (-2))$

The line has slope -1, or $\dfrac{-1}{1}$, and passes through

$(-2, -3)$. We plot $(-2, -3)$ and then find a second point by moving down 1 unit and right 1 unit to $(-1, -4)$. We draw the line through these points.

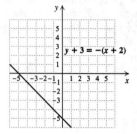

42.

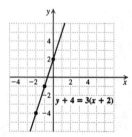

43. $y + 1 = -\dfrac{3}{5}(x + 2)$, or $y - (-1) = -\dfrac{3}{5}(x - (-2))$

The line has slope $-\dfrac{3}{5}$, or $\dfrac{-3}{5}$ and passes through $(-2, -1)$. We plot $(-2, -1)$ and then find a second point by moving down 3 units and right 5 units to $(3, -4)$, and draw the line.

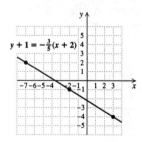

44.

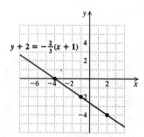

45. $y - 1 = -\dfrac{7}{2}(x + 5)$, or $y - 1 = -\dfrac{7}{2}(x - (-5))$

The line has slope $-\dfrac{7}{2}$, or $\dfrac{-7}{2}$ and passes through $(-5, 1)$. We plot $(-5, 1)$ and then find a second point

by moving down 7 units and right 2 units to $(-3, -6)$, and draw the line.

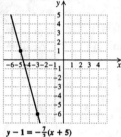

46.

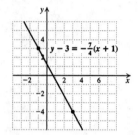

47. ◈

48. ◈

49. $(-5)^3 = (-5)(-5)(-5) = -125$

50. 64

51. $3 \cdot 2^4 - 5 \cdot 2^3$
    $= 3 \cdot 16 - 5 \cdot 8$ Evaluating the exponential expressions
    $= 48 - 40$ Multiplying
    $= 8$ Subtracting

52. 24

53. $(-2)^3(-3)^2 = -8 \cdot 9 = -72$

54. -4

55. ◈

56. ◈

57. $y - 3 = 0(x - 52)$

Observe that the slope is 0. Then this is the equation of a horizontal line that passes through $(52, 3)$. Thus, its graph is a horizontal line 3 units above the x-axis.

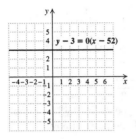

58.

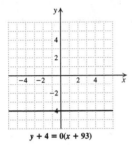

$y + 4 = 0(x + 93)$

59. First find the slope of the line passing through $(1, 2)$ and $(3, 7)$.
$$m = \frac{7 - 2}{3 - 1} = \frac{5}{2}$$
Then write an equation of the line containing $(1, 2)$ and having slope $\frac{5}{2}$.
$$y - 2 = \frac{5}{2}(x - 1)$$
We can also write an equation of the line containing $(3, 7)$ and having slope $\frac{5}{2}$.
$$y - 7 = \frac{5}{2}(x - 3)$$

60. $y - 1 = \frac{1}{2}(x - 3)$; $y - 3 = \frac{1}{2}(x - 7)$

61. First find the slope of the line passing through $(-1, 2)$ and $(3, 8)$.
$$m = \frac{2 - 8}{-1 - 3} = \frac{-6}{-4} = \frac{3}{2}$$
Then write an equation of the line containing $(-1, 2)$ and having slope $\frac{3}{2}$.
$$y - 2 = \frac{3}{2}(x - (-1))$$
We can also write an equation of the line containing $(3, 8)$ and having slope $\frac{3}{2}$.
$$y - 8 = \frac{3}{2}(x - 3)$$

62. $y - 1 = \frac{2}{7}(x - (-3))$; $y - 3 = \frac{2}{7}(x - 4)$

63. First find the slope of the line passing through $(-3, 8)$ and $(1, -2)$.
$$m = \frac{8 - (-2)}{-3 - 1} = \frac{10}{-4} = -\frac{5}{2}$$
Then write an equation of the line containing $(-3, 8)$ and having slope $-\frac{5}{2}$.
$$y - 8 = -\frac{5}{2}(x - (-3))$$
We can also write an equation of the line containing $(1, -2)$ and having slope $-\frac{5}{2}$.
$$y - (-2) = -\frac{5}{2}(x - 1)$$

64. $y - 7 = -\frac{5}{3}(x - (-2))$; $y - (-3) = -\frac{5}{3}(x - 4)$

65. First we find the slope of the line using any two points on the line. We will use $(3, -3)$ and $(4, -1)$.
$$m = \frac{-3 - (-1)}{3 - 4} = \frac{-2}{-1} = 2$$
Then we write an equation of the line in point-slope form using either of the points above.
$$y - (-3) = 2(x - 3)$$
Finally, we find an equivalent equation in slope-intercept form.
$$y - (-3) = 2(x - 3)$$
$$y + 3 = 2x - 6$$
$$y = 2x - 9$$

66. $y = -3x + 7$

67. First we find the slope of the line using any two points on the line. We will use $(2, 5)$ and $(5, 1)$.
$$m = \frac{5 - 1}{2 - 5} = \frac{4}{-3} = -\frac{4}{3}$$
Then we write an equation of the line in point-slope form using either of the points above.
$$y - 5 = -\frac{4}{3}(x - 2)$$
Finally, we find an equivalent equation in slope-intercept form.
$$y - 5 = -\frac{4}{3}(x - 2)$$
$$y - 5 = -\frac{4}{3}x + \frac{8}{3}$$
$$y = -\frac{4}{3}x + \frac{23}{3}$$

68. $y = \frac{5}{3}x + \frac{28}{3}$

69. $(1, 5)$ and $(4, 2)$

First we find the slope.
$$m = \frac{5 - 2}{1 - 4} = \frac{3}{-3} = -1$$
Then we write an equation of the line in point-slope form using either of the points above.
$$y - 5 = -1(x - 1)$$
Finally, we find an equivalent equation in slope-intercept form.
$$y - 5 = -1(x - 1)$$
$$y - 5 = -x + 1$$
$$y = -x + 6$$

70. $y = x + 4$

71. $(-3, 1)$ and $(3, 5)$

First we find the slope.
$$m = \frac{1 - 5}{-3 - 3} = \frac{-4}{-6} = \frac{2}{3}$$
Then we write an equation of the line in point-slope form using either of the points above.
$$y - 5 = \frac{2}{3}(x - 3)$$
Finally, we find an equivalent equation in slope-intercept form.
$$y - 5 = \frac{2}{3}(x - 3)$$
$$y - 5 = \frac{2}{3}x - 2$$
$$y = \frac{2}{3}x + 3$$

72. $y = \frac{1}{2}x + 4$

73. $(5, 0)$ and $(0, -2)$

First we find the slope.
$$m = \frac{0 - (-2)}{5 - 0} = \frac{2}{5}$$
Then we write an equation of the line in point-slope form using either of the points above.
$$y - 0 = \frac{2}{5}(x - 5)$$
Finally, we find an equivalent equation in slope-intercept form.
$$y - 0 = \frac{2}{5}(x - 5)$$
$$y = \frac{2}{5}x - 2$$

74. $y = \frac{3}{2}x + 3$

75. $(-2, -4)$ and $(2, -1)$

First we find the slope.
$$m = \frac{-4 - (-1)}{-2 - 2} = \frac{-4 + 1}{-2 - 2} = \frac{-3}{-4} = \frac{3}{4}$$
Then we write an equation of the line in point-slope form using either of the points above.
$$y - (-4) = \frac{3}{4}(x - (-2))$$
Finally, we find an equivalent equation in slope-intercept form.
$$y - (-4) = \frac{3}{4}(x - (-2))$$
$$y + 4 = \frac{3}{4}(x + 2)$$
$$y + 4 = \frac{3}{4}x + \frac{3}{2}$$
$$y = \frac{3}{4}x - \frac{5}{2}$$

76. $y = -4x - 7$

77. First find the slope of $2x + 3y = 11$.
$$2x + 3y = 11$$
$$3y = -2x + 11$$
$$y = -\frac{2}{3}x + \frac{11}{3}$$
The slope is $-\frac{2}{3}$.

Then write a point-slope equation of the line containing $(-4, 7)$ and having slope $-\frac{2}{3}$.
$$y - 7 = -\frac{2}{3}(x - (-4))$$

78. $y - (-1) = \frac{4}{5}(x - 3)$

79. The slope of $y = 3 - 4x$ is -4. We are given the y-intercept of the line, so we use slope-intercept form. The equation is $y = -4x + 7$.

80. $y = \frac{1}{5}x - 2$

81. First find the slope of the line passing through $(2, 7)$ and $(-1, -3)$.
$$m = \frac{-3 - 7}{-1 - 2} = \frac{-10}{-3} = \frac{10}{3}$$
Now find an equation of the line containing the point $(-1, 5)$ and having slope $\frac{10}{3}$.

$$y - 5 = \frac{10}{3}(x - (-1))$$
$$y - 5 = \frac{10}{3}(x + 1)$$
$$y - 5 = \frac{10}{3}x + \frac{10}{3}$$
$$y = \frac{10}{3}x + \frac{25}{3}$$

82. $y = \frac{1}{2}x + 1$

83. ◈

Chapter 4
Polynomials

1. $r^4 \cdot r^6 = r^{4+6} = r^{10}$

2. 8^7

3. $9^5 \cdot 9^3 = 9^{5+3} = 9^8$

4. n^{23}

5. $a^6 \cdot a = a^6 \cdot a^1 = a^{6+1} = a^7$

6. y^{16}

7. $5^7 \cdot 5^8 = 5^{7+8} = 5^{15}$

8. t^{16}

9. $(3y)^4(3y)^8 = (3y)^{4+8} = (3y)^{12}$

10. $(2t)^{25}$

11. $(5t)(5t)^6 = (5t)^1(5t)^6 = (5t)^{1+6} = (5t)^7$

12. $8x$

13. $(a^2b^7)(a^3b^2) = a^2b^7a^3b^2$ Using an associative law
$\qquad\qquad = a^2a^3b^7b^2$ Using a commutative law
$\qquad\qquad = a^5b^9$ Adding exponents

14. $(m-3)^9$

15. $(x+1)^5(x+1)^7 = (x+1)^{5+7} = (x+1)^{12}$

16. $a^{12}b^4$

17. $r^3 \cdot r^7 \cdot r^0 = r^{3+7+0} = r^{10}$

18. s^{11}

19. $(xy^4)(xy)^3 = (xy^4)(x^3y^3)$
$\qquad\quad = x \cdot x^3 \cdot y^4 \cdot y^3$
$\qquad\quad = x^{1+3}y^{4+3}$
$\qquad\quad = x^4y^7$

20. a^7b^5

21. $\dfrac{7^5}{7^2} = 7^{5-2} = 7^3$ Subtracting exponents

22. 4^4

23. $\dfrac{x^{15}}{x^3} = x^{15-3} = x^{12}$ Subtracting exponents

24. a^8

25. $\dfrac{t^5}{t} = \dfrac{t^5}{t^1} = t^{5-1} = t^4$

26. x^6

27. $\dfrac{(5a)^7}{(5a)^6} = (5a)^{7-6} = (5a)^1 = 5a$

28. $3m$

29. $\dfrac{(x+y)^8}{(x+y)^8}$

Observe that we have an expression divided by itself. Thus, the result is 1.

We could also do this exercise as follows:
$\dfrac{(x+y)^8}{(x+y)^8} = (x+y)^{8-8} = (x+y)^0 = 1$

30. $a-b$

31. $\dfrac{18m^5}{6m^2} = \dfrac{18}{6}m^{5-2} = 3m^3$

32. $5n^4$

33. $\dfrac{a^9b^7}{a^2b} = \dfrac{a^9}{a^2} \cdot \dfrac{b^7}{b^1} = a^{9-2}b^{7-1} = a^7b^6$

34. r^8s^6

35. $\dfrac{m^9n^8}{m^0n^4} = \dfrac{m^9}{m^0} \cdot \dfrac{n^8}{n^4} = m^{9-0}n^{8-4} = m^9n^4$

36. a^8b^{12}

37. When $x = 13$, $x^0 = 13^0 = 1$. (Any nonzero number raised to the 0 power is 1.)

38. 1

39. When $x = -4$, $5x^0 = 5(-4)^0 = 5 \cdot 1 = 5$.

40. 7

41. $8^0 + 5^0 = 1 + 1 = 2$

42. 1

43. $(-3)^1 - (-3)^0 = -3 - 1 = -4$

44. 5

45. $(x^4)^7 = x^{4 \cdot 7} = x^{28}$ Multiplying exponents

46. a^{24}

47. $(5^8)^2 = 5^{8 \cdot 2} = 5^{16}$ Multiplying exponents

48. 2^{15}, or 32,768

49. $(m^7)^5 = m^{7 \cdot 5} = m^{35}$

50. n^{18}

51. $(t^{20})^4 = t^{20 \cdot 4} = t^{80}$

52. t^{27}

53. $(7x)^2 = 7^2 \cdot x^2 = 49x^2$

54. $25a^2$

55. $(-2a)^3 = (-2)^3 a^3 = -8a^3$

56. $-27x^3$

57. $(4m^3)^2 = 4^2(m^3)^2 = 16m^6$

58. $25n^8$

59. $(a^2b)^7 = (a^2)^7(b^7) = a^{14}b^7$

60. $x^9 y^{36}$

61. $(x^3y)^2(x^2y^5) = (x^3)^2 y^2 x^2 y^5 = x^6 y^2 x^2 y^5 = x^8 y^7$

62. $a^{14}b^{11}$

63. $(2x^5)^3(3x^4) = 2^3(x^5)^3(3x^4) = 8x^{15} \cdot 3x^4 = 24x^{19}$

64. $50x^{13}$

65. $\left(\dfrac{a}{4}\right)^3 = \dfrac{a^3}{4^3} = \dfrac{a^3}{64}$ Raising the numerator and the denominator to the third power

66. $\dfrac{81}{x^4}$

67. $\left(\dfrac{7}{5a}\right)^2 = \dfrac{7^2}{(5a)^2} = \dfrac{49}{5^2 a^2} = \dfrac{49}{25a^2}$

68. $\dfrac{125x^3}{8}$

69. $\left(\dfrac{a^4}{b^3}\right)^5 = \dfrac{(a^4)^5}{(b^3)^5} = \dfrac{a^{20}}{b^{15}}$

70. $\dfrac{x^{35}}{y^{14}}$

71. $\left(\dfrac{y^3}{2}\right)^2 = \dfrac{(y^3)^2}{2^2} = \dfrac{y^6}{4}$

72. $\dfrac{a^{15}}{8}$

73. $\left(\dfrac{x^2 y}{z^3}\right)^4 = \dfrac{(x^2 y)^4}{(z^3)^4} = \dfrac{(x^2)^4(y^4)}{z^{12}} = \dfrac{x^8 y^4}{z^{12}}$

74. $\dfrac{x^{15}}{y^{10} z^5}$

75. $\left(\dfrac{a^3}{-2b^5}\right)^4 = \dfrac{(a^3)^4}{(-2b^5)^4} = \dfrac{a^{12}}{(-2)^4(b^5)^4} = \dfrac{a^{12}}{16b^{20}}$

76. $\dfrac{x^{20}}{81y^{12}}$

77. $\left(\dfrac{5x^7 y}{2z^4}\right)^3 = \dfrac{(5x^7 y)^3}{(2z^4)^3} = \dfrac{5^3(x^7)^3 y^3}{2^3(z^4)^3} =$ $\dfrac{125x^{21} y^3}{8z^{12}}$

78. $\dfrac{64a^6 b^3}{27c^{21}}$

79. $\left(\dfrac{4x^3 y^5}{3z^7}\right)^0$

Observe that for $x \neq 0$, $y \neq 0$, and $z \neq 0$, we have a nonzero number raised to the 0 power. Thus, the result is 1.

80. 1

81. ◈

82. ◈

83. $3s - 3r + 3t = 3 \cdot s - 3 \cdot r + 3 \cdot t = 3(s - r + t)$

84. $-7(x - y + z)$

85. $9x + 2y - x - 2y = 9x - x + 2y - 2y =$ $(9 - 1)x + (2 - 2)y = 8x + 0y = 8x$

86. $-3a - 6b$

87. $2y + 3x$

88. $5z + 2xy$

89. ◈

90. ◈

91. ◈

92. ◈

93. Choose any number except 0.

For example, let $a = 1$. Then $(a + 5)^2 = (1 + 5)^2 = 6^2 = 36$, but $a^2 + 5^2 = 1^2 + 5^2 = 1 + 25 = 26$.

94. Choose any number except 0. For example, let $x = 1$.

$$3x^2 = 3 \cdot 1^2 = 3 \cdot 1 = 3, \text{ but}$$
$$(3x)^2 = (3 \cdot 1)^2 = 3^2 = 9.$$

95. Choose any number except $\frac{7}{6}$. For example let $a = 0$.

Then $\frac{0 + 7}{7} = \frac{7}{7} = 1$, but $a = 0$.

96. Choose any number except 0 or 1. For example, let $t = -1$. Then $\frac{t^6}{t^2} = \frac{(-1)^6}{(-1)^2} = \frac{1}{1} = 1$, but $t^3 = (-1)^3 = -1$.

97. $a^{10k} \div a^{2k} = a^{10k-2k} = a^{8k}$

98. y^{6x}

99.

$$\frac{\left(\frac{1}{2}\right)^3 \left(\frac{2}{3}\right)^4}{\left(\frac{5}{6}\right)^3} = \frac{\frac{1}{8} \cdot \frac{16}{81}}{\frac{125}{216}} = \frac{1}{8} \cdot \frac{16}{81} \cdot \frac{216}{125} =$$

$$\frac{1 \cdot 2 \cdot \cancel{8} \cdot \cancel{27} \cdot 8}{\cancel{8} \cdot 3 \cdot \cancel{27} \cdot 125} = \frac{16}{375}$$

100. x^t

101.
$$\frac{t^{26}}{t^x} = t^x$$
$$t^{26-x} = t^x$$
$$26 - x = x \quad \text{Equating exponents}$$
$$26 = 2x$$
$$13 = x$$

The solution is 13.

102. $3^5 > 3^4$

103. Since the bases are the same, the expression with the larger exponent is larger. Thus, $4^2 < 4^3$.

104. $4^3 < 5^3$

105. $4^3 = 64$, $3^4 = 81$, so $4^3 < 3^4$.

106. $9^7 > 3^{13}$

107. $25^8 = (5^2)^8 = 5^{16}$
$125^5 = (5^3)^5 = 5^{15}$
$5^{16} > 5^{15}$, or $25^8 > 125^5$.

108. 16,000; 16,384; 384

109. $2^{22} = 2^{10} \cdot 2^{10} \cdot 2^2 \approx 10^3 \cdot 10^3 \cdot 4 \approx 1000 \cdot 1000 \cdot 4 \approx 4,000,000$

Using a calculator, we find that $2^{22} = 4,194,304$. The difference between the exact value and the approximation is $4,194,304 - 4,000,000$, or $194,304$.

110. 64,000,000; 67,108,864; 3,108,864

111. $2^{31} = 2^{10} \cdot 2^{10} \cdot 2^{10} \cdot 2 \approx 10^3 \cdot 10^3 \cdot 10^3 \cdot 2 \approx 1000 \cdot 1000 \cdot 1000 \cdot 2 = 2,000,000,000$

Using a calculator, we find that $2^{31} = 2,147,483,648$. The difference between the exact value and the approximation is $2,147,483,648 - 2,000,000,000 = 147,483,648$.

112. 57,344 bytes

113. 64 K = $64 \times 1 \times 2^{10}$ bytes = $65,536$ bytes

Exercise Set 4.2

1. $7x^4 + x^3 - 5x + 8 = 7x^4 + x^3 + (-5x) + 8$
The terms are $7x^4$, x^3, $-5x$, and 8.

2. $5a^3$, $4a^2$, $-a$, -7

3. $-t^4 + 7t^3 - 3t^2 + 6 = -t^4 + 7t^3 + (-3t^2) + 6$
The terms are $-t^4$, $7t^3$, $-3t^2$, and 6.

4. n^5, $-4n^3$, $2n$, -8

5. $4x^5 + 7x$

Term	Coefficient	Degree
$4x^5$	4	5
$7x$	7	1

6.

Term	Coefficient	Degree
$9a^3$	9	3
$-4a^2$	-4	2

7. $9t^2 - 3t + 4$

Term	Coefficient	Degree
$9t^2$	9	2
$-3t$	-3	1
4	4	0

8.

Term	Coefficient	Degree
$7x^4$	7	4
$5x$	5	1
-3	-3	0

9. $7a^4 + 9a + a^3$

Term	Coefficient	Degree
$7a^4$	7	4
$9a$	9	1
a^3	1	3

10.

Term	Coefficient	Degree
$6t^5$	6	5
$-3t^2$	-3	2
$-t$	-1	1

11. $x^4 - x^3 + 4x - 3$

Term	Coefficient	Degree
x^4	1	4
$-x^3$	-1	3
$4x$	4	1
-3	-3	0

12.

Term	Coefficient	Degree
$3a^4$	3	4
$-a^3$	-1	3
a	1	1
-9	-9	0

13. $2a^3 + 7a^5 + a^2$

a)

Term	$2a^3$	$7a^5$	a^2
Degree	3	5	2

b) The term of highest degree is $7a^5$. This is the leading term. Then the leading coefficient is 7.

c) Since the term of highest degree is $7a^5$, the degree of the polynomial is 5.

14. a)

Term	$5x$	$-9x^2$	$3x^6$
Degree	1	2	6

b) $3x^6$; 3

c) 6

15. $2t + 3 + 4t^2$

a)

Term	$2t$	3	$4t^2$
Degree	1	0	2

b) The term of highest degree is $4t^2$. This is the leading term. Then the leading coefficient is 4.

c) Since the term of highest degree is $4t^2$, the degree of the polynomial is 2.

16. a)

Term	$3a^2$	-7	$2a^4$
Degree	2	0	4

b) $2a^4$; 2

c) 4

17. $9x^4 + x^2 + x^7 + 4$

a)

Term	$9x^4$	x^2	x^7	4
Degree	4	2	7	0

b) The term of highest degree is x^7. This is the leading term. Then the leading coefficient is 1.

c) Since the term of highest degree is x^7, the degree of the polynomial is 7.

18. a)

Term	8	$6x^2$	$-3x$	$-x^5$
Degree	0	2	1	5

b) $-x^5$; -1

c) 5

19. $9a - a^4 + 3 + 2a^3$

a)

Term	$9a$	$-a^4$	3	$2a^3$
Degree	1	4	0	3

b) The term of highest degree is $-a^4$. This is the leading term. Then the leading coefficient is -1.

c) Since the term of highest degree is $-a^4$, the degree of the polynomial is 4.

20. a)

Term	$-x$	$2x^5$	$-5x^2$	x^6
Degree	1	5	2	6

b) x^6; 1

c) 6

21. $7x^2 + 8x^5 - 4x^3 + 6 - \dfrac{1}{2}x^4$

Term	Coefficient	Degree of Term	Degree of Polynomial
$8x^5$	8	5	
$-\dfrac{1}{2}x^4$	$-\dfrac{1}{2}$	4	
$-4x^3$	-4	3	5
$7x^2$	7	2	
6	6	0	

22.

Term	Coefficient	Degree of Term	Degree of Polynomial
$-3x^4$	-3	4	
$6x^3$	6	3	
$-2x^2$	-2	2	4
$8x$	8	1	
7	7	0	

23. Three monomials are added, so $x^2 - 23x + 17$ is a trinomial.

24. Monomial

25. The polynomial $x^3 - 7x^2 + 2x - 4$ is none of these because it is composed of four monomials.

26. Binomial

27. Two monomials are added, so $8t^2 + 5t$ is a binomial.

28. Trinomial

29. The polynomial 17 is a monomial because it is the product of a constant and a variable raised to a whole number power. (In this case the variable is raised to the power 0.)

30. None of these

31. $7x^2 + 3x + 4x^2 = (7 + 4)x^2 + 3x = 11x^2 + 3x$

32. $7a^2 + 8a$

33. $3a^4 - 2a + 2a + a^4 = (3 + 1)a^4 + (-2 + 2)a = $
$4a^4 + 0a = 4a^4$

34. $7b^5$

35. $2x^2 - 6x + 3x + 4x^2 = (2 + 4)x^2 + (-6 + 3)x = $
$6x^2 - 3x$

36. $4x^4 - 9x$

37. $9x^3 + 2x - 4x^3 + 5 - 3x = (9 - 4)x^3 + (2 - 3)x + 5 = $
$5x^3 - x + 5$

38. x^4

39. $10x^2 + 2x^3 - 3x^3 - 4x^2 - 6x^2 - x^4 = $
$-x^4 + (2 - 3)x^3 + (10 - 4 - 6)x^2 = -x^4 - x^3$

40. $-x^6 + 10x^5$

41. $\dfrac{1}{5}x^4 + 7 - 2x^2 + 3 - \dfrac{2}{15}x^4 + 2x^2 = $
$\left(\dfrac{1}{5} - \dfrac{2}{15}\right)x^4 + (-2 + 2)x^2 + (7 + 3) = $
$\left(\dfrac{3}{15} - \dfrac{2}{15}\right)x^4 + 0x^2 + 10 = \dfrac{1}{15}x^4 + 10$

42. $-\dfrac{1}{6}x^3 + 4x^2 - 3$

43. $5.9x^2 - 2.1x + 6 + 3.4x - 2.5x^2 - 0.5 = $
$(5.9 - 2.5)x^2 + (-2.1 + 3.4)x + (6 - 0.5) = $
$3.4x^2 + 1.3x + 5.5$

44. $9.3x^3 - 8.4x - 1.4$

45. $6t - 9t^3 + 8t^4 + 4t + 2t^4 + 7t - 3t^3 = $
$(8 + 2)t^4 + (-9 - 3)t^3 + (6 + 4 + 7)t = $
$10t^4 - 12t^3 + 17t$

46. $6b^3 + 3b^2 + b$

47. $-7x + 5 = -7 \cdot 3 + 5$
$= -21 + 5$
$= -16$

48. -6

49. $2x^2 - 3x + 7 = 2 \cdot 3^2 - 3 \cdot 3 + 7$
$= 2 \cdot 9 - 3 \cdot 3 + 7$
$= 18 - 9 + 7$
$= 16$

50. 27

51. $5x + 7 = 5(-2) + 7$
$= -10 + 7$
$= -3$

52. 13

53. $x^2 - 3x + 1 = (-2)^2 - 3(-2) + 1$
$$= 4 - 3(-2) + 1$$
$$= 4 + 6 + 1$$
$$= 11$$

54. -15

55. $\quad -3x^3 + 7x^2 - 4x - 5$
$$= -3(-2)^3 + 7(-2)^2 - 4(-2) - 5$$
$$= -3(-8) + 7 \cdot 4 - 4(-2) - 5$$
$$= 24 + 28 + 8 - 5$$
$$= 55$$

56. -5

57. Locate 10 on the horizontal axis. From there move vertically to the graph and then horizontally to the M-axis. This locates an M-value of about 9. Thus, about 9 words were memorized in 10 minutes.

58. About 17

59. Locate 8 on the horizontal axis. From there move vertically to the graph and then horizontally to the M-axis. This locates an M-value of about 6. Thus, the value of $-0.001t^3 + 0.1t^2$ for $t = 8$ is approximately 6.

60. About 13

61. Locate 13 on the horizontal axis. It is halfway between 12 and 14. From there move vertically to the graph and then horizontally to the M-axis. This locates an M-value of about 15. Thus, the value of $-0.001t^3 + 0.1t^2$ when t is 13 is approximately 15.

62. About 4.5

63. $11.12t^2 = 11.12(10)^2 = 11.12(100) = 1112$

A skydiver has fallen approximately 1112 ft 10 seconds after jumping from a plane.

64. 3091 ft

65. $0.4r^2 - 40r + 1039 = 0.4(18)^2 - 40(18) + 1039 =$
$0.4(324) - 720 + 1039 = 129.6 - 720 + 1039 =$
448.6

There are approximately 449 accidents daily involving an 18-year-old driver.

66. 399

67. Evaluate the polynomial for $x = 40$:
$250x - 0.5x^2 = 250(40) - 0.5(40)^2 =$
$10,000 - 800 = 9200$
The total revenue is $9200.

68. $13,200

69. Evaluate the polynomial for $x = 200$:
$4000 + 0.6x^2 = 4000 + 0.6(200)^2 =$
$4000 + 0.6(40,000) = 4000 + 24,000 = 28,000$
The total cost is $28,000.

70. $58,000

71. $2\pi r = 2(3.14)(10)$ Substituting 3.14 for π
$\qquad\qquad\qquad\qquad$ and 10 for r
$\quad = 62.8$
The circumference is 62.8 cm.

72. 31.4 ft

73. $\pi r^2 = 3.14(7)^2$ Substituting 3.14 for π
$\qquad\qquad\qquad\quad$ and 7 for r
$\quad = 3.14(49)$
$\quad = 153.86$
The area is 153.86 m^2.

74. 113.04 ft^2

75. ◈

76. ◈

77. $-19 + 24$ A negative and a positive number. We subtract the absolute values: $24 - 19 = 5$. The positive number has the larger absolute value so the answer is positive.
$\quad -19 + 24 = 5$

78. -9

79. $5x + 15 = 5 \cdot x + 5 \cdot 3 = 5(x + 3)$

80. $7(a - 3)$

81. ***Familiarize.*** Let $x =$ the cost per mile of gasoline in dollars. Then the total cost of the gasoline for the year was $14,800x$.

Translate.

$$\underbrace{\text{Cost of insurance}} + \underbrace{\text{cost of registration and oil}} + \underbrace{\text{cost of gasoline}} = \$2011.$$
$$\downarrow \quad \downarrow \qquad \downarrow \qquad\quad \downarrow \quad\downarrow \qquad \downarrow \quad \downarrow$$
$$972 \quad + \qquad 114 \quad + \quad 14,800x \; = \; 2011$$

Carry out. We solve the equation.
$$972 + 114 + 14,800x = 2011$$
$$1086 + 14,800x = 2011$$
$$14,800x = 925$$
$$x = 0.0625$$

Check. If gasoline cost $0.0625 per mile, then the total cost of the gasoline was $14,800(\$0.0625)$, or $925. Then the total auto expense was $972+$114+ $925, or $2011. The answer checks.

State. Gasoline cost $0.0625, or 6.25¢ per mile.

82. 274 and 275

83. ◈

84. ◈

85. Answers may vary. Use an ax^5-term, where a is an integer, and 3 other terms with different degrees, each less than degree 5, and integer coefficients. Three answers are $-6x^5 + 14x^4 - x^2 + 11$, $x^5 - 8x^3 + 3x + 1$, and $23x^5 + 2x^4 - x^2 + 5x$.

86. Answers may vary; $0.2y^4 - y + \dfrac{5}{2}$

87. $(5m^5)^2 = 5^2 m^{5 \cdot 2} = 25m^{10}$

The degree is 10.

88. Answers may vary; $9y^4$, $-\dfrac{3}{2}y^4$, $4.2y^4$

89.
$$\frac{9}{2}x^8 + \frac{1}{9}x^2 + \frac{1}{2}x^9 + \frac{9}{2}x + \frac{9}{2}x^9 + \frac{8}{9}x^2 +$$
$$\frac{1}{2}x - \frac{1}{2}x^8$$
$$= \left(\frac{1}{2} + \frac{9}{2}\right)x^9 + \left(\frac{9}{2} - \frac{1}{2}\right)x^8 + \left(\frac{1}{9} + \frac{8}{9}\right)x^2 +$$
$$\left(\frac{9}{2} + \frac{1}{2}\right)x$$
$$= \frac{10}{2}x^9 + \frac{8}{2}x^8 + \frac{9}{9}x^2 + \frac{10}{2}x$$
$$= 5x^9 + 4x^8 + x^2 + 5x$$

90. $3x^6$

91. Let c = the coefficient of x^3. Solve:
$$c + (c-3) + 3(c-3) + (c+2) = -4$$
$$c + c - 3 + 3c - 9 + c + 2 = -4$$
$$6c - 10 = -4$$
$$6c = 6$$
$$c = 1$$

Coefficient of x^3, c: 1

Coefficient of x^2, $c - 3$: $1 - 3$, or -2

Coefficient of x, $3(c-3)$: $3(1-3)$, or -6

Coefficient remaining (constant term), $c + 2$: $1 + 2$, or 3

The polynomial is $x^3 - 2x^2 - 6x + 3$.

92.

d	$-0.0064d^2 + 0.8d + 2$
0	2
30	20.24
60	26.96
90	22.16
120	5.84

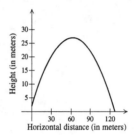

Height (in meters) / Horizontal distance (in meters)

93. We first find q, the quiz average, and t, the test average.
$$q = \frac{60 + 85 + 72 + 91}{4} = \frac{308}{4} = 77$$
$$t = \frac{89 + 93 + 90}{3} = \frac{272}{3} \approx 90.7$$
Now we substitute in the polynomial.
$$A = 0.3q + 0.4t + 0.2f + 0.1h$$
$$= 0.3(77) + 0.4(90.7) + 0.2(84) + 0.1(88)$$
$$= 23.1 + 36.28 + 16.8 + 8.8$$
$$= 84.98$$
$$\approx 85.0$$

94. 84.1

95. Using a calculator, evaluate $0.4r^2 - 40r + 1039$ for $r = 10, 20, 30, 40, 50, 60,$ and 70 and list the values in a table.

Age	Average number of accidents per day
r	$0.4r^2 - 40r + 1039$
10	679
20	399
30	199
40	79
50	39
60	79
70	199

The numbers in the chart increase both below and above age 50. We would assume the number of accidents is the smallest near age 50. Now we evaluate for 49 and 51.

49	39.4
50	39
51	39.4

Again the numbers increase below and above 50.

We conclude that the smallest number of daily accidents occurs at age 50.

96.

t	$-t^2 + 10t - 18$
3	3
4	6
5	7
6	6
7	3

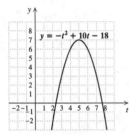

97. When $t = 1$, $-t^2 + 6t - 4 = -1^2 + 6 \cdot 1 - 4 =$
$-1 + 6 - 4 = 1$.
When $t = 2$, $-t^2 + 6t - 4 = -2^2 + 6 \cdot 2 - 4 =$
$-4 + 12 - 4 = 4$.
When $t = 3$, $-t^2 + 6t - 4 = -3^2 + 6 \cdot 3 - 4 =$
$-9 + 18 - 4 = 5$.
When $t = 4$, $-t^2 + 6t - 4 = -4^2 + 6 \cdot 4 - 4 =$
$-16 + 24 - 4 = 4$.
When $t = 5$, $-t^2 + 6t - 4 = -5^2 + 6 \cdot 5 - 4 =$
$-25 + 30 - 4 = 1$.

We complete the table. Then we plot the points and connect them with a smooth curve.

t	$-t^2 + 6t - 4$
1	1
2	4
3	5
4	4
5	1

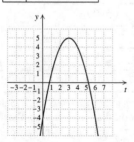

Exercise Set 4.3

1. $(2x + 3) + (-7x + 6) = (2 - 7)x + (3 + 6) = -5x + 9$

2. $-4x + 5$

3. $(-6x + 2) + (x^2 + x - 3) =$
$x^2 + (-6 + 1)x + (2 - 3) = x^2 - 5x - 1$

4. $x^2 + 3x - 5$

5. $(7t^2 - 3t + 6) + (2t^2 + 8t - 9) =$
$(7 + 2)t^2 + (-3 + 8)t + (6 - 9) = 9t^2 + 5t - 3$

6. $15a^2 + a - 6$

7. $(2m^3 - 4m^2 + m - 7) + (4m^3 + 7m^2 - 4m - 2) =$
$(2 + 4)m^3 + (-4 + 7)m^2 + (1 - 4)m + (-7 - 2) =$
$6m^3 + 3m^2 - 3m - 9$

8. $7n^3 - 5n^2 + 7n - 7$

9. $(3 + 6a + 7a^2 + 8a^3) + (4 + 7a - a^2 + 6a^3) =$
$(3 + 4) + (6 + 7)a + (7 - 1)a^2 + (8 + 6)a^3 =$
$7 + 13a + 6a^2 + 14a^3$

10. $9 + 5t + t^2 + 2t^3$

11. $(9x^8 - 7x^4 + 2x^2 + 5) + (8x^7 + 4x^4 - 2x) =$
$9x^8 + 8x^7 + (-7 + 4)x^4 + 2x^2 - 2x + 5 =$
$9x^8 + 8x^7 - 3x^4 + 2x^2 - 2x + 5$

12. $4x^5 + 9x^2 + 1$

13. $\left(\frac{1}{4}x^4 + \frac{2}{3}x^3 + \frac{5}{8}x^2 + 7\right) + \left(-\frac{3}{4}x^4 + \frac{3}{8}x^2 - 7\right) =$
$\left(\frac{1}{4} - \frac{3}{4}\right)x^4 + \frac{2}{3}x^3 + \left(\frac{5}{8} + \frac{3}{8}\right)x^2 + (7 - 7) =$
$-\frac{2}{4}x^4 + \frac{2}{3}x^3 + \frac{8}{8}x^2 + 0 =$
$-\frac{1}{2}x^4 + \frac{2}{3}x^3 + x^2$

14. $\frac{2}{15}x^9 - \frac{2}{5}x^5 + \frac{1}{4}x^4 - \frac{1}{2}x^2 + 7$

15. $(5.3t^2 - 6.4t - 9.1) + (4.2t^3 - 1.8t^2 + 7.3) =$
$4.2t^3 + (5.3 - 1.8)t^2 - 6.4t + (-9.1 + 7.3) =$
$4.2t^3 + 3.5t^2 - 6.4t - 1.8$

16. $4.9a^3 + 5.3a^2 - 8.8a + 4.6$

17. $-3x^4 + 6x^2 + 2x - 1$
$\quad\;\; - 3x^2 + 2x + 1$
___
$\quad -3x^4 + 3x^2 + 4x + 0$
$\quad -3x^4 + 3x^2 + 4x$

18. $-4x^3 + 4x^2 + 6x$

19. Rewrite the problem so the coefficients of like terms have the same number of decimal places.

$\quad\;\; 0.15x^4 + 0.10x^3 - 0.90x^2$
$\quad\qquad\quad\; - 0.01x^3 + 0.01x^2 + x$
$\quad\; 1.25x^4 \qquad\quad + 0.11x^2 \qquad + 0.01$
$\quad\qquad\; 0.27x^3 \qquad\qquad\qquad\; + 0.99$
$\quad -0.35x^4 \qquad\quad\; + 15.00x^2 \qquad - 0.03$
___
$\quad 1.05x^4 + 0.36x^3 + 14.22x^2 + x + 0.97$

20. $1.3x^4 + 0.35x^3 + 9.53x^2 + 2x + 0.96$

21. Two forms of the opposite of $-t^3 + 4t^2 - 9$ are

i) $-(-t^3 + 4t^2 - 9)$ and

ii) $t^3 - 4t^2 + 9$. (Changing the sign of every term)

22. $-(-4x^3 - 5x^2 + 2x),\; 4x^3 + 5x^2 - 2x$

23. Two forms for the opposite of $12x^4 - 3x^3 + 3$ are

i) $-(12x^4 - 3x^3 + 3)$ and

ii) $-12x^4 + 3x^3 - 3$. (Changing the sign of every term)

24. $-(5a^3 + 2a - 17),\; -5a^3 - 2a + 17$

25. We change the sign of every term inside parentheses.
$-(8x - 9) = -8x + 9$

26. $6x - 5$

27. We change the sign of every term inside parentheses.
$-(3a^4 - 5a^2 + 9) = -3a^4 + 5a^2 - 9$

28. $6a^3 - 2a^2 + 7$

29. We change the sign of every term inside parentheses.
$-\left(-4x^4 + 6x^2 + \frac{3}{4}x - 8\right) = 4x^4 - 6x^2 - \frac{3}{4}x + 8$

30. $5x^4 - 4x^3 + x^2 - 0.9$

31. $\quad (7x + 4) - (-2x + 1)$
$= 7x + 4 + 2x - 1 \quad$ Changing the sign of every term inside parentheses
$= 9x + 3$

32. $7x + 2$

33. $(-5t + 4) - (t^2 + 2t - 1) = -5t + 4 - t^2 - 2t + 1 = -t^2 - 7t + 5$

34. $-2a^2 - 7a + 6$

35. $\quad (6x^4 + 3x^3 - 1) - (4x^2 - 3x + 3)$
$= 6x^4 + 3x^3 - 1 - 4x^2 + 3x - 3$
$= 6x^4 + 3x^3 - 4x^2 + 3x - 4$

36. $-3x^3 + x^2 + 2x - 3$

37. $\quad (1.2x^3 + 4.5x^2 - 3.8x) - (-3.4x^3 - 4.7x^2 + 23)$
$= 1.2x^3 + 4.5x^2 - 3.8x + 3.4x^3 + 4.7x^2 - 23$
$= 4.6x^3 + 9.2x^2 - 3.8x - 23$

38. $-1.8x^4 - 0.6x^2 - 1.8x + 4.6$

39. $(7x^3 - 2x^2 + 6) - (7x^3 - 2x^2 + 6)$

Observe that we are subtracting the polynomial $7x^3 - 2x^2 + 6$ from itself. The result is 0.

40. x

41. $\quad (6 + 5a + 3a^2 - a^3) - (2 + 3a - 4a^2 + 2a^3) =$
$6 + 5a + 3a^2 - a^3 - 2 - 3a + 4a^2 - 2a^3 =$
$4 + 2a + 7a^2 - 3a^3$

42. $6 - t - t^2 - 3t^3$

43. $\quad \frac{5}{8}x^3 - \frac{1}{4}x - \frac{1}{3} - \left(-\frac{1}{8}x^3 + \frac{1}{4}x - \frac{1}{3}\right)$
$= \frac{5}{8}x^3 - \frac{1}{4}x - \frac{1}{3} + \frac{1}{8}x^3 - \frac{1}{4}x + \frac{1}{3}$
$= \frac{6}{8}x^3 - \frac{2}{4}x$
$= \frac{3}{4}x^3 - \frac{1}{2}x$

44. $\frac{3}{5}x^3 - \frac{307}{1000}$

45. $(0.07t^3 - 0.03t^2 + 0.01t) - (0.02t^3 + 0.04t^2 - 1) =$
$0.07t^3 - 0.03t^2 + 0.01t - 0.02t^3 - 0.04t^2 + 1 =$
$0.05t^3 - 0.07t^2 + 0.01t + 1$

46. $-0.7a^4 + 0.9a^3 + 0.5a - 4.9$

47. $\quad x^2 + 5x + 6$
$\quad -(x^2 + 2x + 1)$
___
$\quad x^2 + 5x + 6 \quad$ Changing signs and
$\quad -x^2 - 2x - 1 \quad$ removing parentheses
___
$\quad\qquad 3x + 5 \quad$ Adding

48. $2x^2 + 6$

49. $5x^4 + 6x^3 - 9x^2$
            $-(-6x^4 - 6x^3 + x^2)$

 $5x^4 + 6x^3 - 9x^2$ Changing signs and
 $\underline{6x^4 + 6x^3 - x^2}$ removing parentheses
 $11x^4 + 12x^3 - 10x^2$ Adding

50. $-2x^4 - 8x^3 - x^2$

51. a)

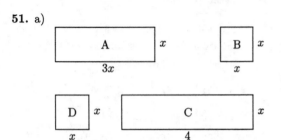

Familiarize. The area of a rectangle is the product of the length and the width.

Translate. The sum of the areas is found as follows:

$$\begin{array}{cccccccc} \text{Area} & & \text{Area} & & \text{Area} & & \text{Area} \\ \text{of } A & + & \text{of } B & + & \text{of } C & + & \text{of } D \\ = 3x \cdot x & + & x \cdot x & + & 4 \cdot x & + & x \cdot x \end{array}$$

Carry out. We collect like terms.

$$3x^2 + x^2 + 4x + x^2 = 5x^2 + 4x$$

Check. We can go over our calculations. We can also assign some value to x, say 2, and carry out the computation of the area in two ways.

 Sum of areas: $3 \cdot 2 \cdot 2 + 2 \cdot 2 + 4 \cdot 2 + 2 \cdot 2 =$
 $12 + 4 + 8 + 4 = 28$

Substituting in the polynomial:
 $5(2)^2 + 4 \cdot 2 = 20 + 8 = 28$

Since the results are the same, our solution is probably correct.

State. A polynomial for the sum of the areas is $5x^2 + 4x$.

b)   For $x = 5$: $5x^2 + 4x = 5 \cdot 5^2 + 4 \cdot 5 =$
 $5 \cdot 25 + 4 \cdot 5 = 125 + 20 = 145$

 When $x = 5$, the sum of the areas is 145 square units.
 For $x = 7$: $5x^2 + 4x = 5 \cdot 7^2 + 4 \cdot 7 =$
 $5 \cdot 49 + 4 \cdot 7 = 245 + 28 = 273$

 When $x = 7$, the sum of the areas is 273 square units.

52. a) $r^2\pi + 13\pi$

 b) $38\pi; = 140.69\pi$

53.

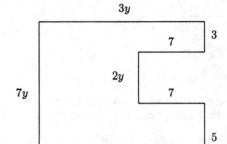

Familiarize. The perimeter is the sum of the lengths of the sides.

Translate. The sum of the lengths is found as follows:

$3y + 7y + (2y + 3) + 5 + 7 + 2y + 7 + 3$

Carry out. We collect like terms.

$(3 + 7 + 2 + 2)y + (3 + 5 + 7 + 7 + 3) = 14y + 25$

Check. We can go over our calculations. We can also assign some value to y, say 3, and carry out the computation of the perimeter in two ways.

Sum of lengths: $3 \cdot 3 + 7 \cdot 3 + (2 \cdot 3 + 3) + 5 + 7 + 2 \cdot 3 + 7 + 3 =$
 $9 + 21 + 9 + 5 + 7 + 6 + 7 + 3 = 67$

Substituting in the polynomial:
 $14 \cdot 3 + 25 = 42 + 25 = 67$

Since the results are the same, our solution is probably correct.

State. A polynomial for the perimeter of the figure is $14y + 25$.

54. $11\frac{1}{2}a + 12$, or $\frac{23}{2}a + 12$

55.

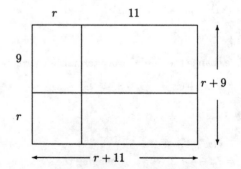

The length and width of the figure can be expressed as $r + 11$ and $r + 9$, respectively. The area of this figure (a rectangle) is the product of the length and width. An algebraic expression for the area is $(r + 11) \cdot (r + 9)$.

The algebraic expressions $9r + 99 + r^2 + 11r$ and $(r + 11) \cdot (r + 9)$ represent the same area.

$$
\begin{array}{c|cc|c}
 & r & 11 & \\
\hline
9 & A & B & 9 \\
\hline
r & C & D & r \\
\hline
 & r & 11 &
\end{array}
$$

The area of the figure can be found by adding the areas of the four rectangles A, B, C, and D. The area of a rectangle is the product of the length and the width.

$$\underset{\text{of } A}{\text{Area}} + \underset{\text{of } B}{\text{Area}} + \underset{\text{of } C}{\text{Area}} + \underset{\text{of } D}{\text{Area}}$$
$$= 9 \cdot r + 11 \cdot 9 + r \cdot r + 11 \cdot r$$
$$= 9r + 99 + r^2 + 11r$$

An algebraic expression for the area of the figure is $9r + 99 + r^2 + 11r$.

56. $(t + 5) \cdot (t + 3); \ t^2 + 5t + 3t + 15$

57.

$$
\begin{array}{c|cc|c}
 & x & 3 & \\
\hline
x & A & B & x \\
\hline
3 & C & D & 3 \\
\hline
 & x & 3 &
\end{array}
$$

The length and width of the figure can each be expressed as $x + 3$. The area can be expressed as $(x + 3) \cdot (x + 3)$, or $(x + 3)^2$. Another way to express the area is to find an expression for the sum of the areas of the four rectangles A, B, C, and D. The area of each rectangle is the product of its length and width.

$$\underset{\text{of } A}{\text{Area}} + \underset{\text{of } B}{\text{Area}} + \underset{\text{of } C}{\text{Area}} + \underset{\text{of } D}{\text{Area}}$$
$$= x \cdot x + 3 \cdot x + 3 \cdot x + 3 \cdot 3$$
$$= x^2 + 3x + 3x + 9$$

The algebraic expressions $(x+3)^2$ and $x^2 + 3x + 3x + 9$ represent the same area.

$$(x + 3)^2 = x^2 + 3x + 3x + 9$$

58. $(x + 10) \cdot (x + 8); \ 8x + 80 + x^2 + 10x$

59.

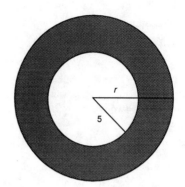

Familiarize. Recall that the area of a circle is the product of π and the square of the radius, r^2.

$$A = \pi r^2$$

Translate.

$$\underset{\text{with radius } r}{\text{Area of circle}} - \underset{\text{with radius } 5}{\text{Area of circle}} = \underset{\text{area}}{\text{Shaded}}$$
$$\pi \cdot r^2 - \pi \cdot 5^2 = \text{Shaded area}$$

Carry out. We simplify the expression.

$$\pi \cdot r^2 - \pi \cdot 5^2 = \pi r^2 - 25\pi$$

Check. We can go over our calculations. We can also assign some value to r, say 7, and carry out the computation in two ways.

Difference of areas: $\pi \cdot 7^2 - \pi \cdot 5^2 = 49\pi - 25\pi = 24\pi$

Substituting in the polynomial: $\pi \cdot 7^2 - 25\pi = 49\pi - 25\pi = 24\pi$

Since the results are the same, our solution is probably correct.

State. A polynomial for the shaded area is $\pi r^2 - 25\pi$.

60. $m^2 - 40$

61. Familiarize. We label the figure with additional information.

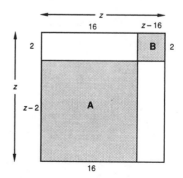

Translate.

Area of shaded sections = Area of A + Area of B

Area of shaded sections = $16(z - 2) + 2(z - 16)$

Carry out. We simplify the expression.

$16(z - 2) + 2(z - 16) = 16z - 32 + 2z - 32 = 18z - 64$

Check. We can go over the calculations. We can also assign some value to z, say 30, and carry out the computation in two ways.

Sum of areas:

$$16 \cdot 28 + 2 \cdot 14 = 448 + 28 = 476$$

Substituting in the polynomial:

$$18 \cdot 30 - 64 = 540 - 64 = 476$$

Since the results are the same, our solution is probably correct.

State. A polynomial for the shaded area is $18z - 64$.

62. $\pi r^2 - 49$

63.

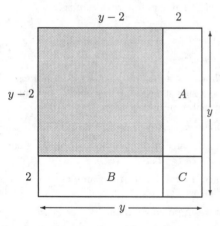

The shaded area is $(y - 2)^2$. We find it as follows:

$$\frac{\text{Shaded}}{\text{area}} = \frac{\text{Area of}}{\text{square}} - \frac{\text{Area}}{\text{of } A} - \frac{\text{Area}}{\text{of } B} - \frac{\text{Area}}{\text{of } C}$$

$(y - 2)^2 = \quad y^2 \quad - 2(y - 2) - 2(y - 2) - \; 2 \cdot 2$

$(y - 2)^2 = y^2 - 2y + 4 - 2y + 4 - 4$

$(y - 2)^2 = y^2 - 4y + 4$

64. $100 - 40x + 4x^2$

65. ◈

66. ◈

67. $5(4 + 3) - 5 \cdot 4 - 5 \cdot 3$

$= 5 \cdot 7 - 5 \cdot 4 - 5 \cdot 3$ Adding inside the parentheses

$= 35 - 20 - 15$ Multiplying

$= 0$ Subtracting

68. 0

69. $2(5t + 7) + 3t = 10t + 14 + 3t = 13t + 14$

70. $14t - 15$

71. $2(x + 3) > 5(x - 3) + 7$

$2x + 6 > 5x - 15 + 7$ Removing parentheses

$2x + 6 > 5x - 8$ Collecting like terms

$2x + 14 > 5x$ Adding 8 to both sides

$14 > 3x$ Adding $-2x$ to both sides

$\dfrac{14}{3} > x$ Dividing both sides by 3

The solution set is $\left\{ x \,\middle|\, \dfrac{14}{3} > x \right\}$, or $\left\{ x \,\middle|\, x < \dfrac{14}{3} \right\}$.

72. $\{x | x \leq 12\}$

73. ◈

74. ◈

75. $(6t^2 - 7t) + (3t^2 - 4t + 5) - (9t - 6)$

$= 6t^2 - 7t + 3t^2 - 4t + 5 - 9t + 6$

$= 9t^2 - 20t + 11$

76. $5x^2 - 9x - 1$

77. $(-8y^2 - 4) - (3y + 6) - (2y^2 - y)$

$= -8y^2 - 4 - 3y - 6 - 2y^2 + y$

$= -10y^2 - 2y - 10$

78. $4x^3 - 5x^2 + 6$

79. $(-y^4 - 7y^3 + y^2) + (-2y^4 + 5y - 2) - (-6y^3 + y^2)$

$= -y^4 - 7y^3 + y^2 - 2y^4 + 5y - 2 + 6y^3 - y^2$

$= -3y^4 - y^3 + 5y - 2$

80. $2 + x + 2x^2 + 4x^3$

81. $(345.099x^3 - 6.178x) - (94.508x^3 - 8.99x)$

$= 345.099x^3 - 6.178x - 94.508x^3 + 8.99x$

$= 250.591x^3 + 2.812x$

82. $36x + 2x^2$

83. *Familiarize.* The surface area is $2lw + 2lh + 2wh$, where $l =$ length, $w =$ width, and $h =$ height of the rectangular solid. Here we have $l = 3$, $w = w$, and $h = 7$.

Translate. We substitute in the formula above.

$$2 \cdot 3 \cdot w + 2 \cdot 3 \cdot 7 + 2 \cdot w \cdot 7$$

Carry out. We simplify the expression.

$$2 \cdot 3 \cdot w + 2 \cdot 3 \cdot 7 + 2 \cdot w \cdot 7$$
$$= 6w + 42 + 14w$$
$$= 20w + 42$$

Check. We can go over the calculations. We can also assign some value to w, say 6, and carry out the computation in two ways.

Using the formula: $2 \cdot 3 \cdot 6 + 2 \cdot 3 \cdot 7 + 2 \cdot 6 \cdot 7 = 36 + 42 + 84 = 162$

Substituting in the polynomial: $20 \cdot 6 + 42 = 120 + 42 = 162$

Since the results are the same, our solution is probably correct.

State. A polynomial for the surface area is $20w + 42$.

84. $22a + 56$

85. *Familiarize.* The surface area is $2lw + 2lh + 2wh$, where $l =$ length, $w =$ width, and $h =$ height of the rectangular solid. Here we have $l = x$, $w = x$, and $h = 5$.

Translate. We substitute in the formula above.

$$2 \cdot x \cdot x + 2 \cdot x \cdot 5 + 2 \cdot x \cdot 5$$

Carry out. We simplify the expression.

$$2 \cdot x \cdot x + 2 \cdot x \cdot 5 + 2 \cdot x \cdot 5$$
$$= 2x^2 + 10x + 10x$$
$$= 2x^2 + 20x$$

Check. We can go over the calculations. We can also assign some value to x, say 3, and carry out the computation in two ways.

Using the formula: $2 \cdot 3 \cdot 3 + 2 \cdot 3 \cdot 5 + 2 \cdot 3 \cdot 5 = 18 + 30 + 30 = 78$

Substituting in the polynomial: $2 \cdot 3^2 + 20 \cdot 3 = 2 \cdot 9 + 60 = 18 + 60 = 78$

Since the results are the same, our solution is probably correct.

State. A polynomial for the surface area is $2x^2 + 20x$.

86. a) $P = -x^2 + 280x - 5000$

b) \$10,375

c) \$13,000

87. ◈

Exercise Set 4.4

1. $(5x^4)6 = (5 \cdot 6)x^4 = 30x^4$

2. $28x^3$

3. $(-x^2)(-x) = (-1 \cdot x^2)(-1 \cdot x) = (-1)(-1)(x^2 \cdot x) = x^3$

4. $-x^7$

5. $(-x^5)(x^3) = (-1 \cdot x^5)(1x^3) = (-1)(1)(x^5 \cdot x^3) = -x^8$

6. x^8

7. $(7t^5)(4t^3) = (7 \cdot 4)(t^5 \cdot t^3) = 28t^8$

8. $30a^4$

9. $(-0.1x^6)(0.2x^4) = (-0.1)(0.2)(x^6 \cdot x^4) = -0.02x^{10}$

10. $-0.12x^9$

11. $\left(-\frac{1}{5}x^3\right)\left(-\frac{1}{3}x\right) = \left(-\frac{1}{5}\right)\left(-\frac{1}{3}\right)(x^3 \cdot x) = \frac{1}{15}x^4$

12. $-\frac{1}{20}x^{12}$

13. $19t^2 \cdot 0 = 0$ Any number multiplied by 0 is 0.

14. $5n^3$

15. $7x^2(-2x^3)(2x^6) = 7(-2)(2)(x^2 \cdot x^3 \cdot x^6) = -28x^{11}$

16. $72y^{10}$

17. $3x(-x + 5) = 3x(-x) + 3x(5)$
$$= -3x^2 + 15x$$

18. $8x^2 - 12x$

19. $4x(x + 1) = 4x(x) + 4x(1)$
$$= 4x^2 + 4x$$

20. $3x^2 + 6x$

21. $(a + 9)3a = a \cdot 3a + 9 \cdot 3a = 3a^2 + 27a$

22. $4a^2 - 28a$

23. $x^2(x^3 + 1) = x^2(x^3) + x^2(1)$
$$= x^5 + x^2$$

24. $-2x^5 + 2x^3$

25. $3x(2x^2 - 6x + 1) = 3x(2x^2) + 3x(-6x) + 3x(1)$
$$= 6x^3 - 18x^2 + 3x$$

26. $-8x^4 + 24x^3 + 20x^2 - 4x$

27. $5t^2(3t+6) = 5t^2(3t) + 5t^2(6) = 15t^3 + 30t^2$

28. $14t^3 + 7t^2$

29. $-6x^2(x^2+x) = -6x^2(x^2) - 6x^2(x)$
$$= -6x^4 - 6x^3$$

30. $-4x^4 + 4x^3$

31. $\quad \dfrac{2}{3}a^4\left(6a^5 - 12a^3 - \dfrac{5}{8}\right)$
$$= \dfrac{2}{3}a^4(6a^5) - \dfrac{2}{3}a^4(12a^3) - \dfrac{2}{3}a^4\left(\dfrac{5}{8}\right)$$
$$= \dfrac{12}{3}a^9 - \dfrac{24}{3}a^7 - \dfrac{10}{24}a^4$$
$$= 4a^9 - 8a^7 - \dfrac{5}{12}a^4$$

32. $6t^{11} - 9t^9 + \dfrac{9}{7}t^5$

33. $\quad (x+6)(x+3) = (x+6)x + (x+6)3$
$$= x \cdot x + 6 \cdot x + x \cdot 3 + 6 \cdot 3$$
$$= x^2 + 6x + 3x + 18$$
$$= x^2 + 9x + 18$$

34. $x^2 + 7x + 10$

35. $\quad (x+5)(x-2) = (x+5)x + (x+5)(-2)$
$$= x \cdot x + 5 \cdot x + x(-2) + 5(-2)$$
$$= x^2 + 5x - 2x - 10$$
$$= x^2 + 3x - 10$$

36. $x^2 + 4x - 12$

37. $\quad (a-6)(a-7) = (a-6)a + (a-6)(-7)$
$$= a \cdot a - 6 \cdot a + a(-7) + (-6)(-7)$$
$$= a^2 - 6a - 7a + 42$$
$$= a^2 - 13a + 42$$

38. $a^2 - 12a - 32$

39. $\quad (x+3)(x-3) = (x+3)x + (x+3)(-3)$
$$= x \cdot x + 3 \cdot x + x(-3) + 3(-3)$$
$$= x^2 + 3x - 3x - 9$$
$$= x^2 - 9$$

40. $x^2 - 36$

41. $\quad (5-x)(5-2x) = (5-x)5 + (5-x)(-2x)$
$$= 5 \cdot 5 - x \cdot 5 + 5(-2x) - x(-2x)$$
$$= 25 - 5x - 10x + 2x^2$$
$$= 25 - 15x + 2x^2$$

42. $18 + 12x + 2x^2$

43. $\quad \left(t + \dfrac{3}{2}\right)\left(t + \dfrac{4}{3}\right) = \left(t + \dfrac{3}{2}\right)t + \left(t + \dfrac{3}{2}\right)\left(\dfrac{4}{3}\right)$
$$= t \cdot t + \dfrac{3}{2} \cdot t + t \cdot \dfrac{4}{3} + \dfrac{3}{2} \cdot \dfrac{4}{3}$$
$$= t^2 + \dfrac{3}{2}t + \dfrac{4}{3}t + 2$$
$$= t^2 + \dfrac{9}{6}t + \dfrac{8}{6}t + 2$$
$$= t^2 + \dfrac{17}{6}t + 2$$

44. $a^2 + \dfrac{21}{10}a - 1$

45. $\quad \left(\dfrac{1}{4}a + 2\right)\left(\dfrac{3}{4}a - 1\right)$
$$= \left(\dfrac{1}{4}a + 2\right)\left(\dfrac{3}{4}a\right) + \left(\dfrac{1}{4}a + 2\right)(-1)$$
$$= \dfrac{1}{4}a\left(\dfrac{3}{4}a\right) + 2 \cdot \dfrac{3}{4}a + \dfrac{1}{4}a(-1) + 2(-1)$$
$$= \dfrac{3}{16}a^2 + \dfrac{3}{2}a - \dfrac{1}{4}a - 2$$
$$= \dfrac{3}{16}a^2 + \dfrac{6}{4}a - \dfrac{1}{4}a - 2$$
$$= \dfrac{3}{16}a^2 + \dfrac{5}{4}a - 2$$

46. $\dfrac{6}{25}t^2 - \dfrac{1}{5}t - 1$

47. Illustrate $x(x+5)$ as the area of a rectangle with width x and length $x + 5$.

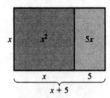

48.

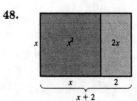

49. Illustrate $(x + 1)(x + 2)$ as the area of a rectangle with width $x + 1$ and length $x + 2$.

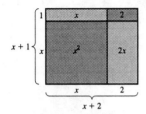

50.

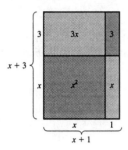

51. Illustrate $(x + 5)(x + 3)$ as the area of a rectangle with length $x + 5$ and width $x + 3$.

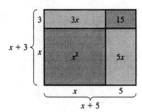

52.

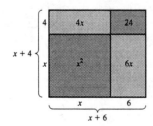

53. Illustrate $(3x + 2)(3x + 2)$ as the area of a square with sides of length $3x + 2$.

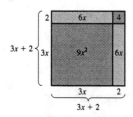

54.

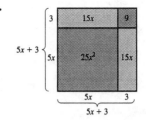

55.
$$(x^2 - x + 5)(x + 1)$$
$$= (x^2 - x + 5)x + (x^2 - x + 5)1$$
$$= x^3 - x^2 + 5x + x^2 - x + 5$$
$$= x^3 + 4x + 5$$

A partial check can be made by selecting a convenient replacement for x, say 1, and comparing the values of the original expression and the result.

$$(1^2 - 1 + 5)(1 + 1) \qquad 1^3 + 4 \cdot 1 + 5$$
$$= (1 - 1 + 5)(1 + 1) \qquad = 1 + 4 + 5$$
$$= 5 \cdot 2 \qquad\qquad\qquad = 10$$
$$= 10$$

Since the value of both expressions is 10, the multiplication is very likely correct.

56. $x^3 + 3x^2 - 5x - 14$

57.
$$(2a + 5)(a^2 - 3a + 2)$$
$$= (2a + 5)a^2 - (2a + 5)(3a) + (2a + 5)2$$
$$= 2a \cdot a^2 + 5 \cdot a^2 - 2a \cdot 3a - 5 \cdot 3a + 2a \cdot 2 + 5 \cdot 2$$
$$= 2a^3 + 5a^2 - 6a^2 - 15a + 4a + 10$$
$$= 2a^3 - a^2 - 11a + 10$$

A partial check can be made as in Exercise 55.

58. $3t^3 - 11t^2 - 17t + 4$

59.
$$(y^2 - 7)(2y^3 + y + 1)$$
$$= (y^2 - 7)(2y^3) + (y^2 - 7)y + (y^2 - 7)(1)$$
$$= y^2 \cdot 2y^3 - 7 \cdot 2y^3 + y^2 \cdot y - 7 \cdot y + y^2 \cdot 1 - 7 \cdot 1$$
$$= 2y^5 - 14y^3 + y^3 - 7y + y^2 - 7$$
$$= 2y^5 - 13y^3 + y^2 - 7y - 7$$

A partial check can be made as in Exercise 55.

60. $5a^5 + 17a^3 - a^2 - 12a - 4$

61.
$$(5x^3 - 7x^2 + 1)(x - 3x^2)$$
$$= (5x^3 - 7x^2 + 1)x - (5x^3 - 7x^2 + 1)(3x^2)$$
$$= 5x^3 \cdot x - 7x^2 \cdot x + 1 \cdot x - 5x^3 \cdot 3x^2 + 7x^2 \cdot 3x^2 - 1 \cdot 3x^2$$
$$= 5x^4 - 7x^3 + x - 15x^5 + 21x^4 - 3x^2$$
$$= -15x^5 + 26x^4 - 7x^3 - 3x^2 + x$$

A partial check can be made in Exercise 55.

62. $8x^5 - 6x^3 - 6x^2 - 5x - 3$

63.

$$
\begin{array}{ll}
x^2 -3x+2 & \text{Line up like terms} \\
x^2 + x +1 & \text{in columns} \\
\hline
x^2 -3x+2 & \text{Multiplying by 1} \\
x^3 -3x^2+2x & \text{Multiplying by } x \\
x^4 -3x^3+2x^2 & \text{Multiplying by } x^2 \\
\hline
x^4 -2x^3 \qquad - x +2 &
\end{array}
$$

A partial check can be made as in Exercise 55.

64. $x^4 + 4x^3 - 3x^2 + 16x - 3$

65.

$$
\begin{array}{ll}
2t^2 -5t-4 & \\
3t^2 - t +1 & \\
\hline
2t^2 -5t-4 & \text{Multiplying by 1} \\
- 2t^3+ 5t^2 +4t & \text{Multiplying by } -t \\
6t^4 -15t^3 -12t^2 & \text{Multiplying by } 3t^2 \\
\hline
6t^4 -17t^3 - 5t^2 - t -4 &
\end{array}
$$

A partial check can be made as in Exercise 55.

66. $10t^4 + 3t^3 - 14t^2 + 4t - 3$

67. We will multiply horizontally while still aligning like terms.

$$(x + 1)(x^3 + 7x^2 + 5x + 4)$$

$$
\begin{array}{ll}
= x^4 + 7x^3 + 5x^2 + 4x & \text{Multiplying by } x \\
\quad + x^3 + 7x^2 + 5x + 4 & \text{Multiplying by 1} \\
\hline
= x^4 + 8x^3 + 12x^2 + 9x + 4 &
\end{array}
$$

A partial check can be made as in Exercise 55.

68. $x^4 + 7x^3 + 19x^2 + 21x + 6$

69. We will multiply horizontally while still aligning like terms.

$$\left(x - \frac{1}{2}\right)\left(2x^3 - 4x^2 + 3x - \frac{2}{5}\right)$$

$$
= 2x^4 - 4x^3 + 3x^2 - \frac{2}{5}x
$$

$$
\quad - x^3 + 2x^2 - \frac{3}{2}x + \frac{1}{5}
$$

$$
\overline{2x^4 - 5x^3 + 5x^2 - \frac{19}{10}x + \frac{1}{5}}
$$

A partial check can be made as in Exercise 55.

70. $6x^4 - 10x^3 - 9x^2 - \dfrac{7}{6}x + \dfrac{1}{6}$

71. ◈

72. ◈

73. $5 - 3 \cdot 2 + 7 = 5 - 6 + 7 = -1 + 7 = 6$

74. 31

75.

$$(8 - 2)(8 + 2) + 2^2 - 8^2$$
$$= 6 \cdot 10 + 2^2 - 8^2$$
$$= 6 \cdot 10 + 4 - 64$$
$$= 60 + 4 - 64$$
$$= 64 - 64$$
$$= 0$$

76. 0

77. ◈

78. ◈

79. The shaded area is the area of the large rectangle, $6y(14y - 5)$ less the area of the unshaded rectangle, $3y(3y + 5)$. We have:

$$6y(14y - 5) - 3y(3y + 5)$$
$$= 84y^2 - 30y - 9y^2 - 15y$$
$$= 75y^2 - 45y$$

80. $78t^2 + 40t$

81. Let $n =$ the missing number. Label the figure with the known areas.

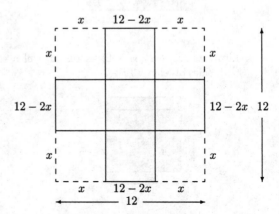

Then the area of the figure is $x^2 + 2x + nx + 2n$. This is equivalent to $x^2 + 7x + 10$, so we have $2x + nx = 7x$ and $2n = 10$. Solving either equation for n, we find that the missing number is 5.

82. 5

83.

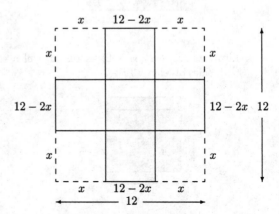

The dimensions, in inches, of the box are $12 - 2x$ by $12 - 2x$ by x. The volume is the product of the dimensions (volume = length × width × height):

$$
\begin{aligned}
\text{Volume} &= (12 - 2x)(12 - 2x)x \\
&= (144 - 48x + 4x^2)x \\
&= 144x - 48x^2 + 4x^3 \text{ in}^3, \text{ or} \\
& \quad 4x^3 - 48x^2 + 144x \text{ in}^3
\end{aligned}
$$

The outside surface area is the sum of the area of the bottom and the areas of the four sides. The dimensions, in inches, of the bottom are $12 - 2x$ by $12 - 2x$, and the dimensions, in inches, of each side are x by $12 - 2x$.

$$
\begin{aligned}
\begin{matrix}\text{Surface} \\ \text{area}\end{matrix} &= \begin{matrix}\text{Area of bottom} + \\ 4 \cdot \text{Area of each side}\end{matrix} \\
&= (12 - 2x)(12 - 2x) + 4 \cdot x(12 - 2x) \\
&= 144 - 24x - 24x + 4x^2 + 48x - 8x^2 \\
&= 144 - 48x + 4x^2 + 48x - 8x^2 \\
&= 144 - 4x^2 \text{ in}^2, \text{ or } -4x^2 + 144 \text{ in}^2
\end{aligned}
$$

84. $x^3 - 5x^2 + 8x - 4 \text{ cm}^3$

85. We have a rectangular solid with dimensions x m by x m by $x+2$ m with a rectangular solid piece with dimensions 6 m by 5 m by 7 m cut out of it.

$$
\begin{aligned}
\text{Volume} &= \begin{matrix}\text{Volume of} \\ \text{large solid}\end{matrix} - \begin{matrix}\text{Volume of} \\ \text{small solid}\end{matrix} \\
&= (x \text{ m})(x \text{ m})(x + 2 \text{ m}) - (6 \text{ m})(5 \text{ m})(7 \text{ m}) \\
&= x^2(x + 2) \text{ m}^3 - 210 \text{ m}^3 \\
&= x^3 + 2x^2 - 210 \text{ m}^3
\end{aligned}
$$

86. $x^3 + 6x^2 + 12x + 8 \text{ cm}^3$

87. Let $x = $ the width of the garden. Then $2x = $ the length of the garden.

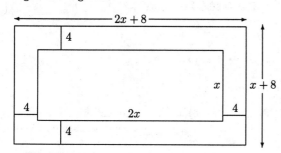

Area of garden and sidewalk together   is   Area of garden alone   + $\underbrace{256 \text{ ft}^2}$

$$(2x + 8)(x + 8) = 2x \cdot x + 256$$

$$
\begin{aligned}
2x^2 + 24x + 64 &= 2x^2 + 256 \\
24x &= 192 \\
x &= 8
\end{aligned}
$$

The dimensions are 8 ft by 16 ft.

88. $2x^2 + 18x + 36$

89. $(x - 2)(x - 7) - (x - 7)(x - 2)$

First observe that, by the commutative law of multiplication, $(x - 2)(x - 7)$ and $(x - 7)(x - 2)$ are equivalent expressions. Then when we subtract $(x - 7)(x - 2)$ from $(x - 2)(x - 7)$, the result is 0.

90. $16x + 16$

91.
$$
\begin{aligned}
&(x - a)(x - b) \cdots (x - x)(x - y)(x - z) \\
&= (x - a)(x - b) \cdots 0 \cdot (x - y)(x - z) \\
&= 0
\end{aligned}
$$

92. ▩

Exercise Set 4.5

1. $(x + 4)(x^2 + 3)$
$$
\begin{aligned}
&\quad\ \text{F}\qquad \text{O}\qquad \text{I}\qquad \text{L} \\
&= x \cdot x^2 + x \cdot 3 + 4 \cdot x^2 + 4 \cdot 3 \\
&= x^3 + 3x + 4x^2 + 12, \text{or } x^3 + 4x^2 + 3x + 12
\end{aligned}
$$

2. $x^3 - x^2 - 3x + 3$

3. $(x^3 + 6)(x + 2)$
$$
\begin{aligned}
&\quad\ \text{F}\qquad \text{O}\qquad \text{I}\qquad \text{L} \\
&= x^3 \cdot x + x^3 \cdot 2 + 6 \cdot x + 6 \cdot 2 \\
&= x^4 + 2x^3 + 6x + 12
\end{aligned}
$$

4. $x^5 + 12x^4 + 2x + 24$

5. $(y + 2)(y - 3)$
$$
\begin{aligned}
&\quad\ \text{F}\qquad \text{O}\qquad\quad \text{I}\qquad \text{L} \\
&= y \cdot y + y \cdot (-3) + 2 \cdot y + 2 \cdot (-3) \\
&= y^2 - 3y + 2y - 6 \\
&= y^2 - y - 6
\end{aligned}
$$

6. $a^2 + 4a + 4$

7. $(3x + 2)(3x + 5)$
$$
\begin{aligned}
&\quad\ \text{F}\qquad\quad \text{O}\qquad\quad \text{I}\qquad \text{L} \\
&= 3x \cdot 3x + 3x \cdot 5 + 2 \cdot 3x + 2 \cdot 5 \\
&= 9x^2 + 15x + 6x + 10 \\
&= 9x^2 + 21x + 10
\end{aligned}
$$

8. $8x^2 + 30x + 7$

9. $(5x - 6)(x + 2)$
$$\quad \text{F}\qquad \text{O}\qquad \text{I}\qquad \text{L}$$
$$= 5x \cdot x + 5x \cdot 2 + (-6) \cdot x + (-6) \cdot 2$$
$$= 5x^2 + 10x - 6x - 12$$
$$= 5x^2 + 4x - 12$$

10. $t^2 - 81$

11. $(1 + 3t)(2 - 3t)$
$$\quad \text{F}\qquad \text{O}\qquad \text{I}\qquad \text{L}$$
$$= 1 \cdot 2 + 1(-3t) + 3t \cdot 2 + 3t(-3t)$$
$$= 2 - 3t + 6t - 9t^2$$
$$= 2 + 3t - 9t^2$$

12. $14 + 19a - 3a^2$

13. $(2x - 7)(x - 1)$
$$\quad \text{F}\qquad \text{O}\qquad \text{I}\qquad \text{L}$$
$$= 2x \cdot x + 2x \cdot (-1) + (-7) \cdot x + (-7) \cdot (-1)$$
$$= 2x^2 - 2x - 7x + 7$$
$$= 2x^2 - 9x + 7$$

14. $6x^2 - x - 1$

15. $\left(p - \dfrac{1}{4}\right)\left(p + \dfrac{1}{4}\right)$
$$\quad \text{F}\qquad \text{O}\qquad \text{I}\qquad \text{L}$$
$$= p \cdot p + p \cdot \frac{1}{4} + \left(-\frac{1}{4}\right) \cdot p + \left(-\frac{1}{4}\right) \cdot \frac{1}{4}$$
$$= p^2 + \frac{1}{4}p - \frac{1}{4}p - \frac{1}{16}$$
$$= p^2 - \frac{1}{16}$$

16. $q^2 + \dfrac{3}{2}q + \dfrac{9}{16}$

17. $(x - 0.1)(x + 0.1)$
$$\quad \text{F}\qquad \text{O}\qquad \text{I}\qquad \text{L}$$
$$= x \cdot x + x \cdot (0.1) + (-0.1) \cdot x + (-0.1)(0.1)$$
$$= x^2 + 0.1x - 0.1x - 0.01$$
$$= x^2 - 0.01$$

18. $x^2 - 0.1x - 0.12$

19. $(2x^2 + 6)(x + 1)$
$$\quad \text{F}\qquad \text{O}\qquad \text{I}\qquad \text{L}$$
$$= 2x^3 + 2x^2 + 6x + 6$$

20. $4x^3 - 2x^2 + 6x - 3$

21. $(-2x + 1)(x + 6)$
$$\quad \text{F}\qquad \text{O}\qquad \text{I}\qquad \text{L}$$
$$= -2x^2 - 12x + x + 6$$
$$= -2x^2 - 11x + 6$$

22. $-2x^2 + 13x - 20$

23. $(a + 9)(a + 9)$
$$\quad \text{F}\qquad \text{O}\qquad \text{I}\qquad \text{L}$$
$$= a^2 + 9a + 9a + 81$$
$$= a^2 + 18a + 81$$

24. $4y^2 + 28y + 49$

25. $(1 + 3t)(1 - 5t)$
$$\quad \text{F}\qquad \text{O}\qquad \text{I}\qquad \text{L}$$
$$= 1 - 5t + 3t - 15t^2$$
$$= 1 - 2t - 15t^2$$

26. $1 + 2t - 3t^2 - 6t^3$

27. $(x^2 + 3)(x^3 - 1)$
$$\quad \text{F}\qquad \text{O}\qquad \text{I}\qquad \text{L}$$
$$= x^5 - x^2 + 3x^3 - 3, \text{ or } x^5 + 3x^3 - x^2 - 3$$

28. $2x^5 + x^4 - 6x - 3$

29. $(3x^2 - 2)(x^4 - 2)$
$$\quad \text{F}\qquad \text{O}\qquad \text{I}\qquad \text{L}$$
$$= 3x^6 - 6x^2 - 2x^4 + 4, \text{ or } 3x^6 - 2x^4 - 6x^2 + 4$$

30. $x^{20} - 9$

31. $(2t^3 + 5)(2t^3 + 3)$
$$\quad \text{F}\qquad \text{O}\qquad \text{I}\qquad \text{L}$$
$$= 4t^6 + 6t^3 + 10t^3 + 15$$
$$= 4t^6 + 16t^3 + 15$$

32. $10t^4 + 17t^2 + 3$

33. $(8x^3 + 5)(x^2 + 2)$
$$\quad \text{F}\qquad \text{O}\qquad \text{I}\qquad \text{L}$$
$$= 8x^5 + 16x^3 + 5x^2 + 10$$

34. $20 - 10x - 8x^2 + 4x^3$

35. $(4x^2 + 3)(x - 3)$
$$\quad \text{F}\qquad \text{O}\qquad \text{I}\qquad \text{L}$$
$$= 4x^3 - 12x^2 + 3x - 9$$

36. $14x^2 - 53x + 14$

37. $(x + 8)(x - 8)$ Product of sum and difference of the same two terms
$$= x^2 - 8^2$$
$$= x^2 - 64$$

38. $x^2 - 1$

39. $(2x + 1)(2x - 1)$ Product of sum and difference of the same two terms
$$= (2x)^2 - 1^2$$
$$= 4x^2 - 1$$

40. $x^4 - 1$

41. $\quad (5m - 2)(5m + 2)$ Product of sum and difference of the same two terms
$= (5m)^2 - 2^2$
$= 25m^2 - 4$

42. $9x^8 - 4$

43. $\quad (2x^2 + 3)(2x^2 - 3)$ Product of sum and difference of the same two terms
$= (2x^2)^2 - 3^2$
$= 4x^4 - 9$

44. $36x^{10} - 25$

45. $\quad (3x^4 - 1)(3x^4 + 1)$
$= (3x^4)^2 - 1^2$
$= 9x^8 - 1$

46. $t^4 - 0.04$

47. $\quad (x^4 + 7)(x^4 - 7)$
$= (x^4)^2 - 7^2$
$= x^8 - 49$

48. $t^6 - 16$

49. $\quad \left(t - \dfrac{3}{4}\right)\left(t + \dfrac{3}{4}\right)$
$= t^2 - \left(\dfrac{3}{4}\right)^2$
$= t^2 - \dfrac{9}{16}$

50. $m^2 - \dfrac{4}{9}$

51. $\quad (x + 2)^2$
$= x^2 + 2 \cdot x \cdot 2 + 2^2$ Square of a binomial
$= x^2 + 4x + 4$

52. $4x^2 - 4x + 1$

53. $\quad (3x^5 + 1)^2$ Square of a binomial
$= (3x^5)^2 + 2 \cdot 3x^5 \cdot 1 + 1^2$
$= 9x^{10} + 6x^5 + 1$

54. $16x^6 + 8x^3 + 1$

55. $\quad \left(a - \dfrac{2}{5}\right)^2$ Square of a binomial
$= a^2 - 2 \cdot a \cdot \dfrac{2}{5} + \left(\dfrac{2}{5}\right)^2$
$= a^2 - \dfrac{4}{5}a + \dfrac{4}{25}$

56. $t^2 - \dfrac{2}{5}t + \dfrac{1}{25}$

57. $\quad = (t^3 + 3)^2$ Square of a binomial
$= (t^3)^2 + 2 \cdot t^3 \cdot 3 + 3^2$
$= t^6 + 6t^3 + 9$

58. $a^8 + 4a^4 + 4$

59. $\quad (2 - 3x^4)^2 = 2^2 - 2 \cdot 2 \cdot 3x^4 + (3x^4)^2$
$= 4 - 12x^4 + 9x^8$

60. $25 - 20t^3 + 4t^6$

61. $\quad (5 + 6t^2)^2 = 5^2 + 2 \cdot 5 \cdot 6t^2 + (6t^2)^2$
$= 25 + 60t^2 + 36t^4$

62. $9p^4 - 6p^3 + p^2$

63. $\quad (7x - 0.3)^2 = (7x)^2 - 2(7x)(0.3) + (0.3)^2$
$= 49x^2 - 4.2x + 0.09$

64. $16a^2 - 4.8a + 0.36$

65. $\quad 5a^3(2a^2 - 1)$
$= 5a^3 \cdot 2a^2 - 5a^3 \cdot 1$ Multiplying each term of
the binomial by the monomial
$= 10a^5 - 5a^3$

66. $18x^5 - 45x^3$

67. $\quad (a - 3)(a^2 + 2a - 4)$

$\begin{aligned} &= a^3 + 2a^2 - 4a \quad \text{Multiplying horizontally} \\ &\underline{ - 3a^2 - 6a + 12} \quad \text{and aligning like terms} \\ &= a^3 - a^2 - 10a + 12 \end{aligned}$

68. $x^4 + x^3 - 6x^2 - 5x + 5$

69. $\quad (3 - 2x^3)^2$
$= 3^2 - 2 \cdot 3 \cdot 2x^3 + (2x^3)^2$ Squaring a binomial
$= 9 - 12x^3 + 4x^6$

70. $x^2 - 8x^4 + 16x^6$

71. $\quad 4x(x^2 + 6x - 3)$
$= 4x \cdot x^2 + 4x \cdot 6x + 4x(-3)$ Multiplying each
term of the trinomial
by the monomial
$= 4x^3 + 24x^2 - 12x$

72. $-8x^6 + 48x^3 + 72x$

73. $\quad (-t^3 + 1)^2$
$= (-t^3)^2 + 2(-t)^3(1) + 1^2$ Squaring a binomial
$= t^6 - 2t^3 + 1$

74. $x^4 - 2x^2 + 1$

75. $3t^2(5t^3 - t^2 + t)$
    $= 3t^2 \cdot 5t^3 + 3t^2(-t^2) + 3t^2 \cdot t$ Multiplying each
                                    term of the trinomial
                                    by the monomial
    $= 15t^5 - 3t^4 + 3t^3$

76. $-5x^5 - 40x^4 + 45x^3$

77. $(6x^4 - 3)^2$ Squaring a binomial
    $= (6x^4)^2 - 2 \cdot 6x^4 \cdot 3 + 3^2$
    $= 36x^8 - 36x^4 + 9$

78. $64a^2 + 80a^3 + 25$

79. $(3x + 2)(4x^2 + 5)$ Product of two
                                binomials; use FOIL
    $= 3x \cdot 4x^2 + 3x \cdot 5 + 2 \cdot 4x^2 + 2 \cdot 5$
    $= 12x^3 + 15x + 8x^2 + 10,$ or
      $12x^3 + 8x^2 + 15x + 10$

80. $6x^4 - 3x^2 - 63$

81. $(5 - 6x^4)^2$ Squaring a binomial
    $= 5^2 - 2 \cdot 5 \cdot 6x^4 + (6x^4)^2$
    $= 25 - 60x^4 + 36x^8$

82. $9 - 24t^5 + 16t^{10}$

83. $(a+1)(a^2 - a + 1)$

    $= a^3 - a^2 + a$ Multiplying horizontally
    $\;\; a^2 - a + 1$ and aligning like terms
    $\overline{a^3 + 1}$

84. $x^3 - 125$

85.

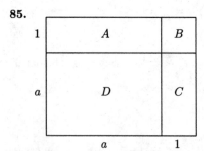

We can find the shaded area in two ways.

Method 1: The figure is a square with side $a + 1$, so the area is $(a + 1)^2 = a^2 + 2a + 1$.

Method 2: We add the areas of A, B, C, and D.

$1 \cdot a + 1 \cdot 1 + 1 \cdot a + a \cdot a = a + 1 + a + a^2 = a^2 + 2a + 1$.

Either way we find that the total shaded area is $a^2 + 2a + 1$.

86. $x^2 + 6x + 9$

87.

We can find the shaded area in two ways.

Method 1: The figure is a rectangle with dimensions $x + 5$ by $x + 2$, so the area is

$(x + 5)(x + 2) = x^2 + 2x + 5x + 10 = x^2 + 7x + 10.$

Method 2: We add the areas of A, B, C, and D.

$5 \cdot x + 2 \cdot 5 + 2 \cdot x + x \cdot x = 5x + 10 + 2x + x^2 = x^2 + 7x + 10.$

Either way, we find that the area is $x^2 + 7x + 10$.

88. $t^2 + 7t + 12$

89.

We can find the shaded area in two ways.

Method 1: The figure is a square with side $x + 7$, so the area is $(x + 7)^2 = x^2 + 14x + 49$.

Method 2: We add the areas of A, B, C, and D.

$x \cdot x + x \cdot 7 + 7 \cdot 7 + 7 \cdot x = x^2 + 7x + 49 + 7x = x^2 + 14x + 49.$

Either way, we find that the total shaded area is $x^2 + 14x + 49$.

90. $a^2 + 10a + 25$

91.

We can find the shaded area in two ways.

Method 1: The figure is a rectangle with dimensions $t + 6$ by $t + 4$, so the area is $(t + 6)(t + 4) = t^2 + 4t + 6t + 24 = t^2 + 10t + 24$.

Method 2: We add the areas of A, B, C, and D.

$t \cdot t + t \cdot 6 + 6 \cdot 4 + 4 \cdot t = t^2 + 6t + 24 + 4t = t^2 + 10t + 24$.

Either way, we find that the total shaded area is $t^2 + 10t + 24$.

92. $x^2 + 10x + 21$

93.

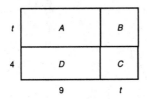

We can find the shaded area in two ways.

Method 1: The figure is a rectangle with dimensions $t + 9$ by $t + 4$, so the area is

$(t + 9)(t + 4) = t^2 + 4t + 9t + 36 = t^2 + 13t + 36$

Method 2: We add the areas of A, B, C, and D.

$9 \cdot t + t \cdot t + 4 \cdot t + 4 \cdot 9 = 9t + t^2 + 4t + 36 = t^2 + 13t + 36$.

Either way, we find that the total shaded area is $t^2 + 13t + 36$.

94. $a^2 + 8a + 7$

95.

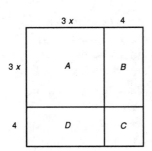

We can find the shaded area in two ways.

Method 1: The figure is a square with side $3x + 4$, so the area is $(3x + 4)^2 = 9x^2 + 24x + 16$.

Method 2: We add the areas of A, B, C, and D.

$3x \cdot 3x + 3x \cdot 4 + 4 \cdot 4 + 3x \cdot 4 = 9x^2 + 12x + 16 + 12x = 9x^2 + 24x + 16$.

Either way, we find that the total shaded area is $9x^2 + 24x + 16$.

96. $25t^2 + 20t + 4$

97. We draw a square with side $x + 5$.

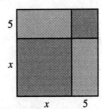

98.

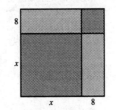

99. We draw a square with side $t + 9$.

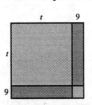

100.

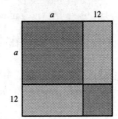

101. We draw a square with side $3 + x$.

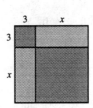

102.

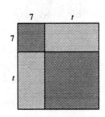

103. ◈

104.

105. *Familiarize*. Let t = the number of watts used by the television set. Then $10t$ = the number of watts used by the lamps, and $40t$ = the number of watts used by the air conditioner.

Translate.

$$\underbrace{\begin{array}{c}\text{Lamp}\\\text{watts}\end{array}}_{} + \underbrace{\begin{array}{c}\text{Air}\\\text{conditioner}\\\text{watts}\end{array}}_{} + \underbrace{\begin{array}{c}\text{Television}\\\text{watts}\end{array}}_{} = \underbrace{\begin{array}{c}\text{Total}\\\text{watts}\end{array}}_{}$$

$$\downarrow \quad \downarrow \qquad \downarrow \qquad \quad \downarrow \qquad \downarrow \quad \downarrow \quad \downarrow$$
$$10t \;+\; 40t \;+\; t \;=\; 2550$$

Solve. We solve the equation.

$$10t + 40t + t = 2550$$
$$51t = 2550$$
$$t = 50$$

The possible solution is:

Television, t: 50 watts

Lamps, $10t$: $10 \cdot 50$, or 500 watts

Air conditioner, $40t$: $40 \cdot 50$, or 2000 watts

Check. The number of watts used by the lamps, 500, is 10 times 50, the number used by the television. The number of watts used by the air conditioner, 2000, is 40 times 50, the number used by the television. Also, $50 + 500 + 2000 = 2550$, the total wattage used.

State. The television uses 50 watts, the lamps use 500 watts, and the air conditioner uses 2000 watts.

106. II

107.
$$5xy = 8$$
$$y = \frac{8}{5x} \qquad \text{Dividing both sides by } 5x$$

108. $a = \dfrac{c}{3b}$

109.
$$ax - b = c$$
$$ax = b + c \qquad \text{Adding } b \text{ to both sides}$$
$$x = \frac{b+c}{a} \qquad \text{Dividing both sides by } a$$

110. $t = \dfrac{u - r}{s}$

111.

112.

113.
$$(4x^2 + 9)(2x + 3)(2x - 3)$$
$$= (4x^2 + 9)(4x^2 - 9)$$
$$= 16x^4 - 81$$

114. $81a^4 - 1$

115.
$$(3t - 2)^2 (3t + 2)^2$$
$$= [(3t - 2)(3t + 2)]^2$$
$$= (9t^2 - 4)^2$$
$$= 81t^4 - 72t^2 + 16$$

116. $625a^4 - 50a^2 + 1$

117.
$$(t^3 - 1)^4 (t^3 + 1)^4$$
$$= [(t^3 - 1)(t^3 + 1)]^4$$
$$= (t^6 - 1)^4$$
$$= [(t^6 - 1)^2]^2$$
$$= (t^{12} - 2t^6 + 1)^2$$
$$= (t^{12} - 2t^6 + 1)(t^{12} - 2t^6 + 1)$$
$$= t^{24} - 2t^{18} + t^{12} - 2t^{18} + 4t^{12} - 2t^6 +$$
$$\qquad t^{12} - 2t^6 + 1$$
$$= t^{24} - 4t^{18} + 6t^{12} - 4t^6 + 1$$

118. $1050.4081x^2 + 348.0834x + 28.8369$

119. $18 \times 22 = (20 - 2)(20 + 2) = 20^2 - 2^2 =$
$400 - 4 = 396$

120. 9951

121.
$$(x + 2)(x - 5) = (x + 1)(x - 3)$$
$$x^2 - 5x + 2x - 10 = x^2 - 3x + x - 3$$
$$x^2 - 3x - 10 = x^2 - 2x - 3$$
$$-3x - 10 = -2x - 3 \qquad \text{Adding } - x^2$$
$$-3x + 2x = 10 - 3 \qquad \text{Adding } 2x \text{ and } 10$$
$$-x = 7$$
$$x = -7$$

The solution is -7.

122. 0

123. If l = the length, then $l + 1$ = the height, and $l - 1 =$ the width. Recall that the volume of a rectangular solid is given by length $\times$ width $\times$ height.

Volume $= l(l - 1)(l + 1) = l(l^2 - 1) = l^3 - l$

124. $w^3 + 3w^2 + 2w$

125.

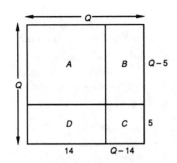

The dimensions of the shaded area, B, are $Q-14$ by $Q-5$, so one expression is $(Q-14)(Q-5)$.

To find another expression we find the area of regions B and C together and subtract the area of region C. The region consisting of B and C together has dimensions Q by $Q-14$, so its area is $Q(Q-14)$. Region C has dimensions 5 by $Q-14$, so its area is $5(Q-14)$. Then another expression for the shaded area, B, is $Q(Q-14)-5(Q-14)$.

It is possible to find other equivalent expressions also.

126. $F^2-(F-7)(F-17)$; $24F-119$; other equivalent expressions are possible.

127.

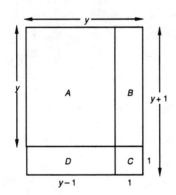

The dimensions of the shaded area, regions A and D together, are $y+1$ by $y-1$ so the area is $(y+1)(y-1)$.

To find another expression we add the areas of regions A and D. The dimensions of region A are y by $y-1$, and the dimensions of region D are $y-1$ by 1, so the sum of the areas is $y(y-1)+(y-1)(1)$, or $y(y-1)+y-1$.

It is possible to find other equivalent expressions also.

128. 10, 11, 12

129.

Exercise Set 4.6

1. We replace x by 5 and y by -2.
$x^2-3y^2+2xy=5^2-3(-2)^2+2\cdot5(-2)=$
$25-12-20=-7$.

2. 85

3. We replace x by 2, y by -3, and z by -4.
$xyz^2-z=2(-3)(-4)^2-(-4)=-96+4=-92$

4. 14

5. Evaluate the polynomial for $h=160$ and $A=50$.
$$0.041h-0.018A-2.69$$
$$=0.041(160)-0.018(50)-2.69$$
$$=6.56-0.9-2.69$$
$$=2.97$$
The woman's lung capacity is 2.97 liters.

6. 3.715 liters

7. Evaluate the polynomial for $h=300$, $v=40$, and $t=2$.
$$h+vt-4.9t^2$$
$$=300+40\cdot2-4.9(2)^2$$
$$=300+80-19.6$$
$$=360.4$$
The rocket will be 360.4 m above the ground 2 seconds after blast off.

8. 250.9 m

9. Evaluate the polynomial for $h=7\frac{1}{2}$, or $\frac{15}{2}$, $r=1\frac{1}{4}$, or $\frac{5}{4}$, and $\pi\approx3.14$.
$$2\pi rh+\pi r^2\approx2(3.14)\left(\frac{5}{4}\right)\left(\frac{15}{2}\right)+(3.14)\left(\frac{5}{4}\right)^2$$
$$\approx2(3.14)\left(\frac{5}{4}\right)\left(\frac{15}{2}\right)+(3.14)\left(\frac{25}{16}\right)$$
$$\approx58.875+4.90625$$
$$\approx63.78125$$
The surface area is about 63.78125 in^2.

10. 20.60625 in^2

11. $x^3y - 2xy + 3x^2 - 5$

Term	Coefficient	Degree	
x^3y	1	4	(Think: $x^3y = x^3y^1$)
$-2xy$	-2	2	(Think: $-2xy =$ $-2x^1y^1$)
$3x^2$	3	2	
-5	-5	0	(Think: $-5 = -5x^0$)

The degree of the polynomial is the degree of the term of highest degree. The term of highest degree is x^3y. Its degree is 4, so the degree of the polynomial is 4.

12. Coefficients: 1, -1, 9, 7

Degrees: 3, 2, 3, 0; 3

13. $17x^2y^3 - 3x^3yz - 7$

Term	Coefficient	Degree	
$17x^2y^3$	17	5	
$-3x^3yz$	-3	5	(Think: $-3x^3yz =$ $-3x^3y^1z^1$)
-7	-7	0	(Think: $-7 = -7x^0$)

The terms of highest degree are $17x^2y^3$ and $-3x^3yz$. Each has degree 5. The degree of the polynomial is 5.

14. Coefficients: 6, -1, 8, -1

Degrees: 0, 2, 4, 5; 5

15. $7a + b - 4a - 3b = (7-4)a + (1-3)b = 3a - 2b$

16. $3r - 3s$

17. $3x^2y - 2xy^2 + x^2 + 5x$

There are <u>no</u> like terms, so none of the terms can be collected.

18. $m^3 + 2m^2n - 3m^2 + 3mn^2$

19. $\quad 2u^2v - 3uv^2 + 6u^2v - 2uv^2 + 7u^2$
$= (2+6)u^2v + (-3-2)uv^2 + 7u^2$
$= 8u^2v - 5uv^2 + 7u^2$

20. $-2x^2 - 4xy + 3y^2$

21. $\quad 5a^2c - 2ab^2 + a^2b - 3ab^2 + a^2c - 2ab^2$
$= (5+1)a^2c + (-2-3-2)ab^2 + a^2b$
$= 6a^2c - 7ab^2 + a^2b$

22. $2s^2t - 6r^2t - st^2$

23. $\quad (4x^2 - xy + y^2) + (-x^2 - 3xy + 2y^2)$
$= (4-1)x^2 + (-1-3)xy + (1+2)y^2$
$= 3x^2 - 4xy + 3y^2$

24. $-3r^3 + 2rs - 9s^2$

25. $\quad (3a^4 - 5ab + 6ab^2) - (9a^4 + 3ab - ab^2)$
$= 3a^4 - 5ab + 6ab^2 - 9a^4 - 3ab + ab^2$
$\qquad\qquad$ Adding the opposite
$= (3-9)a^4 + (-5-3)ab + (6+1)ab^2$
$= -6a^4 - 8ab + 7ab^2$

26. $-5r^2t - 6rt + 6rt^2$

27. $(5r^2 - 4rt + t^2) + (-6r^2 - 5rt - t^2) + (-5r^2 + 4rt - t^2)$

Observe that the polynomials $5r^2 - 4rt + t^2$ and $-5r^2 + 4rt - t^2$ are opposites. Thus, their sum is 0 and the sum in the exercise is the remaining polynomial, $-6r^2 - 5rt - t^2$.

28. $2x^2 - 3xy + y^2$

29. $\quad (x^3 - y^3) - (-2x^3 + x^2y - xy^2 + 2y^3)$
$= x^3 - y^3 + 2x^3 - x^2y + xy^2 - 2y^3$
$= 3x^3 - 3y^3 - x^2y + xy^2, \text{ or}$
$\quad 3x^3 - x^2y + xy^2 - 3y^3$

30. $6a^3 - 2a^2b + ab^2 - 2b^3$

31. $\quad (2y^4x^2 - 5y^3x) + (5y^4x^2 - y^3x) + (3y^4x^2 - 2y^3x)$
$= (2+5+3)y^4x^2 + (-5-1-2)y^3x$
$= 10y^4x^2 - 8y^3x$

32. $15a^2b - 4ab$

33. $\quad (4x + 5y) + (-5x + 6y) - (7x + 3y)$
$= 4x + 5y - 5x + 6y - 7x - 3y$
$= (4-5-7)x + (5+6-3)y$
$= -8x + 8y$

34. $-5b$

35. $\qquad\qquad\qquad\quad\ \ \text{F}\quad\ \ \text{O}\quad\ \ \text{I}\quad\ \ \text{L}$
$(3z - u)(2z + 3u) = 6z^2 + 9zu - 2uz - 3u^2$
$\qquad\qquad\qquad\ = 6z^2 + 7zu - 3u^2$

36. $10x^2 - 13xy - 3y^2$

37. $\qquad\qquad\qquad\quad\ \ \text{F}\quad\ \ \text{O}\quad\ \ \text{I}\quad\ \ \text{L}$
$(xy + 7)(xy - 4) = x^2y^2 - 4xy + 7xy - 28 - 28$
$\qquad\qquad\qquad\ = x^2y^2 + 3xy - 28$

38. $a^2b^2 - 2ab - 15$

39. $\quad (2a - b)(2a + b) \quad [(A+B)(A-B) = A^2 - B^2]$
$= 4a^2 - b^2$

40. $a^2 - 9b^2$

41. $(5rt - 2)(3rt + 1) = \overset{F}{15r^2t^2} + \overset{O}{5rt} - \overset{I}{6rt} - \overset{L}{2}$
$$= 15r^2t^2 - rt - 2$$

42. $12x^2y^2 + 2xy - 2$

43. $(m^3n + 8)(m^3n - 6)$
$$= \overset{F}{m^6n^2} - \overset{O}{6m^3n} + \overset{I}{8m^3n} - \overset{L}{48}$$
$$= m^6n^2 + 2m^3n - 48$$

44. $12 - c^2d^2 - c^4d^4$

45. $(6x - 2y)(5x - 3y)$
$$= \overset{F}{30x^2} - \overset{O}{18xy} - \overset{I}{10xy} + \overset{L}{6y^2}$$
$$= 30x^2 - 28xy + 6y^2$$

46. $35a^2 - 2ab - 24b^2$

47. $(pq + 0.2)(0.4pq - 0.1)$
$$= \overset{F}{0.4p^2q^2} - \overset{O}{0.1pq} + \overset{I}{0.08pq} - \overset{L}{0.02}$$
$$= 0.4p^2q^2 - 0.02pq - 0.02$$

48. $0.2a^2b^2 + 0.18ab - 0.18$

49. $(x + h)^2$
$$= x^2 + 2xh + h^2 \quad [(A + B)^2 = A^2 + 2AB + B^2]$$

50. $r^2 + 2rt + t^2$

51. $(4a + 5b)^2$
$$= 16a^2 + 40ab + 25b^2 \quad [(A+B)^2 = A^2 + 2AB + B^2]$$

52. $9x^2 + 12xy + 4y^2$

53. $(c^2 - d)(c^2 + d) = (c^2)^2 - d^2$
$$= c^4 - d^2$$

54. $p^6 - 25q^2$

55. $(ab + cd^2)(ab - cd^2) = (ab)^2 - (cd^2)^2$
$$= a^2b^2 - c^2d^4$$

56. $x^2y^2 - p^2q^2$

57. $(a + b - c)(a + b + c)$
$$= [(a + b) - c][(a + b) + c]$$
$$= (a + b)^2 - c^2$$
$$= a^2 + 2ab + b^2 - c^2$$

58. $x^2 + 2xy + y^2 - z^2$

59. $[a + b + c][a - (b + c)]$
$$= [a + (b + c)][a - (b + c)]$$
$$= a^2 - (b + c)^2$$
$$= a^2 - (b^2 + 2bc + c^2)$$
$$= a^2 - b^2 - 2bc - c^2$$

60. $a^2 - b^2 - 2bc - c^2$

61. The figure is a rectangle with dimensions $a + b$ by $a + c$. Its area is $(a + b)(a + c) = a^2 + ac + ab + bc$.

62. $x^2 + 2xy + y^2$

63. The figure is a parallelogram with base $x + z$ and height $x - z$. Thus the area is $(x+z)(x-z) = x^2 - z^2$.

64. $\frac{1}{2}a^2b^2 - 2$

65. The figure is a square with side $x + y + z$. Thus the area is
$$(x + y + z)^2$$
$$= [(x + y) + z]^2$$
$$= (x + y)^2 + 2(x + y)(z) + z^2$$
$$= x^2 + 2xy + y^2 + 2xz + 2yz + z^2.$$

66. $a^2 + 2ac + c^2 + ad + cd + ab + bc + bd$

67. The figure is a triangle with base $x + 2y$ and height $x - y$. Thus the area is $\frac{1}{2}(x + 2y)(x - y) = \frac{1}{2}(x^2 + xy - 2y^2) = \frac{1}{2}x^2 + \frac{1}{2}xy - y^2$.

68. $m^2 - n^2$

69. We draw a rectangle with dimensions $r + s$ by $u + v$.

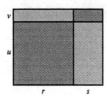

70.

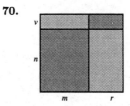

71. We draw a rectangle with dimensions $a + b + c$ by $a + d + f$.

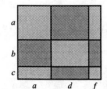

72.

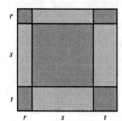

73. ◈

74. ◈

75.
$$5 + \frac{7 + 4 + 2 \cdot 5}{3}$$
$$= 5 + \frac{7 + 4 + 10}{3} \qquad \text{Multiplying}$$
$$= 5 + \frac{21}{3} \qquad \text{Adding in the numerator}$$
$$= 5 + 7 \qquad \text{Dividing}$$
$$= 12 \qquad \text{Adding}$$

76. 5

77.
$$(4 + 3 \cdot 5 + 8) \div 3 \cdot 3$$
$$= (4 + 15 + 8) \div 3 \cdot 3 \qquad \text{Multiplying inside the parentheses}$$
$$= 27 \div 3 \cdot 3 \qquad \text{Adding}$$
$$= 9 \cdot 3 \qquad \text{Dividing}$$
$$= 27 \qquad \text{Multiplying}$$

78. 36

79.
$$[3 \cdot 5 - 4 \cdot 2 + 7(-3)] \div (-2)$$
$$= (15 - 8 - 21) \div (-2) \qquad \text{Multiplying}$$
$$= -14 \div (-2) \qquad \text{Subtracting}$$
$$= 7 \qquad \text{Dividing}$$

80. 5

81. ◈

82. ◈

83. The unshaded region is a circle with radius $a - b$. Then the shaded area is the area of a circle with radius a less the area of a circle with radius $a - b$. Thus, we have:
$$\begin{aligned}\text{Shaded area} &= \pi a^2 - \pi(a - b)^2 \\ &= \pi a^2 - \pi(a^2 - 2ab + b^2) \\ &= \pi a^2 - \pi a^2 + 2\pi ab - \pi b^2 \\ &= 2\pi ab - \pi b^2\end{aligned}$$

84. $4xy - 4y^2$

85. The shaded area is the area of a square with side a less the areas of 4 squares with side b. Thus, the shaded area is $a^2 - 4 \cdot b^2$, or $a^2 - 4b^2$.

86. $\pi x^2 + 2xy$

87. a) The figure is a square with side A, so its area is $A \cdot A$, or A^2. The unshaded square has side B, so its area is $B \cdot B$, or B^2. Then the shaded area is $A^2 - B^2$.

b) In the upper left-hand corner we have a square with side $A - B$, so its area is $(A - B)(A - B)$, or $A^2 - 2AB + B^2$. The rectangles in the upper right-hand and lower left-hand corners have dimensions $A - B$ by B, so each has area $(A - B)(B)$, or $AB - B^2$. The sum of these three areas is $A^2 - 2AB + B^2 + AB - B^2 + AB - B^2$, or $A^2 - B^2$.

88. $2\pi nh + 2\pi mh + 2\pi n^2 - 2\pi m^2$

89. The surface area of the solid consists of the surface area of a rectangular solid with dimensions x by x by h less the areas of 2 circles with radius r plus the lateral surface area of a right circular cylinder with radius r and height h. Thus, we have
$$2x^2 + 2xh + 2xh - 2\pi r^2 + 2\pi rh, \text{ or}$$
$$2x^2 + 4xh - 2\pi r^2 + 2\pi rh.$$

90. ◈

91.
$$(x + a)(x - b)(x - a)(x + b)$$
$$= [(x + a)(x - a)][(x - b)(x + b)]$$
$$= (x^2 - a^2)(x^2 - b^2)$$
$$= x^4 - b^2 x^2 - a^2 x^2 + a^2 b^2$$

92. $P + 2Pr + Pr^2$

93. Replace t with 2 and multiply.
$$P(1 - r)^2$$
$$= P(1 - 2r + r^2)$$
$$= P - 2Pr + Pr^2$$

94. $15,638.03

95. Substitute $90,000 for P, 12.5% or 0.125 for r, and 4 for t.
$$P(1-r)^t$$
$$= \$90,000(1-0.125)^4$$
$$\approx \$52,756.35$$

Exercise Set 4.7

1. $\dfrac{40x^5 - 16x}{8} = \dfrac{40x^5}{8} - \dfrac{16x}{8}$
$$= \dfrac{40}{8}x^5 - \dfrac{16}{8}x \quad \text{Dividing coefficients}$$
$$= 5x^5 - 2x$$

To check, we multiply the quotient by 8:
$$(5x^5 - 2x)8 = 40x^5 - 16x$$
The answer checks.

2. $2a^4 - \dfrac{1}{2}a^2$

3. $\dfrac{u - 2u^2 + u^7}{u}$
$$= \dfrac{u}{u} - \dfrac{2u^2}{u} + \dfrac{u^7}{u}$$
$$= 1 - 2u + u^6$$

Check: We multiply.
$$\begin{array}{r} 1 - 2u + u^6 \\ u \\ \hline u - 2u^2 + u^7 \end{array}$$

4. $50x^4 - 7x^3 + x$

5. $(15t^3 - 24t^2 + 6t) \div (3t)$
$$= \dfrac{15t^3 - 24t^2 + 6t}{3t}$$
$$= \dfrac{15t^3}{3t} - \dfrac{24t^2}{3t} + \dfrac{6t}{3t}$$
$$= 5t^2 - 8t + 2$$

Check: We multiply.
$$\begin{array}{r} 5t^2 - 8t + 2 \\ 3t \\ \hline 15t^3 - 24t^2 + 6t \end{array}$$

6. $4t^2 - 3t + 6$

7. $(25x^6 - 20x^4 - 5x^2) \div (-5x^2)$
$$= \dfrac{25x^6 - 20x^4 - 5x^2}{-5x^2}$$
$$= \dfrac{25x^6}{-5x^2} - \dfrac{20x^4}{-5x^2} - \dfrac{5x^2}{-5x^2}$$
$$= -5x^4 - (-4x^2) - (-1)$$
$$= -5x^4 + 4x^2 + 1$$

Check: We multiply.
$$\begin{array}{r} -5x^4 + 4x^2 + 1 \\ -5x^2 \\ \hline 25x^6 - 20x^4 - 5x^2 \end{array}$$

8. $-2x^4 - 4x^3 + 1$

9. $(24t^5 - 40t^4 + 6t^3) \div (4t^3)$
$$= \dfrac{24t^5 - 40t^4 + 6t^3}{4t^3}$$
$$= \dfrac{24t^5}{4t^3} - \dfrac{40t^4}{4t^3} + \dfrac{6t^3}{4t^3}$$
$$= 6t^2 - 10t + \dfrac{3}{2}$$

Check: We multiply.
$$\begin{array}{r} 6t^2 - 10t + \dfrac{3}{2} \\ 4t^3 \\ \hline 24t^5 - 40t^4 + 6t^3 \end{array}$$

10. $2t^3 - 3t^2 - \dfrac{1}{3}$

11. $\dfrac{6x^2 - 10x + 1}{2}$
$$= \dfrac{6x^2}{2} - \dfrac{10x}{2} + \dfrac{1}{2}$$
$$= 3x^2 - 5x + \dfrac{1}{2}$$

Check: We multiply.
$$\begin{array}{r} 3x^2 - 5x + \dfrac{1}{2} \\ 2 \\ \hline 6x^2 - 10x + 1 \end{array}$$

12. $3x^2 + x - \dfrac{2}{3}$

13. $\dfrac{4x^3 + 6x^2 + 4x}{2x^2}$
$$= \dfrac{4x^3}{2x^2} + \dfrac{6x^2}{2x^2} + \dfrac{4x}{2x^2}$$
$$= 2x + 3 + \dfrac{2}{x}$$

Check: We multiply.

$$2x \ + \ 3 \ + \ \dfrac{2}{x}$$

$$\dfrac{\qquad\qquad 2x^2}{4x^3 \ + \ 6x^2 \ + \ 4x}$$

14. $2x^2 + 3x + \dfrac{1}{x}$

15.
$$\dfrac{9r^2s^2 + 3r^2s - 6rs^2}{-3rs}$$
$$= \dfrac{9r^2s^2}{-3rs} + \dfrac{3r^2s}{-3rs} - \dfrac{6rs^2}{-3rs}$$
$$= -3rs - r + 2s$$

Check: We multiply.

$$-3rs \ - \ r \ + \ 2s$$
$$\dfrac{\qquad\qquad\qquad -3rs}{9r^2s^2 \ + \ 3r^2s \ - \ 6rs^2}$$

16. $1 - 2x^2y + 3x^4y^5$

17.
$$\begin{array}{r} x + 6 \\ x - 2 \overline{\smash{\big)}\, x^2 + 4x - 12} \\ \underline{x^2 - 2x} \\ 6x - 12 \leftarrow (x^2 + 4x) - (x^2 - 2x) = 6x \\ \underline{6x - 12} \\ 0 \leftarrow (6x - 12) - (6x - 12) = 0 \end{array}$$

The answer is $x + 6$.

18. $x - 2$

19.
$$\begin{array}{r} t - 5 \\ t - 5 \overline{\smash{\big)}\, t^2 - 10t - 20} \\ \underline{t^2 - 5t} \\ -5t - 20 \leftarrow (t^2 - 10t) - (t^2 - 5t) = \\ -5t \\ \underline{-5t + 25} \\ -45 \leftarrow (-5t - 20) - (-5t + 25) = \\ -45 \end{array}$$

The answer is $t - 5 + \dfrac{-45}{t-5}$, or $t - 5 - \dfrac{45}{t-5}$.

20. $t + 4 + \dfrac{-31}{t+4}$

21.
$$\begin{array}{r} 2x - 1 \\ x + 6 \overline{\smash{\big)}\, 2x^2 + 11x - 5} \\ \underline{2x^2 + 12x} \\ -x - 5 \leftarrow (2x^2 + 11x) - (2x^2 + 12x) = \\ -x \\ \underline{-x - 6} \\ 1 \leftarrow (-x - 5) - (-x - 6) = 1 \end{array}$$

The answer is $2x - 1 + \dfrac{1}{x+6}$.

22. $3x + 4 + \dfrac{-5}{x-2}$

23.
$$\begin{array}{r} a^2 - 2a + 4 \\ a + 2 \overline{\smash{\big)}\, a^3 + 0a^2 + 0a + 8} \quad \leftarrow \text{Writing in the missing} \\ \text{terms} \\ \underline{a^3 + 2a^2} \\ -2a^2 + 0a \quad \leftarrow a^3 - (a^3 + 2a^2) = -2a^2 \\ \underline{-2a^2 - 4a} \\ 4a + 8 \leftarrow -2a^2 - (-2a^2 - 4a) = 4a \\ \underline{4a + 8} \\ 0 \leftarrow (4a + 8) - (4a + 8) = 0 \end{array}$$

The answer is $a^2 - 2a + 4$.

24. $t^2 - 3t + 9$

25.
$$\begin{array}{r} t + 4 \\ t - 4 \overline{\smash{\big)}\, t^2 + 0t - 15} \quad \leftarrow \text{Writing in the missing term} \\ \underline{t^2 - 4t} \\ 4t - 15 \leftarrow t^2 - (t^2 - 4t) = 4t \\ \underline{4t - 16} \\ 1 \leftarrow (4t - 15) - (4t - 16) = 1 \end{array}$$

The answer is $t + 4 + \dfrac{1}{t-4}$.

26. $a + 5 + \dfrac{2}{a-5}$

27.
$$\begin{array}{r} x + 4 \\ 3x - 1 \overline{\smash{\big)}\, 3x^2 + 11x - 4} \\ \underline{3x^2 - x} \\ 12x - 4 \leftarrow (3x^2 + 11x) - (3x^2 - x) = \\ 12x \\ \underline{12x - 4} \\ 0 \leftarrow (12x - 4) - (12x - 4) = 0 \end{array}$$

The answer is $x + 4$.

28. $2x + 3$

29.
$$\begin{array}{r} 3a + 1 \\ 2a + 5 \overline{\smash{\big)}\, 6a^2 + 17a + 8} \\ \underline{6a^2 + 15a} \\ 2a + 8 \leftarrow (6a^2 + 17a) - (6a^2 + 15a) = \\ 2a + 8 \\ \underline{2a + 5} \\ 3 \leftarrow (2a + 8) - (2a + 5) = 3 \end{array}$$

The answer is $3a + 1 + \dfrac{3}{2a+5}$.

30. $5a + 2 + \dfrac{3}{2a+3}$

31.
$$\begin{array}{r} t^2 - 3t + 1 \\ 2t - 3 \overline{\smash{\big)}\ 2t^3 - 9t^2 + 11t - 3} \\ \underline{2t^3 - 3t^2} \\ -6t^2 + 11t \leftarrow (2t^3 - 9t^2) - (2t^3 - 3t^2) = \\ \underline{-6t^2 + 9t} \qquad -6t^2 \\ 2t - 3 \leftarrow (-6t^2 + 11t) - \\ \underline{2t - 3} \quad (-6t^2 + 9t) = 2t \\ 0 \leftarrow (2t - 3) - (2t - 3) = 0 \end{array}$$

The answer is $t^2 - 3t + 1$.

32. $2t^2 - 7t + 4$

33.
$$\begin{array}{r} t^2 - 2t + 3 \\ t + 1 \overline{\smash{\big)}\ t^3 - t^2 + t - 1} \\ \underline{t^3 + t^2} \\ -2t^2 + t \quad \leftarrow (t^3 - t^2) - (t^3 + t^2) = -2t^2 \\ \underline{-2t^2 - 2t} \\ 3t - 1 \quad \leftarrow (-2t^2 + t) - \\ (-2t^2 - 2t) = 3t \\ \underline{3t + 3} \\ -4 \end{array}$$

The answer is $t^2 - 2t + 3 + \dfrac{-4}{t+1}$, or

$t^2 - 2t + 3 - \dfrac{4}{t+1}$.

34. $x^2 + 1$

35.
$$\begin{array}{r} t^2 \qquad - 1 \\ t^2 + 5 \overline{\smash{\big)}\ t^4 + 0t^3 + 4t^2 + 3t - 6} \leftarrow \text{Writing in the} \\ \underline{t^4 \qquad + 5t^2} \qquad\qquad \text{missing term} \\ -t^2 + 3t - 6 \leftarrow (t^4 + 4t^2) - \\ \underline{-t^2 \qquad -5} \quad (t^4 + 5t^2) = -t^2 \\ 3t - 1 \leftarrow (-t^2 + 3t - 6) - \\ (-t^2 - 5) = 3t - 1 \end{array}$$

The answer is $t^2 - 1 + \dfrac{3t-1}{t^2+5}$.

36. $t^2 + 1 + \dfrac{4t-2}{t^2-3}$

37.
$$\begin{array}{r} 2x^2 \qquad + 1 \\ 2x^2 - 3 \overline{\smash{\big)}\ 4x^4 + 0x^3 - 4x^2 - x - 3} \leftarrow \text{Writing in the} \\ \underline{4x^4 \qquad -6x^2} \qquad\qquad \text{missing term} \\ 2x^2 - x - 3 \leftarrow (4x^4 - 4x^2) - \\ (4x^4 - 6x^2) = 2x^2 \\ \underline{2x^2 \qquad -3} \\ -x \leftarrow (2x^2 - x - 3) - \\ (2x^2 - 3) = -x \end{array}$$

The answer is $2x^2 + 1 + \dfrac{-x}{2x^2-3}$, or

$2x^2 + 1 - \dfrac{x}{2x^2-3}$.

38. $3x^2 - 3 + \dfrac{x-1}{2x^2+1}$

39. ◈

40. ◈

41. $-4 + (-13)$ Two negative numbers.
Add the absolute values,
4 and 13, to get 17. Make
the answer negative.

$-4 + (-13) = -17$

42. -23

43. $-9 - (-7) = -9 + 7 = -2$

44. 5

45. Familiarize. Let $w =$ the width. Then $w + 15 =$ the length. We draw a picture.

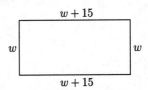

We will use the fact that the perimeter is 640 ft to find w (the width). Then we can find $w + 15$ (the length) and multiply the length and the width to find the area.

Translate.

Width+Width+ Length + Length =Perimeter
$\quad w \quad + \quad w \quad +(w+15)+(w+15)= \quad 640$

Carry out.
$$\begin{aligned} w + w + (w+15) + (w+15) &= 640 \\ 4w + 30 &= 640 \\ 4w &= 610 \\ w &= 152.5 \end{aligned}$$

If the width is 152.5, then the length is $152.5 + 15$, or 167.5.

Check. The length, 167.5 ft, is 15 ft greater than the width, 152.5 ft. The perimeter is $152.5 + 152.5 + 167.5 + 167.5$, or 640 ft. The answer checks.

State. The length is 167.5 ft.

46. 2

47. Graph: $3x - 2y = 12$.

We will graph the equation using intercepts. To find the y-intercept, we let $x = 0$.

$\qquad -2y = 12$ Ignoring the x-term

$\qquad\quad y = -6$

The y-intercept is $(0, -6)$.

To find the x-intercept, we let $y = 0$.

$$3x = 12 \quad \text{Ignoring the } y\text{-term}$$
$$x = 4$$

The x-intercept is $(4, 0)$.

To find a third point, replace x with -2 and solve for y:

$$3(-2) - 2y = 12$$
$$-6 - 2y = 12$$
$$-2y = 18$$
$$y = -9$$

The point $(-2, -9)$ appears to line up with the intercepts, so we draw the graph.

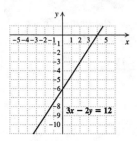

48.

49. ◈

50. ◈

51. $(10x^{9k} - 32x^{6k} + 28x^{3k}) \div (2x^{3k})$

$$= \frac{10x^{9k} - 32x^{6k} + 28x^{3k}}{2x^{3k}}$$

$$= \frac{10x^{9k}}{2x^{3k}} - \frac{32x^{6k}}{2x^{3k}} + \frac{28x^{3k}}{2x^{3k}}$$

$$= 5x^{9k-3k} - 16x^{6k-3k} + 14x^{3k-3k}$$

$$= 5x^{6k} - 16x^{3k} + 14$$

52. $15a^{6k} + 10a^{4k} - 20a^{2k}$

53.
$$
\begin{array}{r}
3t^{2h} + 2t^h - 5 \\
2t^h + 3 \overline{\smash{)}\ 6t^{3h} + 13t^{2h} - 4t^h - 15} \\
\underline{6t^{3h} + 9t^{2h}} \\
4t^{2h} - 4t^h \\
\underline{4t^{2h} + 6t^h} \\
-10t^h - 15 \\
\underline{-10t^h - 15} \\
0
\end{array}
$$

The answer is $3t^{2h} + 2t^h - 5$.

54. $x^3 - ax^2 + a^2x - a^3 + \dfrac{a^2 + a^4}{x + a}$

55.
$$
\begin{array}{r}
a + 3 \\
5a^2 - 7a - 2 \overline{\smash{)}\ 5a^3 + 8a^2 - 23a - 1} \\
\underline{5a^3 - 7a^2 - 2a} \\
15a^2 - 21a - 1 \\
\underline{15a^2 - 21a - 6} \\
5
\end{array}
$$

The answer is $a + 3 + \dfrac{5}{5a^2 - 7a - 2}$.

56. $5y + 2 + \dfrac{-10y + 11}{3y^2 - 5y - 2}$

57.
$$(4x^5 - 14x^3 - x^2 + 3) +$$
$$(2x^5 + 3x^4 + x^3 - 3x^2 + 5x)$$
$$= 6x^5 + 3x^4 - 13x^3 - 4x^2 + 5x + 3$$

$$
\begin{array}{r}
2x^2 + x - 3 \\
3x^3 - 2x - 1 \overline{\smash{)}\ 6x^5 + 3x^4 - 13x^3 - 4x^2 + 5x + 3} \\
\underline{6x^5 \qquad\quad - 4x^3 - 2x^2} \\
3x^4 - 9x^3 - 2x^2 + 5x \\
\underline{3x^4 \qquad\quad - 2x^2 - x} \\
-9x^3 \qquad\quad + 6x + 3 \\
\underline{-9x^3 \qquad\quad + 6x + 3} \\
0
\end{array}
$$

The answer is $2x^2 + x - 3$.

58. $5x^5 + 5x^4 - 8x^2 - 8x + 2$

59.
$$
\begin{array}{r}
x - 3 \\
x - 1 \overline{\smash{)}\ x^2 - 4x + c} \\
\underline{x^2 - x} \\
-3x + c \\
\underline{-3x + 3} \\
c - 3
\end{array}
$$

We set the remainder equal to 0.

$$c - 3 = 0$$
$$c = 3$$

Thus, c must be 3.

60. -2

61.

$$
\begin{array}{r}
c^2x + \ (2c+c^2) \\
x-1\ \overline{\smash{\big)}\ c^2x^2+2cx+1} \\
\underline{c^2x^2-\ c^2x} \\
(2c+c^2)x+1 \\
\underline{(2c+c^2)x-(2c+c^2)} \\
1+(2c+c^2)
\end{array}
$$

We set the remainder equal to 0.

$$c^2 + 2c + 1 = 0$$
$$(c+1)^2 = 0$$
$$c+1 = 0 \quad or \quad c+1 = 0$$
$$c = -1 \quad or \qquad c = -1$$

Thus, c must be -1.

Exercise Set 4.8

1. $5^{-2} = \dfrac{1}{5^2} = \dfrac{1}{25}$

2. $\dfrac{1}{2^4} = \dfrac{1}{16}$

3. $10^{-4} = \dfrac{1}{10^4} = \dfrac{1}{10,000}$

4. $\dfrac{1}{5^3} = \dfrac{1}{125}$

5. $(-2)^{-6} = \dfrac{1}{(-2)^6} = \dfrac{1}{64}$

6. $\dfrac{1}{(-3)^4} = \dfrac{1}{81}$

7. $x^{-8} = \dfrac{1}{x^8}$

8. $\dfrac{1}{t^5}$

9. $xy^{-2} = x \cdot \dfrac{1}{y^2} = \dfrac{x}{y^2}$

10. $\dfrac{b}{a^3}$

11. $r^{-5}t = \dfrac{1}{r^5} \cdot t = \dfrac{t}{r^5}$

12. $\dfrac{x}{y^9}$

13. $\dfrac{1}{t^{-7}} = t^7$

14. z^9

15. $\dfrac{1}{h^{-8}} = h^8$

16. a^{12}

17. $7^{-1} = \dfrac{1}{7^1} = \dfrac{1}{7}$

18. $\dfrac{1}{3}$

19. $\left(\dfrac{2}{5}\right)^{-2} = \left(\dfrac{5}{2}\right)^2 = \dfrac{5^2}{2^2} = \dfrac{25}{4}$

20. $\left(\dfrac{4}{3}\right)^2 = \dfrac{16}{9}$

21. $\left(\dfrac{a}{2}\right)^{-3} = \left(\dfrac{2}{a}\right)^3 = \dfrac{2^3}{a^3} = \dfrac{8}{a^3}$

22. $\left(\dfrac{3}{x}\right)^4 = \dfrac{81}{x^4}$

23. $\left(\dfrac{s}{t}\right)^{-7} = \left(\dfrac{t}{s}\right)^7 = \dfrac{t^7}{s^7}$

24. $\left(\dfrac{v}{r}\right)^5 = \dfrac{v^5}{r^5}$

25. $\dfrac{1}{7^2} = 7^{-2}$

26. 5^{-2}

27. $\dfrac{1}{t^6} = t^{-6}$

28. y^{-2}

29. $\dfrac{1}{a^4} = a^{-4}$

30. t^{-5}

31. $\dfrac{1}{p^8} = p^{-8}$

32. m^{-12}

33. $\dfrac{1}{5} = \dfrac{1}{5^1} = 5^{-1}$

34. 8^{-1}

35. $\dfrac{1}{t} = \dfrac{1}{t^1} = t^{-1}$

36. m^{-1}

37. $2^{-5} \cdot 2^8 = 2^{-5+8} = 2^3$, or 8

38. 5

39. $x^{-2} \cdot x^{-7} = x^{-2+(-7)} = x^{-9} = \dfrac{1}{x^9}$

40. $\dfrac{1}{x^{11}}$

41. $t^{-3} \cdot t = t^{-3} \cdot t^1 = t^{-3+1} = t^{-2} = \dfrac{1}{t^2}$

42. $\dfrac{1}{y^4}$

43. $(a^{-2})^9 = a^{-2 \cdot 9} = a^{-18} = \dfrac{1}{a^{18}}$

44. $\dfrac{1}{x^{30}}$

45. $(t^{-3})^{-6} = t^{-3(-6)} = t^{18}$

46. a^{28}

47. $(t^4)^{-3} = t^{4(-3)} = t^{-12} = \dfrac{1}{t^{12}}$

48. $\dfrac{1}{t^{10}}$

49. $(x^{-2})^{-4} = x^{-2(-4)} = x^8$

50. t^{30}

51. $(ab)^{-3} = \dfrac{1}{(ab)^3} = \dfrac{1}{a^3 b^3}$

52. $\dfrac{1}{x^6 y^6}$

53. $(mn)^{-7} = \dfrac{1}{(mn)^7} = \dfrac{1}{m^7 n^7}$

54. $\dfrac{1}{a^9 b^9}$

55. $(3x^{-4})^2 = 3^2 (x^{-4})^2 = 9x^{-8} = \dfrac{9}{x^8}$

56. $\dfrac{8}{a^{15}}$

57. $(5r^{-4}t^3)^2 = 5^2 (r^{-4})^2 (t^3)^2 = 25 r^{-8} t^6 = \dfrac{25 t^6}{r^8}$

58. $\dfrac{64 x^{15}}{y^{18}}$

59. $\dfrac{t^7}{t^{-3}} = t^{7-(-3)} = t^{10}$

60. x^9

61. $\dfrac{y^{-7}}{y^{-3}} = y^{-7-(-3)} = y^{-4} = \dfrac{1}{y^4}$

62. $\dfrac{1}{z^4}$

63. $\dfrac{y^{-4}}{y^{-9}} = y^{-4-(-9)} = y^5$

64. a^4

65. $\dfrac{x^6}{x} = \dfrac{x^6}{x^1} = x^{6-1} = x^5$

66. x^2

67. $\dfrac{a^{-7}}{b^{-9}} = \dfrac{b^9}{a^7}$

68. $\dfrac{y^{10}}{x^6}$

69. $\dfrac{t^{-7}}{t^{-7}}$

Note that we have an expression divided by itself. Thus, the result is 1. We could also find this result as follows:
$$\dfrac{t^{-7}}{t^{-7}} = t^{-7-(-7)} = t^0 = 1.$$

70. $\dfrac{b^7}{a^5}$

71. $\dfrac{3x^{-5}}{y^{-6} z^{-2}} = \dfrac{3 y^6 z^2}{x^5}$

72. $\dfrac{4 b^5 c^7}{a^6}$

73. $\dfrac{3t^4}{s^{-2} u^{-4}} = 3 s^2 t^4 u^4$

74. $\dfrac{5 y^3}{x^8 z^2}$

75. $(x^4 y^5)^{-3} = (x^4)^{-3} (y^5)^{-3} = x^{-12} y^{-15} = \dfrac{1}{x^{12} y^{15}}$

76. $\dfrac{1}{t^{20} x^{12}}$

77. $(x^{-6} y^{-2})^{-4} = (x^{-6})^{-4} (y^{-2})^{-4} = x^{24} y^8$

78. $x^{10} y^{35}$

79. $(a^{-5} b^7 c^{-2})(a^{-3} b^{-2} c^6) = a^{-5+(-3)} b^{7+(-2)} c^{-2+6} = a^{-8} b^5 c^4 = \dfrac{b^5 c^4}{a^8}$

80. $\dfrac{z^4}{x y^6}$

81. $\left(\dfrac{a^4}{3}\right)^{-2} = \left(\dfrac{3}{a^4}\right)^2 = \dfrac{3^2}{(a^4)^2} = \dfrac{9}{a^8}$

82. $\dfrac{4}{y^4}$

83. $\left(\dfrac{7}{x^{-3}}\right)^2 = (7x^3)^2 = 7^2(x^3)^2 = 49x^6$

84. $27a^6$

85. $\left(\dfrac{m^{-1}}{n^{-4}}\right)^3 = \dfrac{(m^{-1})^3}{(n^{-4})^3} = \dfrac{m^{-3}}{n^{-12}} = \dfrac{n^{12}}{m^3}$

86. $x^6 y^3 z^{15}$

87. $\left(\dfrac{2a^2}{3b^4}\right)^{-3} = \left(\dfrac{3b^4}{2a^2}\right)^3 = \dfrac{(3b^4)^3}{(2a^2)^3} = \dfrac{3^3(b^4)^3}{2^3(a^2)^3} = \dfrac{27b^{12}}{8a^6}$

88. $\dfrac{c^5 d^{15}}{a^{10} b^5}$

89. $\left(\dfrac{5x^{-2}}{3y^{-2}z}\right)^0$

Any nonzero expression raised to the 0 power is equal to 1. Thus, the answer is 1.

90. $\dfrac{4a^3 c^3}{5b^2}$

91. 7.12×10^4

Since the exponent is positive, the decimal point will move to the right.

7.1200.   The decimal point moves right 4 places.

$7.12 \times 10^4 = 71,200$

92. 892

93. 8.92×10^{-3}

Since the exponent is negative, the decimal point will move to the left.

.008.92   The decimal point moves left 3 places.

$8.92 \times 10^{-3} = 0.00892$

94. 0.000726

95. 9.04×10^8

Since the exponent is positive, the decimal point will move to the right.

9.04000000.

⎣_____↑   8 places

$9.04 \times 10^8 = 904,000,000$

96. 13,500,000

97. 2.764×10^{-10}

Since the exponent is negative, the decimal point will move to the left.

0.0000000002.764

↑⎣_____⎦   10 places

$2.764 \times 10^{-10} = 0.0000000002764$

98. 0.009043

99. 4.209×10^9

Since the exponent is positive, the decimal point will move to the right.

4.2090000.

⎣_____↑   7 places

$4.209 \times 10^7 = 42,090,000$

100. 502,900,000

101. $490,000 = 4.9 \times 10^m$

To write 4.9 as 490,000 we move the decimal point 5 places to the right. Thus, m is 5 and

$490,000 = 4.9 \times 10^5$.

102. 7.15×10^4

103. $0.00583 = 5.83 \times 10^m$

To write 5.83 as 0.00583 we move the decimal point 3 places to the left. Thus, m is -3 and

$0.00583 = 5.83 \times 10^{-3}$.

104. 8.14×10^{-2}

105. $78,000,000,000 = 7.8 \times 10^m$

To write 7.8 as 78,000,000,000 we move the decimal point 10 places to the right. Thus, m is 10 and

$78,000,000,000 = 7.8 \times 10^{10}$.

106. 3.7×10^{12}

107. $907,000,000,000,000,000 = 9.07 \times 10^m$

To write 9.07 as 907,000,000,000,000,000 we move the decimal point 17 places to the right. Thus, m is 17 and

$907,000,000,000,000,000 = 9.07 \times 10^{17}$.

108. 1.68×10^{14}

109. $0.000000527 = 5.27 \times 10^m$

To write 5.27 as 0.000000527 we move the decimal point 7 places to the left. Thus, m is -7 and

$0.000000527 = 5.27 \times 10^{-7}$.

110. 6.48×10^{-9}

111. $0.000000018 = 1.8 \times 10^m$

To write 1.8 as 0.000000018 we move the decimal point 8 places to the left. Thus, m is -8 and
$$0.000000018 = 1.8 \times 10^{-8}.$$

112. 2×10^{-11}

113. $1,094,000,000,000,000 = 1.094 \times 10^m$

To write 1.094 as 1,094,000,000,000,000 we move the decimal point 15 places to the right. Thus, m is 15 and
$$1,094,000,000,000,000 = 1.094 \times 10^{15}.$$

114. 1.0302×10^{18}

115.
$$
\begin{aligned}
(4 \times 10^7)(2 \times 10^5) &= (4 \cdot 2) \times (10^7 \cdot 10^5) \\
&= 8 \times 10^{7+5} \quad \text{Adding exponents} \\
&= 8 \times 10^{12}
\end{aligned}
$$

116. 6.46×10^5

117.
$$
\begin{aligned}
(3.8 \times 10^9)(6.5 \times 10^{-2}) &= (3.8 \cdot 6.5) \times (10^9 \cdot 10^{-2}) \\
&= 24.7 \times 10^7
\end{aligned}
$$

The answer is not yet in scientific notation since 24.7 is not a number between 1 and 10. We convert to scientific notation.
$$24.7 \times 10^7 = (2.47 \times 10) \times 10^7 = 2.47 \times 10^8$$

118. 6.106×10^{-11}

119.
$$
\begin{aligned}
&(8.7 \times 10^{-12})(4.5 \times 10^{-5}) \\
&= (8.7 \cdot 4.5) \times (10^{-12} \cdot 10^{-5}) \\
&= 39.15 \times 10^{-17}
\end{aligned}
$$

The answer is not yet in scientific notation since 39.15 is not a number between 1 and 10. We convert to scientific notation.
$$39.15 \times 10^{-17} = (3.915 \times 10) \times 10^{-17} = 3.915 \times 10^{-16}$$

120. 2.914×10^{-6}

121.
$$
\begin{aligned}
\frac{8.5 \times 10^8}{3.4 \times 10^{-5}} &= \frac{8.5}{3.4} \times \frac{10^8}{10^{-5}} \\
&= 2.5 \times 10^{8-(-5)} \\
&= 2.5 \times 10^{13}
\end{aligned}
$$

122. 2.24×10^{-7}

123.
$$
\begin{aligned}
(3.0 \times 10^6) \div (6.0 \times 10^9) &= \frac{3.0 \times 10^6}{6.0 \times 10^9} \\
&= \frac{3.0}{6.0} \times \frac{10^6}{10^9} \\
&= 0.5 \times 10^{6-9} \\
&= 0.5 \times 10^{-3}
\end{aligned}
$$

The answer is not yet in scientific notation because 0.5 is not between 1 and 10. We convert to scientific notation.
$$
\begin{aligned}
0.5 \times 10^{-3} &= (5.0 \times 10^{-1}) \times 10^{-3} = \\
&\quad 5.0 \times 10^{-4}
\end{aligned}
$$

124. 9.375×10^2

125.
$$
\begin{aligned}
\frac{7.5 \times 10^{-9}}{2.5 \times 10^{12}} &= \frac{7.5}{2.5} \times \frac{10^{-9}}{10^{12}} \\
&= 3.0 \times 10^{-9-12} \\
&= 3.0 \times 10^{-21}
\end{aligned}
$$

126. 5×10^{-24}

127. ◈

128. ◈

129.
$$
\begin{aligned}
(3 - 8)(9 - 12) & \\
= (-5)(-3) \quad &\text{Subtracting} \\
= 15 \quad &\text{Multiplying}
\end{aligned}
$$

130. 49

131.
$$
\begin{aligned}
7 \cdot 2 + 8^2 & \\
= 7 \cdot 2 + 64 \quad &\text{Evaluating the exponential expression} \\
= 14 + 64 \quad &\text{Multiplying} \\
= 78 \quad &\text{Adding}
\end{aligned}
$$

132. 6

133. To plot $(-3, 2)$, we start at the origin and move 3 units to the left and then 2 units up. To plot $(4, -1)$, we start at the origin and move 4 units to the right and then 1 unit down. To plot $(5, 3)$, we start at the origin and move 5 units to the right and 3 units up. To plot $(-5, -2)$, we start at the origin and move 5 units to the left and then 2 units down.

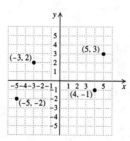

134. $t = \dfrac{r - cx}{b}$

135. ◈

136. ◈

137.
$$\frac{4.2 \times 10^8 [(2.5 \times 10^{-5}) \div (5.0 \times 10^{-9})]}{3.0 \times 10^{-12}}$$
$$= \frac{4.2 \times 10^8 [0.5 \times 10^4]}{3.0 \times 10^{-12}}$$
$$= \frac{2.1 \times 10^{12}}{3.0 \times 10^{-12}}$$
$$= 0.7 \times 10^{24}$$
$$= (7 \times 10^{-1}) \times 10^{24}$$
$$= 7 \times 10^{23}$$

138. 8×10^5

139. $\dfrac{1}{2.5 \times 10^9} = \dfrac{1}{2.5} \times \dfrac{1}{10^9} = 0.4 \times 10^{-9} =$
$(4 \times 10^{-1}) \times 10^{-9} = 4 \times 10^{-10}$

140. 2^{-12}

141. $81^3 \cdot 27 \div 9^2 = (3^4)^3 \cdot 3^3 \div (3^2)^2 = 3^{12} \cdot 3^3 \div 3^4 =$
$3^{15} \div 3^4 = 3^{11}$

142. 7

143. $\dfrac{125^{-4}(25^2)^4}{125} = \dfrac{(5^3)^{-4}((5^2)^2)^4}{5^3} =$
$\dfrac{5^{-12}(5^4)^4}{5^3} = \dfrac{5^{-12} \cdot 5^{16}}{5^3} = \dfrac{5^4}{5^3} = 5^1 = 5$

144. 9

145. a) False; let $x = 2$, $y = 3$, $m = 4$, and $n = 2$:
$$2^4 \cdot 3^2 = 16 \cdot 9 = 144, \text{ but}$$
$$(2 \cdot 3)^{4 \cdot 2} = 6^8 = 1,679,616$$
b) False; let $x = 3$, $y = 4$, and $m = 2$:
$$3^2 \cdot 4^2 = 9 \cdot 16 = 144, \text{ but}$$
$$(3 \cdot 4)^{2 \cdot 2} = 12^4 = 20,736$$
c) False; let $x = 5$, $y = 3$, and $m = 2$:
$$(5 - 3)^2 = 2^2 = 4, \text{ but}$$
$$5^2 - 3^2 = 25 - 9 = 16$$

146. $4.894179894 \times 10^{26}$

147.
$$\frac{5.8 \times 10^{17}}{(4.0 \times 10^{-13})(2.3 \times 10^4)}$$
$$= \frac{5.8}{(4.0 \cdot 2.3)} \times \frac{10^{17}}{(10^{-13} \cdot 10^4)}$$
$$\approx 0.6304347826 \times 10^{17-(-13)-4}$$
$$\approx (6.304347826 \times 10^{-1}) \times 10^{26}$$
$$\approx 6.304347826 \times 10^{25}$$

148. 3.12×10^{43}

149.
$$\frac{(2.5 \times 10^{-8})(6.1 \times 10^{-11})}{1.28 \times 10^{-3}}$$
$$= \frac{(2.5 \cdot 6.1)}{1.28} \times \frac{(10^{-8} \cdot 10^{-11})}{10^{-3}}$$
$$= 11.9140625 \times 10^{-8+(-11)-(-3)}$$
$$= 11.9140625 \times 10^{-16}$$
$$= (1.19140625 \times 10) \times 10^{-16}$$
$$= 1.19140625 \times 10^{-15}$$

150. $\$5.09425 \times 10^{12}$

151. *Familiarize.* Express 1 billion and 2500 in scientific notation:
$$1 \text{ billion} = 1,000,000,000 = 10^9$$
$$2500 = 2.5 \times 10^3$$
Let $b =$ the number of bytes in the network.

Translate. We reword the problem.

$$\underbrace{\text{What}}_{\downarrow} \ \underbrace{\text{is}}_{} \ \underbrace{2500}_{\downarrow} \ \underbrace{\text{times}}_{} \ \underbrace{1 \text{ gigabyte?}}_{\downarrow}$$
$$b \quad = \quad 2.5 \times 10^3 \quad \times \quad 10^9$$

Carry out. We do the computation.
$$b = (2.5 \times 10^3) \times 10^9$$
$$b = 2.5 \times (10^3 \times 10^9)$$
$$b = 2.5 \times 10^{12}$$

Check. We review the computation. Also, the answer seems reasonable since it is larger than 1 billion.

State. There are 2.5×10^{12} bytes in the network.

152. 1.325×10^{14} cubic feet

153. *Familiarize.* We must express both dimensions using the same units. Let's choose centimeters. First, convert 1.5 m to centimeters and express the result in scientific notation.

$$1.5 \text{ m} = 1.5 \times 1 \text{ m} = 1.5 \times 100 \text{ cm} = 1.5 \times 10^2 \text{ cm}$$

Let l represent how many times the DNA is longer than it is wide.

Translate. We reword the problem.

$$\underbrace{\text{The length}}_{\downarrow} \ \underbrace{\text{is}}_{} \ \underbrace{\text{how many}}_{\downarrow} \ \underbrace{\text{times}}_{} \ \underbrace{\text{the width.}}_{\downarrow}$$
$$1.5 \times 10^2 \quad = \quad l \quad \cdot \quad 1.3 \times 10^{-10}$$

Carry out. We solve the equation.
$$1.5 \times 10^2 = l \cdot 1.3 \times 10^{-10}$$
$$\frac{1.5 \times 10^2}{1.3 \times 10^{-10}} = l$$
$$1.15385 \times 10^{12} \approx l$$

Check. Since $(1.15385 \times 10^{12}) \times (1.3 \times 10^{-10}) = 1.498705 \times 10^2 \approx 1.5 \times 10^2$, the answer checks.

State. A strand of DNA is about 1.15385×10^{12} times longer than it is wide.

154. 2×10^{14} gal

Chapter 5

Polynomials and Factoring

1. Answers may vary. $10x^3 = (5x)(2x^2) = (10x^2)(x) = (-2)(-5x^3)$

2. Answers may vary. $(6x)(x^2)$; $(3x^2)(2x)$; $(2x^2)(3x)$

3. Answers may vary. $-15a^4 = (-15)(a^4) = (-5a)(3a^3) = (-3a^2)(5a^2)$

4. Answers may vary. $(-8t)(t^4)$; $(-2t^2)(4t^3)$; $(-4t^3)(2t^2)$

5. Answers may vary. $26x^5 = (2x^4)(13x) = (2x^3)(13x^2) = (-x^2)(-26x^3)$

6. Answers may vary. $(5x^2)(5x^2)$; $(x^3)(25x)$; $(-5x)(-5x^3)$

7. $x^2 + 8x = x \cdot x + x \cdot 8$
   $= x(x + 8)$

8. $x(x + 6)$

9. $10t^2 - 5t = 5t \cdot 2t - 5t \cdot 1$
   $= 5t(2t - 1)$

10. $5a(a - 3)$

11. $x^3 + 6x^2 = x^2 \cdot x + x^2 \cdot 6$
    $= x^2(x + 6)$

12. $x^2(4x^2 + 1)$

13. $8x^4 - 24x^2 = 8x^2 \cdot x^2 - 8x^2 \cdot 3$
    $= 8x^2(x^2 - 3)$

14. $5x^3(x^2 + 2)$

15. $2x^2 + 2x - 8 = 2 \cdot x^2 + 2 \cdot x - 2 \cdot 4$
    $= 2(x^2 + x - 4)$

16. $3(2x^2 + x - 5)$

17. $7a^6 - 10a^4 - 14a^2 = a^2 \cdot 7a^4 - a^2 \cdot 10a^2 - a^2 \cdot 14$
    $= a^2(7a^4 - 10a^2 - 14)$

18. $t^3(10t^2 - 15t + 9)$

19. $\quad 2x^8 + 4x^6 - 8x^4 + 10x^2$
    $= 2x^2 \cdot x^6 + 2x^2 \cdot 2x^4 - 2x^2 \cdot 4x^2 + 2x^2 \cdot 5$
    $= 2x^2(x^6 + 2x^4 - 4x^2 + 5)$

20. $5(x^4 - 3x^3 - 5x - 2)$

21. $\quad x^5y^5 + x^4y^3 + x^3y^3 - x^2y^2$
    $= x^2y^2 \cdot x^3y^3 + x^2y^2 \cdot x^2y + x^2y^2 \cdot xy - x^2y^2 \cdot 1$
    $= x^2y^2(x^3y^3 + x^2y + xy - 1)$

22. $x^3y^3(x^6y^3 - x^4y^2 + xy + 1)$

23. $\quad 5a^3b^4 + 10a^2b^3 - 15a^3b^2$
    $= 5a^2b^2 \cdot ab^2 + 5a^2b^2 \cdot 2b - 5a^2b^2 \cdot 3a$
    $= 5a^2b^2(ab^2 + 2b - 3a)$

24. $7r^3t^4(3r^2 - 2rt^2 + 3t^2)$

25. $\quad y(y - 2) + 7(y - 2)$
    $= (y - 2)(y + 7) \qquad$ Factoring out the common binomial factor $y - 2$

26. $(b + 5)(b + 3)$

27. $\quad x^2(x + 3) - 7(x + 3)$
    $= (x + 3)(x^2 - 7) \quad$ Factoring out the common binomial factor $x + 3$

28. $(2z + 9)(3z^2 + 1)$

29. $y^2(y + 8) + (y + 8) = y^2(y + 8) + 1(y + 8)$
    $= (y + 8)(y^2 + 1) \quad$ Factoring out the common factor

30. $(x - 7)(x^2 - 3)$

31. $\quad x^3 + 3x^2 + 4x + 12$
    $= (x^3 + 3x^2) + (4x + 12)$
    $= x^2(x + 3) + 4(x + 3) \qquad$ Factoring each binomial
    $= (x + 3)(x^2 + 4) \qquad$ Factoring out the common factor $x + 3$

32. $(2z + 1)(3z^2 + 1)$

33. $3a^3 + 9a^2 + 2a + 6$

$= (3a^3 + 9a^2) + (2a + 6)$

$= 3a^2(a+3) + 2(a+3)$ Factoring each binomial

$= (a+3)(3a^2 + 2)$ Factoring out the common factor $a+3$

34. $(3a+2)(a^2+2)$

35. $9x^3 - 12x^2 + 3x - 4$

$= 3x^2(3x-4) + 1(3x-4)$

$= (3x-4)(3x^2+1)$

36. $(2x-5)(5x^2+2)$

37. $4t^3 - 20t^2 + 3t - 15$

$= 4t^2(t-5) + 3(t-5)$

$= (t-5)(4t^2+3)$

38. $(3a-4)(2a^2+3)$

39. $7x^3 + 2x^2 - 14x - 4$

$= x^2(7x+2) - 2(7x+2)$

$= (7x+2)(x^2-2)$

40. $(5x+4)(x^2-2)$

41. $6a^3 - 7a^2 + 6a - 7$

$= a^2(6a-7) + 1(6a-7)$

$= (6a-7)(a^2+1)$

42. $(7t-5)(t^2+1)$

43. $x^3 + 8x^2 - 3x - 24 = x^2(x+8) - 3(x+8)$

$= (x+8)(x^2-3)$

44. $(x+7)(x^2-2)$

45. $2x^3 + 12x^2 - 5x - 30 = 2x^2(x+6) - 5(x+6)$

$= (x+6)(2x^2-5)$

46. $(x+5)(3x^2-5)$

47. $w^3 - 7w^2 + 4w - 28 = w^2(w-7) + 4(w-7)$

$= (w-7)(w^2+4)$

48. Cannot be factored by grouping

49. We try factoring by grouping.

$x^3 - x^2 - 2x + 5 = x^2(x-1) - 1(2x-5)$, or

$x^3 - 2x - x^2 + 5 = x(x^2-2) - (x^2-5)$

Because we cannot find a common binomial factor, this polynomial cannot be factored using factoring by grouping.

50. $(y+8)(y^2-2)$

51. $2x^3 - 8x^2 - 9x + 36 = 2x^2(x-4) - 9(x-4)$

$= (x-4)(2x^2-9)$

52. $(5g-1)(4g^2-5)$

53. ◈

54. ◈

55. $(x+3)(x+5)$

$\quad$ F $\quad$ O $\quad$ I $\quad$ L

$= x \cdot x + x \cdot 5 + 3 \cdot x + 3 \cdot 5$

$= x^2 + 5x + 3x + 15$

$= x^2 + 8x + 15$

56. $x^2 + 9x + 14$

57. $(a-7)(a+3)$

$\quad$ F $\quad$ O $\quad$ I $\quad$ L

$= a \cdot a + a \cdot 3 - 7 \cdot a - 7 \cdot 3$

$= a^2 + 3a - 7a - 21$

$= a^2 - 4a - 21$

58. $a^2 - 3a - 40$

59. $(2x+5)(3x-4)$

$\quad$ F $\quad$ O $\quad$ I $\quad$ L

$= 2x \cdot 3x - 2x \cdot 4 + 5 \cdot 3x - 5 \cdot 4$

$= 6x^2 - 8x + 15x - 20$

$= 6x^2 + 7x - 20$

60. $12t^2 - 13t - 14$

61. $(3t-5)^2$

$= (3t)^2 - 2 \cdot 3t \cdot 5 + 5^2$

$\quad [(A-B)^2 = A^2 - 2AB + B^2]$

$= 9t^2 - 30t + 25$

62. $4t^2 - 36t + 81$

63. ◈

64. ◈

65. $4x^5 + 6x^3 + 6x^2 + 9 = 2x^3(2x^2+3) + 3(2x^2+3)$

$= (2x^2+3)(2x^3+3)$

66. $(x^2+1)(x^4+1)$

67. $x^{12} + x^7 + x^5 + 1 = x^7(x^5+1) + (x^5+1)$

$= (x^5+1)(x^7+1)$

68. Cannot be factored by grouping

69.
$$5x^5 - 5x^4 + x^3 - x^2 + 3x - 3$$
$$= 5x^4(x-1) + x^2(x-1) + 3(x-1)$$
$$= (x-1)(5x^4 + x^2 + 3)$$

We could also do this exercise as follows:
$$5x^5 - 5x^4 + x^3 - x^2 + 3x - 3$$
$$= (5x^5 + x^3 + 3x) - (5x^4 + x^2 + 3)$$
$$= x(5x^4 + x^2 + 3) - 1(5x^4 + x^2 + 3)$$
$$= (5x^4 + x^2 + 3)(x-1)$$

70. $(x^2 + 2x + 3)(a + 1)$

71. Answers may vary. $3x^4y^3 - 9x^3y^3 + 27x^2y^4$

Exercise Set 5.2

1. $x^2 + 6x + 5$

Since the constant term and the coefficient of the middle term are both positive, we look for a factorization of 5 in which both factors are positive. The only possible positive factors are 1 and 5. Their sum is 6, so these are the numbers we want.
$$x^2 + 6x + 5 = (x+1)(x+5).$$

2. $(x+1)(x+6)$

3. $x^2 + 7x + 10$

Since the constant term is positive and the coefficient of the middle term is positive, we look for a factorization of 10 in which both factors are positive. Their sum must be 7.

Pairs of factors	Sums of factors
1, 10	11
2, 5	7

The numbers we want are 2 and 5.
$$x^2 + 7x + 10 = (x+2)(x+5)$$

4. $(x+3)(x+4)$

5. $y^2 + 11y + 28$

Since the constant term is positive and the coefficient of the middle term is positive, we look for a factorization of 28 in which both factors are positive. Their sum must be 11.

Pairs of factors	Sums of factors
1, 28	29
2, 14	16
4, 7	11

The numbers we want are 4 and 7.
$$y^2 + 11y + 28 = (y+4)(y+7)$$

6. $(x-3)(x-3)$, or $(x-3)^2$

7. $a^2 + 11a + 30$

Since the constant term is positive and the coefficient of the middle term is positive, we look for a factorization of 30 in which both factors are positive. Their sum must be 11.

Pairs of factors	Sums of factors
1, 30	31
2, 15	17
3, 10	13
5, 6	11

The numbers we want are 5 and 6.
$$a^2 + 11a + 30 = (a+5)(a+6).$$

8. $(x+2)(x+7)$

9. $x^2 - 5x + 4$

Since the constant term is positive and the coefficient of the middle term is negative, we look for a factorization of 4 in which both factors are negative. Their sum must be -5.

Pairs of factors	Sums of factors
−1, −4	−5
−2, −2	−4

The numbers we want are -1 and -4.
$$x^2 - 5x + 4 = (x-1)(x-4).$$

10. $(b+1)(b+4)$

11. $z^2 - 8z + 7$

Since the constant term is positive and the coefficient of the middle term is negative, we look for a factorization of 7 in which both factors are negative. Their sum must be -8. The only possible negative factors are -1 and -7. Their sum is -8, so these are the numbers we want.
$$z^2 - 8z + 7 = (z-1)(z-7)$$

12. $(a+2)(a-6)$

13. $x^2 - 8x + 15$

Since the constant term is positive and the coefficient of the middle term is negative, we look for a factorization of 15 in which both factors are negative. Their sum must be -8.

Pairs of factors	Sums of factors
−1, −15	−16
−3, −5	−8

The numbers we want are -3 and -5.
$$x^2 - 8x + 15 = (x-3)(x-5).$$

14. $(d-2)(d-5)$

15. $y^2 - 11y + 10$

Since the constant term is positive and the coefficient of the middle term is negative, we look for a factorization of 10 in which both factors are negative. Their sum must be -11.

Pairs of factors	Sums of factors
-1, -10	-11
-2, -5	-7

The numbers we want are -1 and -10.
$y^2 - 11y + 10 = (y-1)(y-10)$.

16. $(x+3)(x-5)$

17. $x^2 + x - 42$

Since the constant term is negative, we look for a factorization of -42 in which one factor is positive and one factor is negative. Their sum must be 1, the coefficient of the middle term, so the positive factor must have the larger absolute value. Thus we consider only pairs of factors in which the positive factor has the larger absolute value.

Pairs of factors	Sums of factors
-1, 42	41
-2, 21	19
-3, 14	11
-6, 7	1

The numbers we need are -6 and 7.
$x^2 + x - 42 = (x-6)(x+7)$.

18. $(x-3)(x+5)$

19. $2x^2 - 14x - 36 = 2(x^2 - 7x - 18)$

After factoring out the common factor, 2, we consider $x^2 - 7x - 18$. Since the constant term is negative, we look for a factorization of -18 in which one factor is positive and one factor is negative. Their sum must be -7, the coefficient of the middle term, so the negative factor must have the larger absolute value. Thus we consider only pairs of factors in which the negative factor has the larger absolute value.

Pairs of factors	Sums of factors
1, -18	-17
2, -9	-7
3, -6	-3

The numbers we want are 2 and -9. The factorization of $x^2 - 7x - 18$ is $(x+2)(x-9)$. We must not forget the common factor, 2. The factorization of $2x^2 - 14x - 36$ is $2(x+2)(x-9)$.

20. $3(y+4)(y-7)$

21. $x^3 - 6x^2 - 16x = x(x^2 - 6x - 16)$

After factoring out the common factor, x, we consider $x^2 - 6x - 16$. Since the constant term is negative, we look for a factorization of -16 in which one factor is positive and one factor is negative. Their sum must be -6, the coefficient of the middle term, so the negative factor must have the larger absolute value. Thus we consider only pairs of factors in which the negative factor has the large absolute value.

Pairs of factors	Sums of factors
1, -16	-15
2, -8	-6

The numbers we want are 2 and -8.
Then $x^2 - 6x - 16 = (x+2)(x-8)$, so $x^3 - 6x^2 - 16x = x(x+2)(x-8)$.

22. $x(x+6)(x-7)$

23. $y^2 + 4y - 45$

The constant term, 45, must be expressed as the product of a negative number and a positive number. Since the sum of those two numbers must be positive, the positive number must have the greater absolute value.

Pairs of factors	Sums of factors
-1, 45	44
-3, 15	12
-5, 9	4

The numbers we need are -5 and 9.
$y^2 + 4y - 45 = (y-5)(y+9)$

24. $(x-5)(x+12)$

25. $-2x - 99 + x^2 = x^2 - 2x - 99$

Since the constant term is negative, we look for a factorization of -99 in which one factor is positive and one factor is negative. Their sum must be -2, the coefficient of the middle term, so the negative factor must have the larger absolute value. Thus we consider only pairs of factors in which the negative factor has the larger absolute value.

Pairs of factors	Sums of factors
1, -99	-98
3, -33	-30
9, -11	-2

The numbers we want are 9 and -11.
$-2x - 99 + x^2 = (x+9)(x-11)$

26. $(x - 6)(x + 12)$

27. $c^4 + c^3 - 56c^2 = c^2(c^2 + c - 56)$

After factoring out the common factor, c^2, we consider $c^2 + c - 56$. Since the constant term is negative, we look for a factorization of -56 in which one factor is positive and one factor is negative. Their sum must be 1, so the positive factor must have the larger absolute value. Thus we consider only pairs of factors in which the positive factor has the larger absolute value.

Pairs of factors	Sums of factors
$-1,\ \ 56$	55
$-2,\ \ 28$	26
$-4,\ \ 14$	10
$-7,\ \ \ 8$	1

The numbers we want are -7 and 8. The factorization of $c^2 + c - 56$ is $(c - 7)(c + 8)$, so $c^4 + c^3 - 56c^2 = c^2(c - 7)(c + 8)$.

28. $5(b - 3)(b + 8)$

29. $2a^2 - 4a - 70 = 2(a^2 - 2a - 35)$

After factoring out the common factor, 2, we consider $a^2 - 2a - 35$. Since the constant term is negative, we look for a factorization of -35 in which one factor is positive and one factor is negative. Their sum must be -2, so the negative factor must have the large absolute value. Thus we consider only pairs of factors in which the negative factor has the larger absolute value.

Pairs of factors	Sums of factors
$1,\ \ -35$	-34
$5,\ \ \ -7$	-2

The numbers we want are 5 and -7. The factorization of $a^2 - 2a - 35$ is $(a + 5)(a - 7)$, so $2a^2 - 4a - 70 = 2(a + 5)(a - 7)$.

30. $x^3(x - 2)(x + 1)$

31. $x^2 + x + 1$

Since the constant term and the coefficient of the middle term are both positive, we look for a factorization of 1 in which both factors are positive. Their sum must be 1. The only possible pair of factors is 1 and 1, but their sum is not 1. Thus, this polynomial is not factorable into polynomials with integer coefficients. It is prime.

32. Prime

33. $7 - 2p + p^2 = p^2 - 2p + 7$

Since the constant term is positive and the coefficient of the middle term is negative, we look for a factorization of 7 in which both factors are negative. Their sum must be -2. The only possible pair of factors is -1 and -7, but their sum is not -2. Thus, this polynomial is not factorable into polynomials with integer coefficients. It is prime.

34. Prime

35. $x^2 + 20x + 100$

We look for two factors, both positive, whose product is 100 and whose sum is 20.

They are 10 and 10: $10 \cdot 10 = 100$ and $10 + 10 = 20$.

$x^2 + 20x + 100 = (x + 10)(x + 10)$, or $(x + 10)^2$.

36. $(x + 9)(x + 11)$

37. $3x^3 - 63x^2 - 300x = 3x(x^2 - 21x - 100)$

After factoring out the common factor, $3x$, we consider $x^2 - 21x - 100$. We look for two factors, one positive and one negative, whose product is -100 and whose sum is -21.

They are 4 and -25: $4 \cdot (-25) = -100$ and $4 + (-25) = -21$.

$x^2 - 21x - 100 = (x + 4)(x - 25)$, so $3x^3 - 63x^2 - 300x = 3x(x + 4)(x - 25)$.

38. $2x(x - 8)(x - 12)$

39. $x^2 - 21x - 72$

We look for two factors, both negative, whose product is -72 and whose sum is -21. They are 3 and -24.

$x^2 - 21x - 72 = (x + 3)(x - 24)$

40. $4(x + 5)^2$

41. $x^2 - 25x + 144$

We look for two factors, both negative, whose product is 144 and whose sum is -25. They are -9 and -16.

$x^2 - 25x + 144 = (x - 9)(x - 16)$

42. $(y - 9)(y - 12)$

43. $a^4 + a^3 - 132a^2 = a^2(a^2 + a - 132)$

After factoring out the common factor, a^2, we consider $a^2 + a - 132$. We look for two factors, one positive and one negative, whose product is -132 and whose sum is 1. They are -11 and 12.

$a^2 + a - 132 = (a - 11)(a + 12)$, so $a^4 + a^3 - 132a^2 = a^2(a - 11)(a + 12)$.

44. $a^4(a-6)(a+15)$

45. $x^2 - \dfrac{2}{5}x + \dfrac{1}{25}$

We look for two factors, both negative, whose product is $\dfrac{1}{25}$ and whose sum is $-\dfrac{2}{5}$. They are $-\dfrac{1}{5}$ and $-\dfrac{1}{5}$.

$$x^2 - \frac{2}{5}x + \frac{1}{25} = \left(x - \frac{1}{5}\right)\left(x - \frac{1}{5}\right), \text{ or } \left(x - \frac{1}{5}\right)^2$$

46. $\left(t + \dfrac{1}{3}\right)\left(t + \dfrac{1}{3}\right)$, or $\left(t + \dfrac{1}{3}\right)^2$

47. $27 + 12y + y^2 = y^2 + 12y + 27$

We look for two factors, both positive, whose product is 27 and whose sum is 12. They are 3 and 9.

$$27 + 12y + y^2 = (y+3)(y+9), \text{ or } (3+y)(9+y)$$

48. $(5+x)(10+x)$

49. $t^2 - 0.3t - 0.10$

We look for two factors, one positive and one negative, whose product is -0.10 and whose sum is -0.3. They are 0.2 and -0.5.

$$t^2 - 0.3t - 0.10 = (t + 0.2)(t - 0.5)$$

50. $(y + 0.2)(y - 0.4)$

51. $p^2 + 3pq - 10q^2 = p^2 + 3qp - 10q^2$

Think of $3q$ as a "coefficient" of p. Then we look for factors of $-10q^2$ whose sum is $3q$. They are $5q$ and $-2q$.

$$p^2 + 3pq - 10q^2 = (p + 5q)(p - 2q).$$

52. $(a - 3b)(a + b)$

53. $m^2 + 5mn + 5n^2 = m^2 + 5nm + 5n^2$

We look for factors of $5n^2$ whose sum is $5n$. The only reasonable possibilities are shown below.

Pairs of factors	Sums of factors
$5n$, n	$6n$
$-5n$, $-n$	$-6n$

There are no factors whose sum is $5n$. Thus, the polynomial is not factorable into polynomials with integer coefficients. It is prime.

54. $(x - 8y)(x - 3y)$

55. $s^2 - 2st - 15t^2 = s^2 - 2ts - 15t^2$

We look for factors of $-15t^2$ whose sum is $-2t$. They are $-5t$ and $3t$.

$$s^2 - 2st - 15t^2 = (s - 5t)(s + 3t)$$

56. $(b + 10c)(b - 2c)$

57. $6a^{10} - 30a^9 - 84a^8 = 6a^8(a^2 - 5a - 14)$

After factoring out the common factor, $6a^8$, we consider $a^2 - 5a - 14$. We look for two factors, one positive and one negative, whose product is -14 and whose sum is -5. They are 2 and -7.

$a^2 - 5a - 14 = (a+2)(a-7)$, so $6a^{10} - 30a^9 - 84a^8 = 6a^8(a+2)(a-7)$.

58. $7x^7(x+1)(x-5)$

59. ◈

60. ◈

61. $3x - 8 = 0$

   $3x = 8$ Adding 8 on both sides

   $x = \dfrac{8}{3}$ Dividing by 3 on both sides

The solution is $\dfrac{8}{3}$.

62. $-\dfrac{7}{2}$

63. $(x+6)(3x+4)$

   $= 3x^2 + 4x + 18x + 24$ Using FOIL

   $= 3x^2 + 22x + 24$

64. $49w^2 + 84w + 36$

65. **Familiarize.** Let $n = $ the number of people arrested the year before.

Translate. We reword the problem.

$$\underbrace{\text{Number arrested the year before}}_{\downarrow \atop n} \underbrace{\text{less}}_{\downarrow \atop -} \underbrace{1.2\% \text{ of}}_{\downarrow \atop 1.2\%} \underbrace{\text{that number}}_{\downarrow \atop n} \underbrace{\text{is}}_{\downarrow \atop =} \underbrace{29,090.}_{\downarrow \atop 29,090}$$

Carry out. We solve the equation.

   $n - 1.2\% \cdot n = 29,090$

   $1 \cdot n - 0.012n = 29,090$

   $0.988n = 29,090$

   $n \approx 29,443$ Rounding

Check. 1.2% of $29,443$ is $0.012(29,443) \approx 353$ and $29,443 - 353 = 29,090$. The answer checks.

State. Approximately 29,443 people were arrested the year before.

66. $100°$, $25°$, $55°$

67.

68.

69. $a^2 + ba - 50$

We look for all pairs of integer factors whose product is -50. The sum of each pair is represented by b.

Pairs of factors whose product is -50	Sums of factors
$-1, \quad 50$	49
$1, -50$	-49
$-2, \quad 25$	23
$2, -25$	-23
$-5, \quad 10$	5
$5, -10$	-5

The polynomial $a^2 + ba - 50$ can be factored if b is $49, -49, 23, -23, 5,$ or -5.

70. $51, -51, 27, -27, 15, -15$

71.
$$\begin{aligned}
30 + 7x - x^2 &= -1(-30) - 1(-7x) - 1 \cdot x^2 \\
&= -1(-30 - 7x + x^2) \\
&= -1(x^2 - 7x - 30)
\end{aligned}$$
We look for factors of -30 whose sum is -7. The numbers we want are -10 and 3.
$$30 + 7x - x^2 = -1(x-10)(x+3), \text{ or } -1(-10+x)(3+x)$$

72. $-1(x-9)(x+5)$, or $-1(-9+x)(5+x)$

73.
$$\begin{aligned}
24 - 10a - a^2 &= -1(-24) - 1(10a) - 1 \cdot a^2 \\
&= -1(-24 + 10a + a^2) \\
&= -1(a^2 + 10a - 24)
\end{aligned}$$
We look for factors of -24 whose sum is 10. The numbers we want are 12 and -2.
$$24 - 10a - a^2 = -1(a+12)(a-2), \text{ or } -1(12+a)(-2+a)$$

74. $-1(a+12)(a-3)$, or $-1(12+a)(-3+a)$

75.
$$\begin{aligned}
84 - 8t - t^2 &= -1(-84) - 1(8t) - 1 \cdot t^2 \\
&= -1(-84 + 8t + t^2) \\
&= -1(t^2 + 8t - 84)
\end{aligned}$$
We look for factors of -84 whose sum is 8. the numbers we want are 14 and -6.
$$84 - 8t - t^2 = -1(t+14)(t-6), \text{ or } -1(14+t)(-6+t)$$

76. $-1(t+12)(t-6)$, or $-1(12+t)(-6+t)$

77. $x^2 + \dfrac{1}{4}x - \dfrac{1}{8}$

We look for two factors, one positive and one negative, whose product is $-\dfrac{1}{8}$ and whose sum is $\dfrac{1}{4}$. They are $\dfrac{1}{2}$ and $-\dfrac{1}{4}$.
$$x^2 + \frac{1}{4}x - \frac{1}{8} = \left(x + \frac{1}{2}\right)\left(x - \frac{1}{4}\right)$$

78. $\left(x + \dfrac{3}{4}\right)\left(x - \dfrac{1}{4}\right)$

79. $\dfrac{1}{3}a^3 - \dfrac{1}{3}a^2 - 2a = \dfrac{1}{3}a(a^2 - a - 6)$

After factoring out the common factor, $\dfrac{1}{3}a$, we consider $a^2 - a - 6$. We look for two factors, one positive and one negative, whose product is -6 and whose sum is -1. They are 2 and -3.
$$a^2 - a - 6 = (a+2)(a-3), \text{ so}$$
$$\frac{1}{3}a^3 - \frac{1}{3}a^2 - 2a = \frac{1}{3}a(a+2)(a-3).$$

80. $a^5(a-5)\left(a + \dfrac{5}{7}\right)$

81. $x^{2m} + 11x^m + 28 = (x^m)^2 + 11x^m + 28$

We look for numbers p and q such that $x^{2m} + 11x^m + 28 = (x^m + p)(x^m + q)$. We find two factors, both positive, whose product is 28 and whose sum is 11. They are 4 and 7.
$$x^{2m} + 11x^m + 28 = (x^m + 4)(x^m + 7)$$

82. $(t^n - 2)(t^n - 5)$

83.
$$\begin{aligned}
&(a+1)x^2 + (a+1)3x + (a+1)2 \\
&= (a+1)(x^2 + 3x + 2)
\end{aligned}$$
After factoring out the common factor $a+1$, we consider $x^2 + 3x + 2$. We look for two factors, whose product is 2 and whose sum is 3. They are 1 and 2.
$$x^2 + 3x + 2 = (x+1)(x+2), \text{ so}$$
$$(a+1)x^2 + (a+1)3x + (a+1)2 = (a+1)(x+1)(x+2).$$

84. $(a-5)(x+9)(x-1)$

85. We first label the drawing with additional information.

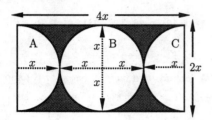

$4x$ represents the length of the rectangle and $2x$ the width. The area of the rectangle is $4x \cdot 2x$, or $8x^2$.

The area of semicircle A is $\frac{1}{2}\pi x^2$.

The area of circle B is πx^2.

The area of semicircle C is $\frac{1}{2}\pi x^2$.

$$\begin{array}{l}
\text{Area of} \\
\text{shaded} = \text{Area of rectangle} - \text{Area of } A - \text{Area of } B - \text{Area of } C \\
\text{region}
\end{array}$$

$$\begin{array}{l}
\text{Area of} \\
\text{shaded} = 8x^2 - \frac{1}{2}\pi x^2 - \pi x^2 - \frac{1}{2}\pi x^2 \\
\text{region}
\end{array}$$

$$= 8x^2 - 2\pi x^2$$
$$= 2x^2(4-\pi)$$

The shaded area can be represented by $2x^2(4-\pi)$.

86. $x^2(\pi - 1)$

87. $6x^2 + 36x + 54 = 6(x^2 + 6x + 9) = 6(x+3)(x+3) = 6(x+3)^2$

Since the surface area of a cube with sides is given by $6s^2$, we know that this cube has side $x + 3$. The volume of a cube with side s is given by s^3, so the volume of this cube is $(x+3)^3$, or $x^3 + 9x^2 + 27x + 27$.

88.

Exercise Set 5.3

1. $2x^2 + 7x - 4$

(1) There is no common factor (other than 1 or -1).

(2) Because $2x^2$ can be factored as $2x \cdot x$, we have this possibility:

$$(2x + \quad)(x + \quad)$$

(3) There are 3 pairs of factors of -4 and they can be listed two ways:

$$-4,1 \quad 4,-1 \quad 2,-2$$
$$\text{and} \quad 1,-4 \quad -1,4 \quad -2,2$$

(4) Look for Outer and Inner products resulting from steps (2) and (3) for which the sum is $7x$. We can immediately reject all possibilities in which a factor has a common factor, such as $(2x - 4)$ or $(2x + 2)$,

because we determined at the outset that there is no common factor other than 1 and -1. We try some possibilities:

$$(2x + 1)(x - 4) = 2x^2 - 7x - 4$$
$$(2x - 1)(x + 4) = 2x^2 + 7x - 4$$

The factorization is $(2x - 1)(x + 4)$.

2. $(3x + 4)(x - 1)$

3. $3t^2 + 4t - 15$

(1) There is no common factor (other than 1 or -1).

(2) Because $3t^2$ can be factored as $3t \cdot t$, we have this possibility:

$$(3t + \quad)(t + \quad)$$

(3) There are 4 pairs of factors of -15 and they can be listed two ways:

$$-15,1 \quad 15,-1 \quad -5,3 \quad 5,-3$$
$$\text{and} \quad 1,-15 \quad -1,15 \quad 3,-5 \quad -3,5$$

(4) Look for Outer and Inner products resulting from steps (2) and (3) for which the sum is $4t$. We can immediately reject all possibilities in which either factor has a common factor, such as $(3t - 15)$ or $(3t + 3)$, because at the outset we determined that there is no common factor other than 1 or -1. We try some possibilities:

$$(3t + 1)(t - 15) = 3t^2 - 44t - 15$$
$$(3t - 5)(t + 3) = 3t^2 + 4t - 15$$

The factorization is $(3t - 5)(t + 3)$.

4. $(5t - 9)(t + 2)$

5. $6x^2 - 23x + 7$

(1) There is no common factor (other than 1 or -1).

(2) Because $6x^2$ can be factored as $6x \cdot x$ or $3x \cdot 2x$, we have these possibilities:

$$(6x + \quad)(x + \quad) \text{ and } (3x + \quad)(2x + \quad)$$

(3) There are 2 pairs of factors of 7 and they can be listed two ways:

$$7,1 \quad -7,-1$$
$$\text{and} \quad 1,7 \quad -1,-7$$

(4) Look for Outer and Inner products resulting from steps (2) and (3) for which the sum is $-23x$. Since the sign of the middle term is negative and the sign of the last term is positive, the factors of 7 must both be negative. We try some possibilities:

$$(6x - 7)(x - 1) = 6x^2 - 13x + 7$$
$$(3x - 7)(2x - 1) = 6x^2 - 17x + 7$$
$$(6x - 1)(x - 7) = 6x^2 - 43x + 7$$
$$(3x - 1)(2x - 7) = 6x^2 - 23x + 7$$

The factorization is $(3x - 1)(2x - 7)$.

6. $(3x - 2)(2x - 3)$

7. $7x^2 + 15x + 2$

(1) There is no common factor (other than 1 or -1).

(2) Because $7x^2$ can be factored as $7x \cdot x$, we have this possibility:

$$(7x + \quad)(x + \quad)$$

(3) There are 2 pairs of factors of 2 and they can be listed two ways:

$$2, 1 \quad -2, -1$$
$$\text{and} \quad 1, 2 \quad -1, -2$$

(4) Look for Outer and Inner products resulting from steps (2) and (3) for which the sum is $15x$. Since all coefficients are positive, we need consider only positive factors of 2. We try some possibilities:

$$(7x + 2)(x + 1) = 7x^2 + 9x + 2$$
$$(7x + 1)(x + 2) = 7x^2 + 15x + 2$$

The factorization is $(7x + 1)(x + 2)$.

8. $(3x + 1)(x + 1)$

9. $9a^2 - 6a - 8$

(1) There is no common factor (other than 1 or -1).

(2) Because $9a^2$ can be factored as $9a \cdot a$ and $3a \cdot 3a$, we have these possibilities:

$$(9a + \quad)(a + \quad) \text{ and } (3a + \quad)(3a + \quad)$$

(3) There are 4 pairs of factors of -8 and they can be listed two ways:

$$-8, 1 \quad 8, -1 \quad -4, 2 \quad 4, -2$$
$$\text{and} \quad 1, -8 \quad -1, 8 \quad 2, -4 \quad -2, 4$$

(4) Look for Outer and Inner products resulting from steps (2) and (3) for which the sum is $-6a$. We try some possibilities:

$$(9a - 8)(a + 1) = 9a^2 + a - 8$$
$$(9a - 4)(a + 2) = 9a^2 + 14a - 8$$
$$(3a + 8)(3a - 1) = 9a^2 + 21a - 8$$
$$(3a - 4)(3a + 2) = 9a^2 - 6a - 8$$

The factorization is $(3a - 4)(3a + 2)$.

10. $(2a - 5)(2a + 3)$

11. $3x^2 - 5x - 2$

(1) There is no common factor (other than 1 or -1).

(2) Because $3x^2$ can be factored as $3x \cdot x$, we have this possibility:

$$(3x + \quad)(x + \quad)$$

(3) There are 2 pairs of factors of -2 and they can be listed two ways:

$$-2, 1 \quad 2, -1$$
$$\text{and} \quad 1, -2 \quad -1, 2$$

(4) Look for Outer and Inner products resulting from steps (2) and (3) for which the sum is $-5x$. We try some possibilities:

$$(3x - 2)(x + 1) = 3x^2 + x - 2$$
$$(3x + 2)(x - 1) = 3x^2 - x - 2$$
$$(3x + 1)(x - 2) = 3x^2 - 5x - 2$$

The factorization is $(3x + 1)(x - 2)$.

12. $(5x + 2)(3x - 5)$

13. $12t^2 - 6t - 6$

(1) We factor out the common factor, 6:

$$6(2t^2 - t - 1).$$

Then we factor the trinomial $2t^2 - t - 1$.

(2) Because $2t^2$ can be factored as $2t \cdot t$, we have this possibility:

$$(2t + \quad)(t + \quad)$$

(3) There are 2 pairs of factors of -1. In this case they can be listed in only one way:

$$-1, 1 \quad 1, -1$$

(4) Look for Outer and Inner products resulting from steps (2) and (3) for which the sum is $-t$. We try some possibilities:

$$(2t - 1)(t + 1) = 2t^2 + t - 1$$
$$(2t + 1)(t - 1) = 2t^2 - t - 1$$

The factorization of $2t^2 - t - 1$ is $(2t + 1)(t - 1)$. We must include the common factor in order to get a factorization of the original trinomial.

$$12t^2 - 6t - 6 = 6(2t + 1)(t - 1)$$

14. $2(3t - 2)(3t + 8)$

15. $18t^2 + 3t - 10$

(1) There is no common factor (other than 1 or -1).

(2) Because $18t^2$ can be factored as $18t \cdot t$, $9t \cdot 2t$, and $6t \cdot 3t$, we have these possibilities:

$$(18t + \quad)(t + \quad) \text{ and } (9t + \quad)(2t + \quad) \text{ and }$$
$$(6t + \quad)(3t + \quad)$$

(3) There are 4 pairs of factors of -10 and they can be listed two ways:

$$-10, 1 \quad 10, -1 \quad -5, 2 \quad 5, -2$$
$$\text{and} \quad 1, -10 \quad -1, 10 \quad 2, -5 \quad -2, 5$$

(4) We can immediately reject all possibilities in which either factor has a common factor, such as

$(18t - 10)$ or $(2t + 2)$, because we determined at the outset that there is no common factor other than 1 or -1. We try some possibilities:

$$(18t - 5)(t + 2) = 18t^2 + 31t - 10$$
$$(9t - 5)(t + 2) = 9t^2 + 13t - 10$$
$$(6t - 5)(3t + 2) = 18t^2 - 3t - 10$$
$$(6t + 5)(3t - 2) = 18t^2 + 3t - 10$$

The factorization is $(6t + 5)(3t - 2)$.

16. $(2t + 1)(t + 2)$

17. $15x^2 + 19x + 6$

(1) There is no common factor (other than 1 or -1).

(2) Because $15x^2$ can be factored as $15x \cdot x$ and $5x \cdot 3x$, we have these possibilities:

$$(15x +)(x +) \text{ and } (5x +)(3x +)$$

(3) Since all coefficients are positive, we need consider only positive pairs of factors of 6. There are 2 such pairs and they can be listed two ways:

$$6, 1 \quad 3, 2$$
$$\text{and} \quad 1, 6 \quad 2, 3$$

(4) We can immediately reject all possibilities in which either factor has a common factor, such as $(15x + 6)$ or $(3x + 3)$, because we determined at the outset that there is no common factor other than 1 or -1. We try some possibilities:

$$(15x + 2)(x + 3) = 15x^2 + 47x + 6$$
$$(5x + 6)(3x + 1) = 15x^2 + 23x + 6$$
$$(5x + 3)(3x + 2) = 15x^2 + 19x + 6$$

The factorization is $(5x + 3)(3x + 2)$.

18. $(4x - 5)(3x - 4)$

19. $35x^2 + 34x + 8$

(1) There is no common factor (other than 1 or -1).

(2) Because $35x^2$ can be factored as $35x \cdot x$ or $7x \cdot 5x$, we have these possibilities:

$$(35x +)(x +) \text{ and } (7x +)(5x +)$$

(3) Since all coefficients are positive, we need consider only positive pairs of factors of 8. There are 2 such pairs and they can be listed two ways:

$$8, 1 \quad 4, 2$$
$$\text{and} \quad 1, 8 \quad 2, 4$$

(4) We try some possibilities:

$$(35x + 8)(x + 1) = 35x^2 + 43x + 8$$
$$(7x + 8)(5x + 1) = 35x^2 + 47x + 8$$
$$(7x + 4)(5x + 2) = 35x^2 + 34x + 8$$

The factorization is $(7x + 4)(5x + 2)$.

20. $2(7x - 1)(2x + 3)$

21. $4 + 6t^2 - 13t = 6t^2 - 13t + 4$

(1) There is no common factor (other than 1 or -1).

(2) Because $6t^2$ can be factored as $6t \cdot t$ or $3t \cdot 2t$, we have these possibilities:

$$(6t +)(t +) \text{ and } (3t +)(2t +)$$

(3) Since the sign of the middle term is negative but the sign of the last term is positive, we need to consider only negative factors of 4. There is only 1 such pair and it can be listed two ways:

$$-4, -1 \text{ and } -1, -4$$

(4) We can immediately reject all possibilities in which either factor has a common factor, such as $(6t - 4)$ or $(2t - 4)$, because we determined at the outset that there is no common factor other than 1 or -1. We try some possibilities:

$$(6t - 1)(t - 4) = 6t^2 - 25t + 4$$
$$(3t - 4)(2t - 1) = 6t^2 - 11t + 4$$

These are the only possibilities that do not contain a common factor. Since neither is the desired factorization, we must conclude that $4 + 6t^2 - 13t$ is prime.

22. $(2t - 3)(4t - 3)$

23. $25x^2 + 40x + 16$

(1) There is no common factor (other than 1 or -1).

(2) Because $25x^2$ can be factored as $25x \cdot x$ or $5x \cdot 5x$, we have these possibilities:

$$(25x +)(x +) \text{ and } (5x +)(5x +)$$

(3) Since all coefficients are positive, we need consider only positive pairs of factors of 16. There are 3 such pairs and two of them can be listed two ways:

$$16, 1 \quad 8, 2 \quad 4, 4$$
$$\text{and} \quad 1, 16 \quad 2, 8$$

(4) We try some possibilities:

$$(25x + 16)(x + 1) = 25x^2 + 41x + 16$$
$$(5x + 8)(5x + 2) = 25x^2 + 50x + 16$$
$$(5x + 4)(5x + 4) = 25x^2 + 40x + 16$$

The factorization is $(5x + 4)(5x + 4)$, or $(5x + 4)^2$.

24. $(7t + 3)(7t + 3)$, or $(7t + 3)^2$

25. $16a^2 + 78a + 27$

(1) There is no common factor (other than 1 or -1).

(2) Because $16a^2$ can be factored as $16a \cdot a$, $8a \cdot 2a$, or $4a \cdot 4a$, we have these possibilities:

$$(16a +)(a +) \text{ and } (8a +)(2a +) \text{ and}$$

$(4a + \quad)(4a + \quad)$

(3) Since all coefficients are positive, we need consider only positive pairs of factors of 27. There are 2 such pairs and two of them can be listed two ways:

$$27, 1 \quad 3, 9$$
$$\text{and} \quad 1, 27 \quad 9, 3$$

(4) We try some possibilities:

$$(16a + 27)(a + 1) = 16a^2 + 43a + 27$$
$$(8a + 3)(2a + 9) = 16a^2 + 78a + 27$$

The factorization is $(8a + 3)(2a + 9)$.

26. $(24x - 1)(x + 2)$

27. $18t^2 + 24t - 10$

(1) Factor out the common factor, 2:

$$2(9t^2 + 12t - 5)$$

Then we factor the trinomial $9t^2 + 12t - 5$.

(2) Because $9t^2$ can be factored as $9t \cdot t$ or $3t \cdot 3t$, we have these possibilities:

$$(9t + \quad)(t + \quad) \text{ and } (3t + \quad)(3t + \quad)$$

(3) There are 2 pairs of factors of -5 and they can be listed two ways:

$$-5, 1 \quad 5, -1$$
$$\text{and} \quad 1, -5 \quad -1, 5$$

(4) We try some possibilities:

$$(9t - 5)(t + 1) = 9t^2 + 4t - 5$$
$$(9t + 1)(t - 5) = 9t^2 - 44t - 5$$
$$(3t + 1)(3t - 5) = 9t^2 - 12t - 5$$
$$(3t - 1)(3t + 5) = 9t^2 + 12t - 5$$

The factorization of $9t^2 + 12t - 5$ is $(3t - 1)(3t + 5)$. We must include the common factor in order to get a factorization of the original trinomial.

$$18t^2 + 24t - 10 = 2(3t - 1)(3t + 5)$$

28. $(7x + 4)(5x - 11)$

29. $2x^2 - 15 - x = 2x^2 - x - 15$

(1) There is no common factor (other than 1 or -1).

(2) Because $2x^2$ can be factored as $2x \cdot x$ we have this possibility:

$$(2x + \quad)(x + \quad)$$

(3) There are 4 pairs of factors of -15 and they can be listed two ways:

$$-15, 1 \quad 15, -1 \quad -5, 3 \quad 5, -3$$
$$\text{and} \quad 1, -15 \quad -1, 15 \quad 3, -5 \quad -3, 5$$

(4) We try some possibilities:

$$(2x - 15)(x + 1) = 2x^2 - 13x - 15$$
$$(2x - 5)(x + 3) = 2x^2 + x - 15$$
$$(2x + 5)(x - 3) = 2x^2 - x - 15$$

The factorization is $(2x + 5)(x - 3)$.

30. Prime

31. $6x^2 + 33x + 15$

(1) Factor out the common factor, 3:

$$3(2x^2 + 11x + 5)$$

Then we factor the trinomial $2x^2 + 11x + 5$.

(2) Because $2x^2$ can be factored as $2x \cdot x$ we have this possibility:

$$(2x + \quad)(x + \quad)$$

(3) Since all coefficients are positive, we need consider only positive pairs of factors of 5. There is one such pair and it can be listed two ways:

$$5, 1 \quad \text{and} \quad 1, 5$$

(4) We try some possibilities:

$$(2x + 5)(x + 1) = 2x^2 + 7x + 5$$
$$(2x + 1)(x + 5) = 2x^2 + 11x + 5$$

The factorization of $2x^2 + 11x + 5$ is $(2x + 1)(x + 5)$. We must include the common factor in order to get a factorization of the original trinomial.

$$6x^2 + 33x + 15 = 3(2x + 1)(x + 5)$$

32. $4(3x - 2)(x + 3)$

33. $20x^2 - 25x + 5$

(1) Factor out the common factor, 5:

$$5(4x^2 - 5x + 1)$$

Then we factor the trinomial $4x^2 - 5x + 1$.

(2) Because $4x^2$ can be factored as $4x \cdot x$ or $2x \cdot 2x$, we have these possibilities:

$$(4x + \quad)(x + \quad) \text{ and } (2x + \quad)(2x + \quad)$$

(3) Since the sign of the middle term is negative but the sign of the last term is positive, we need to consider only negative factors of 1. There is only 1 such pair, $-1, -1$.

(4) We try the possibilities:

$$(4x - 1)(x - 1) = 4x^2 - 5x + 1$$

The factorization of $4x^2 - 5x + 1$ is $(4x - 1)(x - 1)$. We must include the common factor in order to get a factorization of the original trinomial.

$$20x^2 - 25x + 5 = 5(4x - 1)(x - 1)$$

34. $6(5x - 9)(x + 1)$

35. $12x^2 + 68x - 24$

(1) Factor out the common factor, 4:
$$4(3x^2 + 17x - 6)$$
Then we factor the trinomial $3x^2 + 17x - 6$.

(2) Because $3x^2$ can be factored as $3x \cdot x$ we have this possibility:
$$(3x + \quad)(x + \quad)$$

(3) There are 4 pairs of factors of -6 and they can be listed two ways:

$$
\begin{array}{cccc}
6,-1 & -6,1 & 3,-2 & -3,2 \\
\text{and} \quad -1,6 & 1,-6 & -2,3 & 2,-3
\end{array}
$$

(4) We can immediately reject all possibilities in which either factor has a common factor, such as $(3x + 6)$ or $(3x - 3)$, because we determined at the outset that there is no common factor other than 1 or -1. We try some possibilities:
$$(3x - 1)(x + 6) = 3x^2 + 17x - 6$$

The factorization of $3x^2 + 17x - 6$ is $(3x - 1)(x + 6)$. We must include the common factor in order to get a factorization of the original trinomial.
$$12x^2 + 68x - 24 = 4(3x - 1)(x + 6)$$

36. $3(2x + 5)(x + 1)$

37. $4x + 1 + 3x^2 = 3x^2 + 4x + 1$

(1) There is no common factor (other than 1 or -1).

(2) Because $3x^2$ can be factored as $3x \cdot x$ we have this possibility:
$$(3x + \quad)(x + \quad)$$

(3) Since all coefficients are positive, we need consider only positive pairs of factors of 1. There is one such pair: 1,1.

(4) We try the possible factorization:
$$(3x + 1)(x + 1) = 3x^2 + 4x + 1$$
The factorization is $(3x + 1)(x + 1)$.

38. $3(3x - 1)(2x + 3)$

39. $y^2 + 4y - 2y - 8 = y(y + 4) - 2(y + 4)$
$$= (y + 4)(y - 2)$$

40. $(x + 5)(x - 2)$

41. $8t^2 - 6t - 28t + 21 = 2t(4t - 3) - 7(4t - 3)$
$$= (4t - 3)(2t - 7)$$

42. $(7t - 8)(5t + 3)$

43. $6x^2 + 4x + 9x + 6 = 2x(3x + 2) + 3(3x + 2)$
$$= (3x + 2)(2x + 3)$$

44. $(3x - 2)(x + 1)$

45. $2t^2 + 6t - t - 3 = 2t(t + 3) - 1(t + 3)$
$$= (t + 3)(2t - 1)$$

46. $(t + 2)(5t - 1)$

47. $3a^2 - 12a - a + 4 = 3a(a - 4) - 1(a - 4)$
$$= (a - 4)(3a - 1)$$

48. $(a - 5)(2a - 1)$

49. $9t^2 + 14t + 5$

(1) First note that there is no common factor (other than 1 or -1).

(2) Multiply the leading coefficient, 9, and the constant, 5:
$$9 \cdot 5 = 45$$

(3) We look for factors of 45 that add to 14. Since all coefficients are positive, we need to consider only positive factors.

Pairs of factors	Sums of factors
1, 45	46
3, 15	18
5, 9	14

The numbers we need are 5 and 9.

(4) Rewrite the middle term:
$$14t = 5t + 9t$$

(5) Factor by grouping:
$$9t^2 + 14t + 5 = 9t^2 + 5t + 9t + 5$$
$$= t(9t + 5) + 1(9t + 5)$$
$$= (9t + 5)(t + 1)$$

50. $(t + 1)(16t + 7)$

51. $16x^2 + 32x + 7$

(1) First note that there is no common factor (other than 1 or -1).

(2) Multiply the leading coefficient, 16, and the constant, 7:
$$16 \cdot 7 = 112$$

(3) We look for factors of 112 that add to 32. Since all coefficients are positive, we need to consider only positive factors.

Pairs of factors	Sums of factors
1, 112	113
2, 56	58
4, 28	32
7, 16	23
8, 14	22

The numbers we need are 4 and 28.

(4) Rewrite the middle term:

$$32x = 4x + 28x$$

(5) Factor by grouping:

$$16x^2 + 32x + 7 = 16x^2 + 4x + 28x + 7$$
$$= 4x(4x + 1) + 7(4x + 1)$$
$$= (4x + 1)(4x + 7)$$

52. $(3x + 5)(3x + 1)$

53. $10a^2 + 25a - 15$

(1) Factor out the largest common factor, 5:

$$10a^2 + 25a - 15 = 5(2a^2 + 5a - 3)$$

(2) To factor $2a^2 + 5a - 3$ by grouping we first multiply the leading coefficient, 2, and the constant, -3:

$$2(-3) = -6$$

(3) We look for factors of -6 that add to 5.

Pairs of factors	Sums of factors
$-1, 6$	5
$-6, 1$	-5
$-2, 3$	1
$2, -3$	-1

The numbers we need are -1 and 6.

(4) Rewrite the middle term:

$$5a = -a + 6a$$

(5) Factor by grouping:

$$2a^2 + 5a - 3 = 2a^2 - a + 6a - 3$$
$$= a(2a - 1) + 3(2a - 1)$$
$$= (2a - 1)(a + 3)$$

The factorization of $2a^2 + 5a - 3$ is $(2a - 1)(a + 3)$. We must include the common factor in order to get a factorization of the original trinomial:

$$10a^2 + 25a - 15 = 5(2a - 1)(a + 3)$$

54. $(2a - 3)(5a + 6)$

55. $2x^2 - 6x - 14$

(1) Factor out the largest common factor, 2.

$$2x^2 + 6x - 14 = 2(x^2 + 3x - 7)$$

To factor the trinomial $x^2 + 3x - 7$ we must find a pair of factors of -7 whose sum is 3. Since there is no such pair, we conclude that $x^2 + 3x - 7$ is prime. Thus, we have $2x^2 + 6x - 14 = 2(x^2 + 3x - 7)$.

56. $7(x - 2)(2x - 1)$

57. $18x^3 + 21x^2 - 9x$

(1) Factor out the largest common factor, $3x$:

$$18x^3 + 21x^2 - 9x = 3x(6x^2 + 7x - 3)$$

(2) To factor $6x^2 + 7x - 3$ by grouping we first multiply the leading coefficient, 6, and the constant, -3:

$$6(-3) = -18$$

(3) We look for factors of -18 that add to 7.

Pairs of factors	Sums of factors
$-1, 18$	17
$1, -18$	-17
$-2, 9$	7
$2, -9$	-7
$-3, 6$	3
$3, -6$	-3

The numbers we need are -2 and 9.

(4) Rewrite the middle term:

$$7x = -2x + 9x$$

(5) Factor by grouping:

$$6x^2 + 7x - 3 = 6x^2 - 2x + 9x - 3$$
$$= 2x(3x - 1) + 3(3x - 1)$$
$$= (3x - 1)(2x + 3)$$

The factorization of $6x^2 + 7x - 3$ is $(3x - 1)(2x + 3)$. We must include the common factor in order to get a factorization of the original trinomial:

$$18x^3 + 21x^2 - 9x = 3x(3x - 1)(2x + 3)$$

58. $2x(3x - 5)(x + 1)$

59. $89x + 64 + 25x^2 = 25x^2 + 89x + 64$

(1) First note that there is no common factor (other than 1 or -1).

(2) Multiply the leading coefficient, 25, and the constant, 64:

$$25 \cdot 64 = 1600$$

(3) We look for factors of 1600 that add to 89. Since all coefficients are positive, we need to consider only positive factors. The numbers we need are 25 and 64.

(4) Rewrite the middle term:

$$89x = 25x + 64x$$

(5) Factor by grouping:

$$25x^2 + 89x + 64 = 25x^2 + 25x + 64x + 64$$
$$= 25x(x + 1) + 64(x + 1)$$
$$= (x + 1)(25x + 64)$$

60. Prime

61. $168x^3 + 45x^2 + 3x$

(1) Factor out the largest common factor, $3x$:
$$168x^3 + 45x^2 + 3x = 3x(56x^2 + 15x + 1)$$

(2) To factor $56x^2 + 15x + 1$ we first multiply the leading coefficient, 56, and the constant, 1:
$$56 \cdot 1 = 56$$

(3) We look for factors of 56 that add to 15. Since all coefficients are positive, we need to consider only positive factors. The numbers we need are 7 and 8.

(4) Rewrite the middle term:
$$15x = 7x + 8x$$

(5) Factor by grouping:
$$56x^2 + 15x + 1 = 56x^2 + 7x + 8x + 1$$
$$= 7x(8x + 1) + 1(8x + 1)$$
$$= (8x + 1)(7x + 1)$$

The factorization of $56x^2 + 15x + 1$ is $(8x+1)(7x+1)$. We must include the common factor in order to get a factorization of the original trinomial:
$$168x^3 + 45x^2 + 3x = 3x(8x + 1)(7x + 1)$$

62. $24x^3(3x - 2)(2x - 1)$

63. $14t^4 - 19t^3 - 3t^2$

(1) Factor out the largest common factor, t^2:
$$14t^4 - 19t^3 - 3t^2 = t^2(14t^2 - 19t - 3)$$

(2) To factor $14t^2 - 19t - 3$ we first multiply the leading coefficient, 14, and the constant, -3:
$$14(-3) = -42$$

(3) We look for factors of -42 that add to -19. The numbers we need are -21 and 2.

(4) Rewrite the middle term:
$$-19t = -21t + 2t$$

(5) Factor by grouping:
$$14t^2 - 19t - 3 = 14t^2 - 21t + 2t - 3$$
$$= 7t(2t - 3) + 1(2t - 3)$$
$$= (2t - 3)(7t + 1)$$

The factorization of $14t^2 - 19t - 3$ is $(2t-3)(7t+1)$. We must include the common factor in order to get a factorization of the original trinomial:
$$14t^4 - 19t^3 - 3t^2 = t^2(2t - 3)(7t + 1)$$

64. $2a^2(5a - 2)(7a - 4)$

65. $3x + 45x^2 - 18 = 45x^2 + 3x - 18$

(1) Factor out the largest common factor, 3:
$$45x^2 + 3x - 18 = 3(15x^2 + x - 6)$$

(2) To factor $15x^2 + x - 6$ we first multiply the leading coefficient, 15, and the constant, -6:
$$15(-6) = -90$$

(3) We look for factors of -90 that add to 1. The numbers we need are 10 and -9.

(4) Rewrite the middle term:
$$x = 10x - 9x$$

(5) Factor by grouping:
$$15x^2 + x - 6 = 15x^2 + 10x - 9x - 6$$
$$= 5x(3x + 2) - 3(3x + 2)$$
$$= (3x + 2)(5x - 3)$$

The factorization of $15x^2 + x - 6$ is $(3x + 2)(5x - 3)$. We must include the common factor in order to get a factorization of the original trinomial:
$$3x + 45x^2 - 18 = 3(3x + 2)(5x - 3)$$

66. $2(4x - 5)(3x + 4)$

67. $9a^2 + 18ab + 8b^2$

(1) First note that there is no common factor (other than 1 or -1).

(2) Multiply the leading coefficient, 9, and the constant, 8:
$$9 \cdot 8 = 72$$

(3) We look for factors of 72 that add to 18. The numbers we need are 6 and 12.

(4) Rewrite the middle term:
$$18ab = 6ab + 12ab$$

(5) Factor by grouping:
$$9a^2 + 18ab + 8b^2 = 9a^2 + 6ab + 12ab + 8b^2$$
$$= 3a(3a + 2b) + 4b(3a + 2b)$$
$$= (3a + 2b)(3a + 4b)$$

68. $(p - 6q)(3p + 2q)$

69. $35p^2 + 34pq + 8q^2$

(1) First note that there is no common factor (other than 1 or -1).

(2) Multiply the leading coefficient, 35, and the constant, 8:
$$35 \cdot 8 = 280$$

(3) We look for factors of 280 that add to 34. The numbers we need are 14 and 20.

(4) Rewrite the middle term:
$$34pq = 14pq + 20pq$$

(5) Factor by grouping:
$$35p^2 + 34pq + 8q^2 = 35p^2 + 14pq + 20pq + 8q^2$$
$$= 7p(5p + 2q) + 4q(5p + 2q)$$
$$= (5p + 2q)(7p + 4q)$$

70. $2(s+t)(5s-3t)$

71. $18x^2 - 6xy - 24y^2$

(1) Factor out the largest common factor, 6:
$$18x^2 - 6xy - 24y^2 = 6(3x^2 - xy - 4y^2)$$

(2) To factor $3x^2 - xy - 4y^2$, we first multiply the leading coefficient, 3, and the constant, -4:
$$3(-4) = -12$$

(3) We look for factors of -12 that add to -1. The numbers we need are -4 and 3.

(4) Rewrite the middle term:
$$-xy = -4xy + 3xy$$

(5) Factor by grouping:
$$\begin{aligned} 3x^2 - xy - 4y^2 &= 3x^2 - 4xy + 3xy - 4y^2 \\ &= x(3x - 4y) + y(3x - 4y) \\ &= (3x - 4y)(x + y) \end{aligned}$$

The factorization of $3x^2 - xy - 4y^2$ is $(3x-4y)(x+y)$. We must include the common factor in order to get a factorization of the original trinomial:
$$18x^2 - 6xy - 24y^2 = 6(3x - 4y)(x + y)$$

72. $3(5a+2b)(2a+5b)$

73. $24a^2 - 34ab + 12b^2$

(1) Factor out the largest common factor, 2:
$$24a^2 - 34ab + 12b^2 = 2(12a^2 - 17ab + 6b^2)$$

(2) To factor $12a^2 - 17ab + 6b^2$, we first multiply the leading coefficient, 12, and the constant, 6:
$$12 \cdot 6 = 72$$

(3) We look for factors of 72 that add to -17. The numbers we need are -8 and -9.

(4) Rewrite the middle term:
$$-17ab = -8ab - 9ab$$

(5) Factor by grouping:
$$\begin{aligned} 12a^2 - 17ab + 6b^2 &= 12a^2 - 8ab - 9ab + 6b^2 \\ &= 4a(3a - 2b) - 3b(3a - 2b) \\ &= (3a - 2b)(4a - 3b) \end{aligned}$$

The factorization of $12a^2 - 17ab + 6b^2$ is $(3a - 2b)(4a - 3b)$. We must include the common factor in order to get a factorization of the original trinomial:
$$24a^2 - 34ab + 12b^2 = 2(3a - 2b)(4a - 3b)$$

74. $5(a+b)(3a-4b)$

75. $35x^2 + 34x^3 + 8x^4 = 8x^4 + 34x^3 + 35x^2$

(1) Factor out the largest common factor, x^2:
$$x^2(8x^2 + 34x + 35)$$

(2) To factor $8x^2 + 34x + 35$ by grouping we first multiply the leading coefficient, 8, and the constant, 35:
$$8 \cdot 35 = 280$$

(3) We look for factors of 280 that add to 34. The numbers we need are 14 and 20.

(4) Rewrite the middle term:
$$34x = 14x + 20x$$

(5) Factor by grouping:
$$\begin{aligned} 8x^2 + 34x + 35 &= 8x^2 + 14x + 20x + 35 \\ &= 2x(4x + 7) + 5(4x + 7) \\ &= (4x + 7)(2x + 5) \end{aligned}$$

The factorization of $8x^2+34x+35$ is $(4x+7)(2x+5)$. We must include the common factor in order to get a factorization of the original trinomial:
$$35x^2 + 34x^3 + 8x^4 = x^2(4x + 7)(2x + 5)$$

76. $x^2(2x+3)(7x-1)$

77. $18a^7 + 8a^6 + 9a^8 = 9a^8 + 18a^7 + 8a^6$

(1) Factor out the largest common factor, a^6:
$$9a^8 + 18a^7 + 8a^6 = a^6(9a^2 + 18a + 8)$$

(2) To factor $9a^2 + 18a + 8$ we first multiply the leading coefficient, 9, and the constant, 8:
$$9 \cdot 8 = 72$$

(3) Look for factors of 72 that add to 18. The numbers we need are 6 and 12.

(4) Rewrite the middle term:
$$18a = 6a + 12a$$

(5) Factor by grouping:
$$\begin{aligned} 9a^2 + 18a + 8 &= 9a^2 + 6a + 12a + 8 \\ &= 3a(3a + 2) + 4(3a + 2) \\ &= (3a + 2)(3a + 4) \end{aligned}$$

The factorization of $9a^2+18a+8$ is $(3a+2)(3a+4)$. We must include the common factor in order to get a factorization of the original trinomial:
$$18a^7 + 8a^6 + 9a^8 = a^6(3a + 2)(3a + 4)$$

78. $a^7(5a+4)(5a+4)$, or $a^7(5a+4)^2$

79. ◈

80. ◈

81. *Familiarize.* We will use the formula $C = 2\pi r$, where C is circumference and r is radius, to find the radius in kilometers. Then we will multiply that number by 0.62 to find the radius in miles.

Translate.

$$\underbrace{\text{Circumference}}_{40,000} = \underbrace{2 \cdot \pi \cdot}_{\approx} \underbrace{\text{radius}}_{2(3.14)r}$$

Carry out. First we solve the equation.

$$40,000 \approx 2(3.14)r$$
$$40,000 \approx 6.28r$$
$$6369 \approx r$$

Then we multiply to find the radius in miles:

$6369(0.62) \approx 3949$

Check. If $r = 6369$, then $2\pi r = 2(3.14)(6369) \approx 40,000$. We should also recheck the multiplication we did to find the radius in miles. Both values check.

State. The radius of the earth is about 6369 km or 3949 mi. (These values may differ slightly if a different approximation is used for π.)

82. $40°$

83. $(3x+1)^2 = (3x)^2 + 2 \cdot 3x \cdot 1 + 1^2$
$\quad\quad\quad\quad\quad\quad [(A+B)^2 = A^2 + 2AB + B^2]$
$\quad\quad\quad = 9x^2 + 6x + 1$

84. $25x^2 - 20x + 4$

85. $(4t-5)^2 = (4t)^2 - 2 \cdot 4t \cdot 5 + 5^2$
$\quad\quad\quad\quad\quad\quad [(A-B)^2 = A^2 - 2AB + B^2]$
$\quad\quad\quad = 16t^2 - 40t + 25$

86. $49a^2 + 14a + 1$

87. $(5x-2)(5x+2) = (5x)^2 - 2^2$
$\quad\quad\quad\quad\quad\quad [(A+B)(A-B)=A^2-B^2]$
$\quad\quad\quad = 25x^2 - 4$

88. $4x^2 - 9$

89. $(2t+7)(2t-7) = (2t)^2 - 7^2$
$\quad\quad\quad\quad\quad\quad [(A+B)(A-B)=A^2-B^2]$
$\quad\quad\quad = 4t^2 - 49$

90. $16a^2 - 49$

91. ◈

92. ◈

93. $9a^2b^2 - 15ab - 2$

(1) There is no common factor (other than 1 or -1).

(2) Because $9a^2b^2$ can be factored as $9ab \cdot ab$ or $3ab \cdot 3ab$, we have these possibilities:

$(9ab + \quad)(ab + \quad)$ and $(3ab + \quad)(3ab + \quad)$

(3) There are 2 pairs of factors of -2 and they can be listed two ways:

$$-2,1 \quad 2,-1$$
$$\text{and} \quad 1,-2 \quad -1,2$$

(4) We try some possibilities:

$(9ab-2)(ab+1) = 9a^2b^2 + 7ab - 2$
$(9ab+2)(ab-1) = 9a^2b^2 - 7ab - 2$
$(9ab+1)(ab-2) = 9a^2b^2 - 17ab - 2$
$(9ab-1)(ab+2) = 9a^2b^2 + 17ab - 2$
$(3ab-2)(3ab+1) = 9a^2b^2 - 3ab - 2$
$(3ab+2)(3ab-1) = 9a^2b^2 + 3ab - 2$

Since none of the possibilities is the correct factorization, we conclude that $9a^2b^2 - 15ab - 2$ is prime.

94. $(3xy + 2)(6xy - 5)$

95. $8x^2y^3 + 10xy^2 + 2y$

(1) We factor out the common factor, $2y$:

$2y(4x^2y^2 + 5xy + 1)$

Then we factor the trinomial $4x^2y^2 + 5xy + 1$.

(2) Because $4x^2y^2$ can be factored as $4xy \cdot xy$ or $2xy \cdot 2xy$, we have these possibilities:

$(4xy + \quad)(xy + \quad)$ and $(2xy + \quad)(2xy + \quad)$

(3) Since all coefficients are positive, we need consider only positive pairs of factors of 1. The only such pair is 1, 1.

(4) We try some possibilities:

$(4xy + 1)(xy + 1) = 4x^2y^2 + 5xy + 1$

The factorization of $4x^2y^2 + 5xy + 1$ is $(4xy + 1)(xy + 1)$. We must include the common factor in order to get a factorization of the original trinomial.

$8x^2y^3 + 10xy^2 + 2y = 2y(4xy+1)(xy+1)$

96. Prime

97. $9t^{10} + 12t^5 + 4 = 9(t^5)^2 + 12t^5 + 4$

(1) There is no common factor (other than 1 or -1).

(2) Because $9t^{10}$ can be factored as $9t^5 \cdot t^5$ or $3t^5 \cdot 3t^5$, we have these possibilities:

$(9t^5 + \quad)(t^5 + \quad)$ and $(3t^5 + \quad)(3t^5 + \quad)$

(3) Since all coefficients are positive, we need consider only positive pairs of factors of 4. There are two such pairs and one of them can be listed two ways:

$$4, 1 \quad 2, 2$$
$$\text{and} \quad 1, 4$$

(4) We try some possibilities:

$$(9t^5 + 4)(t^5 + 1) = 9t^{10} + 13t^5 + 4$$
$$(3t^5 + 2)(3t^5 + 2) = 9t^{10} + 12t^5 + 4$$

The factorization is $(3t^5 + 2)(3t^5 + 2)$, or $(3t^5 + 2)^2$.

98. $(4t^5 - 1)^2$

99. $-15x^{2m} + 26x^m - 8 = -(15x^{2m} - 26x^m + 8)$

We will factor $15x^{2m} - 26x^m - 8$.

(1) We have factored -1 out of the original trinomial in order to work with a trinomial that has a positive leading term.

(2) Multiply the leading coefficient, 15, and the constant, 8:

$$15 \cdot 8 = 120$$

(3) We look for factors of 120 that add to -26. The numbers we need are -6 and -20.

(4) Rewrite the middle term.

$$-26x^m = -6x^m - 20x^m$$

(5) Factor by grouping:

$$15x^{2m} - 26x^m + 8 = 15x^{2m} - 6x^m - 20x^m + 8$$
$$= 3x^m(5x^m - 2) - 4(5x^m - 2)$$
$$= (5x^m - 2)(3x^m - 4)$$

The factorization of $15x^{2m} - 26x^m + 8$ is $(5x^m - 2)(3x^m - 4)$. We must include the common factor in order to get a factorization of the original trinomial:

$$-15x^{2m} + 26x^m - 8 = -(5x^m - 2)(3x^m - 4)$$

100. $(10x^n + 3)(2x^n + 1)$

101. $a^{2n+1} - 2a^{n+1} + a$

(1) Factor out the largest common factor, a:

$$a^{2n+1} - 2a^{n+1} + a = a(a^{2n} - 2a^n + 1)$$

(2) Multiply the leading coefficient, 1, and the constant, 1:

$$1 \cdot 1 = 1$$

(3) Look for factors of 1 that add to -2. The numbers we need are -1 and -1.

(4) Rewrite the middle term.

$$-2a^n = -a^n - a^n$$

(5) Factor by grouping:

$$a^{2n} - 2a^n + 1 = a^{2n} - a^n - a^n + 1$$
$$= a^n(a^n - 1) - 1(a^n - 1)$$
$$= (a^n - 1)(a^n - 1), \text{ or } (a^n - 1)^2$$

The factorization of $a^{2n} - 2a^n + 1$ is $(a^n - 1)^2$. We must include the common factor in order to get a factorization of the original trinomial:

$$a^{2n+1} - 2a^{n+1} + a = a(a^n - 1)^2$$

102. $(3a^{3n} + 1)(a^{3n} - 1)$

103.
$$3(a + 1)^{n+1}(a + 3)^2 - 5(a + 1)^n(a + 3)^3$$
$$= (a + 1)^n(a + 3)^2[3(a + 1) - 5(a + 3)]$$

Removing the common factors

$$= (a + 1)^n(a + 3)^2[3a + 3 - 5a - 15] \text{ Simplifying inside the brackets}$$

$$= (a + 1)^n(a + 3)^2(-2a - 12)$$

$$= (a + 1)^n(a + 3)^2(-2)(a + 6) \text{ Removing the common factor}$$

$$= -2(a + 1)^n(a + 3)^2(a + 6) \text{ Rearranging}$$

104. $[7(t - 3)^n - 2][(t - 3)^n + 1]$

Exercise Set 5.4

1. $x^2 - 18x + 81$

(1) Two terms, x^2 and 81, are squares.

(2) There is no minus sign before x^2 or 81.

(3) Twice the product of the square roots, $2 \cdot x \cdot 9$, is $18x$, the opposite of the remaining term, $-18x$.

Thus, $x^2 - 18x + 81$ is a perfect-square trinomial.

2. Yes

3. $x^2 + 16x - 64$

(1) Two terms, x^2 and 64, are squares.

(2) There is a minus sign before 64, so $x^2 + 16x - 64$ is not a perfect-square trinomial.

4. No

5. $x^2 - 3x + 9$

(1) Two terms, x^2 and 9, are squares.

(2) There is no minus sign before x^2 or 9.

(3) Twice the product of the square roots, $2 \cdot x \cdot 3$, is $6x$. This is neither the remaining term nor its opposite, so $x^2 - 3x + 9$ is not a perfect-square trinomial.

6. No

7. $9x^2 - 36x + 24$

(1) Only one term, $9x^2$, is a square. Thus, $9x^2 - 36x + 24$ is not a perfect-square trinomial.

8. No

9. $\begin{aligned} & x^2 - 16x + 64 \\ = {} & x^2 - 2 \cdot x \cdot 8 + 8^2 \ = \ (x - 8)^2 \\ & \ \uparrow \quad \uparrow \quad \uparrow \quad \uparrow \qquad \uparrow \\ = {} & A^2 - 2 \ \ A \ \ B + B^2 = (A - B)^2 \end{aligned}$

10. $(x - 7)^2$

11. $\begin{aligned} & x^2 + 14x + 49 \\ = {} & x^2 + 2 \cdot x \cdot 7 + 7^2 \ = \ (x + 7)^2 \\ & \ \uparrow \quad \uparrow \quad \uparrow \quad \uparrow \qquad \uparrow \\ = {} & A^2 + 2 \ \ A \ \ B + B^2 = (A + B)^2 \end{aligned}$

12. $(x + 8)^2$

13. $\begin{aligned} 3x^2 - 6x + 3 &= 3(x^2 - 2x + 1) \\ &= 3(x^2 - 2 \cdot x \cdot 1 + 1^2) \\ &= 3(x - 1)^2 \end{aligned}$

14. $5(x - 1)^2$

15. $\begin{aligned} 4 + 4x + x^2 &= 2^2 + 2 \cdot 2 \cdot x + x^2 \\ &= (2 + x)^2, \text{ or } (x + 2)^2 \end{aligned}$

16. $(x - 2)^2$

17. $\begin{aligned} 18x^2 - 12x + 2 &= 2(9x^2 - 6x + 1) \\ &= 2[(3x)^2 - 2 \cdot 3x \cdot 1 + 1^2] \\ &= 2(3x - 1)^2 \end{aligned}$

18. $(5x + 1)^2$

19. $\begin{aligned} 49 + 56y + 16y^2 &= 16y^2 + 56y + 49 \\ &= (4y)^2 + 2 \cdot 4y \cdot 7 + 7^2 \\ &= (4y + 7)^2 \end{aligned}$

We could also factor as follows:
$\begin{aligned} 49 + 56y + 16y^2 &= 7^2 + 2 \cdot 7 \cdot 4y + (4y)^2 \\ &= (7 + 4y)^2 \end{aligned}$

20. $3(4m + 5)^2$

21. $\begin{aligned} x^5 - 18x^4 + 81x^3 &= x^3(x^2 - 18x + 81) \\ &= x^3(x^2 - 2 \cdot x \cdot 9 + 9^2) \\ &= x^3(x - 9)^2 \end{aligned}$

22. $2(x - 10)^2$

23. $\begin{aligned} 2x^3 - 4x^2 + 2x &= 2x(x^2 - 2x + 1) \\ &= 2x(x^2 - 2 \cdot x \cdot 1 + 1^2) \\ &= 2x(x - 1)^2 \end{aligned}$

24. $x(x + 12)^2$

25. $\begin{aligned} 20x^2 + 100x + 125 &= 5(4x^2 + 20x + 25) \\ &= 5[(2x)^2 + 2 \cdot 2x \cdot 5 + 5^2] \\ &= 5(2x + 5)^2 \end{aligned}$

26. $3(2x + 3)^2$

27. $49 - 42x + 9x^2 = 7^2 - 2 \cdot 7 \cdot 3x + (3x)^2 = (7 - 3x)^2$, or $(3x - 7)^2$

28. $(8 - 7x)^2$, or $(7x - 8)^2$

29. $16x^2 + 24x + 9 = (4x)^2 + 2 \cdot 4x \cdot 3 + 3^2 = (4x + 3)^2$

30. $2(a + 7)^2$

31. $\begin{aligned} 2 + 20x + 50x^2 &= 2(1 + 10x + 25x^2) \\ &= 2[1^2 + 2 \cdot 1 \cdot 5x + (5x)^2] \\ &= 2(1 + 5x)^2, \text{ or } 2(5x + 1)^2 \end{aligned}$

32. $(3x + 5)^2$

33. $\begin{aligned} 4p^2 + 12pq + 9q^2 &= (2p)^2 + 2 \cdot 2p \cdot 3q + (3q)^2 \\ &= (2p + 3q)^2 \end{aligned}$

34. $(5m + 2n)^2$

35. $a^2 - 12ab + 49b^2$

This is not a perfect square trinomial because $-2 \cdot a \cdot 7b = -14ab \neq -12ab$. Nor can it be factored using the methods of Sections 5.2 and 5.3. Thus, it is prime.

36. Prime

37. $\begin{aligned} 64m^2 + 16mn + n^2 &= (8m)^2 + 2 \cdot 8m \cdot n + n^2 \\ &= (8m + n)^2 \end{aligned}$

38. $(9p - q)^2$

39. $\begin{aligned} 32s^2 - 80st + 50t^2 &= 2(16s^2 - 40st + 25t^2) \\ &= 2[(4s)^2 - 2 \cdot 4s \cdot 5t + (5t)^2] \\ &= 2(4s - 5t)^2 \end{aligned}$

40. $4(3a + 4b)^2$

41. $x^2 - 100$

    (1) The first expression is a square: x^2

        The second expression is a square: $100 = 10^2$

    (2) The terms have different signs.

    Thus, $x^2 - 100$ is a difference of squares, $x^2 - 10^2$.

42. Yes

43. $x^2 + 36$

    (1) The first expression is a square: x^2

        The second expression is a square: $36 = 6^2$

    (2) The terms do not have different signs.

    Thus, $x^2 + 36$ is not a difference of squares.

44. No

45. $9t^2 - 32$

    (1) The expression 32 is not a square.

    Thus, $9t^2 - 32$ is not a difference of squares.

46. No

47. $-25 + 4t^2$

    (1) The expressions 25 and $4t^2$ are squares:

        $25 = 5^2$ and $4t^2 = (2t)^2$.

    (2) The terms have different signs.

    Thus, $-25 + 4t^2$ is a difference of squares,

    $(2t)^2 - 5^2$.

48. Yes

49. $y^2 - 4 = y^2 - 2^2 = (y + 2)(y - 2)$

50. $(x + 6)(x - 6)$

51. $p^2 - 9 = p^2 - 3^2 = (p + 3)(p - 3)$

52. Prime

53. $-49 + t^2 = t^2 - 49 = t^2 - 7^2 = (t + 7)(t - 7)$, or $(7 + t)(-7 + t)$

54. $(m + 8)(m - 8)$, or $(8 + m)(-8 + m)$

55. $6a^2 - 54 = 6(a^2 - 9) = 6(a^2 - 3^2) = 6(a + 3)(a - 3)$

56. $(x - 4)^2$

57. $49x^2 - 14x + 1 = (7x)^2 - 2 \cdot 7x \cdot 1 + 1^2 = (7x - 1)^2$

58. $3(t + 2)(t - 2)$

59. $200 - 2t^2 = 2(100 - t^2) = 2(10^2 - t^2) =$
    $2(10 + t)(10 - t)$

60. $2(7 + 2w)(7 - 2w)$

61. $80a^2 - 45 = 5(16a^2 - 9) = 5[(4a^2) - 3^2] =$
    $5(4a + 3)(4a - 3)$

62. $(5x + 2)(5x - 2)$

63. $5t^2 - 80 = 5(t^2 - 16) = 5(t^2 - 4^2) =$
    $5(t + 4)(t - 4)$

64. $4(t + 4)(t - 4)$

65. $8x^2 - 98 = 2(4x^2 - 49) = 2[(2x)^2 - 7^2] =$
    $2(2x + 7)(2x - 7)$

66. $6(2x + 3)(2x - 3)$

67. $36x - 49x^3 = x(36 - 49x^2) = x[6^2 - (7x)^2] =$
    $x(6 + 7x)(6 - 7x)$

68. $x(4 + 9x)(4 - 9x)$

69. $49a^4 - 20$

    There is no common factor (other than 1 or -1). Since 20 is not a square, this is not a difference of squares. Thus, the polynomial is prime.

70. $(5a^2 + 3)(5a^2 - 3)$

71. $t^4 - 1$

    $= (t^2)^2 - 1^2$

    $= (t^2 + 1)(t^2 - 1)$

    $= (t^2 + 1)(t + 1)(t - 1)$ Factoring further;

               $t^2 - 1$ is a difference of squares

72. $(x^2 + 4)(x + 2)(x - 2)$

73. $3x^3 - 24x^2 + 48x = 3x(x^2 - 8x + 16)$

                   $= 3x(x^2 - 2 \cdot x \cdot 4 + 4^2)$

                   $= 3x(x - 4)^2$

74. $2a^2(a - 9)^2$

75. $48t^2 - 27 = 3(16t^2 - 9)$

              $= 3[(4t)^2 - 3^2]$

              $= 3(4t + 3)(4t - 3)$

76. $5(5t + 3)(5t - 3)$

77. $a^8 - 2a^7 + a^6 = a^6(a^2 - 2a + 1)$

                   $= a^6(a^2 - 2 \cdot a \cdot 1 + 1^2)$

                   $= a^6(a - 1)^2$

78. $x^6(x - 4)^2$

79. $7a^2 - 7b^2 = 7(a^2 - b^2)$
$$= 7(a + b)(a - b)$$

80. $6(p + q)(p - q)$

81. $25x^2 - 4y^2 = (5x)^2 - (2y)^2$
$$= (5x + 2y)(5x - 2y)$$

82. $(4a + 3b)(4a - 3b)$

83. $1 - a^4b^4 = 1^2 - (a^2b^2)^2$
$$= (1 + a^2b^2)(1 - a^2b^2)$$
$$= (1 + a^2b^2)[1^2 - (ab)^2]$$
$$= (1 + a^2b^2)(1 + ab)(1 - ab)$$

84. $3(5 + m^2n^2)(5 - m^2n^2)$

85. $18t^2 - 8s^2 = 2(9t^2 - 4s^2)$
$$= 2[(3t)^2 - (2s)^2]$$
$$= 2(3t + 2s)(3t - 2s)$$

86. $(7x + 4y)(7x - 4y)$

87. ◈

88. ◈

89. *Familiarize*. Let a = the amount of oxygen, in liters, that can be dissolved in 100 L of water at 20° C.

Translate. We reword the problem.

$$\underbrace{5\text{ L}}\ \underbrace{\text{is}}\ \underbrace{1.6}\ \underbrace{\text{times}}\ \underbrace{\text{amount } a.}$$
$$\ \ \downarrow\ \ \ \ \downarrow\ \ \downarrow\ \ \ \ \downarrow\ \ \ \ \ \ \ \downarrow$$
$$\ \ 5\ \ = 1.6\ \ \ \cdot\ \ \ \ \ \ \ a$$

Carry out. We solve the equation.

$$5 = 1.6a$$
$$3.125 = a \qquad \text{Dividing both sides by 1.6}$$

Check. Since 1.6 times 3.125 is 5, the answer checks.

State. 3.125 L of oxygen can be dissolved in 100 L of water at 20° C.

90. Scores of 77 or higher

91. $(x^3y^5)(x^9y^7) = x^{3+9}y^{5+7} = x^{12}y^{12}$

92. $25a^4b^6$

93. Graph: $y = \dfrac{3}{2}x - 3$

Because the equation is in the form $y = mx + b$, we know the y-intercept is $(0, -3)$. We find two other points on the line, substituting multiples of 2 for x to avoid fractions.

When $x = -2$, $y = \dfrac{3}{2}(-2) - 3 = -3 - 3 = -6$.

When $x = 4$, $y = \dfrac{3}{2} \cdot 4 - 3 = 6 - 3 = 3$.

x	y
0	-3
-2	-6
4	3

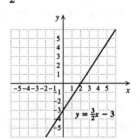

94.

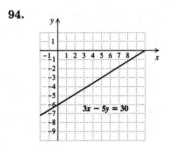

95. ◈

96. ◈

97. $x^8 - 2^8 = (x^4 + 2^4)(x^4 - 2^4)$
$$= (x^4 + 2^4)(x^2 + 2^2)(x^2 - 2^2)$$
$$= (x^4 + 2^4)(x^2 + 2^2)(x + 2)(x - 2), \text{ or}$$
$$(x^4 + 16)(x^2 + 4)(x + 2)(x - 2)$$

98. $3\left(x + \dfrac{1}{3}\right)\left(x - \dfrac{1}{3}\right); \dfrac{1}{3}(3x + 1)(3x - 1)$

99. $18x^3 - \dfrac{8}{25}x = 2x\left(9x^2 - \dfrac{4}{25}\right) =$
$$2x\left(3x + \dfrac{2}{5}\right)\left(3x - \dfrac{2}{5}\right)$$

100. $p(0.7 + p)(0.7 - p)$

101. $0.64x^2 - 1.21 = (0.8x)^2 - (1.1)^2 =$
$$(0.8x + 1.1)(0.8x - 1.1)$$

102. $x(x + 6)(x^2 + 6x + 18)$

103. $(y - 5)^4 - z^8$
$$= [(y - 5)^2 + z^4][(y - 5)^2 - z^4]$$
$$= [(y - 5)^2 + z^4][y - 5 + z^2][y - 5 - z^2]$$
$$= (y^2 - 10y + 25 + z^4)(y - 5 + z^2)(y - 5 - z^2)$$

104. $\left(x + \dfrac{1}{x}\right)\left(x - \dfrac{1}{x}\right)$

105. $a^{2n} - 49b^{2n} = (a^n)^2 - (7b^n)^2 =$
$(a^n + 7b^n)(a^n - 7b^n)$

106. $(9 + b^{2k})(3 + b^k)(3 - b^k)$

107. $x^4 - 8x^2 - 9 = (x^2 - 9)(x^2 + 1)$
$\qquad = (x + 3)(x - 3)(x^2 + 1)$

108. $(3b^n + 2)^2$

109. $16x^4 - 96x^2 + 144 = 16(x^4 - 6x^2 + 9)$
$\qquad\qquad\qquad\qquad = 16(x^2 - 3)^2$

110. $(y + 4)^2$

111. $49(x + 1)^2 - 42(x + 1) + 9 = [7(x + 1) - 3]^2 =$
$(7x + 7 - 3)^2 = (7x + 4)^2$

112. $(3x - 7)^2(3x + 7)$

113. $x^2(x + 1)^2 - (x^2 + 1)^2$
$= x^2(x^2 + 2x + 1) - (x^4 + 2x^2 + 1)$
$= x^4 + 2x^3 + x^2 - x^4 - 2x^2 - 1$
$= 2x^3 + x^2 - 2x^2 - 1$
$= (2x^3 - 2x^2) + (x^2 - 1)$
$= 2x^3 - x^2 - 1$

114. $(a + 4)(a - 2)$

115. $y^2 + 6y + 9 - x^2 - 8x - 16$
$= (y^2 + 6y + 9) - (x^2 + 8x + 16)$
$= (y + 3)^2 - (x + 4)^2$
$= [(y + 3) + (x + 4)][(y + 3) - (x + 4)]$
$= (y + 3 + x + 4)(y + 3 - x - 4)$
$= (y + x + 7)(y - x - 1)$

116. 9

117. For $c = a^2$, $2 \cdot a \cdot 3 = 24$. Then $a = 4$, so $c = 4^2 = 16$.

118. 0, 2

119. $(x + 1)^2 - x^2$
$= [(x + 1) + x][(x + 1) - x]$
$= 2x + 1$

Exercise Set 5.5

1. $\quad 5x^2 - 45$
$= 5(x^2 - 9) \qquad$ 5 is a common factor.
$= 5(x + 3)(x - 3) \quad$ Factoring the difference of squares

2. $10(a + 8)(a - 8)$

3. $\quad a^2 + 25 + 10a$
$= a^2 + 10a + 25 \qquad$ Perfect-square trinomial
$= (a + 5)^2$

4. $(y - 7)^2$

5. $8t^2 - 18t - 5$

There is no common factor (other than 1). This polynomial has three terms, but it is not a perfect-square trinomial. Multiply the leading coefficient and the constant, 8 and -5: $8(-5) = -40$. Try to factor -40 so that the sum of the factors is -18. The numbers we want are -20 and 2: $-20 \cdot 2 = -40$ and $-20 + 2 = -18$. Split the middle term and factor by grouping.

$8t^2 - 18t - 5 = 8t^2 - 20t + 2t - 5$
$\qquad\qquad = 4t(2t - 5) + 1(2t - 5)$
$\qquad\qquad = (2t - 5)(4t + 1)$

6. $(2t + 3)(t + 4)$

7. $\quad x^3 - 24x^2 + 144x$
$= x(x^2 - 24x + 144) \qquad$ x is a common factor.
$= x(x^2 - 2 \cdot x \cdot 12 + 12^2) \quad$ Perfect-square trinomial
$= x(x - 12)^2$

8. $x(x - 9)^2$

9. $\quad x^3 + 3x^2 - 4x - 12$
$= x^2(x + 3) - 4(x + 3) \qquad$ Factoring by grouping
$= (x + 3)(x^2 - 4)$
$= (x + 3)(x + 2)(x - 2) \quad$ Factoring the difference of squares

10. $(x + 5)(x - 5)^2$

11. $\quad 98t^2 - 18$
$= 2(49t^2 - 9) \qquad$ 2 is a common factor.
$= 2[(7t)^2 - 3^2] \quad$ Difference of squares
$= 2(7t + 3)(7t - 3)$

12. $3t(3t+1)(3t-1)$

13. $20x^3 - 4x^2 - 72x$

$= 4x(5x^2 - x - 18)$ $4x$ is a common factor.

$= 4x(5x+9)(x-2)$ Factoring the trinomial us-
                                    ing trial and error

14. $3x(x+3)(3x-5)$

15. $x^2 + 4$

The polynomial has no common factor and is not a
difference of squares. It is prime.

16. Prime

17. $a^4 + 8a^2 + 8a^3 + 64a$

$= a(a^3 + 8a + 8a^2 + 64)$ a is a common
                                         factor.

$= a[a(a^2+8) + 8(a^2+8)]$ Factoring by
                                          grouping

$= a(a^2+8)(a+8)$

18. $t(t^2+7)(t-3)$

19. $x^5 - 14x^4 + 49x^3$

$= x^3(x^2 - 14x + 49)$ x^3 is a common factor.

$= x^3(x^2 - 2 \cdot x \cdot 7 + 7^2)$ Trinomial square

$= x^3(x-7)^2$

20. $2x^4(x+2)^2$

21. $20 - 6x - 2x^2$

$= -2x^2 - 6x + 20$ Rewriting

$= -2(x^2 + 3x - 10)$ -2 is a common factor.

$= -2(x+5)(x-2)$ Using trial and error

We could also express this result as $2(5+x)(2-x)$.

22. $-3(2x-5)(x+3)$, or $3(5-2x)(3+x)$

23. $t^2 - 7t - 6$

There is no common factor (other than 1). This is
not a trinomial square, because only t^2 is a square.
We try factoring by trial and error. We look for two
factors whose product is -6 and whose sum is -7.
There are none. The polynomial cannot be factored.
It is prime.

24. Prime

25. $4x^4 - 64$

$= 4(x^4 - 16)$ 4 is a common factor.

$= 4[(x^2)^2 - 4^2]$ Difference of squares

$= 4(x^2+4)(x^2-4)$ Difference of squares

$= 4(x^2+4)(x+2)(x-2)$

26. $5x(x^2+4)(x+2)(x-2)$

27. $9 + t^8$

There is no common factor (other than 1). Although
both 9 and t^8 are squares, this is not a difference of
squares because the terms do not have different signs.
This polynomial cannot be factored. It is prime.

28. $(t^2+3)(t^2-3)$

29. $x^5 - 4x^4 + 3x^3$

$= x^3(x^2 - 4x + 3)$ x^3 is a common factor.

$= x^3(x-3)(x-1)$ Factoring the trinomial us-
                                 ing trial and error

30. $x^4(x^2 - 2x + 7)$

31. $x^2 - y^2$ Difference of squares

$= (x+y)(x-y)$

32. $(pq+r)(pq-r)$

33. $12n^2 + 24n^3 = 12n^2(1+2n)$

34. $a(x^2 + y^2)$

35. $ab^2 - a^2b = ab(b-a)$

36. $9mn(4 - mn)$

37. $2\pi rh + 2\pi r^2 = 2\pi r(h+r)$

38. $2\pi r(2r+1)$

39. $(a+b)(x-3) + (a+b)(x+4)$

$= (a+b)[(x-3) + (x+4)]$ $(a+b)$ is a com-
                                            mon factor.

$= (a+b)(2x+1)$

40. $(a^3+b)(5c-1)$

41. $n^2 + 2n + np + 2p$

$= n(n+2) + p(n+2)$ Factoring by grouping

$= (n+2)(n+p)$

42. $(x+1)(x+y)$

43. $2x^2 - 4x + xz - 2z$

$= 2x(x-2) + z(x-2)$ Factoring by grouping

$= (x-2)(2x+z)$

44. $(a-3)(a+y)$

45. $x^2 + y^2 + 2xy$

$= x^2 + 2xy + y^2$ Perfect-square trinomial

$= (x+y)^2$

46. $(3x - 2y)(x + 5y)$

47. $\quad 9c^2 - 6cd + d^2$

$\quad = (3c)^2 - 2 \cdot 3c \cdot d + d^2 \quad$ Perfect-square trinomial

$\quad = (3c - d)^2$

48. $(2b - a)^2$, or $(a - 2b)^2$

49. $\quad 7p^4 - 7q^4$

$\quad = 7(p^4 - q^4) \qquad\qquad$ 7 is a common factor.

$\quad = 7(p^2 + q^2)(p^2 - q^2) \qquad$ Difference of squares

$\quad = 7(p^2 + q^2)(p + q)(p - q) \quad$ Difference of squares

50. $(4x + 3y)^2$

51. $\quad 25z^2 + 10zy + y^2$

$\quad = (5z)^2 + 2 \cdot 5z \cdot y + y^2 \quad$ Perfect-square trinomial

$\quad = (5z + y)^2$

52. $(2xy + 3z)^2$

53. $\quad a^5 - 4a^4b - 5a^3b^2$

$\quad = a^3(a^2 - 4ab - 5b^2) \quad a^3$ is a common factor.

$\quad = a^3(a - 5b)(a + b) \quad$ Using trial and error

54. $(a^2b^2 + 4)(ab + 2)(ab - 2)$

55. $a^2 + ab + 2b^2$

There is no common factor (other than 1). This is not a perfect-square trinomial because only a^2 is a square. We try factoring by trial and error. We look for two factors whose product is 2 and whose sum is 1. There are none. This polynomial cannot be factored. It is prime.

56. $p(4pq - q^2 + 4p^2)$

57. $\quad 2mn - 360n^2 + m^2$

$\quad = m^2 + 2mn - 360n^2 \quad$ Rewriting

$\quad = (m + 20n)(m - 18n) \quad$ Using trial and error

58. $(3b - a)(b + 6a)$

59. $\quad m^2n^2 - 4mn - 32$

$\quad = (mn - 8)(mn + 4) \quad$ Using trial and error

60. $(xy + 2)(xy + 6)$

61. $\quad a^5b^2 + 3a^4b - 10a^3$

$\quad = a^3(a^2b^2 + 3ab - 10) \quad a^3$ is a common factor.

$\quad = a^3(ab + 5)(ab - 2) \quad$ Using trial and error

62. $(pq + 1)(pq + 6)$

63. $\quad 8x^2 - 36x + 40$

$\quad = 4(2x^2 - 9x + 10) \quad$ 4 is a common factor.

$\quad = 4(2x - 5)(x - 2) \quad$ Using trial and error

64. $b^4(ab + 8)(ab - 4)$

65. $\quad 2s^6t^2 + 10s^3t^3 + 12t^4$

$\quad = 2t^2(s^6 + 5s^3t + 6t^2) \quad 2t^2$ is a common factor.

$\quad = 2t^2(s^3 + 3t)(s^3 + 2t) \quad$ Using trial and error

66. $x^4(x + 2y)(x - y)$

67. $\quad a^2 + 2a^2bc + a^2b^2c^2$

$\quad = a^2(1 + 2bc + b^2c^2) \quad a^2$ is a common factor.

$\quad = a^2[1^2 + 2 \cdot 1 \cdot bc + (bc)^2] \quad$ Perfect-square trinomial

$\quad = a^2(1 + bc)^2$

68. $\left(6a - \dfrac{5}{4}\right)^2$

69. $\quad \dfrac{1}{81}x^2 - \dfrac{8}{27}x + \dfrac{16}{9}$

$\quad = \left(\dfrac{1}{9}x\right)^2 - 2 \cdot \dfrac{1}{9}x \cdot \dfrac{4}{3} + \left(\dfrac{4}{3}\right)^2 \quad$ Perfect-square trinomial

$\quad = \left(\dfrac{1}{9}x - \dfrac{4}{3}\right)^2$

If we had factored out $\dfrac{1}{9}$ at the outset, the final result would have been $\dfrac{1}{9}\left(\dfrac{1}{3}x - 4\right)^2$.

70. $\left(\dfrac{1}{2}a + \dfrac{1}{3}b\right)^2$

71. $\quad 1 - 16x^{12}y^{12}$

$\quad = (1 + 4x^6y^6)(1 - 4x^6y^6) \quad$ Difference of squares

$\quad = (1 + 4x^6y^6)(1 + 2x^3y^3)(1 - 2x^3y^3) \quad$ Difference of squares

72. $a(b^2 + 9a^2)(b + 3a)(b - 3a)$

73. ◈

74. ◈

75. $\quad \dfrac{y = -4x + 7}{}$

$\qquad 11 \ \ ? \ \ -4(-1) + 7$

$\qquad\qquad \big| \ \ 4 + 7$

$\qquad 11 \ \ \big| \ \ 11 \qquad\qquad$ TRUE

Since $11 = 11$ is true, $(-1, 11)$ is a solution.

$$y = -4x + 7$$

$$\begin{array}{c|c} 7 \; ? \; -4 \cdot 0 + 7 & \\ & 0 + 7 \\ \hline 7 & 7 \qquad\qquad \text{TRUE} \end{array}$$

Since $7 = 7$ is true, $(0, 7)$ is a solution.

$$y = -4x + 7$$

$$\begin{array}{c|c} -5 \; ? \; -4 \cdot 3 + 7 & \\ & -12 + 7 \\ \hline -5 & -5 \qquad\qquad \text{TRUE} \end{array}$$

Since $-5 = -5$ is true, $(3, -5)$ is a solution.

76. $\dfrac{4}{5}$

77. $3x + 7 = 0$

$\qquad 3x = -7 \qquad$ Subtracting 7 from both sides

$\qquad x = -\dfrac{7}{3} \qquad$ Dividing both sides by 3

The solution is $-\dfrac{7}{3}$.

78. $-\dfrac{9}{2}$

79. $4x - 9 = 0$

$\qquad 4x = 9 \qquad$ Adding 9 to both sides

$\qquad x = \dfrac{9}{4} \qquad$ Dividing both sides by 4

The solution is $\dfrac{9}{4}$.

80.

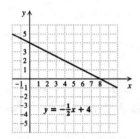

$$y = -\tfrac{1}{2}x + 4$$

81.

82. ◈

83. $-(x^5 + 7x^3 - 18x)$

$\qquad = -x(x^4 + 7x^2 - 18)$

$\qquad = -x(x^2 + 9)(x^2 - 2)$

84. $(a - 2)(a + 3)(a - 3)$

85. $3a^4 - 15a^2 + 12$

$\qquad = 3(a^4 - 5a^2 + 4)$

$\qquad = 3(a^2 - 1)(a^2 - 4)$

$\qquad = 3(a + 1)(a - 1)(a + 2)(a - 2)$

86. $(x^2 + 2)(x + 3)(x - 3)$

87. $y^2(y + 1) - 4y(y + 1) - 21(y + 1)$

$\qquad = (y + 1)(y^2 - 4y - 21)$

$\qquad = (y + 1)(y - 7)(y + 3)$

88. $(y - 1)^3$

89. $6(x - 1)^2 + 7y(x - 1) - 3y^2$

$\qquad = [2(x - 1) + 3y][3(x - 1) - y]$

$\qquad = (2x + 3y - 2)(3x - y - 3)$

90. $(y + 4 + x)^2$

91. $2(a + 3)^4 - (a + 3)^3(b - 2) - (a + 3)^2(b - 2)^2$

$\qquad = (a + 3)^2[2(a + 3)^2 - (a + 3)(b - 2) - (b - 2)^2]$

$\qquad = (a + 3)^2[2(a + 3) + (b - 2)][(a + 3) - (b - 2)]$

$\qquad = (a + 3)^2(2a + 6 + b - 2)(a + 3 - b + 2)$

$\qquad = (a + 3)^2(2a + b + 4)(a - b + 5)$

92. $(t - 1)^3(5t - s - 4)(t - s)$

Exercise Set 5.6

1. $(x + 5)(x + 6) = 0$

We use the principle of zero products.

$\qquad x + 5 = 0 \quad or \quad x + 6 = 0$

$\qquad\qquad x = -5 \quad or \qquad\quad x = -6$

Check:

For -5:

$$\begin{array}{c|c} \multicolumn{2}{l}{(x + 5)(x + 6) = 0} \\ \hline (-5 + 5)(-5 + 6) \; ? \; 0 & \\ 0 \cdot 1 & \\ 0 & 0 \qquad \text{TRUE} \end{array}$$

For -6:

$$\begin{array}{c|c} \multicolumn{2}{l}{(x + 5)(x + 6) = 0} \\ \hline (-6 + 5)(-6 + 6) \; ? \; 0 & \\ -1 \cdot 0 & \\ 0 & 0 \qquad \text{TRUE} \end{array}$$

The solutions are -5 and -6.

2. $-1, -2$

3. $(x-3)(x+7) = 0$

$x - 3 = 0 \quad or \quad x + 7 = 0$

$x = 3 \quad or \qquad x = -7$

Check:

For 3:

$$\frac{(x-3)(x+5) = 0}{(3-3)(3+5) \; ? \; 0}$$
$$\begin{array}{c|c} 0 \cdot 8 & \\ 0 & 0 \quad \text{TRUE} \end{array}$$

For -7:

$$\frac{(x-3)(x+7) = 0}{(-7-3)(-7+7) \; ? \; 0}$$
$$\begin{array}{c|c} -10 \cdot 0 & \\ 0 & 0 \quad \text{TRUE} \end{array}$$

The solutions are 3 and -7.

4. $-9, 3$

5. $(2x-9)(x+4) = 0$

$2x - 9 = 0 \quad or \quad x + 4 = 0$

$2x = 9 \quad or \qquad x = -4$

$x = \dfrac{9}{2} \quad or \qquad x = -4$

The solutions are $\dfrac{9}{2}$ and -4.

6. $\dfrac{5}{3}, -1$

7. $(10x-9)(4x+7) = 0$

$10x - 9 = 0 \quad or \quad 4x + 7 = 0$

$10x = 9 \quad or \qquad 4x = -7$

$x = \dfrac{9}{10} \quad or \qquad x = -\dfrac{7}{4}$

The solutions are $\dfrac{9}{10}$ and $-\dfrac{7}{4}$.

8. $\dfrac{7}{2}, -\dfrac{4}{3}$

9. $x(x+6) = 0$

$x = 0 \quad or \quad x + 6 = 0$

$x = 0 \quad or \qquad x = -6$

The solutions are 0 and -6.

10. $0, -9$

11. $\left(\dfrac{2}{3}x - \dfrac{12}{11}\right)\left(\dfrac{7}{4}x - \dfrac{1}{12}\right) = 0$

$\dfrac{2}{3}x - \dfrac{12}{11} = 0 \qquad or \qquad \dfrac{7}{4}x - \dfrac{1}{12} = 0$

$\dfrac{2}{3}x = \dfrac{12}{11} \qquad or \qquad \dfrac{7}{4}x = \dfrac{1}{12}$

$x = \dfrac{3}{2} \cdot \dfrac{12}{11} \quad or \qquad x = \dfrac{4}{7} \cdot \dfrac{1}{12}$

$x = \dfrac{18}{11} \qquad or \qquad x = \dfrac{1}{21}$

The solutions are $\dfrac{18}{11}$ and $\dfrac{1}{21}$.

12. $-\dfrac{1}{10}, \dfrac{1}{27}$

13. $5x(2x+9) = 0$

$5x = 0 \quad or \quad 2x + 9 = 0$

$x = 0 \quad or \qquad 2x = -9$

$x = 0 \quad or \qquad x = -\dfrac{9}{2}$

The solutions are 0 and $-\dfrac{9}{2}$.

14. $0, -\dfrac{5}{4}$

15. $(20x - 0.4x)(7 - 0.1x) = 0$

$20 - 0.4x = 0 \quad or \quad 7 - 0.1x = 0$

$-0.4x = -20 \quad or \qquad -0.1x = -7$

$x = 50 \quad or \qquad x = 70$

The solutions are 50 and 70.

16. $3.\overline{3}$, or $\dfrac{10}{3}$; 20

17. $\quad x^2 + 7x + 6 = 0$

$(x+6)(x+1) = 0 \quad$ Factoring

$x + 6 = 0 \quad or \quad x + 1 = 0 \quad$ Using the principle of zero products

$x = -6 \quad or \qquad x = -1$

The solutions are -6 and -1.

18. $1, 5$

19. $\quad x^2 - 4x - 21 = 0$

$(x+3)(x-7) = 0 \quad$ Factoring

$x + 3 = 0 \quad or \quad x - 7 = 0 \quad$ Using the principle of zero products

$x = -3 \quad or \qquad x = 7$

The solutions are -3 and 7.

20. $-2, 9$

21. $x^2 + 9x + 14 = 0$

$(x + 7)(x + 2) = 0$

$x + 7 = 0 \quad or \quad x + 2 = 0$

$x = -7 \quad or \qquad x = -2$

The solutions are -7 and -2.

22. $-5, -3$

23. $x^2 - 6x = 0$

$x(x - 6) = 0$

$x = 0 \quad or \quad x - 6 = 0$

$x = 0 \quad or \qquad x = 6$

The solutions are 0 and 6.

24. $-8, 0$

25. $7t + t^2 = 0$

$t(7 + t) = 0$

$t = 0 \quad or \quad 7 + t = 0$

$t = 0 \quad or \qquad t = -7$

The solutions are 0 and -7.

26. $-4, 0$

27. $9x^2 = 4$

$9x^2 - 4 = 0$

$(3x + 2)(3x - 2) = 0$

$3x + 2 = 0 \quad or \quad 3x - 2 = 0$

$3x = -2 \quad or \qquad 3x = 2$

$x = -\dfrac{2}{3} \quad or \qquad x = \dfrac{2}{3}$

The solutions are $-\dfrac{2}{3}$ and $\dfrac{2}{3}$.

28. $-\dfrac{5}{2}, \dfrac{5}{2}$

29. $0 = 25 + x^2 + 10x$

$0 = x^2 + 10x + 25$ Writing in descending order

$0 = (x + 5)(x + 5)$

$x + 5 = 0 \quad or \quad x + 5 = 0$

$x = -5 \quad or \qquad x = -5$

The solution is -5.

30. -3

31. $1 + x^2 = 2x$

$x^2 - 2x + 1 = 0$

$(x - 1)(x - 1) = 0$

$x - 1 = 0 \quad or \quad x - 1 = 0$

$x = 1 \quad or \qquad x = 1$

The solution is 1.

32. 4

33. $8x^2 = 5x$

$8x^2 - 5x = 0$

$x(8x - 5) = 0$

$x = 0 \quad or \quad 8x - 5 = 0$

$x = 0 \quad or \qquad 8x = 5$

$x = 0 \quad or \qquad x = \dfrac{5}{8}$

The solutions are 0 and $\dfrac{5}{8}$.

34. $0, \dfrac{7}{3}$

35. $3x^2 - 7x = 20$

$3x^2 - 7x - 20 = 0$

$(3x + 5)(x - 4) = 0$

$3x + 5 = 0 \quad or \quad x - 4 = 0$

$3x = -5 \quad or \qquad x = 4$

$x = -\dfrac{5}{3} \quad or \qquad x = 4$

The solutions are $-\dfrac{5}{3}$ and 4.

36. $-1, \dfrac{5}{3}$

37. $2y^2 + 12y = -10$

$2y^2 + 12y + 10 = 0$

$2(y^2 + 6y + 5) = 0$

$2(y + 5)(y + 1) = 0$

$y + 5 = 0 \quad or \quad y + 1 = 0$

$y = -5 \quad or \qquad y = -1$

The solutions are -5 and -1.

38. $-\dfrac{1}{4}, \dfrac{2}{3}$

39. $(x - 7)(x + 1) = -16$

$x^2 - 6x - 7 = -16$

$x^2 - 6x + 9 = 0$

$(x - 3)(x - 3) = 0$

$x - 3 = 0 \quad or \quad x - 3 = 0$

$x = 3 \quad or \qquad x = 3$

The solution is 3.

40. 1, 4

41.
$$y(3y + 1) = 2$$
$$3y^2 + y = 2$$
$$3y^2 + y - 2 = 0$$
$$(3y - 2)(y + 1) = 0$$
$$3y - 2 = 0 \quad or \quad y + 1 = 0$$
$$3y = 2 \quad or \quad y = -1$$
$$y = \frac{2}{3} \quad or \quad y = -1$$
The solutions are $\frac{2}{3}$ and -1.

42. $-2, 7$

43.
$$81x^2 - 5 = 20$$
$$81x^2 - 25 = 0$$
$$(9x + 5)(9x - 5) = 0$$
$$9x + 5 = 0 \quad or \quad 9x - 5 = 0$$
$$9x = -5 \quad or \quad 9x = 5$$
$$x = -\frac{5}{9} \quad or \quad x = \frac{5}{9}$$
The solutions are $-\frac{5}{9}$ and $\frac{5}{9}$.

44. $-\frac{7}{6}, \frac{7}{6}$

45.
$$(x - 1)(5x + 4) = 2$$
$$5x^2 - x - 4 = 2$$
$$5x^2 - x - 6 = 0$$
$$(5x - 6)(x + 1) = 0$$
$$5x - 6 = 0 \quad or \quad x + 1 = 0$$
$$5x = 6 \quad or \quad x = -1$$
$$x = \frac{6}{5} \quad or \quad x = -1$$
The solutions are $\frac{6}{5}$ and -1.

46. $-4, -\frac{2}{3}$

47.
$$x^2 - 2x = 18 + 5x$$
$$x^2 - 7x - 18 = 0 \qquad \text{Subtracting 18 and } 5x$$
$$(x - 9)(x + 2) = 0$$
$$x - 9 = 0 \quad or \quad x + 2 = 0$$
$$x = 9 \quad or \quad x = -2$$
The solutions are 9 and -2.

48. $-3, 1$

49.
$$(6a + 1)(a + 1) = 21$$
$$6a^2 + 7a + 1 = 21$$
$$6a^2 + 7a - 20 = 0$$
$$(3a - 4)(2a + 5) = 0$$
$$3a - 4 = 0 \quad or \quad 2a + 5 = 0$$
$$3a = 4 \quad or \quad 2a = -5$$
$$a = \frac{4}{3} \quad or \quad a = -\frac{5}{2}$$
The solutions are $\frac{4}{3}$ and $-\frac{5}{2}$.

50. $-\frac{3}{2}, \frac{5}{4}$

51. The solutions of the equation are the first coordinates of the x-intercepts of the graph. From the graph we see that the x-intercepts are $(-1, 0)$ and $(4, 0)$, so the solutions of the equation are -1 and 4.

52. $-3, 2$

53. The solutions of the equation are the first coordinates of the x-intercepts of the graph. From the graph we see that the x-intercepts are $(-1, 0)$ and $(3, 0)$, so the solutions of the equation are -1 and 3.

54. $-3, 2$

55. We let $y = 0$ and solve for x.
$$0 = x^2 + 3x - 4$$
$$0 = (x + 4)(x - 1)$$
$$x + 4 = 0 \quad or \quad x - 1 = 0$$
$$x = -4 \quad or \quad x = 1$$
The x-intercepts are $(-4, 0)$ and $(1, 0)$.

56. $(-2, 0), (3, 0)$

57. We let $y = 0$ and solve for x.
$$0 = x^2 - 2x - 15$$
$$0 = (x - 5)(x + 3)$$
$$x - 5 = 0 \quad or \quad x + 3 = 0$$
$$x = 5 \quad or \quad x = -3$$
The x-intercepts are $(5, 0)$ and $(-3, 0)$.

58. $(-4, 0), (2, 0)$

59. We let $y = 0$ and solve for x
$$0 = 2x^2 + x - 10$$
$$0 = (2x + 5)(x - 2)$$
$$2x + 5 = 0 \quad or \quad x - 2 = 0$$
$$2x = -5 \quad or \quad x = 2$$
$$x = -\frac{5}{2} \quad or \quad x = 2$$

The x-intercepts are $\left(-\dfrac{5}{2}, 0\right)$ and $(2, 0)$.

60. $(-3, 0), \left(\dfrac{3}{2}, 0\right)$

61. ◈

62. ◈

63. $(a+b)^2$

64. $a^2 + b^2$

65. Let x represent the smaller integer; $x + (x+1)$

66. Let x represent the number; $2x + 5 < 19$

67. Let x represent the number; $\dfrac{1}{2}x - 7 > 24$

68. Let n represent the number; $n - 3 \geq 34$

69. ◈

70. ◈

71. $(2x - 5)(x + 7)(3x + 8) = 0$

$\quad 2x - 5 = 0 \quad or \quad x + 7 = 0 \quad or \quad 3x + 8 = 0$

$\quad\quad 2x = 5 \quad or \quad\quad x = -7 \quad or \quad\quad 3x = -8$

$\quad\quad\quad x = \dfrac{5}{2} \quad or \quad\quad x = -7 \quad or \quad\quad x = -\dfrac{8}{3}$

The solutions are $\dfrac{5}{2}$, -7, and $-\dfrac{8}{3}$.

72. $-\dfrac{9}{4}, \dfrac{2}{3}, -\dfrac{1}{5}$

73. a) $\quad\quad x = -4 \quad or \quad\quad x = 5$

$\quad\quad x + 4 = 0 \quad or \quad x - 5 = 0$

$\quad (x + 4)(x - 5) = 0 \quad\quad$ Principle of zero products

$\quad x^2 - x - 20 = 0 \quad\quad$ Multiplying

b) $\quad\quad x = -1 \quad or \quad\quad x = 7$

$\quad\quad x + 1 = 0 \quad or \quad x - 7 = 0$

$\quad (x + 1)(x - 7) = 0$

$\quad x^2 - 6x + -7 = 0$

c) $\quad\quad x = \dfrac{1}{4} \quad or \quad\quad x = 3$

$\quad\quad x - \dfrac{1}{4} = 0 \quad or \quad x - 3 = 0$

$\quad \left(x - \dfrac{1}{4}\right)(x - 3) = 0$

$\quad x^2 - \dfrac{13}{4}x + \dfrac{3}{4} = 0$

$\quad 4\left(x^2 - \dfrac{13}{4}x + \dfrac{3}{4}\right) = 4 \cdot 0 \quad$ Multiplying both sides by 4

$\quad\quad 4x^2 - 13x + 3 = 0$

d) $\quad\quad x = \dfrac{1}{2} \quad or \quad\quad x = \dfrac{1}{3}$

$\quad x - \dfrac{1}{2} = 0 \quad or \quad x - \dfrac{1}{3} = 0$

$\quad \left(x - \dfrac{1}{2}\right)\left(x - \dfrac{1}{3}\right) = 0$

$\quad\quad x^2 - \dfrac{5}{6}x + \dfrac{1}{6} = 0$

$\quad 6x^2 - 5x + 1 = 0 \quad$ Multiplying by 6

e) $\quad\quad x = \dfrac{2}{3} \quad or \quad\quad x = \dfrac{3}{4}$

$\quad x - \dfrac{2}{3} = 0 \quad or \quad x - \dfrac{3}{4} = 0$

$\quad \left(x - \dfrac{2}{3}\right)\left(x - \dfrac{3}{4}\right) = 0$

$\quad\quad x^2 - \dfrac{17}{12}x + \dfrac{1}{2} = 0$

$\quad 12x^2 - 17x + 6 = 0 \quad$ Multiplying by 12

f) $\quad x = -1 \quad or \quad\quad x = 2 \; or \quad\quad x = 3$

$\quad x + 1 = 0 \quad or \quad x - 2 = 0 \; or \quad x - 3 = 0$

$\quad (x + 1)(x - 2)(x - 3) = 0$

$\quad (x^2 - x - 2)(x - 3) = 0$

$\quad\quad x^3 - 4x^2 + x + 6 = 0$

74. 4

75. $\quad\quad a(9 + a) = 4(2a + 5)$

$\quad\quad 9a + a^2 = 8a + 20$

$\quad a^2 + a - 20 = 0 \quad\quad$ Subtracting $8a$ and 20

$\quad (a + 5)(a - 4) = 0$

$\quad\quad a + 5 = 0 \quad or \quad a - 4 = 0$

$\quad\quad\quad a = -5 \quad or \quad\quad a = 4$

The solutions are -5 and 4.

76. 3, 5

77.
$$x^2 - \frac{9}{25} = 0$$
$$\left(x - \frac{3}{5}\right)\left(x + \frac{3}{5}\right) = 0$$
$$x - \frac{3}{5} = 0 \quad or \quad x + \frac{3}{5} = 0$$
$$x = \frac{3}{5} \quad or \qquad x = -\frac{3}{5}$$

The solutions are $\frac{3}{5}$ and $-\frac{3}{5}$.

78. $-\dfrac{5}{6}, \dfrac{5}{6}$

79. $(t+1)^2 = 9$

Observe that $t + 1$ is a number which yields 9 when it is squared. Thus, we have
$$t + 1 = -3 \quad or \quad t + 1 = 3$$
$$t = -4 \quad or \qquad t = 2$$

The solutions are -4 and 2.

We could also do this exercise as follows:
$$(t+1)^2 = 9$$
$$t^2 + 2t + 1 = 9$$
$$t^2 + 2t - 8 = 0$$
$$(t+4)(t-2) = 0$$
$$t + 4 = 0 \quad or \quad t - 2 = 0$$
$$t = -4 \quad or \qquad t = 2$$

Again we see that the solutions are -4 and 2.

80. $-\dfrac{5}{9}, \dfrac{5}{9}$

81. a) $2(x^2 + 10x - 2) = 2 \cdot 0$ Multiplying (a) by 2
$$2x^2 + 20x - 4 = 0$$
(a) and $2x^2 + 20x - 4 = 0$ are equivalent.

b) $(x - 6)(x + 3) = x^2 - 3x - 18$ Multiplying
(b) and $x^2 - 3x - 18 = 0$ are equivalent.

c) $5x^2 - 5 = 5(x^2 - 1) = 5(x + 1)(x - 1) = (x + 1)5(x - 1) = (x + 1)(5x - 5)$
(c) and $(x + 1)(5x - 5) = 0$ are equivalent.

d) $2(2x - 5)(x + 4) = 2 \cdot 0$ Multiplying (d) by 2
$$2(x + 4)(2x - 5) = 0$$
$$(2x + 8)(2x - 5) = 0$$
(d) and $(2x + 8)(2x - 5) = 0$ are equivalent.

e) $4(x^2 + 2x + 9) = 4 \cdot 0$ Multiplying (e) by 4
$$4x^2 + 8x + 36 = 0$$
(e) and $4x^2 + 8x + 36 = 0$ are equivalent.

f) $3(3x^2 - 4x + 8) = 3 \cdot 0$ Multiplying (f) by 3
$$9x^2 - 12x + 24 = 0$$
(f) and $9x^2 - 12x + 24 = 0$ are equivalent.

82. ◈

83. ◈

84. $-3.45, 1.65$

85. $2.33, 6.77$

86. $-0.25, 0.88$

87. $-4.59, -9.15$

88. $4.55, -3.23$

89. $0, 2.74$

90. $-3.76, 0$

Exercise Set 5.7

1. *Familiarize*. Let $x =$ the number (or numbers).

***Translate*.** We reword the problem.

$$\underbrace{\text{The square of a number}}_{\displaystyle x^2} \quad \underset{\displaystyle -}{\text{minus}} \quad \underbrace{\text{the number}}_{\displaystyle x} \quad \underset{\displaystyle =}{\text{is}} \quad \underset{\displaystyle 6}{6}$$

***Carry out*.** We solve the equation.
$$x^2 - x = 6$$
$$x^2 - x - 6 = 0$$
$$(x - 3)(x + 2) = 0$$

$$x - 3 = 0 \quad or \quad x + 2 = 0$$
$$x = 3 \quad or \qquad x = -2$$

***Check*.** For 3: The square of 3 is 3^2, or 9, and $9 - 3 = 6$.

For -2: The square of -2 is $(-2)^2$, or 4, and $4 - (-2) = 4 + 2 = 6$. Both numbers check.

***State*.** The numbers are 3 and -2.

2. $-1, 2$

3. *Familiarize*. Let $x =$ the length of the shorter leg, in cm. Then $x + 3 =$ the length of the longer leg.

***Translate*.** we use the Pythagorean theorem.
$$a^2 + b^2 = c^2$$
$$x^2 + (x + 3)^2 = 15^2$$

***Carry out*.** We solve the equation.

$$x^2 + (x+3)^2 = 15^2$$
$$x^2 + x^2 + 6x + 9 = 225$$
$$2x^2 + 6x + 9 = 225$$
$$2x^2 + 6x - 216 = 0$$
$$2(x^2 + 3x - 108) = 0$$
$$2(x+12)(x-9) = 0$$

$$x + 12 = 0 \quad or \quad x - 9 = 0$$
$$x = -12 \ or \qquad x = 9$$

Check. The number -12 cannot be the length of a side because it is negative. When $x = 9$, then $x + 3 = 12$, and $9^2 + 12^2 = 81 + 144 = 225 = 15^2$, so the number 9 checks.

State. The lengths of the sides are 9 cm, 12 cm, and 15 cm.

4. 6 cm, 8 cm, 10 cm

5. **Familiarize**. The page numbers on facing pages are consecutive integers. Let $x =$ the smaller integer. Then $x + 1 =$ the larger integer.

Translate. We reword the problem.

Smaller integer times larger integer is 110.
$$x \qquad \cdot \qquad (x+1) \qquad = 110$$

Carry out. We solve the equation.
$$x(x+1) = 110$$
$$x^2 + x = 110$$
$$x^2 + x - 110 = 0$$
$$(x+11)(x-10) = 0$$
$$x + 11 = 0 \quad or \quad x - 10 = 0$$
$$x = -11 \quad or \qquad x = 10$$

Check. The solutions of the equation are -11 and 10. Since a page number cannot be negative, -11 cannot be a solution of the original problem. We only need to check 10. When $x = 10$, then $x + 1 = 11$, and $10 \cdot 11 = 110$. This checks.

State. The page numbers are 10 and 11.

6. 14, 15

7. **Familiarize**. Let $x =$ the smaller odd integer. Then $x + 2 =$ the larger odd integer.

Translate. We reword the problem.

Smaller odd integer times larger odd integer is 255.
$$x \qquad \cdot \qquad (x+2) \qquad = 255$$

Carry out.
$$x(x+2) = 255$$
$$x^2 + 2x = 255$$
$$x^2 + 2x - 255 = 0$$
$$(x+17)(x-15) = 0$$
$$x + 17 = 0 \quad or \quad x - 15 = 0$$
$$x = -17 \quad or \qquad x = 15$$

Check. The solutions of the equation are -17 and 15. When x is -17, then $x + 2$ is -15 and $-17(-15) = 225$. The numbers -17 and -15 are consecutive odd integers which are solutions of the problem. When x is 15, then $x + 2 = 17$ and $15 \cdot 17 = 255$. The numbers 15 and 17 are also consecutive odd integers which are solutions of the problem.

State. We have two solutions, each of which consists of a pair of numbers: -17 and -15 or 15 and 17.

8. 14 and 16; -16 and -14

9. **Familiarize**. Let $w =$ the width of the frame, in inches. Then $2w =$ the length of the frame. Recall that the area of a rectangle is Length $\cdot$ Width.

Translate.

The area of the rectangle is 288 in².
$$2w \cdot w \qquad = \qquad 288$$

Carry out. We solve the equation.
$$2w \cdot w = 288$$
$$2w^2 = 288$$
$$2w^2 - 288 = 0$$
$$2(w^2 - 144) = 0$$
$$2(w+12)(w-12) = 0$$
$$w + 12 = 0 \quad or \quad w - 12 = 0$$
$$w = -12 \quad or \qquad w = 12$$

Check. Since the width must be positive, -12 cannot be a solution. If the width is 12 in., then the length is $2 \cdot 12$ in, or 24 in., and the area is 12 in. $\cdot$ 24 in. $= 288$ in². Thus, 12 checks.

State. The frame is 12 in. wide and 24 in. long.

10. Length: 12 ft; width: 2 ft

11. **Familiarize**. We make a drawing. Let $w =$ the width, in cm. Then $2w + 2 =$ the length, in cm.

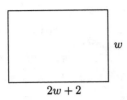

$$2w + 2$$

Recall that the area of a rectangle is length times width.

Translate. We reword the problem.

Length times width is $\underline{144 \text{ cm}^2}$.

$$\downarrow \qquad \downarrow \qquad \downarrow \quad \downarrow \qquad \downarrow$$
$$(2w + 2) \quad \cdot \quad w \quad = \quad 144$$

Carry out. We solve the equation.

$$(2w + 2)w = 144$$
$$2w^2 + 2w = 144$$
$$2w^2 + 2w - 144 = 0$$
$$2(w^2 + w - 72) = 0$$
$$2(w + 9)(w - 8) = 0$$
$$w + 9 = 0 \quad or \quad w - 8 = 0$$
$$w = -9 \quad or \qquad w = 8$$

Check. Since the width must be positive, -9 cannot be a solution. If the width is 8 cm, then the length is $2 \cdot 8 + 2$, or 18 cm, and the area is $8 \cdot 18$, or 144 cm². Thus, 8 checks.

State. The width is 8 cm, and the length is 18 cm.

12. Length: 12 m; width: 8 m

13. *Familiarize*. Using the labels shown on the drawing in the text, we let $h =$ the height, in cm, and $h + 10 =$ the base, in cm. Recall that the formula for the area of a triangle is $\frac{1}{2} \cdot$ (base) $\cdot$ (height).

Translate.

$\frac{1}{2}$ times base times height is $\underline{28 \text{ cm}^2}$.

$$\downarrow \quad \downarrow \qquad \downarrow \qquad \downarrow \quad \downarrow \quad \downarrow \qquad \downarrow$$
$$\frac{1}{2} \quad \cdot \quad (h + 10) \quad \cdot \quad h \quad = \quad 28$$

Carry out.

$$\frac{1}{2}(h + 10)h = 28$$
$$(h + 10)h = 56 \quad \text{Multiplying by 2}$$
$$h^2 + 10h = 56$$
$$h^2 + 10h - 56 = 0$$
$$(h + 14)(h - 4) = 0$$
$$h + 14 = 0 \quad or \quad h - 4 = 0$$
$$h = -14 \quad or \qquad h = 4$$

Check. Since the height of the triangle must be positive, -14 cannot be a solution. If the height is 4 cm, then the base is $4 + 10$, or 14 cm, and the area is $\frac{1}{2} \cdot 14 \cdot 4$, or 28 cm². Thus, 4 checks.

State. The height of the triangle is 4 cm, and the base is 14 cm.

14. Bse: 10 cm; height: 7 cm

15. *Familiarize*. Using the labels show on the drawing in the text, we let $x =$ the length of the foot of the sail, in ft, and $x + 5 =$ the height of the sail, in ft. Recall that the formula for the area of a triangle is $\frac{1}{2} \cdot$ (base) $\cdot$ (height).

Translate.

$\frac{1}{2}$ times base times height is $\underline{42 \text{ ft}^2}$.

$$\downarrow \qquad \downarrow \qquad \downarrow \qquad \downarrow \qquad \downarrow \quad \downarrow \qquad \downarrow$$
$$\frac{1}{2} \quad \cdot \qquad x \qquad \cdot \quad (x + 5) \quad = \qquad 42$$

Carry out.

$$\frac{1}{2}x(x + 5) = 42$$
$$x(x + 5) = 84 \quad \text{Multiplying by 2}$$
$$x^2 + 5x = 84$$
$$x^2 + 5x - 84 = 0$$
$$(x + 12)(x - 7) = 0$$
$$x + 12 = 0 \quad or \quad x - 7 = 0$$
$$x = -12 \quad or \qquad x = 7$$

Check. The solutions of the equation are -12 and 7. The length of the base of a triangle cannot be negative, so -12 cannot be a solution. Suppose the length of the foot of the sail is 7 ft. Then the height is $7 + 5$, or 12 ft, and the area is $\frac{1}{2} \cdot 7 \cdot 12$, or 42 ft². These numbers check.

State. The length of the foot of the sail is 7 ft, and the height is 12 ft.

16. Base: 8 m; hieght: 16 m

17. *Familiarize*. We make a drawing. Let $l =$ the length of the cable, in ft.

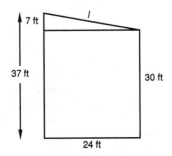

Note that we have a right triangle with hypotenuse l and legs of 24 ft and $37 - 30$, or 7 ft.

Translate. We use the Pythagorean theorem.
$$a^2 + b^2 = c^2$$
$$7^2 + 24^2 = l^2 \quad \text{Substituting}$$

Carry out.
$$7^2 + 24^2 = l^2$$
$$49 + 576 = l^2$$
$$625 = l^2$$
$$0 = l^2 - 625$$
$$0 = (l + 25)(l - 25)$$
$$l + 25 = 0 \quad or \quad l - 25 = 0$$
$$l = -25 \quad or \quad l = 25$$

Check. The integer -25 cannot be the length of the cable, because it is negative. When $l = 25$, we have $7^2 + 24^2 = 25^2$. This checks.

State. The cable is 25 ft long.

18. 8000 ft

19. Familiarize. We will use the formula $n^2 - n = N$.

Translate. Substitute 20 for n.
$$20^2 - 20 = N$$

Carry out. We do the computation on the left.
$$20^2 - 20 = N$$
$$400 - 20 = N$$
$$380 = N$$

Check. We can recheck the computation or we can solve $n^2 - n = 380$. The answer checks.

State. 380 games will be played.

20. 182

21. Familiarize. We will use the formula $n^2 - n = N$.

Translate. Substitute 132 for N.
$$n^2 - n = 132$$

Carry out.
$$n^2 - n = 132$$
$$n^2 - n - 132 = 0$$
$$(n - 12)(n + 11) = 0$$
$$n - 12 = 0 \quad or \quad n + 11 = 0$$
$$n = 12 \quad or \quad n = -11$$

Check. The solutions of the equation are 12 and -11. Since the number of teams cannot be negative, -11 cannot be a solution. But 12 checks since $12^2 - 12 = 144 - 12 = 132$.

State. There are 12 teams in the league.

22. 10

23. Familiarize. Let $h =$ the vertical height to which each brace reaches, in feet. We have a right triangle with hypotenuse 15 ft and legs 12 ft and h.

Translate. We use the Pythagorean theorem.
$$a^2 + b^2 = c^2$$
$$12^2 + h^2 = 15^2$$

Carry out. We solve the equation.
$$12^2 + h^2 = 15^2$$
$$144 + h^2 = 225$$
$$h^2 - 81 = 0$$
$$(h + 9)(h - 9) = 0$$
$$h + 9 = 0 \quad or \quad h - 9 = 0$$
$$h = -9 \quad or \quad h = 9$$

Check. Since the vertical height must be positive, -9 cannot be a solution. If the height is 9 ft, then we have $12^2 + 9^2 = 144 + 81 = 225 = 15^2$. The number 9 checks.

State. Each brace reaches 9 ft vertically.

24. 24 ft

25. Familiarize. We will use the formula
$$N = \frac{1}{2}(n^2 - n).$$

Translate. Substitute 15 for n.
$$N = \frac{1}{2}(15^2 - 15)$$

Carry out. We do the computation on the right.
$$N = \frac{1}{2}(15^2 - 15)$$
$$N = \frac{1}{2}(225 - 15)$$
$$N = \frac{1}{2}(210)$$
$$N = 105$$

Check. We can recheck the computation, or we can solve the equation $105 = \frac{1}{2}(n^2 - n)$. The answer checks.

State. 105 handshakes are possible.

26. 435

27. Familiarize. We will use the formula $N = \frac{1}{2}(n^2 - n)$, since "high fives" can be substituted for handshakes.

Translate. Substitute 66 for N.
$$66 = \frac{1}{2}(n^2 - n)$$

Carry out.

$$66 = \frac{1}{2}(n^2 - n)$$
$$132 = n^2 - n \qquad \text{Multiplying by 2}$$
$$0 = n^2 - n - 132$$
$$0 = (n - 12)(n + 11)$$
$$n - 12 = 0 \quad or \quad n + 11 = 0$$
$$n = 12 \ or \qquad n = -11$$

Check. The solutions of the equation are 12 and -11. Since the number of people cannot be negative, -11 cannot be a solution. However, 12 checks since $\frac{1}{2}(12^2 - 12) = \frac{1}{2}(144 - 12) = \frac{1}{2}(132) = 66$.

State. 12 people were on the team.

28. 20

29. Familiarize. We label the drawing. Let $x =$ the length of a side of the dining room, in ft. Then the dining room has dimensions x by x and the kitchen has dimensions x by 10. The entire rectangular space has dimension x by $x + 10$. Recall that we multiply these dimensions to find the area of the rectangle.

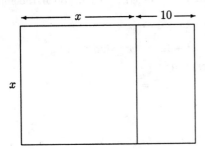

Translate.

The area of the rectangular space is 264 ft².

$$x(x + 10) = 264$$

Carry out. We solve the equation.

$$x(x + 10) = 264$$
$$x^2 + 10x = 264$$
$$x^2 + 10x - 264 = 0$$
$$(x + 22)(x - 12) = 0$$
$$x + 22 = 0 \quad or \quad x - 12 = 0$$
$$x = -22 \ or \qquad x = 12$$

Check. Since the length of a side of the dining room must be positive, -22 cannot be a solution. If x is 12 ft, then $x + 10$ is 22 ft, and the area of the space is $12 \cdot 22$, or 264 ft². The number 12 checks.

State. The dining room is 12 ft by 12 ft, and the kitchen is 12 ft by 10 ft.

30. 4 m

31. Familiarize. We will use the formula $h = 48t - 16t^2$.

Translate. Substitute $\frac{1}{2}$ for t.

$$h = 48 \cdot \frac{1}{2} - 16\left(\frac{1}{2}\right)^2$$

Carry out. We do the computation on the right.

$$h = 48 \cdot \frac{1}{2} - 16\left(\frac{1}{2}\right)^2$$
$$h = 48 \cdot \frac{1}{2} - 16 \cdot \frac{1}{4}$$
$$h = 24 - 4$$
$$h = 20$$

Check. We can recheck the computation, or we can solve the equation $20 = 48t - 16t^2$. The answer checks.

State. The rocket is 20 ft high $\frac{1}{2}$ sec after it is launched.

32. 36 ft

33. Familiarize. We will use the formula $h = 48t - 16t^2$.

Translate. Substitute 32 for h.

$$32 = 48t - 16t^2$$

Carry out. We solve the equation.

$$32 = 48t - 16t^2$$
$$0 = -16t^2 + 48t - 32$$
$$0 = -16(t^2 - 3t + 2)$$
$$0 = -16(t - 1)(t - 2)$$
$$t - 1 = 0 \quad or \quad t - 2 = 0$$
$$t = 1 \quad or \qquad t = 2$$

Check. When $t = 1$, $h = 48 \cdot 1 - 16 \cdot 1^2 = 48 - 16 = 32$. When $t = 2$, $h = 48 \cdot 2 - 16 \cdot 2^2 = 96 - 64 = 32$. Both numbers check.

State. The rocket will be exactly 32 ft above the ground at 1 sec and at 2 sec after it is launched.

34. 3 sec after launch

35. ◈

36. ◈

37. $-\frac{2}{3} \cdot \frac{4}{7} = -\frac{2 \cdot 4}{3 \cdot 7} = -\frac{8}{21}$

38. $-\frac{8}{45}$

39. $\frac{5}{6}\left(\frac{-7}{9}\right) = \frac{5(-7)}{6\cdot 9} = \frac{-35}{54}$, or $-\frac{35}{54}$

40. $-\frac{5}{16}$

41. $-\frac{2}{3} + \frac{4}{7} = -\frac{2}{3}\cdot\frac{7}{7} + \frac{4}{7}\cdot\frac{3}{3}$

$= -\frac{14}{21} + \frac{12}{21}$

$= -\frac{2}{21}$

42. $-\frac{26}{45}$

43. $\frac{5}{6} + \frac{-7}{9} = \frac{5}{6}\cdot\frac{3}{3} + \frac{-7}{9}\cdot\frac{2}{2}$

$= \frac{15}{18} + \frac{-14}{18}$

$= \frac{1}{18}$

44. $-\frac{11}{24}$

45. ◈

46. ◈

47. *Familiarize*. First we can use the Pythagorean theorem to find x, in ft. Then the height of the telephone pole is $x + 5$.

Translate. We use the Pythagorean theorem.

$$a^2 + b^2 = c^2$$
$$\left(\frac{1}{2}x + 1\right)^2 + x^2 = 34^2$$

Carry out. We solve the equation.

$$\left(\frac{1}{2}x + 1\right)^2 + x^2 = 34^2$$
$$\frac{1}{4}x^2 + x + 1 + x^2 = 1156$$
$$x^2 + 4x + 4 + 4x^2 = 4624 \quad \text{Multiplying by 4}$$
$$5x^2 + 4 + 4 = 4624$$
$$5x^2 + 4x - 4620 = 0$$
$$(5x + 154)(x - 30) = 0$$
$$5x + 154 = 0 \quad or \quad x - 30 = 0$$
$$5x = -154 \quad or \quad x = 30$$
$$x = -30.8 \quad or \quad x = 30$$

Check. Since the length x must be positive, -30.8 cannot be a solution. If x is 30 ft, then $\frac{1}{2}x + 1$ is $\frac{1}{2}\cdot 30 + 1$, or 16 ft. Since $16^2 + 30^2 = 1156 = 34^2$,

the number 30 checks. When x is 30 ft, then $x + 5$ is 35 ft.

State. The height of the telephone pole is 35 ft.

48. $1200

49. *Familiarize*. From the drawing in the text we see that the length of each half of the roof is 32 ft. Next we need to find the width w of each half of the roof, in ft. Then we will find the area of the roof and determine how many squares of shingles are needed. We make a drawing.

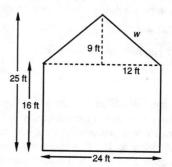

Translate. We use the Pythagorean theorem to find w.

$$a^2 + b^2 = c^2$$
$$9^2 + 12^2 = w^2 \quad \text{Substituting}$$

Carry out.

$$9^2 + 12^2 = w^2$$
$$81 + 144 = w^2$$
$$225 = w^2$$
$$0 = w^2 - 225$$
$$0 = (w + 15)(w - 15)$$
$$w + 15 = 0 \quad or \quad w - 15 = 0$$
$$w = -15 \quad or \quad w = 15$$

Since the width of the roof cannot be negative, we use $w = 15$ ft. The roof consists of two rectangles, each of which has dimensions 15 ft by 32 ft. We find the area of the roof:

$$2\cdot 32\cdot 15 = 960$$

Since a square of shingles covers 100 ft^2, we divide 960 by 100 to find the number of squares needed: $960 \div 100 = 9.6$. Assuming it is not possible to buy a fraction of a square, we round up, finding that 10 squares would be needed.

Check. Recheck the calculations.

State. 10 squares of shingles will be needed.

50. 39 cm

51. *Familiarize*. Let $y =$ the ten's digit. Then $y + 4 =$ the one's digit and $10y + y + 4$, or $11y + 4$, represents the number.

***Translate*.**

$$\underbrace{\text{The number}}_{} \text{ plus } \underbrace{\begin{array}{c}\text{the product}\\\text{of the digits}\end{array}}_{} \text{ is } 58.$$

$$11y + 4 \quad + \quad y(y + 4) \quad = \quad 58$$

***Carry out*.** We solve the equation.

$$11y + 4 + y(y + 4) = 58$$
$$11y + 4 + y^2 + 4y = 58$$
$$y^2 + 15y + 4 = 58$$
$$y^2 + 15y - 54 = 0$$
$$(y + 18)(y - 3) = 0$$
$$y + 18 = 0 \quad or \quad y - 3 = 0$$
$$y = -18 \quad or \quad y = 3$$

***Check*.** Since -18 cannot be a digit of the number, we only need to check 3. When $y = 3$, then $y + 4 = 7$ and the number is 37. We see that $37 + 3 \cdot 7 = 37 + 21$, or 58. The result checks.

***State*.** The number is 37.

52. 5 ft

53. *Familiarize*. Let $w =$ the width of the piece of cardboard, in cm. Then $2w =$ the length, in cm. The length and width of the base of the box are $2x - 8$ and $x - 8$, respectively, and its height is 4.

Recall that the formula for the volume of a rectangular solid is given by length $\cdot$ width $\cdot$ height.

***Translate*.**

$$\underbrace{\text{The volume}}_{} \text{ is } \underbrace{616 \text{ cm}^3}_{}.$$

$$(2w - 8)(w - 8)(4) = \quad 616$$

***Carry out*.** We solve the equation.

$$(2w - 8)(w - 8)(4) = 616$$
$$(2w^2 - 24w + 64)(4) = 616$$
$$8w^2 - 96 + 256 = 616$$
$$8w^2 - 96w - 360 = 0$$
$$8(w^2 - 12w - 45) = 0$$
$$w^2 - 12w - 45 = 0 \quad \text{Dividing by 8}$$
$$(w - 15)(w + 3) = 0$$
$$w - 15 = 0 \quad or \quad w + 3 = 0$$
$$w = 15 \quad or \quad w = -3$$

***Check*.** The width cannot be negative, so we only need to check 15. When $w = 15$, then $2w = 30$ and the dimensions of the box are $30 - 8$ by $15 - 8$ by 4, or 22 by 7 by 4. The volume is $22 \cdot 7 \cdot 4$, or 616.

***State*.** The original dimension of the cardboard are 15 cm by 30 cm.

54. 7 m

55. *Familiarize*. We make a drawing. Let $x =$ the depth of the gutter, in inches.

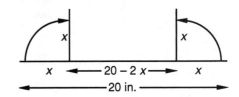

The cross-section has dimensions $20 - 2x$ by x.

***Translate*.**

$$\underbrace{\text{Area of cross-section}}_{} \text{ is } \underbrace{48 \text{ in}^2}_{}.$$

$$(20 - 2x)(x) = \quad 48$$

***Carry out*.**

$$(20 - 2x)(x) = 48$$
$$20x - 2x^2 = 48$$
$$-2x^2 + 20x - 48 = 0$$
$$-2(x^2 - 10x + 24) = 0$$
$$-2(x - 4)(x - 6) = 0$$
$$x - 4 = 0 \quad or \quad x - 6 = 0$$
$$x = 4 \quad or \quad x = 6$$

***Check*.** If the depth of the gutter is 4 in., then the cross-section has dimension $20 - 2 \cdot 4$, or 12 in. by 4 in., and its area is $12 \cdot 4 = 48 \text{ in}^2$. If the depth of the gutter is 6 in., then the cross-section has dimensions $20 - 2 \cdot 6$, or 8 in. by 6 in., and its area is $8 \cdot 6$, or 48 in^2. Both answers check.

***State*.** The gutter is 4 in. deep or 6 in. deep.

56. 100 cm^2; 225 cm^2

Chapter 6

Rational Expressions and Equations

Exercise Set 6.1

1. $\dfrac{25}{-7x}$

We find the real number(s) that make the denominator 0. To do so we set the denominator equal to 0 and solve for x:

$$-7x = 0$$
$$x = 0$$

The expression is undefined for $x = 0$.

2. 0

3. $\dfrac{t - 3}{t + 8}$

Set the denominator equal to 0 and solve for t:

$$t + 8 = 0$$
$$t = -8$$

The expression is undefined for $t = -8$.

4. -7

5. $\dfrac{a - 4}{3a - 12}$

Set the denominator equal to 0 and solve for a:

$$3a - 12 = 0$$
$$3a = 12$$
$$a = 4$$

The expression is undefined for $a = 4$.

6. 3

7. $\dfrac{x^2 - 16}{x^2 - 3x - 28}$

Set the denominator equal to 0 and solve for x:

$$x^2 - 3x - 28 = 0$$
$$(x - 7)(x + 4) = 0$$
$$x - 7 = 0 \quad or \quad x + 4 = 0$$
$$x = 7 \quad or \quad x = -4$$

The expression is undefined for $x = 7$ and $x = -4$.

8. 2, 5

9. $\dfrac{m^3 - 2m}{m^2 - 25}$

Set the denominator equal to 0 and solve for m:

$$m^2 - 25 = 0$$
$$(m + 5)(m - 5) = 0$$
$$m + 5 = 0 \quad or \quad m - 5 = 0$$
$$m = -5 \quad or \quad m = 5$$

The expression is undefined for $m = -5$ and $m = 5$.

10. -7, 7

11. $\dfrac{60a^2 b}{40ab^3}$

$= \dfrac{3a \cdot 20ab}{2b^2 \cdot 20ab}$ Factoring the numerator and denominator. Note the common factor of $20ab$.

$= \dfrac{3a}{2b^2} \cdot \dfrac{20ab}{20ab}$ Rewriting as a product of two rational expressions

$= \dfrac{3a}{2b^2} \cdot 1$ $\qquad \dfrac{20ab}{20ab} = 1$

$= \dfrac{3a}{2b^2}$ Removing the factor 1

12. $\dfrac{5y}{x^2}$

13. $\dfrac{35x^2 y}{14x^3 y^5} = \dfrac{5 \cdot 7x^2 y}{2xy^4 \cdot 7x^2 y}$

$= \dfrac{5}{2xy^4} \cdot \dfrac{7x^2 y}{7x^2 y}$

$= \dfrac{5}{2xy^4} \cdot 1$

$= \dfrac{5}{2xy^4}$

14. $\dfrac{2a^2 b^5}{3}$

15. $\dfrac{9x + 15}{12x + 20} = \dfrac{3(3x + 5)}{4(3x + 5)}$

$= \dfrac{3}{4} \cdot \dfrac{3x + 5}{3x + 5}$

$= \dfrac{3}{4} \cdot 1$

$= \dfrac{3}{4}$

16. $\dfrac{7}{5}$

17. $\dfrac{a^2 - 9}{a^2 + 4a + 3} = \dfrac{(a+3)(a-3)}{(a+3)(a+1)}$

$\qquad\qquad\quad = \dfrac{a+3}{a+3} \cdot \dfrac{a-3}{a+1}$

$\qquad\qquad\quad = 1 \cdot \dfrac{a-3}{a+1}$

$\qquad\qquad\quad = \dfrac{a-3}{a+1}$

18. $\dfrac{a+2}{a-3}$

19. $\dfrac{36x^6}{24x^9} = \dfrac{3 \cdot 12x^6}{2x^3 \cdot 12x^6}$

$\qquad\quad = \dfrac{3}{2x^3} \cdot \dfrac{12x^6}{12x^6}$

$\qquad\quad = \dfrac{3}{2x^3} \cdot 1$

$\qquad\quad = \dfrac{3}{2x^3}$

Check: Let $x = 1$.

$\dfrac{36x^6}{24x^9} = \dfrac{36 \cdot 1^6}{24 \cdot 1^9} = \dfrac{36}{24} = \dfrac{3}{2}$

$\dfrac{3}{2x^3} = \dfrac{3}{2 \cdot 1^3} = \dfrac{3}{2}$

The answer is probably correct.

20. $\dfrac{3a^2}{2}$

21. $\dfrac{-2y+6}{-8y} = \dfrac{-2(y-3)}{-2 \cdot 4y}$

$\qquad\qquad = \dfrac{-2}{-2} \cdot \dfrac{y-3}{4y}$

$\qquad\qquad = 1 \cdot \dfrac{y-3}{4y}$

$\qquad\qquad = \dfrac{y-3}{4y}$

Check: Let $x = 2$.

$\dfrac{-2y+6}{-8y} = \dfrac{-2 \cdot 2 + 6}{-8 \cdot 2} = \dfrac{2}{-16} = -\dfrac{1}{8}$

$\dfrac{y-3}{4y} = \dfrac{2-3}{4 \cdot 2} = \dfrac{-1}{8} = -\dfrac{1}{8}$

The answer is probably correct.

22. $\dfrac{2(x-3)}{3x}$

23. $\dfrac{6a^2 - 3a}{7a^2 - 7a} = \dfrac{3a(2a-1)}{7a(a-1)}$

$\qquad\qquad\quad = \dfrac{a}{a} \cdot \dfrac{3(2a-1)}{7(a-1)}$

$\qquad\qquad\quad = 1 \cdot \dfrac{3(2a-1)}{7(a-1)}$

$\qquad\qquad\quad = \dfrac{3(2a-1)}{7(a-1)}$

Check: Let $a = 2$.

$\dfrac{6a^2 - 3a}{7a^2 - 7a} = \dfrac{6 \cdot 2^2 - 3 \cdot 2}{7 \cdot 2^2 - 7 \cdot 2} = \dfrac{18}{14} = \dfrac{9}{7}$

$\dfrac{3(2a-1)}{7(a-1)} = \dfrac{3(2 \cdot 2 - 1)}{7(2-1)} = \dfrac{3 \cdot 3}{7 \cdot 1} = \dfrac{9}{7}$

The answer is probably correct.

24. $\dfrac{m+1}{2m+3}$

25. $\dfrac{t^2 - 16}{t^2 + t - 20} = \dfrac{(t+4)(t-4)}{(t+5)(t-4)}$

$\qquad\qquad\quad = \dfrac{t+4}{t+5} \cdot \dfrac{t-4}{t-4}$

$\qquad\qquad\quad = \dfrac{t+4}{t+5} \cdot 1$

$\qquad\qquad\quad = \dfrac{t+4}{t+5}$

Check: Let $t = 1$.

$\dfrac{t^2 - 16}{t^2 + t - 20} = \dfrac{1^2 - 16}{1^2 + 1 - 20} = \dfrac{-15}{-18} = \dfrac{5}{6}$

$\dfrac{t+4}{t+5} = \dfrac{1+4}{1+5} = \dfrac{5}{6}$

The answer is probably correct.

26. $\dfrac{a-2}{a+3}$

27. $\dfrac{3a^2 + 9a - 12}{6a^2 - 30a + 24} = \dfrac{3(a^2 + 3a - 4)}{6(a^2 - 5a + 4)}$

$\qquad\qquad\qquad\quad = \dfrac{3(a+4)(a-1)}{3 \cdot 2(a-4)(a-1)}$

$\qquad\qquad\qquad\quad = \dfrac{3(a-1)}{3(a-1)} \cdot \dfrac{a+4}{2(a-4)}$

$\qquad\qquad\qquad\quad = 1 \cdot \dfrac{a+4}{2(a-4)}$

$\qquad\qquad\qquad\quad = \dfrac{a+4}{2(a-4)}$

Check: Let $a = 2$.

$\dfrac{3a^2 + 9a - 12}{6a^2 - 30a + 24} = \dfrac{3 \cdot 2^2 + 9 \cdot 2 - 12}{6 \cdot 2^2 - 30 \cdot 2 + 24} = \dfrac{18}{-12} = -\dfrac{3}{2}$

$\dfrac{a+4}{2(a-4)} = \dfrac{2+4}{2(2-4)} = \dfrac{6}{-4} = -\dfrac{3}{2}$

The answer is probably correct.

28. $\dfrac{t-2}{2(t+4)}$

29. $\dfrac{x^2 + 8x + 16}{x^2 - 16} = \dfrac{(x+4)(x+4)}{(x+4)(x-4)}$

$\qquad = \dfrac{x+4}{x+4} \cdot \dfrac{x+4}{x-4}$

$\qquad = 1 \cdot \dfrac{x+4}{x-4}$

$\qquad = \dfrac{x+4}{x-4}$

Check: Let $x = 1$.

$\dfrac{x^2 + 8x + 16}{x^2 - 16} = \dfrac{1^2 + 8 \cdot 1 + 16}{1^2 - 16} = \dfrac{25}{-15} = -\dfrac{5}{3}$

$\dfrac{x+4}{x-4} = \dfrac{1+4}{1-4} = \dfrac{5}{-3} = -\dfrac{5}{3}$

The answer is probably correct.

30. $\dfrac{x+5}{x-5}$

31. $\dfrac{t^2 - 1}{t + 1} = \dfrac{(t+1)(t-1)}{t+1}$

$\qquad = \dfrac{t+1}{t+1} \cdot \dfrac{t-1}{1}$

$\qquad = 1 \cdot \dfrac{t-1}{1}$

$\qquad = t - 1$

Check: Let $t = 2$.

$\dfrac{t^2 - 1}{t+1} = \dfrac{2^2 - 1}{2+1} = \dfrac{3}{3} = 1$

$t - 1 = 2 - 1 = 1$

The answer is probably correct.

32. $a + 1$

33. $\dfrac{y^2 + 4}{y + 2}$ cannot be simplified.

Neither the numerator nor the denominator can be factored.

34. $\dfrac{x^2 + 1}{x + 1}$

35. $\dfrac{5x^2 - 20}{10x^2 - 40} = \dfrac{5(x^2 - 4)}{10(x^2 - 4)}$

$\qquad = \dfrac{1 \cdot \cancel{5} \cdot \cancel{(x^2 - 4)}}{2 \cdot \cancel{5} \cdot \cancel{(x^2 - 4)}}$

$\qquad = \dfrac{1}{2}$

Check: Let $x = 1$.

$\dfrac{5x^2 - 20}{10x^2 - 40} = \dfrac{5 \cdot 1^2 - 20}{10 \cdot 1^2 - 40} = \dfrac{-15}{-30} = \dfrac{1}{2}$

$\dfrac{1}{2} = \dfrac{1}{2}$

The answer is probably correct.

36. $\dfrac{3}{2}$

37. $\dfrac{5y + 5}{y^2 + 7y + 6} = \dfrac{5(y+1)}{(y+1)(y+6)}$

$\qquad = \dfrac{y+1}{y+1} \cdot \dfrac{5}{y+6}$

$\qquad = 1 \cdot \dfrac{5}{y+6}$

$\qquad = \dfrac{5}{y+6}$

Check: Let $x = 1$.

$\dfrac{5y + 5}{y^2 + 7y + 6} = \dfrac{5 \cdot 1 + 5}{1^2 + 7 \cdot 1 + 6} = \dfrac{10}{14} = \dfrac{5}{7}$

$\dfrac{5}{y+6} = \dfrac{5}{1+6} = \dfrac{5}{7}$

The answer is probably correct.

38. $\dfrac{6}{t - 3}$

39. $\dfrac{y^2 + 3y - 18}{y^2 + 2y - 15} = \dfrac{(y+6)(y-3)}{(y+5)(y-3)}$

$\qquad = \dfrac{y+6}{y+5} \cdot \dfrac{y-3}{y-3}$

$\qquad = \dfrac{y+6}{y+5} \cdot 1$

$\qquad = \dfrac{y+6}{y+5}$

Check: Let $y = 1$.

$\dfrac{y^2 + 3y - 18}{y^2 + 2y - 15} = \dfrac{1^2 + 3 \cdot 1 - 18}{1^2 + 2 \cdot 1 - 15} = \dfrac{-14}{-12} = \dfrac{7}{6}$

$\dfrac{y+6}{y+5} = \dfrac{1+6}{1+5} = \dfrac{7}{6}$

The answer is probably correct.

40. $\dfrac{a+3}{a+4}$

41. $\dfrac{(a-3)^2}{a^2 - 9} = \dfrac{(a-3)(a-3)}{(a+3)(a-3)}$

$\qquad = \dfrac{a-3}{a+3} \cdot \dfrac{a-3}{a-3}$

$\qquad = \dfrac{a-3}{a+3} \cdot 1$

$\qquad = \dfrac{a-3}{a+3}$

Check: Let $a = 2$.

$\dfrac{(a-3)^2}{a^2 - 9} = \dfrac{(2-3)^2}{2^2 - 9} = \dfrac{1}{-5} = -\dfrac{1}{5}$

$\dfrac{a-3}{a+3} = \dfrac{2-3}{2+3} = \dfrac{-1}{5} = -\dfrac{1}{5}$

The answer is probably correct.

42. $\dfrac{t-2}{t+2}$

43.
$$\begin{aligned}
\frac{x-8}{8-x} &= \frac{x-8}{-(x-8)} \\
&= \frac{1}{-1} \cdot \frac{x-8}{x-8} \\
&= \frac{1}{-1} \cdot 1 \\
&= -1
\end{aligned}$$
Check: Let $x = 2$.
$$\frac{x-8}{8-x} = \frac{2-8}{8-2} = \frac{-6}{6} = -1$$
The answer is probably correct.

44. -1

45.
$$\begin{aligned}
\frac{7t-14}{2-t} &= \frac{7(t-2)}{-(t-2)} \\
&= \frac{7}{-1} \cdot \frac{t-2}{t-2} \\
&= \frac{7}{-1} \cdot 1 \\
&= -7
\end{aligned}$$
Check: Let $t = 1$.
$$\frac{7t-14}{2-t} = \frac{7 \cdot 1 - 14}{2-1} = \frac{-7}{1} = -7$$
The answer is probably correct.

46. -4

47.
$$\begin{aligned}
\frac{a-b}{3b-3a} &= \frac{a-b}{-3(a-b)} \\
&= \frac{1}{-3} \cdot \frac{a-b}{a-b} \\
&= \frac{1}{-3} \cdot 1 \\
&= -\frac{1}{3}
\end{aligned}$$
Check: Let $a = 2$ and $b = 1$.
$$\frac{a-b}{3b-3a} = \frac{2-1}{3 \cdot 1 - 3 \cdot 2} = \frac{1}{-3} = -\frac{1}{3}$$
The answer is probably correct.

48. $-\dfrac{1}{2}$

49.
$$\begin{aligned}
\frac{3x^2 - 3y^2}{2y^2 - 2x^2} &= \frac{3(x^2 - y^2)}{2(y^2 - x^2)} \\
&= \frac{3(x^2 - y^2)}{2(-1)(x^2 - y^2)} \\
&= \frac{3}{2(-1)} \cdot \frac{x^2 - y^2}{x^2 - y^2} \\
&= \frac{3}{2(-1)} \cdot 1 \\
&= -\frac{3}{2}
\end{aligned}$$
Check: Let $x = 1$ and $y = 2$.
$$\frac{3x^2 - 3y^2}{2y^2 - 2x^2} = \frac{3 \cdot 1^2 - 3 \cdot 2^2}{2 \cdot 2^2 - 2 \cdot 1^2} = \frac{-9}{6} = -\frac{3}{2}$$
$$-\frac{3}{2} = -\frac{3}{2}$$
The answer is probably correct.

50. $-\dfrac{7}{3}$

51. $\dfrac{7s^2 - 28t^2}{28t^2 - 7s^2}$

Note that the numerator and denominator are opposites. Thus, we have an expression divided by its opposite, so the result is -1.

52. -1

53. ▧

54. ▧

55.
$$\begin{aligned}
-\frac{2}{3} \cdot \frac{6}{7} &= -\frac{2 \cdot 6}{3 \cdot 7} \\
&= -\frac{2 \cdot 2 \cdot 3}{3 \cdot 7} \\
&= -\frac{4}{7}
\end{aligned}$$

56. $-\dfrac{10}{33}$

57.
$$\begin{aligned}
\frac{5}{8} \div \left(-\frac{1}{6} \right) &= \frac{5}{8} \cdot (-6) \\
&= -\frac{5 \cdot 6}{8} \\
&= -\frac{5 \cdot 2 \cdot 3}{2 \cdot 4} \\
&= -\frac{15}{4}
\end{aligned}$$

58. $-\dfrac{21}{16}$

59. $\dfrac{7}{9} - \dfrac{2}{3} \cdot \dfrac{6}{7} = \dfrac{7}{9} - \dfrac{4}{7} = \dfrac{7}{9} \cdot \dfrac{7}{7} - \dfrac{4}{7} \cdot \dfrac{9}{2} =$

$\dfrac{49}{63} - \dfrac{36}{63} = \dfrac{13}{63}$

60. $\dfrac{5}{48}$

61. ◈

62. ◈

63. $\dfrac{x^4 - y^4}{(y-x)^4} = \dfrac{(x^2+y^2)(x^2-y^2)}{[-(x-y)]^4}$

$= \dfrac{(x^2+y^2)(x+y)(x-y)}{(-1)^4(x-y)(x-y)^3}$

$= \dfrac{x^2+y^2)(x+y)}{(x-y)^3}$

64. $-(2y + x)$

65. $\dfrac{(x-1)(x^4-1)(x^2-1)}{(x^2+1)(x-1)^2(x^4-2x^2+1)} =$

$\dfrac{(x-1)(x^4-1)(x^2-1)}{(x^2+1)(x-1)^2(x^2-1)^2} =$

$\dfrac{(x-1)(x^2+1)(x^2-1)(x+1)(x-1)}{(x^2+1)(x-1)(x-1)(x^2-1)(x^2-1)} =$

$\dfrac{(x-1)(x^2+1)(x^2-1)(x+1)(x-1) \cdot 1}{(x^2+1)\,(x-1)(x-1)(x^2-1)(x+1)\,(x-1)} =$

$\dfrac{1}{x-1}$

66. $\dfrac{x^3 + 4}{(x^3 + 2)(x^2 + 2)}$

67. $\dfrac{10t^4 - 8t^3 + 15t - 12}{8 - 10t + 12t^2 - 15t^3}$

$= \dfrac{2t^3(5t-4) + 3(5t-4)}{2(4-5t) + 3t^2(4-5t)}$

$= \dfrac{(5t-4)(2t^3+3)}{(4-5t)(2+3t^2)}$

$= \dfrac{(5t-4)(2t^3+3)}{(-1)(5t-4)(2+3t^2)}$

$= -\dfrac{2t^3+3}{2+3t^2}, \text{ or } \dfrac{-2t^3-3}{2+3t^2}, \text{ or } \dfrac{2t^3+3}{-2-3t^2}$

68. $\dfrac{(t-1)(t-9)^2}{(t^2+9)(t+1)}$

69. $\dfrac{(t+2)^3(t^2+2t+1)(t+1)}{(t+1)^3(t^2+4t+4)(t+2)} =$

$\dfrac{(t+2)^3(t+1)^2(t+1)}{(t+1)^3(t+2)^2(t+2)} = \dfrac{(t+2)^3(t+1)^3}{(t+1)^3(t+2)^3} = 1$

70. $\dfrac{(x-y)^3}{(x+y)^2(x-5y)}$

71. ◈

Exercise Set 6.2

1. $\dfrac{9x}{4} \cdot \dfrac{x-5}{2x+1} = \dfrac{9x(x-5)}{4(2x+1)}$

2. $\dfrac{3x(5x+2)}{4(x-1)}$

3. $\dfrac{a-4}{a+6} \cdot \dfrac{a+2}{a+6} = \dfrac{(a-4)(a+2)}{(a+6)(a+6)}, \text{ or } \dfrac{(a-4)(a+2)}{(a+6)^2}$

4. $\dfrac{(a+3)^2}{(a+6)(a-1)}$

5. $\dfrac{2x+3}{4} \cdot \dfrac{x+1}{x-5} = \dfrac{(2x+3)(x+1)}{4(x-5)}$

6. $\dfrac{4(x+2)}{(3x-4)(5x+6)}$

7. $\dfrac{a-5}{a^2+1} \cdot \dfrac{a+2}{a^2-1} = \dfrac{(a-5)(a+2)}{(a^2+1)(a^2-1)}$

8. $\dfrac{(t+3)^2}{(t^2-2)(t^2-4)}$

9. $\dfrac{x+4}{2+x} \cdot \dfrac{x-1}{x+1} = \dfrac{(x+4)(x-1)}{(2+x)(x+1)}$

10. $\dfrac{(m+4)(2+m)}{(m+8)(m+5)}$

11. $\dfrac{5a^4}{6a} \cdot \dfrac{2}{a}$

$= \dfrac{5a^4 \cdot 2}{6a \cdot a}$ Multiplying the numerators and the denominators

$= \dfrac{5 \cdot a \cdot a \cdot a \cdot a \cdot 2}{2 \cdot 3 \cdot a \cdot a}$ Factoring the numerator and the denominator

$= \dfrac{5 \cdot a \cdot a \cdot a \cdot a \cdot 2}{2 \cdot 3 \cdot a \cdot a}$ Removing a factor equal to 1

$= \dfrac{5a^2}{3}$ Simplifying

12. $\dfrac{6}{5t^6}$

13. $\dfrac{3c}{d^2} \cdot \dfrac{8d}{6c^3}$

$= \dfrac{3c \cdot 8d}{d^2 \cdot 6c^3}$ Multiplying the numerators and the denominators

$= \dfrac{3 \cdot c \cdot 2 \cdot 4 \cdot d}{d \cdot d \cdot 3 \cdot 2 \cdot c \cdot c \cdot c}$ Factoring the numerator and the denominator

$= \dfrac{\cancel{3} \cdot \cancel{c} \cdot \cancel{2} \cdot 4 \cdot \cancel{d}}{\cancel{d} \cdot d \cdot \cancel{3} \cdot \cancel{2} \cdot \cancel{c} \cdot c \cdot c}$

$= \dfrac{4}{dc^2}$

14. $\dfrac{6x}{y^2}$

15. $\dfrac{x^2 - 3x - 10}{(x-2)^2} \cdot \dfrac{x-2}{x-5} = \dfrac{(x^2-3x-10)(x-2)}{(x-2)^2(x-5)}$

$= \dfrac{(x-5)(x+2)(x-2)}{(x-2)(x-2)(x-5)}$

$= \dfrac{(\cancel{x-5})(x+2)(\cancel{x-2})}{(\cancel{x-2})(x-2)(\cancel{x-5})}$

$= \dfrac{x+2}{x-2}$

16. $\dfrac{t-3}{t+2}$

17. $\dfrac{a^2+25}{a^2-4a+3} \cdot \dfrac{a-5}{a+5} = \dfrac{(a^2+25)(a-5)}{(a^2-4a+3)(a+5)}$

$= \dfrac{(a^2+25)(a-5)}{(a-3)(a-1)(a+5)}$

(No simplification is possible.)

18. $\dfrac{(x+3)(x+4)(x+1)}{(x^2+9)(x+9)}$

19. $\dfrac{a^2-9}{a^2} \cdot \dfrac{5a}{a^2+a-12} = \dfrac{(a+3)(a-3) \cdot 5 \cdot a}{a \cdot a(a+4)(a-3)}$

$= \dfrac{(a+3)(\cancel{a-3}) \cdot 5 \cdot \cancel{a}}{\cancel{a} \cdot a(a+4)(\cancel{a-3})}$

$= \dfrac{5(a+3)}{a(a+4)}$

20. $\dfrac{x^2(x-1)}{5}$

21. $\dfrac{4a^2}{3a^2-12a+12} \cdot \dfrac{3a-6}{2a}$

$= \dfrac{4a^2(3a-6)}{(3a^2-12a+12)2a}$

$= \dfrac{2 \cdot 2 \cdot a \cdot a \cdot 3 \cdot (a-2)}{3 \cdot (a-2) \cdot (a-2) \cdot 2 \cdot a}$

$= \dfrac{\cancel{2} \cdot 2 \cdot \cancel{a} \cdot a \cdot \cancel{3} \cdot (\cancel{a-2})}{\cancel{3} \cdot (\cancel{a-2}) \cdot (a-2) \cdot \cancel{2} \cdot \cancel{a}}$

$= \dfrac{2a}{a-2}$

22. $\dfrac{10(v-2)}{v-1}$

23. $\dfrac{t^2+2t-3}{t^2+4t-5} \cdot \dfrac{t^2-3t-10}{t^2+5t+6}$

$= \dfrac{(t^2+2t-3)(t^2-3t-10)}{(t^2+4t-5)(t^2+5t+6)}$

$= \dfrac{(t+3)(t-1)(t-5)(t+2)}{(t+5)(t-1)(t+3)(t+2)}$

$= \dfrac{(\cancel{t+3})(\cancel{t-1})(t-5)(\cancel{t+2})}{(t+5)(\cancel{t-1})(\cancel{t+3})(\cancel{t+2})}$

$= \dfrac{t-5}{t+5}$

24. $\dfrac{x+4}{x-4}$

25. $\dfrac{5a^2-180}{10a^2-10} \cdot \dfrac{20a+20}{2a-12}$

$= \dfrac{(5a^2-180)(20a+20)}{(10a^2-10)(2a-12)}$

$= \dfrac{5(a+6)(a-6)(2)(10)(a+1)}{10(a+1)(a-1)(2)(a-6)}$

$= \dfrac{5(a+6)(\cancel{a-6})\cancel{(2)}\cancel{(10)}(\cancel{a+1})}{\cancel{10}(\cancel{a+1})(a-1)\cancel{(2)}(\cancel{a-6})}$

$= \dfrac{5(a+6)}{a-1}$

26. $\dfrac{t+7}{4(t-1)}$

27. $\dfrac{x^2+4x+4}{(x-1)^2} \cdot \dfrac{x^2-2x+1}{(x+2)^2} = \dfrac{(x+2)^2(x-1)^2}{(x-1)^2(x+2)^2} = 1$

28. $\dfrac{1}{x+2}$

29. $\dfrac{t^2+8t+16}{(t+4)^3} \cdot \dfrac{(t+2)^3}{t^2+4t+4} = \dfrac{(t+4)^2(t+2)^3}{(t+4)^3(t+2)^2}$

$\qquad = \dfrac{(t+4)^2(t+2)^2(t+2)}{(t+4)^2(t+4)(t+2)^2}$

$\qquad = \dfrac{(t+4)^2(t+2)^2}{(t+4)^2(t+2)^2} \cdot \dfrac{t+2}{t+4}$

$\qquad = 1 \cdot \dfrac{t+2}{t+4}$

$\qquad = \dfrac{t+2}{t+4}$

30. $\dfrac{y-1}{y-2}$

31. The reciprocal of $\dfrac{3x}{7}$ is $\dfrac{7}{3x}$ because $\dfrac{3x}{7} \cdot \dfrac{7}{3x} = 1$.

32. $\dfrac{x^2+4}{3-x}$

33. The reciprocal of $a^3 - 8a$ is $\dfrac{1}{a^3 - 8a}$ because

$\dfrac{a^3 - 8a}{1} \cdot \dfrac{1}{a^3 - 8a} = 1$.

34. $\dfrac{a^2 - b^2}{7}$

35. The reciprocal of $\dfrac{x^2 + 2x - 5}{x^2 - 4x + 7}$ is $\dfrac{x^2 - 4x + 7}{x^2 + 2x - 5}$ because

$\dfrac{x^2 + 2x - 5}{x^2 - 4x + 7} \cdot \dfrac{x^2 - 4x + 7}{x^2 + 2x - 5} = 1$.

36. $\dfrac{x^2 + 7xy - y^2}{x^2 - 3xy + y^2}$

37. $\dfrac{3}{8} \div \dfrac{5}{2}$

$= \dfrac{3}{8} \cdot \dfrac{2}{5}$ Multiplying by the reciprocal of the divisor

$= \dfrac{3 \cdot 2}{8 \cdot 5}$

$= \dfrac{3 \cdot 2}{2 \cdot 4 \cdot 5}$ Factoring the denominator

$= \dfrac{2}{2} \cdot \dfrac{3}{4 \cdot 5}$ Factoring the fractional expression

$= \dfrac{3}{20}$ Simplifying

38. $\dfrac{35}{18}$

39. $\dfrac{x}{4} \div \dfrac{5}{x}$

$= \dfrac{x}{4} \cdot \dfrac{x}{5}$ Multiplying by the reciprocal of the divisor

$= \dfrac{x \cdot x}{4 \cdot 5}$

$= \dfrac{x^2}{20}$

40. $\dfrac{60}{x^2}$

41. $\dfrac{a^5}{b^4} \div \dfrac{a^2}{b} = \dfrac{a^5}{b^4} \cdot \dfrac{b}{a^2}$

$\qquad = \dfrac{a^5 \cdot b}{b^4 \cdot a^2}$

$\qquad = \dfrac{a^2 \cdot a^3 \cdot b}{b \cdot b^3 \cdot a^2}$

$\qquad = \dfrac{a^2 b}{a^2 b} \cdot \dfrac{a^3}{b^3}$

$\qquad = \dfrac{a^3}{b^3}$

42. $\dfrac{x^3}{y}$

43. $\dfrac{y+5}{4} \div \dfrac{y}{2} = \dfrac{y+5}{4} \cdot \dfrac{2}{y}$

$\qquad = \dfrac{(y+5)(2)}{4 \cdot y}$

$\qquad = \dfrac{(y+5)(\cancel{2})}{\cancel{2} \cdot 2y}$

$\qquad = \dfrac{y+5}{2y}$

44. $\dfrac{(a+2)(a+3)}{(a-3)(a-1)}$

45. $\dfrac{4y-8}{y+2} \div \dfrac{y-2}{y^2-4} = \dfrac{4y-8}{y+2} \cdot \dfrac{y^2-4}{y-2}$

$\qquad = \dfrac{(4y-8)(y^2-4)}{(y+2)(y-2)}$

$\qquad = \dfrac{4(\cancel{y-2})(\cancel{y+2})(y-2)}{(\cancel{y+2})(\cancel{y-2})(1)}$

$\qquad = 4(y-2)$

46. $\dfrac{(x-1)^2}{x}$

47.
$$\frac{a}{a-b} \div \frac{b}{b-a} = \frac{a}{a-b} \cdot \frac{b-a}{b}$$
$$= \frac{a(b-a)}{(a-b)(b)}$$
$$= \frac{a(-1)(a-b)}{(a-b)(b)}$$
$$= \frac{-a}{b} = -\frac{a}{b}$$

48. $-\dfrac{1}{2}$

49.
$$(y^2 - 9) \div \frac{y^2 - 2y - 3}{y^2 + 1} = \frac{(y^2 - 9)}{1} \cdot \frac{y^2 + 1}{y^2 - 2y - 3}$$
$$= \frac{(y^2 - 9)(y^2 + 1)}{y^2 - 2y - 3}$$
$$= \frac{(y+3)(y-3)(y^2+1)}{(y-3)(y+1)}$$
$$= \frac{(y+3)(y-3)(y^2+1)}{(y-3)(y+1)}$$
$$= \frac{(y+3)(y^2+1)}{y+1}$$

50. $\dfrac{(x-6)(x+6)}{x-1}$

51.
$$\frac{5x-5}{16} \div \frac{x-1}{6} = \frac{5x-5}{16} \cdot \frac{6}{x-1}$$
$$= \frac{(5x-5)\cdot 6}{16(x-1)}$$
$$= \frac{5(x-1)\cdot 2 \cdot 3}{2 \cdot 8(x-1)}$$
$$= \frac{5(x-1)\cdot 2 \cdot 3}{2 \cdot 8(x-1)}$$
$$= \frac{15}{8}$$

52. $\dfrac{2}{5}$

53.
$$\frac{-6+3x}{5} \div \frac{4x-8}{25} = \frac{-6+3x}{5} \cdot \frac{25}{4x-8}$$
$$= \frac{(-6+3x)\cdot 25}{5(4x-8)}$$
$$= \frac{3(x-2)\cdot 5 \cdot 5}{5 \cdot 4(x-2)}$$
$$= \frac{3(x-2)\cdot 5 \cdot 5}{5 \cdot 4(x-2)}$$
$$= \frac{15}{4}$$

54. 1

55.
$$\frac{a+2}{a-1} \div \frac{3a+6}{a-5} = \frac{a+2}{a-1} \cdot \frac{a-5}{3a+6}$$
$$= \frac{(a+2)(a-5)}{(a-1)(3a+6)}$$
$$= \frac{(a+2)(a-5)}{(a-1)\cdot 3 \cdot (a+2)}$$
$$= \frac{(a+2)(a-5)}{(a-1)\cdot 3 \cdot (a+2)}$$
$$= \frac{a-5}{3(a-1)}$$

56. $\dfrac{t+1}{4(t+2)}$

57.
$$(2x-1) \div \frac{2x^2 - 11x + 5}{4x^2 - 1}$$
$$= \frac{2x-1}{1} \cdot \frac{4x^2 - 1}{2x^2 - 11x + 5}$$
$$= \frac{(2x-1)(4x^2-1)}{1\cdot(2x^2-11x+5)}$$
$$= \frac{(2x-1)(2x+1)(2x-1)}{1\cdot(2x-1)(x-5)}$$
$$= \frac{(2x-1)(2x+1)(2x-1)}{1\cdot(2x-1)(x-5)}$$
$$= \frac{(2x-1)(2x+1)}{x-5}$$

58. $\dfrac{(a+7)(a+1)}{3a-7}$

59.
$$\frac{x-5}{x+5} \div \frac{2x^2 - 50}{x^2 + 25} = \frac{x-5}{x+5} \cdot \frac{x^2 + 25}{2x^2 - 50}$$
$$= \frac{(x-5)(x^2+25)}{(x+5)(2)(x^2-25)}$$
$$= \frac{(x-5)(x^2+25)}{(x+5)(2)(x+5)(x-5)}$$
$$= \frac{x^2+25}{2(x+5)^2}$$

60. $\dfrac{3(x-3)^2}{x^2+1}$

61.
$$\frac{a^2 - 10a + 25}{a^2 + 7a + 12} \div \frac{a^2 - a - 20}{a^2 + 6a + 9}$$
$$= \frac{a^2 - 10a + 25}{a^2 + 7a + 12} \cdot \frac{a^2 + 6a + 9}{a^2 - a - 20}$$
$$= \frac{(a-5)(a-5)(a+3)(a+3)}{(a+3)(a+4)(a-5)(a+4)}$$
$$= \frac{(a-5)(a+3)}{(a+4)^2}$$

62. $\dfrac{(a+1)^2(a-6)}{(a-1)^2(a+4)}$

63.
$$\dfrac{c^2+10c+21}{c^2-2c-15} \div (c^2+2c-35)$$
$$= \dfrac{c^2+10c+21}{c^2-2c-25} \cdot \dfrac{1}{c^2+2c-35}$$
$$= \dfrac{(c^2+10c+21)\cdot 1}{(c^2-2c-15)(c^2+2c-35)}$$
$$= \dfrac{(c+7)(c+3)}{(c-5)(c+3)(c+7)(c-5)}$$
$$= \dfrac{(c+7)(c+3)}{(c+7)(c+3)} \cdot \dfrac{1}{(c-5)(c-5)}$$
$$= \dfrac{1}{(c-5)^2}$$

64. $\dfrac{1}{1+2z-z^2}$

65.
$$\dfrac{x-y}{x^2+2xy+y^2} \div \dfrac{x^2-y^2}{x^2-5xy+4y^2}$$
$$= \dfrac{x-y}{x^2+2xy+y^2} \cdot \dfrac{x^2-5xy+4y^2}{x^2-y^2}$$
$$= \dfrac{(x-y)(x-y)(x-4y)}{(x+y)(x+y)(x+y)(x-y)}$$
$$= \dfrac{(x-y)(x-4y)}{(x+y)^3}$$

66. $\dfrac{a+b}{(a-2b)^2}$

67. ◈

68. ◈

69.
$$\dfrac{3}{4} + \dfrac{5}{6} = \dfrac{3}{4}\cdot\dfrac{3}{3} + \dfrac{5}{6}\cdot\dfrac{2}{2}$$
$$= \dfrac{9}{12} + \dfrac{10}{12}$$
$$= \dfrac{19}{12}$$

70. $\dfrac{41}{24}$

71.
$$\dfrac{2}{9} - \dfrac{1}{6} = \dfrac{2}{9}\cdot\dfrac{2}{2} - \dfrac{1}{6}\cdot\dfrac{3}{3}$$
$$= \dfrac{4}{18} - \dfrac{3}{18}$$
$$= \dfrac{1}{18}$$

72. $-\dfrac{1}{6}$

73. $\dfrac{2}{5} - \left(\dfrac{3}{2}\right)^2 = \dfrac{2}{5} - \dfrac{9}{4} = \dfrac{8}{20} - \dfrac{45}{20} = -\dfrac{37}{20}$

74. $\dfrac{49}{45}$

75. ◈

76. ◈

77. $\dfrac{3x-y}{2x+y} \div \dfrac{3x-y}{2x+y}$

We have the rational expression $\dfrac{3x-y}{2x+y}$ divided by itself. Thus, the result is 1.

78. $\dfrac{a}{(c-3d)(2a+5b)}$

79.
$$(x-2a) \div \dfrac{a^2x^2-4a^4}{a^2x+2a^3} = \dfrac{x-2a}{1} \cdot \dfrac{a^2x+2a^3}{a^2x^2-4a^4}$$
$$= \dfrac{(x-2a)(a^2)(x+2a)}{a^2(x+2a)(x-2a)}$$
$$= 1$$

80. $\dfrac{1}{b^3(a-3b)}$

81.
$$\dfrac{3x^2-2xy-y^2}{x^2-y^2} \div (3x^2+4xy+y^2)^2 =$$
$$\dfrac{3x^2-2xy-y^2}{x^2-y^2} \cdot \dfrac{1}{(3x^2+4xy+y^2)^2} =$$
$$\dfrac{(3x+y)(x-y)\cdot 1}{(x+y)(x-y)(3x+y)(3x+y)(x+y)(x+y)} =$$
$$\dfrac{1}{(x+y)^3(3x+y)}$$

82. $\dfrac{x^3y^2}{4}$

83.
$$\dfrac{a^2-3b}{a^2+2b} \cdot \dfrac{a^2-2b}{a^2+3b} \cdot \dfrac{a^2+2b}{a^2-3b}$$

Note that $\dfrac{a^2-3b}{a^2+2b} \cdot \dfrac{a^2+2b}{a^2-3b}$ is the product of reciprocals and thus is equal to 1. Then the product in the original exercise is the remaining factor, $\dfrac{a^2-2b}{a^2+3b}$.

84. $\dfrac{(z+4)^3}{3(z-4)^2}$

85. $\dfrac{x^2 - x + xy - y}{x^2 + 6x - 7} \div \dfrac{x^2 + 2xy + y^2}{4x + 4y}$

$= \dfrac{x^2 - x + xy - y}{x^2 + 6x - 7} \cdot \dfrac{4x + 4y}{x^2 + 2xy + y^2}$

$= \dfrac{x(x - 1) + y(x - 1)}{x^2 + 6x - 7} \cdot \dfrac{4x + 4y}{x^2 + 2xy + y^2}$

$= \dfrac{(x - 1)(x + y) \cdot 4(x + y)}{(x + 7)(x - 1)(x + y)(x + y)}$

$= \dfrac{(x - 1)(x + y)(x + y)}{(x - 1)(x + y)(x + y)} \cdot \dfrac{4}{x + 7}$

$= \dfrac{4}{x + 7}$

86. $\dfrac{x(x^2 + 1)}{3(x + y - 1)}$

87. $\dfrac{(t + 2)^3}{(t + 1)^3} \div \dfrac{t^2 + 4t + 4}{t^2 + 2t + 1} \cdot \dfrac{t + 1}{t + 2}$

$= \dfrac{(t + 2)^3}{(t + 1)^3} \cdot \dfrac{t^2 + 2t + 1}{t^2 + 4t + 4} \cdot \dfrac{t + 1}{t + 2}$

$= \dfrac{(t + 2)(t + 2)(t + 2)(t + 1)(t + 1)(t + 1)}{(t + 1)(t + 1)(t + 1)(t + 2)(t + 2)(t + 2)}$

$= 1$

88. $\dfrac{3(y + 2)^3}{y(y - 1)}$

89. $\dfrac{6y - 4x}{(2x + 3y)^2} \cdot \dfrac{2x - 3y}{x^2 - 9y^2} \div \dfrac{4x^2 - 12xy + 9y^2}{9y^2 + 12xy + 4x^2}$

$= \dfrac{6y - 4x}{(2x + 3y)^2} \cdot \dfrac{2x - 3y}{x^2 - 9y^2} \cdot \dfrac{9y^2 + 12xy + 4x^2}{4x^2 - 12xy + 9y^2}$

$= \dfrac{2(3y - 2x)(2x - 3y)(3y + 2x)(3y + 2x)}{(2x + 3y)(2x + 3y)(x + 3y)(x - 3y)(2x - 3y)(2x - 3y)}$

$= \dfrac{2(-1)(2x - 3y)(2x - 3y)(2x + 3y)(2x + 3y)}{(2x + 3y)(2x + 3y)(x + 3y)(x - 3y)(2x - 3y)(2x - 3y)}$

$= \dfrac{-2}{(x + 3y)(x - 3y)}, \text{ or } -\dfrac{2}{(x + 3y)(x - 3y)}$

90. $\dfrac{a - 3b}{c}$

Exercise Set 6.3

1. $\dfrac{3}{x} + \dfrac{9}{x} = \dfrac{12}{x}$ Adding numerators

2. $\dfrac{13}{a^2}$

3. $\dfrac{x}{15} + \dfrac{2x + 5}{15} = \dfrac{3x + 5}{15}$ Adding numerators

4. $\dfrac{4a - 4}{7}$

5. $\dfrac{4}{a + 3} + \dfrac{5}{a + 3} = \dfrac{9}{a + 3}$

6. $\dfrac{13}{x + 2}$

7. $\dfrac{9}{a + 2} - \dfrac{3}{a + 2} = \dfrac{6}{a + 2}$ Subtracting numerators

8. $\dfrac{6}{x + 7}$

9. $\dfrac{3y + 8}{2y} - \dfrac{y + 1}{2y}$

$= \dfrac{3y + 8 - (y + 1)}{2y}$

$= \dfrac{3y + 8 - y - 1}{2y}$ Removing parentheses

$= \dfrac{2y + 7}{2y}$

10. $\dfrac{t + 4}{4t}$

11. $\dfrac{7x + 8}{x + 1} + \dfrac{4x + 3}{x + 1}$

$= \dfrac{11x + 11}{x + 1}$ Adding numerators

$= \dfrac{11(x + 1)}{x + 1}$ Factoring

$= \dfrac{11(x + 1)}{x + 1}$ Removing a factor equal to 1

$= 11$

12. 5

13. $\dfrac{7x + 8}{x + 1} - \dfrac{4x + 3}{x + 1} = \dfrac{7x + 8 - (4x + 3)}{x + 1}$

$= \dfrac{7x + 8 - 4x - 3}{x + 1}$

$= \dfrac{3x + 5}{x + 1}$

14. $\dfrac{a + 6}{a + 4}$

15. $\dfrac{a^2}{a - 4} + \dfrac{a - 20}{a - 4} = \dfrac{a^2 + a - 20}{a - 4}$

$= \dfrac{(a + 5)(a - 4)}{a - 4}$

$= \dfrac{(a + 5)(a - 4)}{a - 4}$

$= a + 5$

16. $x + 2$

17. $\dfrac{x^2}{x-2} - \dfrac{6x-8}{x-2} = \dfrac{x^2-(6x-8)}{x-2}$

$ \qquad\qquad = \dfrac{x^2-6x+8}{x-2}$

$ \qquad\qquad = \dfrac{(x-4)(x-2)}{x-2}$

$ \qquad\qquad = \dfrac{(x-4)(x\!-\!2)}{x\!-\!2}$

$ \qquad\qquad = x - 4$

18. $a - 5$

19. $\dfrac{t^2-5t}{t-1} + \dfrac{5t-t^2}{t-1}$

Note that the numerators are opposites, so their sum is 0. Then we have $\dfrac{0}{t-1}$, or 0.

20. $y + 6$

21. $\dfrac{x-4}{x^2+5x+6} + \dfrac{7}{x^2+5x+6} = \dfrac{x+3}{x^2+5x+6}$

$ \qquad\qquad\qquad\qquad = \dfrac{x+3}{(x+3)(x+2)}$

$ \qquad\qquad\qquad\qquad = \dfrac{x\!+\!3}{(x\!+\!3)(x+2)}$

$ \qquad\qquad\qquad\qquad = \dfrac{1}{x+2}$

22. $\dfrac{1}{x-1}$

23. $\dfrac{3a^2+14}{a^2+5a-6} - \dfrac{13a}{a^2+5a-6} = \dfrac{3a^2-13a+14}{a^2+5a-6}$

$ \qquad\qquad\qquad\qquad = \dfrac{(3a-7)(a-2)}{(a+6)(a-1)}$

(No simplification is possible.)

24. $\dfrac{2a-5}{a-4}$

25. $\dfrac{t^2-3t}{t^2+6t+9} + \dfrac{2t-12}{t^2+6t+9} = \dfrac{t^2-t-12}{t^2+6t+9}$

$ \qquad\qquad\qquad\qquad = \dfrac{(t-4)(t+3)}{(t+3)^2}$

$ \qquad\qquad\qquad\qquad = \dfrac{(t-4)(t\!+\!3)}{(t+3)(t\!+\!3)}$

$ \qquad\qquad\qquad\qquad = \dfrac{t-4}{t+3}$

26. $\dfrac{y-5}{y+4}$

27. $\dfrac{2x^2+x}{x^2-8x+12} - \dfrac{x^2-2x+10}{x^2-8x+12}$

$ = \dfrac{2x^2+x-(x^2-2x+10)}{x^2-8x+12}$

$ = \dfrac{2x^2+x-x^2+2x-10}{x^2-8x+12}$

$ = \dfrac{x^2+3x-10}{x^2-8x+12}$

$ = \dfrac{(x+5)(x-2)}{(x-6)(x-2)}$

$ = \dfrac{(x+5)(x\!-\!2)}{(x-6)(x\!-\!2)}$

$ = \dfrac{x+5}{x-6}$

28. 0

29. $\dfrac{3-2x}{x^2-6x+8} + \dfrac{7-3x}{x^2-6x+8}$

$ = \dfrac{10-5x}{x^2-6x+8}$

$ = \dfrac{5(2-x)}{(x-4)(x-2)}$

$ = \dfrac{5(-1)(x-2)}{(x-4)(x-2)}$

$ = \dfrac{5(-1)(x\!-\!2)}{(x-4)(x\!-\!2)}$

$ = \dfrac{-5}{x-4}$, or $-\dfrac{5}{x-4}$, or $\dfrac{5}{4-x}$

30. $-\dfrac{5}{t-4}$, or $\dfrac{5}{4-t}$

31. $\dfrac{x-7}{x^2+3x-4} - \dfrac{2x-3}{x^2+3x-4}$

$ = \dfrac{x-7-(2x-3)}{x^2+3x-4}$

$ = \dfrac{x-7-2x+3}{x^2+3x-4}$

$ = \dfrac{-x-4}{x^2+3x-4}$

$ = \dfrac{-(x+4)}{(x+4)(x-1)}$

$ = \dfrac{-1(x\!+\!4)}{(x\!+\!4)(x-1)}$

$ = \dfrac{-1}{x-1}$, or $-\dfrac{1}{x-1}$, or $\dfrac{1}{1-x}$

32. $-\dfrac{4}{x-1}$, or $\dfrac{4}{1-x}$

33. $15 = 3 \cdot 5$

$27 = 3 \cdot 3 \cdot 3$

$\text{LCM} = 3 \cdot 3 \cdot 3 \cdot 5, \text{ or } 135$

34. 30

35. $8 = 2 \cdot 2 \cdot 2$

$9 = 3 \cdot 3$

$\text{LCM} = 2 \cdot 2 \cdot 2 \cdot 3 \cdot 3, \text{ or } 72$

36. 60

37. $6 = 2 \cdot 3$

$9 = 3 \cdot 3$

$21 = 3 \cdot 7$

$\text{LCM} = 2 \cdot 3 \cdot 3 \cdot 7, \text{ or } 126$

38. 360

39. $12x^2 = 2 \cdot 2 \cdot 3 \cdot x \cdot x$

$6x^3 = 2 \cdot 3 \cdot x \cdot x \cdot x$

$\text{LCM} = 2 \cdot 2 \cdot 3 \cdot x \cdot x \cdot x, \text{ or } 12x^3$

40. $10t^4$

41. $15a^4b^7 = 3 \cdot 5 \cdot a \cdot a \cdot a \cdot a \cdot b \cdot b \cdot b \cdot b \cdot b \cdot b \cdot b$

$10a^2b^8 = 2 \cdot 5 \cdot a \cdot a \cdot b \cdot b \cdot b \cdot b \cdot b \cdot b \cdot b \cdot b$

$\text{LCM} = 2 \cdot 3 \cdot 5 \cdot a \cdot a \cdot a \cdot a \cdot b \cdot b \cdot b \cdot b \cdot b \cdot b \cdot b \cdot b,$

$\quad\quad \text{or } 30a^4b^8$

42. $18a^5b^7$

43. $2(y - 3) = 2 \cdot (y - 3)$

$6(y - 3) = 2 \cdot 3 \cdot (y - 3)$

$\text{LCM} = 2 \cdot 3 \cdot (y - 3), \text{ or } 6(y - 3)$

44. $8(x - 1)$

45. $x^2 - 4 = (x + 2)(x - 2)$

$x^2 + 5x + 6 = (x + 3)(x + 2)$

$\text{LCM} = (x + 2)(x - 2)(x + 3)$

46. $(x + 2)(x + 1)(x - 2)$

47. $t^3 + 4t^2 + 4t = t(t^2 + 4t + 4) = t(t + 2)(t + 2)$

$t^2 - 4t = t(t - 4)$

$\text{LCM} = t(t + 2)(t + 2)(t - 4) = t(t + 2)^2(t - 4)$

48. $y^2(y + 1)(y - 1)$

49. $10x^2y = 2 \cdot 5 \cdot x \cdot x \cdot y$

$6y^2z = 2 \cdot 3 \cdot y \cdot y \cdot z$

$5xz^3 = 5 \cdot x \cdot z \cdot z \cdot z$

$\text{LCM} = 2 \cdot 3 \cdot 5 \cdot x \cdot x \cdot y \cdot y \cdot z \cdot z \cdot z = 30x^2y^2z^3$

50. $24x^3y^5z^2$

51. $a + 1 = a + 1$

$(a - 1)^2 = (a - 1)(a - 1)$

$a^2 - 1 = (a + 1)(a - 1)$

$\text{LCM} = (a + 1)(a - 1)(a - 1) = (a + 1)(a - 1)^2$

52. $(x + 3)(x - 3)^2$

53. $m^2 - 5m + 6 = (m - 3)(m - 2)$

$m^2 - 4m + 4 = (m - 2)(m - 2)$

$\text{LCM} = (m - 3)(m - 2)(m - 2) = (m - 3)(m - 2)^2$

54. $(2x + 1)(x + 2)(x - 1)$

55. $t - 3, t + 3, (t^2 - 9)^2$

Note that $(t^2 - 9)^2 = [(t + 3)(t - 3)]^2$, so this expression is a multiple of each of the other expressions. Thus, the LCM is $(t^2 - 9)^2$.

56. $(a^2 - 10a + 25)^2$

57. $6x^3 - 24x^2 + 18x = 6x(x^2 - 4x + 3) =$

$2 \cdot 3 \cdot x(x - 1)(x - 3)$

$4x^5 - 24x^4 + 20x^3 = 4x^3(x^2 - 6x + 5) =$

$2 \cdot 2 \cdot x \cdot x \cdot x(x - 1)(x - 5)$

$\text{LCM} = 2 \cdot 2 \cdot 3 \cdot x \cdot x \cdot x(x - 1)(x - 3)(x - 5) =$

$12x^3(x - 1)(x - 3)(x - 5)$

58. $18x^3(x - 2)^2(x + 1)$

59. $6x^5 = 2 \cdot 3 \cdot x \cdot x \cdot x \cdot x \cdot x$

$12x^3 = 2 \cdot 2 \cdot 3 \cdot x \cdot x \cdot x$

The LCD is $2 \cdot 2 \cdot 3 \cdot x \cdot x \cdot x \cdot x \cdot x$, or $12x^5$.

The factor of the LCD that is missing from the first denominator is 2. We multiply by 1 using 2/2:

$$\frac{5}{6x^5} \cdot \frac{2}{2} = \frac{10}{12x^5}$$

The second denominator is missing two factors of x, or x^2. We multiply by 1 using x^2/x^2:

$$\frac{y}{12x^3} \cdot \frac{x^2}{x^2} = \frac{x^2y}{12x^5}$$

60. $\dfrac{3a^3}{10a^6}; \dfrac{2b}{10a^6}$

61. $2a^2b = 2 \cdot a \cdot a \cdot b$

$8ab^2 = 2 \cdot 2 \cdot 2 \cdot a \cdot b \cdot b$

The LCD is $2 \cdot 2 \cdot 2 \cdot a \cdot a \cdot b \cdot b$, or $8a^2b^2$.

We multiply the first expression by $\dfrac{4b}{4b}$ to obtain the LCD:

$$\frac{3}{2a^2b} \cdot \frac{4b}{4b} = \frac{12b}{8a^2b^2}$$

We multiply the second expression by a/a to obtain the LCD:

$$\frac{7}{8ab^2} \cdot \frac{a}{a} = \frac{7a}{8a^2b^2}$$

62. $\dfrac{21y}{9x^4y^3}; \dfrac{4x^3}{9x^4y^3}$

63. The LCD is $(x+2)(x-2)(x+3)$. (See Exercise 45.)

$$\frac{2x}{x^2-4} = \frac{2x}{(x+2)(x-2)} \cdot \frac{x+3}{x+3}$$

$$= \frac{(2x)(x+3)}{(x+2)(x-2)(x+3)}$$

$$\frac{4x}{x^2+5x+6} = \frac{4x}{(x+3)(x+2)} \cdot \frac{x-2}{x-2}$$

$$= \frac{4x(x-2)}{(x+3)(x+2)(x-2)}$$

64. $\dfrac{5x(x+8)}{(x+3)(x-3)(x+8)}; \dfrac{2x(x-3)}{(x+3)(x+8)(x-3)}$

65. ◈

66. ◈

67. $\dfrac{7}{-9} = -\dfrac{7}{9} = \dfrac{-7}{9}$

68. $\dfrac{-3}{2}, \dfrac{3}{-2}$

69. $\dfrac{5}{18} - \dfrac{7}{12} = \dfrac{5}{18} \cdot \dfrac{2}{2} - \dfrac{7}{12} \cdot \dfrac{3}{3}$

$$= \frac{10}{36} - \frac{21}{36}$$

$$= -\frac{11}{36}$$

70. $-\dfrac{7}{60}$

71. The shaded area has dimensions $x-6$ by $x-3$. Then the area is $(x-6)(x-3)$, or $x^2-9x+18$.

72. $s^2 - \pi r^2$

73. ◈

74. ◈

75. $\dfrac{6x-1}{x-1} + \dfrac{3(2x+5)}{x-1} + \dfrac{3(2x-3)}{x-1}$

$$= \frac{6x-1+6x+15+6x-9}{x-1}$$

$$= \frac{18x+5}{x-1}$$

76. $\dfrac{30}{(x-3)(x+4)}$

77. $\dfrac{x^2}{3x^2-5x-2} - \dfrac{2x}{3x+1} \cdot \dfrac{1}{x-2}$

$$= \frac{x^2}{(3x+1)(x-2)} - \frac{2x}{(3x+1)(x-2)}$$

$$= \frac{x^2-2x}{(3x+1)(x-2)}$$

$$= \frac{x(x-2)}{(3x+1)(x-2)}$$

$$= \frac{x}{3x+1}$$

78. 0

79. The smallest number of strands that can be used is the LCM of 10 and 3.

$10 = 2 \cdot 5$

$3 = 3$

LCM $= 2 \cdot 5 \cdot 3 = 30$

80. 24

81. If the number of strands must also be a multiple of 4, we find the smallest multiple of 30 that is also a multiple of 4.

$1 \cdot 30 = 30$, not a multiple of 4

$2 \cdot 30 = 60 = 15 \cdot 4$, a multiple of 4

The smallest number of strands that can be used is 60.

82. 1440

83. $8x^2 - 8 = 8(x^2-1) = 2 \cdot 2 \cdot 2(x+1)(x-1)$

$6x^2 - 12x + 6 = 6(x^2-2x+1) = 2 \cdot 3(x-1)(x-1)$

$10 - 10x = 10(1-x) = 2 \cdot 5(1-x)$

Note that $x-1$ and $1-x$ and opposites.

LCM $= 2 \cdot 2 \cdot 2 \cdot 3 \cdot 5(x+1)(x-1)(x-1) = 120(x+1)(x-1)^2$

$\Big($We could also express the LCM as $120(x+1)(x-1)(1-x)$. It is not necessary to include both a factor and its opposite in the LCM since $\dfrac{a}{-b} = \dfrac{-a}{b} = -\dfrac{a}{b}.\Big)$

84. $(3x+4)(3x-4)^2(2x-3)$

85. The time it takes Kim and Jed to meet again at the starting place is the LCM of the times it takes them to complete one round of the course.

$6 = 2 \cdot 3$

$8 = 2 \cdot 2 \cdot 2$

$\text{LCM} = 2 \cdot 2 \cdot 2 \cdot 3$, or 24

It takes 24 min.

86. 7:55 A.M.

87. The number of years after 2002 in which all three appliances will need to be replaced at once is the LCM of the average numbers of years each will last.

$10 = 2 \cdot 5$

$14 = 2 \cdot 7$

$20 = 2 \cdot 2 \cdot 5$

$\text{LCM} = 2 \cdot 2 \cdot 5 \cdot 7 = 140$

All three appliances will need to be replaced 140 years after 2002, or in 2142.

88. ◈

89. ◈

Exercise Set 6.4

1. $\dfrac{3}{x} + \dfrac{7}{x^2} = \dfrac{3}{x} + \dfrac{7}{x \cdot x}$ $\text{LCD} = x \cdot x$, or x^2

$\qquad = \dfrac{3}{x} \cdot \dfrac{x}{x} + \dfrac{7}{x \cdot x}$

$\qquad = \dfrac{3x+7}{x^2}$

2. $\dfrac{5x+6}{x^2}$

3. $\left.\begin{array}{l} 6r = 2 \cdot 3 \cdot r \\ 8r = 2 \cdot 2 \cdot 2 \cdot r \end{array}\right\}\text{LCD} = 2 \cdot 2 \cdot 2 \cdot 3 \cdot r$, or $24r$

$\dfrac{1}{6r} - \dfrac{3}{8r} = \dfrac{1}{6r} \cdot \dfrac{4}{4} - \dfrac{3}{8r} \cdot \dfrac{3}{3}$

$\qquad = \dfrac{4-9}{24r}$

$\qquad = \dfrac{-5}{24r}$, or $-\dfrac{5}{24r}$

4. $-\dfrac{13}{18t}$

5. $\left.\begin{array}{l} xy^2 = x \cdot y \cdot y \\ x^2y = x \cdot x \cdot y \end{array}\right\}\text{LCD} = x \cdot x \cdot y \cdot y$, or x^2y^2

$\dfrac{4}{xy^2} + \dfrac{2}{x^2y} = \dfrac{4}{xy^2} \cdot \dfrac{x}{x} + \dfrac{2}{x^2y} \cdot \dfrac{y}{y}$

$\qquad\qquad = \dfrac{4x + 2y}{x^2y^2}$

6. $\dfrac{2d^2 + 7c}{c^2d^3}$

7. $\left.\begin{array}{l} 9t^3 = 3 \cdot 3 \cdot t \cdot t \cdot t \\ 6t^2 = 2 \cdot 3 \cdot t \cdot t \end{array}\right\}\text{LCD} = 2 \cdot 3 \cdot 3 \cdot t \cdot t \cdot t$, or $18t^3$

$\dfrac{8}{9t^3} - \dfrac{5}{6t^2} = \dfrac{8}{9t^3} \cdot \dfrac{2}{2} - \dfrac{5}{6t^2} \cdot \dfrac{3t}{3t}$

$\qquad\qquad = \dfrac{16 - 15t}{18t^3}$

8. $\dfrac{-2xy - 18}{3x^2y^3}$

9. $\text{LCD} = 24$ (See Example 1.)

$\dfrac{x+5}{8} + \dfrac{x-3}{12} = \dfrac{x+5}{8} \cdot \dfrac{3}{3} + \dfrac{x-3}{12} \cdot \dfrac{2}{2}$

$\qquad\qquad = \dfrac{3(x+5)}{24} + \dfrac{2(x-3)}{24}$

$\qquad\qquad = \dfrac{3x+15}{24} + \dfrac{2x-6}{24}$

$\qquad\qquad = \dfrac{5x+9}{24}$ Adding numerators

10. $\dfrac{5x+7}{18}$

11. $\left.\begin{array}{l} 2 = 2 \\ 4 = 2 \cdot 2 \end{array}\right\} \text{LCD} = 4$

$\dfrac{a+2}{2} - \dfrac{a-4}{4} = \dfrac{a+2}{2} \cdot \dfrac{2}{2} - \dfrac{a-4}{4}$

$\qquad\qquad = \dfrac{2a+4}{4} - \dfrac{a-4}{4}$

$\qquad\qquad = \dfrac{2a+4 - (a-4)}{4}$

$\qquad\qquad = \dfrac{2a+4 - a+4}{4}$

$\qquad\qquad = \dfrac{a+8}{4}$

12. $\dfrac{-x-4}{6}$

13. $\left.\begin{array}{l} 3a^2 = 3 \cdot a \cdot a \\ 9a = 3 \cdot 3 \cdot a \end{array}\right\}\text{LCD} = 3 \cdot 3 \cdot a \cdot a$, or $9a^2$

$$\frac{2a-1}{3a^2}+\frac{5a+1}{9a}=\frac{2a-1}{3a^2}\cdot\frac{3}{3}+\frac{5a+1}{9a}\cdot\frac{a}{a}$$
$$=\frac{6a-3}{9a^2}+\frac{5a^2+a}{9a^2}$$
$$=\frac{5a^2+7a-3}{9a^2}$$

14. $\dfrac{a^2+16a+16}{16a^2}$

15. $\left.\begin{array}{l}4x=4\cdot x\\ x=x\end{array}\right\}$ LCD $=4x$

$$\frac{x-1}{4x}-\frac{2x+3}{x}=\frac{x-1}{4x}-\frac{2x+3}{x}\cdot\frac{4}{4}$$
$$=\frac{x-1}{4x}-\frac{8x+12}{4x}$$
$$=\frac{x-1-(8x+12)}{4x}$$
$$=\frac{x-1-8x-12}{4x}$$
$$=\frac{-7x-13}{4x}$$

16. $\dfrac{7z-12}{12z}$

17. $\left.\begin{array}{l}c^2d=c\cdot c\cdot d\\ cd^2=c\cdot d\cdot d\end{array}\right\}$ LCD $=c\cdot c\cdot d\cdot d$, or c^2d^2

$$\frac{2c-d}{c^2d}+\frac{c+d}{cd^2}=\frac{2c-d}{c^2d}\cdot\frac{d}{d}+\frac{c+d}{cd^2}\cdot\frac{c}{c}$$
$$=\frac{d(2c-d)+c(c+d)}{c^2d^2}$$
$$=\frac{2cd-d^2+c^2+cd}{c^2d^2}$$
$$=\frac{c^2+3cd-d^2}{c^2d^2}$$

18. $\dfrac{x^2+4xy+y^2}{x^2y^2}$

19. $\left.\begin{array}{l}2x^2y=2\cdot x\cdot x\cdot y\\ xy^2=x\cdot y\cdot y\end{array}\right\}$ LCD $=2\cdot x\cdot x\cdot y\cdot y$, or $2x^2y^2$

$$\frac{5x+3y}{2x^2y}-\frac{3x+4y}{xy^2}=\frac{5x+3y}{2x^2y}\cdot\frac{y}{y}-\frac{3x+4y}{xy^2}\cdot\frac{2x}{2x}$$
$$=\frac{5xy+3y^2}{2x^2y^2}-\frac{6x^2+8xy}{2x^2y^2}$$
$$=\frac{5xy+3y^2-(6x^2+8xy)}{2x^2y^2}$$
$$=\frac{5xy+3y^2-6x^2-8xy}{2x^2y^2}$$
$$=\frac{3y^2-3xy-6x^2}{2x^2y^2}$$

(Although $3y^2-3xy-6x^2$ can be factored, doing so will not enable us to simplify the result further.)

20. $\dfrac{4x^2-13xt+9t^2}{3x^2t^2}$

21. The denominators cannot be factored, so the LCD is their product, $(x-1)(x+1)$.

$$\frac{5}{x-1}+\frac{5}{x+1}=\frac{5}{x-1}\cdot\frac{x+1}{x+1}+\frac{5}{x+1}\cdot\frac{x-1}{x-1}$$
$$=\frac{5(x+1)+5(x-1)}{(x-1)(x+1)}$$
$$=\frac{5x+5+5x-5}{(x-1)(x+1)}$$
$$=\frac{10x}{(x-1)(x+1)}$$

22. $\dfrac{6x}{(x-2)(x+2)}$

23. The denominators cannot be factored, so the LCD is their product, $(z-1)(z+1)$.

$$\frac{4}{z-1}-\frac{2}{z+1}=\frac{4}{z-1}\cdot\frac{z+1}{z+1}-\frac{2}{z+1}\cdot\frac{z-1}{z-1}$$
$$=\frac{4z+4}{(z-1)(z+1)}-\frac{2z-2}{(z-1)(z+1)}$$
$$=\frac{4z+4-(2z-2)}{(z-1)(z+1)}$$
$$=\frac{4z+4-2z+2}{(z-1)(z+1)}$$
$$=\frac{2z+6}{(z-1)(z+1)}$$

(Although $2z+6$ can be factored, doing so will not enable us to simplify the result further.)

24. $\dfrac{2x-40}{(x+5)(x-5)}$

25. $\left.\begin{array}{l}x+5=x+5\\ 4x=4\cdot x\end{array}\right\}$ LCD $=4x(x+5)$

$$\frac{2}{x+5}+\frac{3}{4x}=\frac{2}{x+5}\cdot\frac{4x}{4x}+\frac{3}{4x}\cdot\frac{x+5}{x+5}$$
$$=\frac{2\cdot4x+3(x+5)}{4x(x+5)}$$
$$=\frac{8x+3x+15}{4x(x+5)}$$
$$=\frac{11x+15}{4x(x+5)}$$

26. $\dfrac{11x+2}{3x(x+1)}$

27. $3t^2 - 15t = 3t(t-5)$
$\quad\,\, 2t - 10 = 2(t-5)$ $\bigg\}$ LCD $= 6t(t-5)$

$\qquad \dfrac{8}{3t(t-5)} - \dfrac{3}{2(t-5)}$

$= \dfrac{8}{3t(t-5)} \cdot \dfrac{2}{2} - \dfrac{3}{2(t-5)} \cdot \dfrac{3t}{3t}$

$= \dfrac{16}{6t(t-5)} - \dfrac{9t}{6t(t-5)}$

$= \dfrac{16 - 9t}{6t(t-5)}$

28. $\dfrac{3 - 5t}{2t(t-1)}$

29. $\dfrac{4x}{x^2 - 25} + \dfrac{x}{x+5}$

$= \dfrac{4x}{(x+5)(x-5)} + \dfrac{x}{x+5}$ $\quad$ LCD $= (x+5)(x-5)$

$= \dfrac{4x + x(x-5)}{(x+5)(x-5)}$

$= \dfrac{4x + x^2 - 5x}{(x+5)(x-5)}$

$= \dfrac{x^2 - x}{(x+5)(x-5)}$

(Although $x^2 - x$ can be factored, doing so will not enable us to simplify the result further.)

30. $\dfrac{x^2 + 6x}{(x+4)(x-4)}$

31. $\dfrac{t}{t-3} - \dfrac{5}{4t - 12}$

$= \dfrac{t}{t-3} - \dfrac{5}{4(t-3)}$ $\quad$ LCD $= 4(t-3)$

$= \dfrac{t}{t-3} \cdot \dfrac{4}{4} - \dfrac{5}{4(t-3)}$

$= \dfrac{4t - 5}{4(t-3)}$

32. $\dfrac{16}{3(z+4)}$

33. $\dfrac{2}{x+3} + \dfrac{4}{(x+3)^2}$ $\quad$ LCD $= (x+3)^2$

$= \dfrac{2}{x+3} \cdot \dfrac{x+3}{x+3} + \dfrac{4}{(x+3)^2}$

$= \dfrac{2(x+3) + 4}{(x+3)^2}$

$= \dfrac{2x + 6 + 4}{(x+3)^2}$

$= \dfrac{2x + 10}{(x+3)^2}$

(Although $2x + 10$ can be factored, doing so will not enable us to simplify the result further.)

34. $\dfrac{3x - 1}{(x-1)^2}$

35. $\dfrac{3}{x+2} - \dfrac{8}{x^2 - 4}$

$= \dfrac{3}{x+2} - \dfrac{8}{(x+2)(x-2)}$ $\quad$ LCD $= (x+2)(x-2)$

$= \dfrac{3}{x+2} \cdot \dfrac{x-2}{x-2} - \dfrac{8}{(x+2)(x-2)}$

$= \dfrac{3(x-2) - 8}{(x+2)(x-2)}$

$= \dfrac{3x - 6 - 8}{(x+2)(x-2)}$

$= \dfrac{3x - 14}{(x+2)(x-2)}$

36. $\dfrac{-t - 9}{(t+3)(t-3)}$

37. $\dfrac{3a}{4a - 20} + \dfrac{9a}{6a - 30}$

$= \dfrac{3a}{2 \cdot 2(a-5)} + \dfrac{9a}{2 \cdot 3(a-5)}$

$\qquad\qquad$ LCD $= 2 \cdot 2 \cdot 3(a-5)$

$= \dfrac{3a}{2 \cdot 2(a-5)} \cdot \dfrac{3}{3} + \dfrac{9a}{2 \cdot 3(a-5)} \cdot \dfrac{2}{2}$

$= \dfrac{9a + 18a}{2 \cdot 2 \cdot 3(a-5)}$

$= \dfrac{27a}{2 \cdot 2 \cdot 3(a-5)}$

$= \dfrac{\cancel{3} \cdot 9 \cdot a}{2 \cdot 3 \cdot \cancel{3}(a-5)}$

$= \dfrac{9a}{4(a-5)}$

38. $\dfrac{11a}{10(a-2)}$

39. $\dfrac{x}{x-5} + \dfrac{x}{5-x} = \dfrac{x}{x-5} + \dfrac{x}{5-x} \cdot \dfrac{-1}{-1}$

$\qquad\qquad\qquad\quad = \dfrac{x}{x-5} + \dfrac{-x}{x-5}$

$\qquad\qquad\qquad\quad = 0$

40. $\dfrac{2x^2 + 8x + 16}{x(x+4)}$

41.
$$\frac{7}{a^2 + a - 2} + \frac{5}{a^2 - 4a + 3}$$

$$= \frac{7}{(a+2)(a-1)} + \frac{5}{(a-3)(a-1)}$$

$$\text{LCD} = (a+2)(a-1)(a-3)$$

$$= \frac{7}{(a+2)(a-1)} \cdot \frac{a-3}{a-3} + \frac{5}{(a-3)(a-1)} \cdot \frac{a+2}{a+2}$$

$$= \frac{7(a-3) + 5(a+2)}{(a+2)(a-1)(a-3)}$$

$$= \frac{7a - 21 + 5a + 10}{(a+2)(a-1)(a-3)}$$

$$= \frac{12a - 11}{(a+2)(a-1)(a-3)}$$

42. $\dfrac{x^2 + 5x + 1}{(x+1)^2(x+4)}$

43.
$$\frac{x}{x^2 + 9x + 20} - \frac{4}{x^2 + 7x + 12}$$

$$= \frac{x}{(x+4)(x+5)} - \frac{4}{(x+3)(x+4)}$$

$$\text{LCD} = (x+3)(x+4)(x+5)$$

$$= \frac{x}{(x+4)(x+5)} \cdot \frac{x+3}{x+3} - \frac{4}{(x+3)(x+4)} \cdot \frac{x+5}{x+5}$$

$$= \frac{x(x+3) - 4(x+5)}{(x+3)(x+4)(x+5)}$$

$$= \frac{x^2 + 3x - 4x - 20}{(x+3)(x+4)(x+5)}$$

$$= \frac{x^2 - x - 20}{(x+3)(x+4)(x+5)}$$

$$= \frac{(x+4)(x-5)}{(x+3)(x+4)(x+5)}$$

$$= \frac{x-5}{(x+3)(x+5)}$$

44. $\dfrac{x-3}{(x+3)(x+1)}$

45.
$$\frac{3z}{z^2 - 4x + 4} + \frac{10}{z^2 + z - 6}$$

$$= \frac{3z}{(z-2)^2} + \frac{10}{(z-2)(z+3)},$$

$$\text{LCD} = (z-2)^2(z+3)$$

$$= \frac{3z}{(z-2)^2} \cdot \frac{z+3}{z+3} + \frac{10}{(z-2)(z+3)} \cdot \frac{z-2}{z-2}$$

$$= \frac{3z(z+3) + 10(z-2)}{(x-2)^2(z+3)}$$

$$= \frac{3z^2 + 9z + 10z - 20}{(z-2)^2(z+3)}$$

$$= \frac{3z^2 + 19z - 20}{(z-2)^2(z+3)}$$

46. $\dfrac{5x + 12}{(x+3)(x-3)(x+2)}$

47. $\dfrac{-5}{x^2 + 17x + 16} - \dfrac{0}{x^2 + 9x + 8}$

Note that $\dfrac{0}{x^2 + 9x + 8} = 0$, so the difference is

$\dfrac{-5}{x^2 + 17x + 16}.$

48. $\dfrac{x^2 + 5x - 8}{(x+7)(x+8)(x+6)}$

49.
$$\frac{2x}{5} - \frac{x-3}{-5} = \frac{2x}{5} - \frac{x-3}{-5} \cdot \frac{-1}{-1}$$

$$= \frac{2x}{5} - \frac{3-x}{5}$$

$$= \frac{2x - (3-x)}{5}$$

$$= \frac{2x - 3 + x}{5}$$

$$= \frac{3x - 3}{5}$$

(Although $3x - 3$ can be factored, doing so will not enable us to simplify the result further.)

50. $\dfrac{4x - 5}{4}$

51.
$$\frac{y^2}{y-3} + \frac{9}{3-y} = \frac{y^2}{y-3} + \frac{9}{3-y} \cdot \frac{-1}{-1}$$

$$= \frac{y^2}{y-3} + \frac{-9}{-3+y}$$

$$= \frac{y^2 - 9}{y-3}$$

$$= \frac{(y+3)(y-3)}{y-3}$$

$$= y + 3$$

52. $t + 2$

53.
$$\frac{b-7}{b^2 - 16} + \frac{7-b}{16 - b^2} = \frac{b-7}{b^2 - 16} + \frac{7-b}{16 - b^2} \cdot \frac{-1}{-1}$$

$$= \frac{b-7}{b^2 - 16} + \frac{b-7}{b^2 - 16}$$

$$= \frac{2b - 14}{b^2 - 16}$$

(Although both $2b - 14$ and $b^2 - 16$ can be factored, doing so will not enable us to simplify the result further.)

54. 0

55.
$$\frac{y+2}{y-7} + \frac{3-y}{49-y^2}$$
$$= \frac{y+2}{y-7} + \frac{3-y}{(7+y)(7-y)}$$
$$= \frac{y+2}{y-7} + \frac{3-y}{(7+y)(7-y)} \cdot \frac{-1}{-1}$$
$$= \frac{y+2}{y-7} + \frac{y-3}{(y+7)(y-7)} \quad \text{LCD} = (y+7)(y-7)$$
$$= \frac{y+2}{y-7} \cdot \frac{y+7}{y+7} + \frac{y-3}{(y+7)(y-7)}$$
$$= \frac{y^2+9y+14+y-3}{(y+7)(y-7)}$$
$$= \frac{y^2+10y+11}{(y+7)(y-7)}$$

56. $\dfrac{p^2+7p+1}{(p+5)(p-5)}$

57.
$$\frac{5x}{x^2-9} - \frac{4}{3-x}$$
$$= \frac{5x}{(x+3)(x-3)} - \frac{4}{3-x}$$
$$= \frac{5x}{(x+3)(x-3)} - \frac{4}{3-x} \cdot \frac{-1}{-1}$$
$$= \frac{5x}{(x+3)(x-3)} - \frac{-4}{x-3} \quad \text{LCD} = (x+3)(x-3)$$
$$= \frac{5x}{(x+3)(x-3)} - \frac{-4}{x-3} \cdot \frac{x+3}{x+3}$$
$$= \frac{5x-(-4)(x+3)}{(x+3)(x-3)}$$
$$= \frac{5x+4x+12}{(x+3)(x-3)}$$
$$= \frac{9x+12}{(x+3)(x-3)}$$

(Although $9x+12$ can be factored, doing so will not enable us to simplify the result further.)

58. $\dfrac{13x+20}{(4+x)(4-x)}$, or $\dfrac{-13x-20}{(x+4)(x-4)}$

59.
$$\frac{3x+2}{3x+6} + \frac{x}{4-x^2}$$
$$= \frac{3x+2}{3(x+2)} + \frac{x}{(2+x)(2-x)}$$
$$\qquad\qquad \text{LCD} = 3(x+2)(2-x)$$
$$= \frac{3x+2}{3(x+2)} \cdot \frac{2-x}{2-x} + \frac{x}{(2+x)(2-x)} \cdot \frac{3}{3}$$
$$= \frac{(3x+2)(2-x) + x \cdot 3}{3(x+2)(2-x)}$$
$$= \frac{-3x^2+4x+4+3x}{3(x+2)(2-x)}$$
$$= \frac{-3x^2+7x+4}{3(x+2)(2-x)}, \text{ or}$$
$$\frac{3x^2-7x-4}{3(x+2)(x-2)}$$

60. $\dfrac{-a-2}{(a+1)(a-1)}$, or $\dfrac{a+2}{(1+a)(1-a)}$

61.
$$\frac{4-a^2}{a^2-9} - \frac{a-2}{3-a}$$
$$= \frac{4-a^2}{(a+3)(a-3)} - \frac{a-2}{3-a}$$
$$= \frac{4-a^2}{(a+3)(a-3)} - \frac{a-2}{3-a} \cdot \frac{-1}{-1}$$
$$= \frac{4-a^2}{(a+3)(a-3)} - \frac{2-a}{a-3} \quad \text{LCD} = (a+3)(a-3)$$
$$= \frac{4-a^2}{(a+3)(a-3)} - \frac{2-a}{a-3} \cdot \frac{a+3}{a+3}$$
$$= \frac{4-a^2-(2a+6-a^2-3a)}{(a+3)(a-3)}$$
$$= \frac{4-a^2-2a-6+a^2+3a}{(a+3)(a-3)}$$
$$= \frac{a-2}{(a+3)(a-3)}$$

62. $\dfrac{10x+6y}{(x+y)(x-y)}$

63.
$$\frac{x-3}{2-x} - \frac{x+3}{x+2} + \frac{x+6}{4-x^2}$$

$$= \frac{x-3}{2-x} - \frac{x+3}{x+2} + \frac{x+6}{(2+x)(2-x)}$$

$$\qquad\qquad \text{LCD} = (2+x)(2-x)$$

$$= \frac{x-3}{2-x} \cdot \frac{2+x}{2+x} - \frac{x+3}{x+2} \cdot \frac{2-x}{2-x} + \frac{x+6}{(2+x)(2-x)}$$

$$= \frac{(x-3)(2+x) - (x+3)(2-x) + (x+6)}{(2+x)(2-x)}$$

$$= \frac{x^2 - x - 6 - (-x^2 - x + 6) + x + 6}{(2+x)(2-x)}$$

$$= \frac{x^2 - x - 6 + x^2 + x - 6 + x + 6}{(2+x)(2-x)}$$

$$= \frac{2x^2 + x - 6}{(2+x)(2-x)}$$

$$= \frac{(2x-3)(x+2)}{(2+x)(2-x)}$$

$$= \frac{2x-3}{2-x}$$

64. $\dfrac{-2t^2 + 2t + 11}{(t+1)(t-1)}$

65.
$$\frac{x+5}{x+3} + \frac{x+7}{x+2} - \frac{7x+19}{(x+3)(x+2)}$$

$$\qquad\qquad \text{LCD is } (x+3)(x+2)$$

$$= \frac{x+5}{x+3} \cdot \frac{x+2}{x+2} + \frac{x+7}{x+2} \cdot \frac{x+3}{x+3} - \frac{7x+19}{(x+3)(x+2)}$$

$$= \frac{(x+5)(x+2) + (x+7)(x+3) - (7x+19)}{(x+3)(x+2)}$$

$$= \frac{x^2 + 7x + 10 + x^2 + 10x + 21 - 7x - 19}{(x+3)(x+2)}$$

$$= \frac{2x^2 + 10x + 12}{(x+3)(x+2)}$$

$$= \frac{2(x^2 + 5x + 6)}{(x+3)(x+2)}$$

$$= \frac{2(x+3)(x+2)}{(x+3)(x+2)}$$

$$= 2$$

66. 3

67.
$$\frac{t}{s+t} - \frac{t}{s-t} \qquad \text{LCD} = (s+t)(s-t)$$

$$= \frac{t}{s+t} \cdot \frac{s-t}{s-t} - \frac{t}{s-t} \cdot \frac{s+t}{s+t}$$

$$= \frac{t(s-t) - t(s+t)}{(s+t)(s-t)}$$

$$= \frac{st - t^2 - st - t^2}{(s+t)(s-t)}$$

$$= \frac{-2t^2}{(s+t)(s-t)}$$

68. $\dfrac{a^2 + 2ab - b^2}{(b-a)(b+a)}$

69.
$$\frac{1}{x+y} + \frac{1}{x-y} - \frac{2x}{x^2 - y^2}$$

$$\qquad\qquad \text{LCD} = (x+y)(x-y)$$

$$= \frac{1}{x+y} \cdot \frac{x-y}{x-y} + \frac{1}{x-y} \cdot \frac{x+y}{x+y} - \frac{2x}{(x+y)(x-y)}$$

$$= \frac{(x-y) + (x+y) - 2x}{(x+y)(x-y)}$$

$$= 0$$

70. $\dfrac{2}{r+s}$

71. ◈

72. ◈

73.
$$-\frac{3}{7} \div \frac{6}{13} = -\frac{3}{7} \cdot \frac{13}{6}$$

$$= -\frac{3 \cdot 13}{7 \cdot 2 \cdot 3}$$

$$= -\frac{13}{14}$$

74. $-\dfrac{5}{9}$

75.
$$\frac{\frac{2}{9}}{\frac{5}{3}} = \frac{2}{9} \div \frac{5}{3}$$

$$= \frac{2}{9} \cdot \frac{3}{5}$$

$$= \frac{2 \cdot 3}{3 \cdot 3 \cdot 5}$$

$$= \frac{2}{15}$$

76. $\dfrac{7}{6}$

77. Graph: $y = -\dfrac{1}{2}x - 5$

Since the equation is in the form $y = mx + b$, we know the y−intercept is $(0, -5)$. We find two other solutions, substituting multiples of 2 for x to avoid fractions.

When $x = -2$, $y = -\dfrac{1}{2}(-2) - 5 = 1 - 5 = -4$.

When $x = -4$, $y = -\dfrac{1}{2}(-4) - 5 = 2 - 5 = -3$.

x	y
0	-5
-2	-4
-4	-3

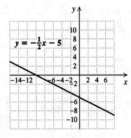

$y = -\frac{1}{2}x - 5$

78.

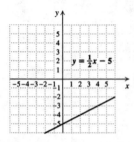

$y = \frac{1}{2}x - 5$

79. ◈

80. ◈

81. $P = 2\left(\dfrac{3}{x+4}\right) + 2\left(\dfrac{2}{x-5}\right)$

$= \dfrac{6}{x+4} + \dfrac{4}{x-5}$ LCD $= (x+4)(x-5)$

$= \dfrac{6}{x+4} \cdot \dfrac{x-5}{x-5} + \dfrac{4}{x-5} \cdot \dfrac{x+4}{x+4}$

$= \dfrac{6x - 30 + 4x + 16}{(x+4)(x-5)}$

$= \dfrac{10x-14}{(x+4)(x-5)}$, or $\dfrac{10x-14}{x^2-x-20}$

$A = \left(\dfrac{3}{x+4}\right)\left(\dfrac{2}{x-5}\right) = \dfrac{6}{(x+4)(x-5)}$, or

$\dfrac{6}{x^2-x-20}$

82. $\dfrac{4x^2+18x}{(x+4)(x+5)}$; $\dfrac{x^2}{(x+4)(x+5)}$

83. $\dfrac{2x+11}{x-3} \cdot \dfrac{3}{x+4} + \dfrac{2x+1}{4+x} \cdot \dfrac{3}{3-x}$

$= \dfrac{6x+33}{(x-3)(x+4)} + \dfrac{6x+3}{(4+x)(3-x)}$

$= \dfrac{6x+33}{(x-3)(x+4)} + \dfrac{6x+3}{(4+x)(3-x)} \cdot \dfrac{-1}{-1}$

$= \dfrac{6x+33}{(x-3)(x+4)} + \dfrac{-6x-3}{(x+4)(x-3)}$

$= \dfrac{6x+33-6x-3}{(x-3)(x+4)}$

$= \dfrac{30}{(x-3)(x+4)}$

84. $\dfrac{x}{3x+1}$

85. $\left(\dfrac{x}{x+7} - \dfrac{3}{x+2}\right)\left(\dfrac{x}{x+7} + \dfrac{3}{x+2}\right)$

$= \dfrac{x^2}{(x+7)^2} - \dfrac{9}{(x+2)^2}$ LCD $= (x+7)^2(x+2)^2$

$= \dfrac{x^2}{(x+7)^2} \cdot \dfrac{(x+2)^2}{(x+2)^2} - \dfrac{9}{(x+2)^2} \cdot \dfrac{(x+7)^2}{(x+7)^2}$

$= \dfrac{x^2(x+2)^2 - 9(x+7)^2}{(x+7)^2(x+2)^2}$

$= \dfrac{x^2(x^2+4x+4) - 9(x^2+14x+49)}{(x+7)^2(x+2)^2}$

$= \dfrac{x^4 + 4x^3 + 4x^2 - 9x^2 - 126x - 441}{(x+7)^2(x+2)^2}$

$= \dfrac{x^4 + 4x^3 - 5x^2 - 126x - 441}{(x+7)^2(x+2)^2}$

86. $\dfrac{-3xy - 3a + 6x}{(y-3)^2(a+2x)(a-2x)}$

87. $\dfrac{2x^2+5x-3}{2x^2-9x+9} + \dfrac{x+1}{3-2x} + \dfrac{4x^2+8x+3}{x-3} \cdot \dfrac{x+3}{9-4x^2}$

$= \dfrac{2x^2+5x-3}{(2x-3)(x-3)} + \dfrac{x+1}{3-2x} +$

$\qquad\qquad \dfrac{(4x^2+8x+3)(x+3)}{(x-3)(3+2x)(3-2x)}$

$= \dfrac{2x^2+5x-3}{(2x-3)(x-3)} \cdot \dfrac{-1}{-1} + \dfrac{x+1}{3-2x} +$

$\qquad\qquad \dfrac{4x^3+20x^2+27x+9}{(x-3)(3+2x)(3-2x)}$

$= \dfrac{-2x^2-5x+3}{(3-2x)(x-3)} + \dfrac{x+1}{3-2x} + \dfrac{4x^3+20x^2+27x+9}{(x-3)(3+2x)(3-2x)}$

$\qquad\qquad \text{LCD} = (x-3)(3+2x)(3-2x)$

$= \dfrac{-2x^2-5x+3}{(3-2x)(x-3)} \cdot \dfrac{3+2x}{3+2x} + \dfrac{x+1}{3-2x} \cdot \dfrac{(x-3)(3+2x)}{(x-3)(3+2x)} +$

$\qquad\qquad \dfrac{4x^3+20x^2+27x+9}{(x-3)(3+2x)(3-2x)}$

$= [(-4x^3-16x^2-9x+9+2x^3-x^2-12x-9+$

$\qquad\quad 4x^3+20x^2+27x+9)]/$

$\qquad\qquad [(x-3)(3+2x)(3-2x)]$

$= \dfrac{2x^3+3x^2+6x+9}{(x-3)(3+2x)(3-2x)}$

$= \dfrac{x^2(2x+3)+3(2x+3)}{(x-3)(3+2x)(3-2x)}$

$= \dfrac{(2x+3)(x^2+3)}{(x-3)(3+2x)(3-2x)}$

$= \dfrac{x^2+3}{(x-3)(3-2x)}, \text{ or } \dfrac{-x^2-3}{(x-3)(2x-3)}$

88. $\dfrac{5(a^2+2ab-b^2)}{(a-b)(3a+b)(3a-b)}$

89. Answers may vary. $\dfrac{a}{a-b} + \dfrac{3b}{b-a}$

Exercise Set 6.5

1. $\dfrac{1+\dfrac{1}{2}}{1+\dfrac{1}{4}}$ \qquad LCD is 4

$= \dfrac{1+\dfrac{1}{2}}{1+\dfrac{1}{4}} \cdot \dfrac{4}{4}$ \qquad Multiplying by $\dfrac{4}{4}$

$= \dfrac{\left(1+\dfrac{1}{2}\right)4}{\left(1+\dfrac{1}{4}\right)4}$ \qquad Multiplying numerator and denominator by 4

$= \dfrac{1\cdot 4 + \dfrac{1}{2}\cdot 4}{1\cdot 4 + \dfrac{1}{4}\cdot 4}$

$= \dfrac{4+2}{4+1}$

$= \dfrac{6}{5}$

2. $\dfrac{7}{6}$

3. $\dfrac{1+\dfrac{1}{3}}{5-\dfrac{5}{27}}$

$= \dfrac{1\cdot\dfrac{3}{3}+\dfrac{1}{3}}{5\cdot\dfrac{27}{27}-\dfrac{5}{27}}$ \qquad Getting a common denominator in numerator and in denominator

$= \dfrac{\dfrac{3}{3}+\dfrac{1}{3}}{\dfrac{135}{27}-\dfrac{5}{27}}$

$= \dfrac{\dfrac{4}{3}}{\dfrac{130}{27}}$ \qquad Adding in the numerator; subtracting in the denominator

$= \dfrac{4}{3}\cdot\dfrac{27}{130}$ \qquad Multiplying by the reciprocal of the divisor

$= \dfrac{2\cdot 2\cdot 3\cdot 9}{3\cdot 2\cdot 65}$

$= \dfrac{2\cdot 2\cdot 3\cdot 9}{3\cdot 2\cdot 65}$

$= \dfrac{18}{65}$

4. 8

5. $\dfrac{\dfrac{s}{3}+s}{\dfrac{3}{s}+s}$ LCD is $3s$

$=\dfrac{\dfrac{s}{3}+s}{\dfrac{3}{s}+s}\cdot\dfrac{3s}{3s}$

$=\dfrac{\left(\dfrac{s}{3}+s\right)(3s)}{\left(\dfrac{3}{s}+s\right)(3s)}$

$=\dfrac{\dfrac{5}{3}\cdot 3s+s\cdot 3s}{\dfrac{3}{s}\cdot 3s+s\cdot 3s}$

$=\dfrac{s^2+3s^2}{9+3s^2}$

$=\dfrac{4s^2}{9+3s^2}$

6. $\dfrac{1-5x}{1+3x}$

7. $\dfrac{\dfrac{2}{x}}{\dfrac{3}{x}+\dfrac{1}{x^2}}$ LCD is x^2

$=\dfrac{\dfrac{2}{x}}{\dfrac{3}{x}+\dfrac{1}{x^2}}\cdot\dfrac{x^2}{x^2}$

$=\dfrac{\dfrac{2}{x}\cdot x^2}{\left(\dfrac{3}{x}+\dfrac{1}{x^2}\right)x^2}$

$=\dfrac{2x}{\dfrac{3}{x}\cdot x^2+\dfrac{1}{x^2}\cdot x^2}$

$=\dfrac{2x}{3x+1}$

8. $\dfrac{4x-1}{2x}$

9. $\dfrac{\dfrac{2a-5}{3a}}{\dfrac{a-1}{6a}}$

$=\dfrac{2a-5}{3a}\cdot\dfrac{6a}{a-1}$ Multiplying by the reciprocal of the divisor

$=\dfrac{(2a-5)\cdot 2\cdot 3a}{3a\cdot(a-1)}$

$=\dfrac{(2a-5)\cdot 2\cdot 3\cancel{a}}{3\cancel{a}\cdot(a-1)}$

$=\dfrac{2(2a-5)}{a-1}$

$=\dfrac{4a-10}{a-1}$

10. $\dfrac{3a+12}{a^2-2a}$

11. $\dfrac{\dfrac{x}{4}-\dfrac{4}{x}}{\dfrac{1}{4}+\dfrac{1}{x}}$ LCD is $4x$

$=\dfrac{\dfrac{x}{4}-\dfrac{4}{x}}{\dfrac{1}{4}+\dfrac{1}{x}}\cdot\dfrac{4x}{4x}$

$=\dfrac{\dfrac{x}{4}\cdot 4x-\dfrac{4}{x}\cdot 4x}{\dfrac{1}{4}\cdot 4x+\dfrac{1}{x}\cdot 4x}$

$=\dfrac{x^2-16}{x+4}$

$=\dfrac{(x+4)(x-4)}{x+4}$

$=\dfrac{\cancel{(x+4)}(x-4)}{\cancel{(x+4)}\cdot 1}$

$=x-4$

12. $\dfrac{24+3x}{x^2-24}$

13. $\dfrac{\dfrac{1}{5} + \dfrac{1}{x}}{\dfrac{5+x}{5}}$ LCD is $5x$

$= \dfrac{\dfrac{1}{5} + \dfrac{1}{x}}{\dfrac{5+x}{5}} \cdot \dfrac{5x}{5x}$

$= \dfrac{\dfrac{1}{5} \cdot 5x + \dfrac{1}{x} \cdot 5x}{\left(\dfrac{5+x}{5}\right)(5x)}$

$= \dfrac{x+5}{x(5+x)}$

$= \dfrac{(x+5) \cdot 1}{x(5+x)}$ $(5+x = x+5)$

$= \dfrac{1}{x}$

14. $-\dfrac{1}{a}$

15. $\dfrac{\dfrac{1}{t^2} + 1}{\dfrac{1}{t} - 1}$ LCD is t^2

$= \dfrac{\dfrac{1}{t^2} + 1}{\dfrac{1}{t} - 1} \cdot \dfrac{t^2}{t^2}$

$= \dfrac{\dfrac{1}{t^2} \cdot t^2 + 1 \cdot t^2}{\dfrac{1}{t} \cdot t^2 - 1 \cdot t^2}$

$= \dfrac{1 + t^2}{t - t^2}$

(Although the denominator can be factored, doing so will not enable us to simplify further.)

16. $\dfrac{2x^2 + x}{2x^2 - 1}$

17. $\dfrac{\dfrac{x^2}{x^2 - y^2}}{\dfrac{x}{x+y}}$

$= \dfrac{x^2}{x^2 - y^2} \cdot \dfrac{x+y}{x}$ Multiplying by the reciprocal of the divisor

$= \dfrac{x^2(x+y)}{(x^2 - y^2)(x)}$

$= \dfrac{x \cdot x \cdot (x+y)}{(x+y)(x-y)(x)}$

$= \dfrac{x \cdot x \cdot (x+y)}{(x+y)(x-y)(x)}$

$= \dfrac{x}{x-y}$

18. $\dfrac{a^2 + 3a}{2}$

19. $\dfrac{\dfrac{2}{a} + \dfrac{4}{a^2}}{\dfrac{5}{a^3} - \dfrac{3}{a}}$ LCD is a^3

$= \dfrac{\dfrac{2}{a} + \dfrac{4}{a^2}}{\dfrac{5}{a^3} - \dfrac{3}{a}} \cdot \dfrac{a^3}{a^3}$

$= \dfrac{\dfrac{2}{a} \cdot a^3 + \dfrac{4}{a^2} \cdot a^3}{\dfrac{5}{a^3} \cdot a^3 - \dfrac{3}{a} \cdot a^3}$

$= \dfrac{2a^2 + 4a}{5 - 3a^2}$

(Although the numerator can be factored, doing so will not enable us to simplify further.)

20. $\dfrac{5 - x}{2x^2 + 3x}$

21. $\dfrac{\dfrac{2}{7a^4} - \dfrac{1}{14a}}{\dfrac{3}{5a^2} + \dfrac{2}{15a}} = \dfrac{\dfrac{2}{7a^4} \cdot \dfrac{2}{2} - \dfrac{1}{14a} \cdot \dfrac{a^3}{a^3}}{\dfrac{3}{5a^2} \cdot \dfrac{3}{3} + \dfrac{2}{15a} \cdot \dfrac{a}{a}}$

$= \dfrac{\dfrac{4 - a^3}{14a^4}}{\dfrac{9 + 2a}{15a^2}}$

$= \dfrac{4 - a^3}{14a^4} \cdot \dfrac{15a^2}{9 + 2a}$

$= \dfrac{15 \cdot a^2(4 - a^3)}{14a^2 \cdot a^2(9 + 2a)}$

$= \dfrac{15(4 - a^3)}{14a^2(9 + 2a)}$, or $\dfrac{60 - 15a^3}{126a^2 + 28a^3}$

22. $\dfrac{10 - 3x^2}{12x^2 + 6}$

23. $\dfrac{\dfrac{x}{5y^3} + \dfrac{3}{10y}}{\dfrac{3}{10y} + \dfrac{x}{5y^3}}$

Observe that, by the commutative law of addition, the numerator and denominator are equivalent, so the result is 1.

24. $\dfrac{3a + 8b}{15b^2 - 2}$

25.
$$\frac{\dfrac{5}{ab^4} + \dfrac{2}{a^3b}}{\dfrac{5}{a^3b} - \dfrac{3}{ab}} = \frac{\dfrac{5}{ab^4} \cdot \dfrac{a^2}{a^2} + \dfrac{2}{a^3b} \cdot \dfrac{b^3}{b^3}}{\dfrac{5}{a^3b} - \dfrac{3}{ab} \cdot \dfrac{a^2}{a^2}}$$

$$= \frac{\dfrac{5a^2 + 2b^3}{a^3b^4}}{\dfrac{5 - 3a^2}{a^3b}}$$

$$= \frac{5a^2 + 2b^3}{a^3b^4} \cdot \frac{a^3b}{5 - 3a^2}$$

$$= \frac{\cancel{a^3b}(5a^2 + 2b^3)}{\cancel{a^3b} \cdot b^3(5 - 3a^2)}$$

$$= \frac{5a^2 + 2b^3}{b^3(5 - 3a^2)}, \text{ or } \frac{5a^2 + 2b^3}{5b^3 - 3a^2b^3}$$

26. 1

27.
$$\frac{2 - \dfrac{3}{x^2}}{2 + \dfrac{3}{x^4}} = \frac{2 - \dfrac{3}{x^2}}{2 + \dfrac{3}{x^4}} \cdot \frac{x^4}{x^4}$$

$$= \frac{2 \cdot x^4 - \dfrac{3}{x^2} \cdot x^4}{2 \cdot x^4 + \dfrac{3}{x^4} \cdot x^4}$$

$$= \frac{2x^4 - 3x^2}{2x^4 + 3}$$

28. $\dfrac{3a^4 - 2}{2a^4 + 3a}$

29.
$$\frac{t - \dfrac{2}{t}}{t + \dfrac{5}{t}} = \frac{t \cdot \dfrac{t}{t} - \dfrac{2}{t}}{t \cdot \dfrac{t}{t} + \dfrac{5}{t}}$$

$$= \frac{\dfrac{t^2 - 2}{t}}{\dfrac{t^2 + 5}{t}}$$

$$= \frac{t^2 - 2}{t} \cdot \frac{t}{t^2 + 5}$$

$$= \frac{\cancel{t}(t^2 - 2)}{\cancel{t}(t^2 + 5)}$$

$$= \frac{t^2 - 2}{t^2 + 5}$$

30. $\dfrac{x^2 + 3}{x^2 - 2}$

31.
$$\frac{3 + \dfrac{4}{ab^3}}{\dfrac{3 + a}{a^2b}} = \frac{3 + \dfrac{4}{ab^3}}{\dfrac{3 + a}{a^2b}} \cdot \frac{a^2b^3}{a^2b^3}$$

$$= \frac{3 \cdot a^2b^3 + \dfrac{4}{ab^3} \cdot a^2b^3}{\dfrac{3 + a}{a^2b} \cdot a^2b^3}$$

$$= \frac{3a^2b^3 + 4a}{b^2(3 + a)}, \text{ or } \frac{3a^2b^3 + 4a}{3b^2 + ab^2}$$

32. $\dfrac{5x^3y + 3x}{3 + x}$

33.
$$\frac{\dfrac{x + 5}{x^2}}{\dfrac{2}{x} - \dfrac{3}{x^2}} = \frac{\dfrac{x + 5}{x^2}}{\dfrac{2}{x} \cdot \dfrac{x}{x} - \dfrac{3}{x^2}}$$

$$= \frac{\dfrac{x + 5}{x^2}}{\dfrac{2x - 3}{x^2}}$$

$$= \frac{x + 5}{x^2} \cdot \frac{x^2}{2x - 3}$$

$$= \frac{\cancel{x^2}(x + 5)}{\cancel{x^2}(2x - 3)}$$

$$= \frac{x + 5}{2x - 3}$$

34. $\dfrac{a + 6}{2a + 3a^2}$

35.
$$\frac{x - 3 + \dfrac{2}{x}}{x - 4 + \dfrac{3}{x}} = \frac{x \cdot \dfrac{x}{x} - 3 \cdot \dfrac{x}{x} + \dfrac{2}{x}}{x \cdot \dfrac{x}{x} - 4 \cdot \dfrac{x}{x} + \dfrac{3}{x}}$$

$$= \frac{\dfrac{x^2 - 3x + 2}{x}}{\dfrac{x^2 - 4x + 3}{x}}$$

$$= \frac{x^2 - 3x + 2}{x} \cdot \frac{x}{x^2 - 4x + 3}$$

$$= \frac{(x - 2)(x - 1)}{x} \cdot \frac{x}{(x - 3)(x - 1)}$$

$$= \frac{x(x - 1)}{x(x - 1)} \cdot \frac{x - 2}{x - 3}$$

$$= \frac{x - 2}{x - 3}$$

36. $\dfrac{x - 1}{x - 4}$

37. ◈

38. ◈

39. $3x - 5 + 2(4x - 1) = 12x - 3$
$$3x - 5 + 8x - 2 = 12x - 3$$
$$11x - 7 = 12x - 3$$
$$-7 = x - 3$$
$$-4 = x$$

The solution is -4.

40. -4

41. $\dfrac{3}{4}x - \dfrac{5}{8} = \dfrac{3}{8}x + \dfrac{7}{4}$ LCD is 8

$$8\left(\dfrac{3}{4}x - \dfrac{5}{8}\right) = 8\left(\dfrac{3}{8}x + \dfrac{7}{4}\right)$$
$$8 \cdot \dfrac{3}{4}x - 8 \cdot \dfrac{5}{8} = 8 \cdot \dfrac{3}{8}x + 8 \cdot \dfrac{7}{4}$$
$$6x - 5 = 3x + 14$$
$$3x - 5 = 14$$
$$3x = 19$$
$$x = \dfrac{19}{3}$$

The solution is $\dfrac{19}{3}$.

42. $-\dfrac{14}{27}$

43. $x^2 - 7x - 30 = 0$
$$(x - 10)(x + 3) = 0$$
$$x - 10 = 0 \quad or \quad x + 3 = 0$$
$$x = 10 \quad or \quad x = -3$$

The solutions are 10 and -3.

44. $-10, 2$

45. ◈

46. ◈

47. $\dfrac{\dfrac{x-5}{x-6}}{\dfrac{x-7}{x-8}}$

This expression is undefined for any value of x that makes a denominator 0. We see that $x - 6 = 0$ when $x = 6$, $x - 7 = 0$ when $x = 7$, and $x - 8 = 0$ when $x = 8$, so the expression is undefined for the x-values 6, 7, and 8.

48. $-2, -3, -4$

49. $\dfrac{\dfrac{2x+3}{5x+4}}{\dfrac{3}{7} - \dfrac{2x}{9}}$

This expression is undefined for any value of x that makes a denominator 0. First we find the value of x for which $5x + 4 = 0$.

$$5x + 4 = 0$$
$$5x = -4$$
$$x = -\dfrac{4}{5}$$

Then we find the value of x for which $\dfrac{3}{7} - \dfrac{2x}{9} = 0$:

$$\dfrac{3}{7} - \dfrac{2x}{9} = 0$$
$$63\left(\dfrac{3}{7} - \dfrac{2x}{9}\right) = 63 \cdot 0$$
$$63 \cdot \dfrac{3}{7} - 63 \cdot \dfrac{2x}{9} = 0$$
$$27 - 14x = 0$$
$$27 = 14x$$
$$\dfrac{27}{14} = x$$

The expression is undefined for the x-values $-\dfrac{4}{5}$ and $\dfrac{27}{14}$.

50. $\dfrac{25}{24}, \dfrac{7}{2}$

51.

$$\cfrac{\cfrac{P\left(1+\dfrac{i}{12}\right)^2}{\left(1+\dfrac{1}{12}\right)^2-1}}{\dfrac{i}{12}} = \cfrac{\cfrac{P\left(1+\dfrac{i}{6}+\dfrac{i^2}{144}\right)}{\left(1+\dfrac{i}{6}+\dfrac{i^2}{144}\right)-1}}{\dfrac{i}{12}}$$

$$= \cfrac{P\left(1+\dfrac{i}{6}+\dfrac{i^2}{144}\right)}{\cfrac{\dfrac{i}{6}+\dfrac{i^2}{144}}{\dfrac{i}{12}}}$$

$$= \cfrac{P\left(1+\dfrac{i}{6}+\dfrac{i^2}{144}\right)}{\left(\dfrac{i}{6}+\dfrac{i^2}{144}\right)\left(\dfrac{12}{i}\right)}$$

$$= \cfrac{P\left(1+\dfrac{i}{6}+\dfrac{i^2}{144}\right)}{2+\dfrac{i}{12}}$$

$$= \cfrac{P\left(1+\dfrac{i}{6}+\dfrac{i^2}{144}\right)}{2+\dfrac{i}{12}} \cdot \dfrac{144}{144}$$

$$= \cfrac{144P\left(1+\dfrac{i}{6}+\dfrac{i^2}{144}\right)}{144\left(2+\dfrac{i}{12}\right)}$$

$$= \frac{P(144+24i+i^2)}{288+12i}$$

$$= \frac{P(12+i)^2}{12(24+i)}, \text{ or}$$

$$\frac{P(i+12)^2}{12(i+24)}$$

52. $\dfrac{(x-1)(3x-2)}{5x-3}$

53. $\dfrac{\dfrac{5}{x+2}-\dfrac{3}{x-2}}{\dfrac{x}{x-1}+\dfrac{x}{x+1}} = \dfrac{\dfrac{5}{x+2}\cdot\dfrac{x-2}{x-2}-\dfrac{3}{x-2}\cdot\dfrac{x+2}{x+2}}{\dfrac{x}{x-1}\cdot\dfrac{x+1}{x+1}+\dfrac{x}{x+1}\cdot\dfrac{x-1}{x-1}}$

$$= \dfrac{\dfrac{5(x-2)-3(x+2)}{(x+2)(x-2)}}{\dfrac{x(x+1)+x(x-1)}{(x+1)(x-1)}}$$

$$= \dfrac{\dfrac{5x-10-3x-6}{(x+2)(x-2)}}{\dfrac{x^2+x+x^2-x}{(x+1)(x-1)}}$$

$$= \dfrac{\dfrac{2x-16}{(x+2)(x-2)}}{\dfrac{2x^2}{(x+1)(x-1)}}$$

$$= \dfrac{2x-16}{(x+2)(x-2)} \cdot \dfrac{(x+1)(x-1)}{2x^2}$$

$$= \dfrac{2(x-8)(x+1)(x-1)}{2 \cdot x^2(x+2)(x-2)}$$

$$= \dfrac{(x-8)(x+1)(x-1)}{x^2(x+2)(x-2)}$$

54. $\dfrac{x^2+5x+5}{-x^2+10}$

55. $\left[\dfrac{\dfrac{x-1}{x-1}-1}{\dfrac{x+1}{x-1}+1}\right]^5$

Consider the numerator of the complex rational expression:

$$\frac{x-1}{x-1}-1 = 1-1 = 0$$

Since the denominator, $\dfrac{x+1}{x-1}+1$ is not equal to 0, the simplified form of the original expression is 0.

56. $\dfrac{3x+2}{2x+1}$

57.
$$\dfrac{\dfrac{z}{1-\dfrac{z}{2+2z}}-2z}{\dfrac{2z}{5z-2}-3} = \dfrac{\dfrac{z}{\dfrac{2+2z-z}{2+2z}}-2z}{\dfrac{2z-15z+6}{5z-2}}$$

$$= \dfrac{\dfrac{z}{\dfrac{2+z}{2+2z}}-2z}{\dfrac{-13z+6}{5z-2}}$$

$$= \dfrac{z\cdot\dfrac{2+2z}{2+z}-2z}{\dfrac{-13z+6}{5z-2}}$$

$$= \dfrac{\dfrac{z(2+2z)-2z(2+z)}{2+z}}{\dfrac{-13z+6}{5z-2}}$$

$$= \dfrac{\dfrac{2z+2z^2-4z-2z^2}{2+z}}{\dfrac{-13z+6}{5z-2}}$$

$$= \dfrac{\dfrac{-2z}{2+z}}{\dfrac{-13z+6}{5z-2}}$$

$$= \dfrac{-2z}{2+z}\cdot\dfrac{5z-2}{-13z+6}$$

$$= \dfrac{-2z(5z-2)}{(2+z)(-13z+6)}, \text{ or}$$

$$\dfrac{2z(5z-2)}{(2+z)(13z-6)}$$

58. ◈

59. ▨

Exercise Set 6.6

1. Because no variable appears in a denominator, no restrictions exist.

$$\frac{5}{8}-\frac{4}{5}=\frac{x}{20}, \text{ LCD}=40$$

$$40\left(\frac{5}{8}-\frac{4}{5}\right)=40\cdot\frac{x}{20}$$

$$40\cdot\frac{5}{8}-40\cdot\frac{4}{5}=40\cdot\frac{x}{20}$$

$$25-32=2x$$

$$-7=2x$$

$$-\frac{7}{2}=x$$

Check:

$$\dfrac{\dfrac{5}{8}-\dfrac{4}{5}=\dfrac{x}{20}}{}$$

$$\dfrac{5}{8}-\dfrac{4}{5} \;?\; \dfrac{-\dfrac{7}{2}}{20}$$

$$\dfrac{25}{40}-\dfrac{32}{40} \;\Big|\; -\dfrac{7}{2}\cdot\dfrac{1}{20}$$

$$-\dfrac{7}{40} \;\Big|\; -\dfrac{7}{40} \qquad \text{TRUE}$$

This checks, so the solution is $-\dfrac{7}{2}$.

2. $\dfrac{6}{5}$

3. Note that x cannot be 0.

$$\frac{1}{3}+\frac{5}{6}=\frac{1}{x}, \text{ LCD}=6x$$

$$6x\left(\frac{1}{3}+\frac{5}{6}\right)=6x\cdot\frac{1}{x}$$

$$6x\cdot\frac{1}{3}+6x\cdot\frac{5}{6}=6x\cdot\frac{1}{x}$$

$$2x+5x=6$$

$$7x=6$$

$$x=\frac{6}{7}$$

Check:

$$\dfrac{\dfrac{1}{3}+\dfrac{5}{6}=\dfrac{1}{x}}{}$$

$$\dfrac{1}{3}+\dfrac{5}{6} \;?\; \dfrac{1}{\dfrac{6}{7}}$$

$$\dfrac{2}{6}+\dfrac{5}{6} \;\Big|\; 1\cdot\dfrac{7}{6}$$

$$\dfrac{7}{6} \;\Big|\; \dfrac{7}{6} \qquad \text{TRUE}$$

This checks, so the solution is $\dfrac{6}{7}$.

4. $\dfrac{40}{29}$

5. Note that t cannot be 0.

$$\frac{1}{6}+\frac{1}{8}=\frac{1}{t}, \text{ LCD}=24t$$

$$24t\left(\frac{1}{6}+\frac{1}{8}\right)=24t\cdot\frac{1}{t}$$

$$24t\cdot\frac{1}{6}+24t\cdot\frac{1}{8}=24t\cdot\frac{1}{t}$$

$$4t+3t=24$$

$$7t=24$$

$$t=\frac{24}{7}$$

Check:

$$\frac{1}{6} + \frac{1}{8} = \frac{1}{t}$$

$$\frac{1}{6} + \frac{1}{8} \ ? \ \frac{1}{\frac{24}{7}}$$

$$\frac{4}{24} + \frac{3}{24} \ \Big| \ 1 \cdot \frac{7}{24}$$

$$\frac{7}{24} \ \Big| \ \frac{7}{24} \qquad \text{TRUE}$$

This checks, so the solution is $\dfrac{24}{7}$.

6. $\dfrac{40}{9}$

7. Note that x cannot be 0.

$$x + \frac{5}{x} = -6, \ \text{LCD} = x$$

$$x\left(x + \frac{5}{x}\right) = -6 \cdot x$$

$$x \cdot x + x \cdot \frac{5}{x} = -6 \cdot x$$

$$x^2 + 5 = -6x$$

$$x^2 + 6x + 5 = 0$$

$$(x + 5)(x + 1) = 0$$

$$x + 5 = 0 \quad or \quad x + 1 = 0$$

$$x = -5 \quad or \qquad x = -1$$

Check:

$$x + \frac{5}{x} = -6 \qquad\qquad x + \frac{5}{x} = -6$$

$$-5 + \frac{5}{-5} \ ? \ -6 \qquad -1 + \frac{5}{-1} \ ? \ -6$$

$$-5 - 1 \ \Big| \qquad\qquad -1 - 5 \ \Big|$$

$$-6 \ \Big| \ -6 \ \text{TRUE} \qquad -6 \ \Big| \ -6 \ \text{TRUE}$$

Both of these check, so the two solutions are -5 and -1.

8. $-6, -1$

9. Note that x cannot be 0.

$$\frac{x}{6} - \frac{6}{x} = 0, \ \text{LCD} = 6x$$

$$6x\left(\frac{x}{6} - \frac{6}{x}\right) = 6x \cdot 0$$

$$6x \cdot \frac{x}{6} - 6x \cdot \frac{6}{x} = 6x \cdot 0$$

$$x^2 - 36 = 0$$

$$(x + 6)(x - 6) = 0$$

$$x + 6 = 0 \quad or \quad x - 6 = 0$$

$$x = -6 \quad or \qquad x = 6$$

Check:

$$\frac{x}{6} - \frac{6}{x} = 0 \qquad\qquad \frac{x}{6} - \frac{6}{x} = 0$$

$$\frac{-6}{6} - \frac{6}{-6} \ ? \ 0 \qquad \frac{6}{6} - \frac{6}{6} \ ? \ 0$$

$$-1 + 1 \ \Big| \qquad\qquad 1 - 1 \ \Big|$$

$$0 \ \Big| \ 0 \ \text{TRUE} \qquad\qquad 0 \ \Big| \ 0 \ \text{TRUE}$$

Both of these check, so the two solutions are -6 and 6.

10. $-7, 7$

11. Note that x cannot be 0.

$$\frac{5}{x} = \frac{6}{x} - \frac{1}{3}, \ \text{LCD} = 3x$$

$$3x \cdot \frac{5}{x} = 3x\left(\frac{6}{x} - \frac{1}{3}\right)$$

$$3x \cdot \frac{5}{x} = 3x \cdot \frac{6}{x} - 3x \cdot \frac{1}{3}$$

$$15 = 18 - x$$

$$-3 = -x$$

$$3 = x$$

Check:

$$\frac{5}{x} = \frac{6}{x} - \frac{1}{3}$$

$$\frac{5}{3} \ ? \ \frac{6}{3} - \frac{1}{3}$$

$$\frac{5}{3} \ \Big| \ \frac{5}{3} \qquad \text{TRUE}$$

This checks, so the solution is 3.

12. 2

13. Note that t cannot be 0.

$$\frac{5}{3t} + \frac{3}{t} = 1, \ \text{LCD} = 3t$$

$$3t\left(\frac{5}{3t} + \frac{3}{t}\right) = 3t \cdot 1$$

$$3t \cdot \frac{5}{3t} + 3t \cdot \frac{3}{t} = 3t \cdot 1$$

$$5 + 9 = 3t$$

$$14 = 3t$$

$$\frac{14}{3} = t$$

Check:

$$\frac{5}{3t} + \frac{3}{t} = 1$$

$$\frac{5}{3 \cdot \frac{14}{3}} + \frac{3}{\frac{14}{3}} \;?\; 1$$

$$\frac{5}{14} + \frac{9}{14}$$

$$\frac{14}{14}$$

$$1 \;\Big|\; 1 \quad \text{TRUE}$$

This checks, so the solution is $\dfrac{14}{3}$.

14. $\dfrac{23}{4}$

15. To avoid division by 0, we must have $x + 3 \neq 0$, or $x \neq -3$.

$$\frac{x-8}{x+3} = \frac{1}{4}, \; \text{LCD} = 4(x+3)$$

$$4(x+3) \cdot \frac{x-8}{x+3} = 4(x+3) \cdot \frac{1}{4}$$

$$4(x-8) = x+3$$

$$4x - 32 = x + 3$$

$$3x = 35$$

$$x = \frac{35}{3}$$

Check:

$$\frac{x-8}{x+3} = \frac{1}{4}$$

$$\frac{\frac{35}{3} - 8}{\frac{35}{3} + 3} \;?\; \frac{1}{4}$$

$$\frac{\frac{35}{3} - \frac{24}{3}}{\frac{35}{3} + \frac{9}{3}}$$

$$\frac{\frac{11}{3}}{\frac{44}{3}}$$

$$\frac{11}{3} \cdot \frac{3}{44}$$

$$\frac{1}{4} \;\Big|\; \frac{1}{4} \quad \text{TRUE}$$

This checks, so the solution is $\dfrac{35}{3}$.

16. $\dfrac{47}{5}$

17. To avoid division by 0, we must have $x + 1 \neq 0$ and $x - 2 \neq 0$, or $x \neq -1$ and $x \neq 2$.

$$\frac{2}{x+1} = \frac{1}{x-2},$$

$$\text{LCD} = (x+1)(x-2)$$

$$(x+1)(x-2) \cdot \frac{2}{x+1} = (x+1)(x-2) \cdot \frac{1}{x-2}$$

$$2(x-2) = x+1$$

$$2x - 4 = x + 1$$

$$x = 5$$

This checks, so the solution is 5.

18. $-\dfrac{13}{2}$

19. Because no variable appears in a denominator, no restrictions exist.

$$\frac{a}{6} - \frac{a}{10} = \frac{1}{6}, \; \text{LCD} = 30$$

$$30\left(\frac{a}{6} - \frac{a}{10}\right) = 30 \cdot \frac{1}{6}$$

$$30 \cdot \frac{a}{6} - 30 \cdot \frac{a}{10} = 30 \cdot \frac{1}{6}$$

$$5a - 3a = 5$$

$$2a = 5$$

$$a = \frac{5}{2}$$

This checks, so the solution is $\dfrac{5}{2}$.

20. 3

21. Because no variable appears in a denominator, no restrictions exist.

$$\frac{x+1}{3} - 1 = \frac{x-1}{2}, \; \text{LCD} = 6$$

$$6\left(\frac{x+1}{3} - 1\right) = 6 \cdot \frac{x-1}{2}$$

$$6 \cdot \frac{x+1}{3} - 6 \cdot 1 = 6 \cdot \frac{x-1}{2}$$

$$2(x+1) - 6 = 3(x-1)$$

$$2x + 2 - 6 = 3x - 3$$

$$2x - 4 = 3x - 3$$

$$-1 = x$$

This checks, so the solution is -1.

22. -2

23. To avoid division by 0, we must have $t - 5 \neq 0$, or $t \neq 5$.

$$\frac{4}{t-5} = \frac{t-1}{t-5}, \text{ LCD} = t-5$$

$$(t-5) \cdot \frac{4}{t-5} = (t-5) \cdot \frac{t-1}{t-5}$$

$$4 = t-1$$

$$5 = t$$

Because of the restriction $t \neq 5$, the number 5 must be rejected as a solution. The equation has no solution.

24. No solution

25. To avoid division by 0, we must have $x + 4 \neq 0$ and $x \neq 0$, or $x \neq -4$ and $x \neq 0$.

$$\frac{3}{x+4} = \frac{5}{x}, \text{ LCD} = x(x+4)$$

$$x(x+4) \cdot \frac{3}{x+4} = x(x+4) \cdot \frac{5}{x}$$

$$3x = 5(x+4)$$

$$3x = 5x + 20$$

$$-2x = 20$$

$$x = -10$$

This checks, so the solution is -10.

26. $-\dfrac{21}{5}$

27. To avoid division by 0, we must have $a - 1 \neq 0$ and $a - 2 \neq 0$, or $a \neq 1$ and $a \neq 2$.

$$\frac{a-4}{a-1} = \frac{a+2}{a-2}, \text{ LCD} = (a-1)(a-2)$$

$$(a-1)(a-2) \cdot \frac{a-4}{a-1} = (a-1)(a-2) \cdot \frac{a+2}{a-2}$$

$$(a-2)(a-4) = (a-1)(a+2)$$

$$a^2 - 6a + 8 = a^2 + a - 2$$

$$-6a + 8 = a - 2$$

$$10 = 7a$$

$$\frac{10}{7} = a$$

This checks, so the solution is $\dfrac{10}{7}$.

28. $\dfrac{5}{3}$

29. To avoid division by 0, we must have $x - 3 \neq 0$ and $x + 3 \neq 0$, or $x \neq 3$ and $x \neq -3$.

$$\frac{4}{x-3} + \frac{2x}{x^2-9} = \frac{1}{x+3},$$
$$\text{LCD} = (x-3)(x+3)$$

$$(x-3)(x+3)\left(\frac{4}{x-3} + \frac{2x}{(x+3)(x-3)} \right) =$$
$$(x-3)(x+3) \cdot \frac{1}{x+3}$$

$$4(x+3) + 2x = x - 3$$

$$4x + 12 + 2x = x - 3$$

$$6x + 12 = x - 3$$

$$5x = -15$$

$$x = -3$$

Because of the restriction of $x \neq -3$, we must reject the number -3 as a solution. The equation has no solution.

30. No solution.

31. To avoid division by 0, we must have $y - 3 \neq 0$ and $y + 3 \neq 0$, or $y \neq 3$ and $y \neq -3$.

$$\frac{5}{y-3} - \frac{30}{y^2-9} = 1$$

$$\frac{5}{y-3} - \frac{30}{(y+3)(y-3)} = 1,$$
$$\text{LCD} = (y-3)(y+3)$$

$$(y-3)(y+3)\left(\frac{5}{y-3} - \frac{30}{(y+3)(y-3)} \right) =$$
$$(y-3)(y+3) \cdot 1$$

$$5(y+3) - 30 = (y+3)(y-3)$$

$$5y + 15 - 30 = y^2 - 9$$

$$0 = y^2 - 5y + 6$$

$$0 = (y-3)(y-2)$$

$$y - 3 = 0 \quad or \quad y - 2 = 0$$

$$y = 3 \quad or \quad y = 2$$

Because of the restriction $y \neq 3$, we must reject the number 3 as a solution. The number 2 checks, so it is the solution.

32. $\dfrac{1}{2}$

33. To avoid division by 0, we must have $8 - a \neq 0$ (or equivalently $a - 8 \neq 0$), or $a \neq 8$.

$$\frac{4}{8-a} = \frac{4-a}{a-8}$$

$$\frac{-1}{-1} \cdot \frac{4}{8-a} = \frac{4-a}{a-8}$$

$$\frac{-4}{a-8} = \frac{4-a}{a-8}, \text{ LCD} = a-8$$

$$(a-8) \cdot \frac{-4}{a-8} = (a-8) \cdot \frac{4-a}{a-8}$$

$$-4 = 4 - a$$

$$-8 = -a$$

$$8 = a$$

Because of the restriction $a \neq 8$, we must reject the number 8 as a solution. The equation has no solution.

34. -13

35. $\dfrac{-2}{x+2} = \dfrac{x}{x+2}$

To avoid division by 0, we must have $x + 2 \neq 0$, or $x \neq -2$. Now observe that the denominators are the same, so the numerators must be the same. Thus, we have $-2 = x$, but because of the restriction $x \neq -2$ this cannot be a solution. The equation has no solution.

36. No solution

37. ◈

38. ◈

39. *Familiarize*. Let $x =$ the first odd integer. Then $x + 2 =$ the next odd integer.

Translate.

$$\underbrace{\text{The sum of two consecutive odd integers}}_{x + (x+2)} \;\; \underset{=}{\text{is}} \;\; \underset{276}{276}.$$

Carry out. We solve the equation.

$$x + (x + 2) = 276$$
$$2x + 2 = 276$$
$$2x = 274$$
$$x = 137$$

When $x = 137$, then $x + 2 = 137 + 2 = 139$.

Check. The numbers 137 and 139 are consecutive odd integers and $137 + 139 = 276$. These numbers check.

State. The integers are 137 and 139.

40. 14 yd

41. *Familiarize*. Let $b =$ the base of the triangle, in cm. Then $b + 3 =$ the height. Recall that the area of a triangle is given by $\frac{1}{2} \times$ base $\times$ height.

Translate.

$$\underbrace{\text{The area of the triangle}}_{\frac{1}{2} \cdot b \cdot (b+3)} \;\; \underset{=}{\text{is}} \;\; \underset{54}{54 \text{ cm}^2}.$$

Carry out. We solve the equation.

$$\frac{1}{2}b(b + 3) = 54$$
$$2 \cdot \frac{1}{2}b(b + 3) = 2 \cdot 54$$
$$b(b + 3) = 108$$
$$b^2 + 3b = 108$$
$$b^2 + 3b - 108 = 0$$
$$(b - 9)(b + 12) = 0$$
$$b - 9 = 0 \;\; or \;\; b + 12 = 0$$
$$b = 9 \;\; or \;\;\;\;\;\; b = -12$$

Check. The length of the base cannot be negative so we need to check only 9. If the base is 9 cm, then the height is $9+3$, or 12 cm, and the area is $\frac{1}{2} \cdot 9 \cdot 12$, or 54 cm^2. The answer checks.

State. The base measures 9 cm, and the height measures 12 cm.

42. -8, -6; 6, and 8

43. To find the rate, in centimeters per day, we divide the amount of growth by the number of days. From June 9 to June 24 is $24 - 9 = 15$ days.

$$\text{Rate, in cm per day} = \frac{0.9 \text{ cm}}{15 \text{ days}}$$
$$= 0.06 \text{ cm/day}$$
$$= 0.06 \text{ cm per day}$$

44. 0.28 in. per day

45. ◈

46. ◈

47. To avoid division by 0, we must have $x - 3 \neq 0$, or $x \neq 3$.

$$1 + \frac{x-1}{x-3} = \frac{2}{x-3} - x, \;\; \text{LCD} = x - 3$$
$$(x-3)\left(1 + \frac{x-1}{x-3}\right) = (x-3)\left(\frac{2}{x-3} - x\right)$$
$$(x-3)\cdot 1 + (x-3)\cdot\frac{x-1}{x-3} = (x-3)\cdot\frac{2}{x-3} - (x-3)x$$
$$x - 3 + x - 1 = 2 - x^2 + 3x$$
$$2x - 4 = 2 - x^2 + 3x$$
$$x^2 - x - 6 = 0$$
$$(x - 3)(x + 2) = 0$$
$$x - 3 = 0 \;\; or \;\; x + 2 = 0$$
$$x = 3 \;\; or \;\;\;\;\;\; x = -2$$

Because of the restriction $x \neq 3$, we must reject the number 3 as a solution. The number -2 checks, so it is the solution.

48. 7

49. To avoid division by 0, we must have $x + 4 \neq 0$ and $x - 1 \neq 0$ and $x + 2 \neq 0$, or $x \neq -4$ and $x \neq 1$ and $x \neq -2$.

$$\frac{x}{x^2 + 3x - 4} + \frac{x}{x^2 + 6x + 8} =$$
$$\frac{2x}{x^2 + x - 2} - \frac{1}{x^2 + 6x + 8}$$
$$\frac{x}{(x+4)(x-1)} + \frac{x}{(x+2)(x+4)} =$$
$$\frac{2x}{(x+2)(x-1)} - \frac{1}{(x+2)(x+4)},$$
$$\text{LCD} = (x + 4)(x - 1)(x + 2)$$

$$(x+4)(x-1)(x+2)\left(\frac{x}{(x+4)(x-1)} + \frac{x}{(x+2)(x+4)}\right) =$$
$$(x+4)(x-1)(x+2)\left(\frac{2x}{(x+2)(x-1)} - \frac{1}{(x+2)(x+4)}\right)$$
$$x(x + 2) + x(x - 1) = 2x(x + 4) - (x - 1)$$
$$x^2 + 2x + x^2 - x = 2x^2 + 8x - x + 1$$
$$2x^2 + x = 2x^2 + 7x + 1$$
$$x = 7x + 1$$
$$-6x = 1$$
$$x = -\frac{1}{6}$$

This checks, so the solution is $-\dfrac{1}{6}$.

50. 3

51. To avoid division by 0, we must have $x + 2 \neq 0$ and $x - 2 \neq 0$, or $x \neq -2$ and $x \neq 2$.

$$\frac{x^2}{x^2 - 4} = \frac{x}{x + 2} - \frac{2x}{2 - x}$$
$$\frac{x^2}{x^2 - 4} = \frac{x}{x + 2} - \frac{2x}{2 - x} \cdot \frac{-1}{-1}$$
$$\frac{x^2}{(x + 2)(x - 2)} = \frac{x}{x + 2} - \frac{-2x}{x - 2},$$
$$\text{LCD} = (x + 2)(x - 2)$$

$$(x+2)(x-2) \cdot \frac{x^2}{(x+2)(x-2)} =$$
$$(x+2)(x-2)\left(\frac{x}{x+2} - \frac{-2x}{x-2}\right)$$
$$x^2 = x(x - 2) - (-2x)(x + 2)$$
$$x^2 = x^2 - 2x + 2x^2 + 4x$$
$$x^2 = 3x^2 + 2x$$
$$0 = 2x^2 + 2x$$
$$0 = 2x(x + 1)$$

$$2x = 0 \quad or \quad x + 1 = 0$$
$$x = 0 \quad or \qquad x = -1$$

Both of these check, so the solutions are -1 and 0.

52. 4

53. To avoid division by 0, we must have $x - 1 \neq 0$, or $x \neq 1$.

$$\frac{1}{x-1} + x - 5 = \frac{5x-4}{x-1} - 6, \ \text{LCD} = x - 1$$
$$(x-1)\left(\frac{1}{x-1} + x - 5\right) = (x-1)\left(\frac{5x-4}{x-1} - 6\right)$$
$$1 + x(x - 1) - 5(x - 1) = 5x - 4 - 6(x - 1)$$
$$1 + x^2 - x - 5x + 5 = 5x - 4 - 6x + 6$$
$$x^2 - 6x + 6 = -x + 2$$
$$x^2 - 5x + 4 = 0$$
$$(x - 1)(x - 4) = 0$$
$$x - 1 = 0 \quad or \quad x - 4 = 0$$
$$x = 1 \quad or \qquad x = 4$$

Because of the restriction $x \neq 1$, we must reject the number 1 as a solution. The number 4 checks, so it is the solution.

54. -6

55.

56.

Exercise Set 6.7

1. ***Familiarize.*** Let $x =$ the number.

Translate.

A number	minus	four times its reciprocal	is	3.
↓	↓	↓	↓	↓
x	$-$	$4 \cdot \dfrac{1}{x}$	$=$	3

Carry out.

$$x - \frac{4}{x} = 3$$
$$x\left(x - \frac{4}{x}\right) = x \cdot 3 \quad \text{Multiplying by the LCD}$$
$$x^2 - 4 = 3x$$
$$x^2 - 3x - 4 = 0$$
$$(x - 4)(x + 1) = 0$$
$$x - 4 = 0 \quad or \quad x + 1 = 0$$
$$x = 4 \quad or \qquad x = -1$$

Check. Four times the reciprocal of 4 is $4 \cdot \dfrac{1}{4}$, or 1. Since $4 - 1 = 3$, the number 4 is a solution. Four times the reciprocal of -1 is $4(-1)$, or -4. Since $-1 - (-4) = -1 + 4$, or 3, the number -1 is a solution.

State. The solutions are 4 and -1.

2. -1, 5

3. *Familiarize*. Let x = the number.

Translate. We reword the problem.

$\underbrace{\text{A number}}$ plus $\underbrace{\text{its reciprocal}}$ is $\ 2$.

$\quad\quad\downarrow\quad\quad\quad\quad\downarrow\quad\quad\quad\quad\downarrow\quad\quad\quad\downarrow\ \downarrow$

$\quad\quad x\quad\quad +\quad\quad \dfrac{1}{x}\quad\quad =\ 2$

Carry out. We solve the equation.

$$x + \frac{1}{x} = 2$$

$$x\left(x + \frac{1}{x}\right) = x \cdot 2 \quad \text{Multiplying by the LCD}$$

$$x^2 + 1 = 2x$$

$$x^2 - 2x + 1 = 0$$

$$(x-1)(x-1) = 0$$

$$x - 1 = 0 \quad or \quad x - 1 = 0$$

$$x = 1 \quad or \quad\quad x = 1$$

Check. The reciprocal of 1 is 1. Since $1 + 1 = 2$, the number 1 is a solution.

State. The number is 1.

4. 1, 5

5. *Familiarize*. The job takes Fontella 4 hours working alone and Omar 5 hours working alone. Then in 1 hour Fontella does $\frac{1}{4}$ of the job and Omar does $\frac{1}{5}$ of the job. Working together, they can do $\frac{1}{4} + \frac{1}{5}$, or $\frac{9}{20}$ of the job in 1 hour. In two hours, Fontella does $2\left(\frac{1}{4}\right)$ of the job and Omar does $2\left(\frac{1}{5}\right)$ of the job. Working together they can do $2\left(\frac{1}{4}\right) + 2\left(\frac{1}{5}\right)$, or $\frac{9}{10}$ of the job in 2 hours. In 3 hours they can do $3\left(\frac{1}{4}\right) + 3\left(\frac{1}{5}\right)$, or $1\frac{7}{20}$ of the job which is more of the job then needs to be done. The answer is somewhere between 2 hr and 3 hr.

Translate. If they work together t hours, then Fontella does $t\left(\frac{1}{4}\right)$ of the job and Omar does $t\left(\frac{1}{5}\right)$ of the job. We want some number t such that

$$\left(\frac{1}{4} + \frac{1}{5}\right)t = 1, \text{ or } \frac{9}{20} \cdot t = 1.$$

Carry out. We solve the equation.

$$\frac{9}{20} \cdot t = 1$$

$$\frac{20}{9} \cdot \frac{9}{20} \cdot t = \frac{20}{9} \cdot 1$$

$$t = \frac{20}{9}, \text{ or } 2\frac{2}{9}$$

Check. The check can be done by repeating the computations. We also have a partial check in that we expected from our familiarization step that the answer would be between 2 hr and 3 hr.

State. Working together, it takes them $2\frac{2}{9}$ hr to complete the job.

6. $6\frac{6}{7}$ hr

7. *Familiarize*. The job takes Vern 45 min working alone and Nina 60 min working alone. Then in 1 minute Vern does $\frac{1}{45}$ of the job and Nina does $\frac{1}{60}$ of the job. Working together, they can do $\frac{1}{45} + \frac{1}{60}$, or $\frac{7}{180}$ of the job in 1 minute. In 20 minutes, Vern does $\frac{20}{45}$ of the job and Nina does $\frac{20}{60}$ of the job. Working together, they can do $\frac{20}{45} + \frac{20}{60}$, or $\frac{7}{9}$ of the job. In 30 minutes, they can do $\frac{30}{45} + \frac{30}{60}$, or $\frac{7}{6}$ of the job which is more of the job than needs to be done. The answer is somewhere between 20 minutes and 30 minutes.

Translate. If they work together t minutes, then Vern does $t\left(\frac{1}{45}\right)$ of the job and Nina does $t\left(\frac{1}{60}\right)$ of the job. We want some number t such that

$$\left(\frac{1}{45} + \frac{1}{60}\right)t = 1, \text{ or } \frac{7}{180} \cdot t = 1.$$

Carry out. We solve the equation.

$$\frac{7}{180} \cdot t = 1$$

$$\frac{180}{7} \cdot \frac{7}{180} \cdot t = \frac{180}{7} \cdot 1$$

$$t = \frac{180}{7}, \text{ or } 25\frac{5}{7}$$

Check. The check can be done by repeating the computations. We also have a partial check in that we expected from our familiarization step that the answer would be between 20 minutes and 30 minutes.

State. It would take them $25\frac{5}{7}$ minutes to complete the job working together.

8. $1\frac{5}{7}$ hr

9. *Familiarize*. The job takes Kenny Dewitt 8 hours working alone and Betty Wohat 6 hours working alone. Then in 1 hour Kenny does $\frac{1}{8}$ of the job and Betty does $\frac{1}{6}$ of the job. Working together they can do $\frac{1}{8} + \frac{1}{6}$, or $\frac{7}{24}$ of the job in 1 hour. In two hours, Kenny does $2\left(\frac{1}{8}\right)$ of the job and Betty does $2\left(\frac{1}{6}\right)$ of the job. Working together they can and do $2\left(\frac{1}{8}\right) + 2\left(\frac{1}{6}\right)$, or $\frac{7}{12}$ of the job in two hours. In five hours they can do $5\left(\frac{1}{8}\right) + 5\left(\frac{1}{6}\right)$, or $\frac{35}{24}$, or $1\frac{11}{24}$ of the job which is more of the job than needs to be done. The answer is somewhere between 2 hr and 5 hr.

Translate. If they work together t hours, Kenny does $t\left(\frac{1}{8}\right)$ of the job and Betty does $t\left(\frac{1}{6}\right)$ of the job. We want some number t such that

$$\left(\frac{1}{8} + \frac{1}{6}\right)t = 1, \text{ or } \frac{7}{24} \cdot t = 1.$$

Carry out. We solve the equation.

$$\frac{7}{24} \cdot t = 1$$
$$\frac{24}{7} \cdot \frac{7}{24} \cdot t = \frac{24}{7} \cdot 1$$
$$t = \frac{24}{7}, \text{ or } 3\frac{3}{7}$$

Check. The check can be done by repeating the computations. We also have a partial check in that we expected from our familiarization step that the answer would be between 2 hr and 5 hr.

State. Working together, it takes them $3\frac{3}{7}$ hr to complete the job.

10. $20\frac{4}{7}$ hr

11. *Familiarize*. Let t = the number of minutes it takes Nicole and Glen to weed the garden, working together.

Translate. We use the work principle.

$$\left(\frac{1}{50} + \frac{1}{40}\right)t = 1, \text{ or } \frac{9}{200} \cdot t = 1$$

Carry out. We solve the equation.

$$\frac{9}{200} \cdot t = 1$$
$$\frac{200}{9} \cdot \frac{9}{200} \cdot t = \frac{200}{9} \cdot 1$$
$$t = \frac{200}{9}, \text{ or } 22\frac{2}{9}$$

Check. In $\frac{200}{9}$ min, the portion of the job done is $\frac{1}{50} \cdot \frac{200}{9} + \frac{1}{40} \cdot \frac{200}{9} = \frac{4}{9} + \frac{5}{9} = 1$. The answer checks.

State. It would take $22\frac{2}{9}$ min to weed the garden if Nicole and Glen worked together.

12. $11\frac{1}{9}$ min

13. *Familiarize*. Let t = the number of minutes it would take the two machines to copy the dissertation, working together.

Translate. We use the work principle.

$$\left(\frac{1}{12} + \frac{1}{20}\right)t = 1, \text{ or } \frac{8}{60} \cdot t = 1, \text{ or } \frac{2}{15} \cdot t = 1$$

Carry out. We solve the equation.

$$\frac{2}{15} \cdot 1 = 1$$
$$\frac{15}{2} \cdot \frac{2}{15} \cdot t = \frac{15}{2} \cdot 1$$
$$t = \frac{15}{2}, \text{ or } 7.5$$

Check. In $\frac{15}{2}$ min, the portion of the job done is $\frac{1}{12} \cdot \frac{15}{2} + \frac{1}{20} \cdot \frac{15}{2} = \frac{5}{8} + \frac{3}{8} = 1$. The answer checks.

State. It would take the two machines 7.5 min to copy the dissertation, working together.

14. $4\frac{4}{9}$ min

15. *Familiarize*. We complete the table shown in the text.

	Distance	Speed	Time
			$d = r \cdot t$
Truck	350	r	$\frac{350}{r}$
Train	150	$r - 40$	$\frac{150}{r-40}$

Translate. Since the times must be the same for both vehicles, we have the equation

$$\frac{350}{r} = \frac{150}{r - 40}.$$

Carry out. We first multiply by the LCD, $r(r-40)$.

$$r(r-40) \cdot \frac{350}{r} = r(r-40) \cdot \frac{150}{r-40}$$

$$350(r-40) = 150r$$

$$350r - 14{,}000 = 150r$$

$$-14{,}000 = -200r$$

$$70 = r$$

If the truck's speed is 70 mph, then the speed of the train is $70 - 40$, or 30 mph.

Check. First note that the speed of the truck, 70 mph, is 40 mph faster than the speed of the train, 30 mph. If the truck travels 350 mi at 70 mph, it travels for 350/70, or 5 hr. If the train travels 150 mi at 30 mph, it travels for 150/30, or 5 hr. Since the times are the same, the speeds check.

State. The speed of the truck is 70 mph, and the speed of the train is 30 mph.

16. AMTRAK: 80 km/h; B & M: 66 km/h

17. *Familiarize*. Let r = Kelly's speed, in km/h, and t = the time the bicyclists travel, in hours. Organize the information in a table.

	Distance	Speed	Time
Hank	42	$r-5$	t
Kelly	57	r	t

Translate. We can replace the t's in the table above using the formula $r = d/t$.

	Distance	Speed	Time
Hank	42	$r-5$	$\dfrac{42}{r-5}$
Kelly	57	r	$\dfrac{57}{r}$

Since the times are the same for both bicyclists, we have the equation

$$\frac{42}{r-5} = \frac{57}{r}.$$

Carry out. We first multiply by the LCD, $r(r-5)$.

$$r(r-5) \cdot \frac{42}{r-5} = r(r-5) \cdot \frac{57}{r}$$

$$42r = 57(r-5)$$

$$42r = 57r - 285$$

$$-15r = -285$$

$$r = 19$$

If $r = 19$, then $r - 5 = 14$.

Check. If Hank's speed is 14 km/h and Kelly's speed is 19 km/h, then Hank bicycles 5 km/h slower than

Kelly. Hank's time is 42/14, or 3 hr. Kelly's time is 57/19, or 3 hr. Since the times are the same, the answer checks.

State. Hank travels at 14 km/h, and Kelly travels at 19 km/h.

18. Bill: 50 mph; Hillary: 80 mph

19. *Familiarize*. Let r = Ralph's speed, in km/h. Then Bonnie's speed is $r + 3$. Also set t = the time, in hours, that Ralph and Bonnie walk. We organize the information in a table.

	Distance	Speed	Time
Ralph	7.5	r	t
Bonnie	12	$r+3$	t

Translate. We can replace the t's in the table shown above using the formula $r = d/t$.

	Distance	Speed	Time
Ralph	7.5	r	$\dfrac{7.5}{r}$
Bonnie	12	$r+3$	$\dfrac{12}{r+3}$

Since the times are the same for both walkers, we have the equation

$$\frac{7.5}{r} = \frac{12}{r+3}.$$

Carry out. We first multiply by the LCD, $r(r+3)$.

$$r(r+3) \cdot \frac{7.5}{r} = r(r+3) \cdot \frac{12}{r+3}$$

$$7.5(r+3) = 12r$$

$$7.5r + 22.5 = 12r$$

$$22.5 = 4.5r$$

$$5 = r$$

If $r = 5$, then $r + 3 = 8$.

Check. If Ralph's speed is 5 km/h and Bonnie's speed is 8 km/h, then Bonnie walks 3 km/h faster than Ralph. Ralph's time is 7.5/5, or 1.5 hr. Bonnie's time is 12/8, or 1.5 hr. Since the times are the same, the answer checks.

State. Ralph's speed is 5 km/h, and Bonnie's speed is 8 km/h.

20. Sally: 12 km/h; Gerard: 16 km/h

21. *Familiarize*. Let t = the time it takes Caledonia to drive to town and organize the given information in a table.

	Distance	Speed	Time
Caledonia	15	r	t
Manley	20	r	$t+1$

Translate. We can replace the r's in the table above using the formula $r = d/t$.

	Distance	Speed	Time
Caledonia	15	$\dfrac{15}{t}$	t
Manley	20	$\dfrac{20}{t+1}$	$t+1$

Since the speeds are the same for both riders, we have the equation

$$\frac{15}{t} = \frac{20}{t+1}.$$

Carry out. We multiply by the LCD, $t(t+1)$.

$$t(t+1) \cdot \frac{15}{t} = t(t+1) \cdot \frac{20}{t+1}$$

$$15(t+1) = 20t$$

$$15t + 15 = 20t$$

$$15 = 5t$$

$$3 = t$$

If $t = 3$, then $t + 1 = 3 + 1$, or 4.

Check. If Caledonia's time is 3 hr and Manley's time is 4 hr, then Manley's time is 1 hr more than Caledonia's. Caledonia's speed is 15/3, or 5 mph. Manley's speed is 20/4, or 5 mph. Since the speeds are the same, the answer checks.

State. It takes Caledonia 3 hr to drive to town.

22. $1\frac{1}{3}$ hr

23. We write a proportion and then solve it.

$$\frac{b}{6} = \frac{7}{4}$$

$$b = \frac{7}{4} \cdot 6$$

$$b = \frac{42}{4}, \text{ or } 10.5$$

$\left(\text{Note that the proportions } \dfrac{6}{b} = \dfrac{4}{7}, \dfrac{b}{7} = \dfrac{6}{4}, \text{ or } \dfrac{7}{b} = \dfrac{4}{6}\right.$
could also be used.$\Big)$

24. 6.75

25. We write a proportion and then solve it.

$$\frac{4}{f} = \frac{6}{4}$$

$$4f \cdot \frac{4}{f} = 4f \cdot \frac{6}{4}$$

$$16 = 6f$$

$$\frac{8}{3} = f \qquad \text{Simplifying}$$

$\left(\text{One of the following proportions could also be used:}\right.$
$\dfrac{f}{4} = \dfrac{4}{6}, \dfrac{4}{f} = \dfrac{9}{6}, \dfrac{f}{4} = \dfrac{6}{9}, \dfrac{4}{9} = \dfrac{f}{6}, \dfrac{9}{4} = \dfrac{6}{f}\Big)$

26. 7.5

27. We write a proportion and then solve it.

$$\frac{4}{10} = \frac{6}{l}$$

$$10l \cdot \frac{4}{10} = 10l \cdot \frac{6}{l}$$

$$4l = 60$$

$$l = 15 \text{ ft}$$

$\left(\text{One of the following proportions could also be used:}\right.$
$\dfrac{4}{6} = \dfrac{10}{l}, \dfrac{10}{4} = \dfrac{l}{6}, \text{ or } \dfrac{6}{4} = \dfrac{l}{10}\Big)$

28. 4.5 ft

29.
$$\frac{a}{b} = \frac{c}{d}$$

$$\frac{8}{5} = \frac{6}{d}$$

$$5d \cdot \frac{8}{5} = 5d \cdot \frac{6}{d}$$

$$8d = 30$$

$$d = \frac{30}{8} = \frac{15}{4} \text{ cm, or } 3.75 \text{ cm}$$

30. $\dfrac{90}{7}$ cm

31. Let $c = b + 2$ and $d = b - 2$.

$$\frac{a}{b} = \frac{c}{d}$$

$$\frac{15}{b} = \frac{b+2}{b-2}$$

$$b(b-2) \cdot \frac{15}{b} = b(b-2) \cdot \frac{b+2}{b-2}$$

$$15(b-2) = b(b+2)$$

$$15b - 30 = b^2 + 2b$$

$$0 = b^2 - 13b + 30$$

$$0 = (b-3)(b-10)$$

$b - 3 = 0$ *or* $b - 10 = 0$

$b = 3$ *or* $b = 10$

If $b = 3$ m, then $c = 3 + 2$, or 5 m and $d = 3 - 2$, or 1 m.

If $b = 10$ m, then $c = 10 + 2$, or 12 m and $d = 10 - 2$, or 8 m.

32. $b = 3$ m, $c = 6$ m, $d = 1$ m; $b = 12$ m, $c = 15$, m, $d = 10$ m

33. *Familiarize.* The coffee beans from 14 trees are required to produce 7.7 kilograms of coffee, and we wish to find how many trees are required to produce 308 kilograms of coffee. We can set up ratios:

$$\frac{T}{308} \qquad \frac{14}{7.7}$$

Translate. Assuming the two ratios are the same, we can translate to a proportion.

$$\begin{array}{l} \text{Trees} \rightarrow \\ \text{Kilograms} \rightarrow \end{array} \frac{T}{308} = \frac{14}{7.7} \begin{array}{l} \leftarrow \text{Trees} \\ \leftarrow \text{Kilograms} \end{array}$$

Carry out. We solve the proportion.

$$308 \cdot \frac{T}{308} = 308 \cdot \frac{14}{7.7}$$

$$T = \frac{4312}{7.7}$$

$$T = 560$$

Check. $\dfrac{560}{308} = 1.8\overline{1} \qquad \dfrac{14}{7.7} = 1.8\overline{1}$

The ratios are the same.

State. 560 trees are required to produce 308 kg of coffee.

34. 702 km

35. *Familiarize.* 10 cm^3 of human blood contains 1.2 grams of hemoglobin, and we wish to find how many grams of hemoglobin are contained in 16 cm^3 of the same blood. We can set up ratios:

$$\frac{H}{16} \qquad \frac{1.2}{10}$$

Translate. Assuming the two ratios are the same, we can translate to a proportion.

$$\begin{array}{l} \text{Grams} \rightarrow \\ \text{cm}^3 \rightarrow \end{array} \frac{H}{16} = \frac{1.2}{10} \begin{array}{l} \leftarrow \text{Grams} \\ \leftarrow \text{cm}^3 \end{array}$$

Carry out. We solve the proportion. We multiply by 16 to get H alone.

$$16 \cdot \frac{H}{16} = 16 \cdot \frac{1.2}{10}$$

$$H = \frac{19.2}{10}$$

$$H = 1.92$$

Check.

$$\frac{1.92}{16} = 0.12 \qquad \frac{1.2}{10} = 0.12$$

The ratios are the same.

State. 16 cm^3 of the same blood would contain 1.92 grams of hemoglobin.

36. $21\dfrac{2}{3}$ cups

37. *Familiarize.* U.S. women earn 77 cents for each dollar earned by a man. This gives us one ratio, expressed in dollars: $\dfrac{0.77}{1}$. If a male sales manager earns \$42,000, we want to find how much a female would earn for comparable work. This gives us a second ratio, also expressed in dollars: $\dfrac{F}{42,000}$.

Translate. Assuming the two ratios are the same, we can translate to a proportion.

$$\begin{array}{l} \text{Female's} \\ \text{earnings} \\ \text{Male's earnings} \end{array} \begin{array}{l} \rightarrow \\ \rightarrow \end{array} \frac{0.77}{1} = \frac{F}{42,000} \begin{array}{l} \leftarrow \\ \leftarrow \end{array} \begin{array}{l} \text{Female's} \\ \text{earnings} \\ \text{Male's earnings} \end{array}$$

Carry out. We solve the proportion.

$$42,000 \cdot \frac{0.77}{1} = 42,000 \cdot \frac{F}{42,000}$$

$$32,340 = F$$

Check.

$$\frac{0.77}{1} = 0.77 \qquad \frac{32,340}{42,000} = 0.77$$

The ratios are the same.

State. If a male sales manager earns \$42,000, a female would earn \$32,340 for comparable work.

38. $1\dfrac{11}{39}$ kg

39. *Familiarize.* The ratio of deer tagged to the total number of deer in the preserve, D, is $\dfrac{318}{D}$. Of the 168 deer caught later, 56 are tagged. The ratio of tagged deer to deer caught is $\dfrac{56}{168}$.

Translate. We translate to a proportion.

$$\begin{array}{l} \text{Deer originally} \\ \text{tagged} \\ \text{Deer} \\ \text{in preserve} \end{array} \begin{array}{l} \rightarrow \\ \rightarrow \end{array} \frac{318}{D} = \frac{56}{168} \begin{array}{l} \leftarrow \\ \leftarrow \end{array} \begin{array}{l} \text{Tagged deer} \\ \text{caught later} \\ \text{Deer} \\ \text{caught later} \end{array}$$

Carry out. We solve the proportion. We multiply by the LCD, 168D.

$$168D \cdot \frac{318}{D} = 168D \cdot \frac{56}{168}$$
$$168 \cdot 318 = D \cdot 56$$
$$\frac{166 \cdot 318}{56} = D$$
$$954 = D$$

Check.
$$\frac{318}{954} = 0.\overline{3} \qquad \frac{56}{168} = 0.\overline{3}$$
The ratios are the same.

State. We estimate that there are 954 deer in the preserve.

40. 184

41. Familiarize. Let D = the number of defective bulbs you would expect in a sample of 1288 bulbs. We set up two ratios:
$$\frac{6}{184} \qquad \frac{D}{1288}$$
Translate. Assuming the ratios are the same, we can translate to a proportion.

$$\begin{array}{c} \text{Defective} \rightarrow \\ \text{Sample} \rightarrow \end{array} \frac{6}{184} = \frac{D}{1288} \begin{array}{c} \leftarrow \text{Defective} \\ \leftarrow \text{Sample} \end{array}$$

Carry out. We solve the proportion.
$$184(1288) \cdot \frac{6}{184} = 184(1288) \cdot \frac{D}{1288}$$
$$1288 \cdot 6 = 184 \cdot D$$
$$\frac{1288 \cdot 6}{184} = D$$
$$42 = D$$

Check.
$$\frac{6}{184} \approx 0.0326 \qquad \frac{42}{1288} \approx 0.0326$$
The ratios are the same.

State. You would expect 42 defective bulbs in a sample of 1288 bulbs.

42. 287

43. Familiarize. Let M = the number of miles Emmanuel will drive in 4 years if he continues to drive at the current rate. We set up two ratios:
$$\frac{16,000}{1\frac{1}{2}} \qquad \frac{M}{4}$$
Translate. We write a proportion.

$$\begin{array}{c} \text{Miles} \rightarrow \\ \text{Years} \rightarrow \end{array} \frac{16,000}{1.5} = \frac{M}{4} \begin{array}{c} \leftarrow \text{Miles} \\ \leftarrow \text{Years} \end{array}$$

Carry out. We solve the proportion.
$$1.5(4) \cdot \frac{16,000}{1.5} = 1.5(4) \cdot \frac{M}{4}$$
$$64,000 = 1.5M$$
$$42,666.\overline{6} = M$$
If this possible answer is correct, Emmanuel will not exceed the 45,000 miles allowed for four years.

Check.
$$\frac{16,000}{1.5} = 10,666.\overline{6} \qquad \frac{42,666.\overline{6}}{4} = 10,666.\overline{6}$$
The ratios are the same.

State. At this rate, Emmanuel will not exceed the mileage allowed for four years.

44. 20

45. Familiarize. The ratio of foxes tagged to the total number of foxes in the county, F, is $\frac{25}{F}$. Of the 36 foxes caught later, 4 had tags. The ratio of tagged foxes to foxes caught is $\frac{4}{36}$.

Translate. Assuming the two ratios are the same, we can translate to a proportion.

$$\begin{array}{c} \text{Foxes tagged} \\ \text{originally} \rightarrow \\ \text{Foxes} \rightarrow \\ \text{in county} \end{array} \frac{25}{F} = \frac{4}{36} \begin{array}{c} \text{Tagged foxes} \\ \leftarrow \text{caught later} \\ \leftarrow \text{Foxes} \\ \text{caught later} \end{array}$$

Carry out. We solve the proportion.
$$36F \cdot \frac{25}{F} = 36F \cdot \frac{4}{36}$$
$$900 = 4F$$
$$225 = F$$

Check.
$$\frac{25}{225} = \frac{1}{9} \qquad \frac{4}{36} = \frac{1}{9}$$
The ratios are the same.

State. We estimate that there are 225 foxes in the county.

46. a) 1.92 tons

b) 28.8 lb

47. Familiarize. The ratio of the weight of an object on Mars to the weight of an object on the earth is 0.4 to 1.

a) We wish to find how much a 12-ton rocket would weigh on Mars.

b) We wish to find how much a 120-lb astronaut would weigh on Mars.

We can set up ratios.

$$\frac{0.4}{1} \qquad \frac{T}{12} \qquad \frac{P}{120}$$

Translate. Assuming the ratios are the same, we can translate to proportions.

a)
$$\begin{array}{ll} \text{Weight} & \text{Weight} \\ \text{on Mars} \rightarrow \dfrac{0.4}{1} = \dfrac{T}{12} \leftarrow \text{on Mars} \\ \text{Weight} \rightarrow & \leftarrow \text{Weight} \\ \text{on earth} & \text{on earth} \end{array}$$

b)
$$\begin{array}{ll} \text{Weight} & \text{Weight} \\ \text{on Mars} \rightarrow \dfrac{0.4}{1} = \dfrac{P}{120} \leftarrow \text{on Mars} \\ \text{Weight} \rightarrow & \leftarrow \text{Weight} \\ \text{on earth} & \text{on earth} \end{array}$$

Carry out. We solve each proportion.

a) $\dfrac{0.4}{1} = \dfrac{T}{12}$ b) $\dfrac{0.4}{1} = \dfrac{P}{120}$

$12(0.4) = T$ $120(0.4) = P$

$4.8 = T$ $48 = P$

Check. $\dfrac{0.4}{1} = 0.4$, $\dfrac{4.8}{12} = 0.4$, and $\dfrac{48}{120} = 0.4$.

The ratios are the same.

State.

a) A 12-ton rocket would weigh 4.8 tons on Mars.

b) A 120-lb astronaut would weigh 48 lb on Mars.

48. $\dfrac{36}{68}$

49. ◈

50. ◈

51. Graph: $y = 2x - 6$.

We select some x-values and compute y-values.

If $x = 1$, then $y = 2 \cdot 1 - 6 = -4$.

If $x = 3$, then $y = 2 \cdot 3 - 6 = 0$.

If $x = 5$, then $y = 2 \cdot 5 - 6 = 4$.

x	y	(x, y)
1	-4	$(1, -4)$
3	0	$(3, 0)$
5	4	$(5, 4)$

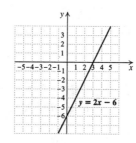

52.

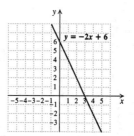

53. Graph: $3x + 2y = 12$.

We can replace either variable with a number and then calculate the other coordinate. We will find the intercepts and one other point.

If $y = 0$, we have:

$$3x + 2 \cdot 0 = 12$$
$$3x = 12$$
$$x = 4$$

The x-intercept is $(4, 0)$.

If $x = 0$, we have:

$$3 \cdot 0 + 2y = 12$$
$$2y = 12$$
$$y = 6$$

The y-intercept is $(0, 6)$.

If $y = -3$, we have:

$$3x + 2(-3) = 12$$
$$3x - 6 = 12$$
$$3x = 18$$
$$x = 6$$

The point $(6, -3)$ is on the graph.

We plot these points and draw a line through them.

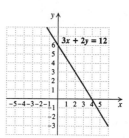

54.

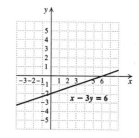

55. Graph: $y = -\dfrac{3}{4}x + 2$

We select some x-values and compute y-values. We use multiples of 4 to avoid fractions.

If $x = -4$, then $y = -\dfrac{3}{4}(-4) + 2 = 5$.

If $x = 0$, then $y = -\dfrac{3}{4} \cdot 0 + 2 = 2$.

If $x = 4$, then $y = -\dfrac{3}{4} \cdot 4 + 2 = -1$.

x	y	(x, y)
-4	5	$(-4, 5)$
0	2	$(0, 2)$
4	-1	$(4, -1)$

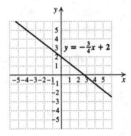

56.

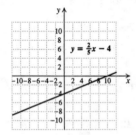

57.

58. ◈

59. *Familiarize*. Let $t =$ the time, in hours, it takes Michelle to wax the car alone. Then $\dfrac{t}{2} =$ Sal's time alone, and $t - 2 =$ Kristen's time alone. In 1 hr they do $\dfrac{1}{t} + \dfrac{1}{\frac{t}{2}} + \dfrac{1}{t-2}$, or $\dfrac{1}{t} + \dfrac{2}{t} + \dfrac{1}{t-2}$ of the job working together. The entire job takes 1 hr and 20 min, or $\dfrac{4}{3}$ hr.

***Translate*.** To get an entire job, we multiply the amount of work done in 1 hr by the number of hours required to complete the job.

$$\frac{4}{3}\left(\frac{1}{t} + \frac{2}{t} + \frac{1}{t-2}\right) = 1$$

***Carry out*.** We solve the equation.

$$3 \cdot \frac{4}{3}\left(\frac{1}{t} + \frac{2}{t} + \frac{1}{t-2}\right) = 3 \cdot 1$$

$$4\left(\frac{1}{t} + \frac{2}{t} + \frac{1}{t-2}\right) = 3$$

$$4\left(\frac{3}{t} + \frac{1}{t-2}\right) = 3 \quad \text{Adding:} \frac{1}{t} + \frac{2}{t} = \frac{3}{t}$$

$$\frac{12}{t} + \frac{4}{t-2} = 3$$

$$t(t-2)\left(\frac{12}{t} + \frac{4}{t-2}\right) = t(t-2)(3)$$

$$12(t-2) + t \cdot 4 = 3t(t-2)$$

$$12t - 24 + 4t = 3t^2 - 6t$$

$$16t - 24 = 3t^2 - 6t$$

$$0 = 3t^2 - 22t + 24$$

$$0 = (3t - 4)(t - 6)$$

$$3t - 4 = 0 \quad or \quad t - 6 = 0$$

$$3t = 4 \quad or \qquad t = 6$$

$$t = \frac{4}{3} \quad or \qquad t = 6$$

***Check*.** If $t = \dfrac{4}{3}$, then $t - 2 = -\dfrac{2}{3}$. Since time cannot be negative in this problem, $\dfrac{4}{3}$ cannot be a solution. If $t = 6$, then $t/2 = 3$ and $t - 2 = 4$. If Michelle, Sal, and Kristen can do the job in 6 hr, 3 hr, and 4 hr, respectively, then in one hour they do $\dfrac{1}{6} + \dfrac{1}{3} + \dfrac{1}{4}$, or $\dfrac{3}{4}$ of the job working together and in $\dfrac{4}{3}$ hr they do $\dfrac{4}{3} \cdot \dfrac{3}{4}$, or 1 entire job. The answer checks.

***State*.** Working alone, the job would take Michelle 6 hr, Sal 3 hr, and Kristen 4 hr.

60. $9\dfrac{3}{13}$ days

61. *Familiarize*. Let $t =$ the number of hours it takes to wire one house, working together. We want to find the number of hours it takes to wire two houses, working together.

***Translate*.** We write an equation.

$$\frac{t}{28} + \frac{t}{34} = 2$$

***Carry out*.** We solve the equation.

$$\frac{t}{28} + \frac{t}{34} = 2, \text{ LCD} = 476$$

$$476\left(\frac{t}{28} + \frac{t}{34}\right) = 476 \cdot 2$$

$$17t + 14t = 952$$

$$31t = 952$$

$$t = \frac{952}{31}, \text{ or } 32\frac{22}{31}$$

Check. If $30\frac{22}{31}$ hr, Janet does $\frac{952}{31} \cdot \frac{1}{28} = \frac{34}{31}$ of one complete job and Linus does $\frac{952}{31} \cdot \frac{1}{34}$, or $\frac{28}{31}$ of one complete job. Together they do $\frac{34}{31} + \frac{28}{31}$, or $\frac{62}{31}$, or 2 complete jobs. The answer checks.

State. It will take Janet and Linus $30\frac{22}{31}$ hr to wire two houses, working together.

62. Ann: 6 hr; Betty: 12 hr

63. Familiarize. We will begin by finding how long it will take Alma and Kevin to grade a batch of exams, working together. Then we will find what percentage of the job was done by Alma.

Translate. We use the work principle to find how long it will take Alma and Kevin to do the job, working together.

$$\left(\frac{1}{3} + \frac{1}{4}\right)t = 1, \text{ or } \frac{7}{12} \cdot t = 1$$

Carry out. We solve the equation.

$$\frac{7}{12} \cdot t = 1$$

$$\frac{12}{7} \cdot \frac{7}{12} \cdot t = \frac{12}{7} \cdot 1$$

$$t = \frac{12}{7}$$

Now, since Alma can do the job alone in 3 hr, she does $\frac{1}{3}$ of the job in 1 hr and in $\frac{12}{7}$ hr she does $\frac{12}{7} \cdot \frac{1}{3} \approx 0.57 \approx 57\%$ of the job.

Check. We can repeat the calculations. The answer checks.

State. About 57% of the exams will have been graded by Alma.

64. 12 hr

65. Familiarize. We organize the information in a table. Let $r =$ the speed of the current and $t =$ the time it takes to travel upstream.

Translate.

	Distance	Speed	Time
Upstream	24	$10-r$	t
Downstream	24	$10+r$	$5-t$

From the rows of the table we get two equations:

$$24 = (10-r)t$$
$$24 = (10+r)(5-t)$$

We solve each equation for t and set the results equal:

Solving $24 = (10-r)t$ for t: $t = \dfrac{24}{10-r}$

Solving $24 = (10+r)(5-t)$ for t: $t = 5 - \dfrac{24}{10+r}$

Then $\dfrac{24}{10-r} = 5 - \dfrac{24}{10+r}$.

Carry out. We first multiply on both sides of the equation by the LCD, $(10-r)(10+r)$:

$$(10-r)(10+r) \cdot \frac{24}{10-r} = (10-r)(10+r)\left(5 - \frac{24}{10+r}\right)$$

$$24(10+r) = 5(10-r)(10+r) - 24(10-r)$$

$$240 + 24r = 500 - 5r^2 - 240 + 24r$$

$$240 + 24r = 260 - 5r^2 + 24r$$

$$5r^2 - 20 = 0$$

$$5(r^2 - 4) = 0$$

$$5(r+2)(r-2) = 0$$

$$r + 2 = 0 \quad or \quad r - 2 = 0$$

$$r = -2 \quad or \quad\quad r = 2$$

Check. We only check 2 since the speed of the current cannot be negative. If $r = 2$, then the speed upstream is $10 - 2$, or 8 mph and the time is $\frac{24}{8}$, or 3 hours. If $r = 2$, then the speed downstream is $10 + 2$, or 12 mph and the time is $\frac{24}{12}$, or 2 hours. The sum of 3 hr and 2 hr is 5 hr. This checks.

State. The speed of the current is 2 mph.

66. 270

67. Familiarize. We organize the information in a table. Let $r =$ the speed on the first part of the trip and $t =$ the time driven at that speed.

	Distance	Speed	Time
First part	30	r	t
Second part	30	$r + 15$	$1 - t$

Translate. From the rows of the table we obtain two equations:

$$30 = rt$$
$$30 = (r + 15)(1 - t)$$

We solve each equation for t and set the results equal:

Solving $30 = rt$ for t: $t = \dfrac{30}{r}$

Solving $20 = (r + 15)(1 - t)$ for t: $t = 1 - \dfrac{20}{r + 15}$

Then $\dfrac{30}{r} = 1 - \dfrac{20}{r + 15}$.

Carry out. We first multiply the equation by the LCD, $r(r + 15)$.

$$r(r+15) \cdot \frac{30}{r} = r(r+15)\left(1 - \frac{20}{r+15}\right)$$

$$30(r+15) = r(r+15) - 20r$$

$$30r + 450 = r^2 + 15r - 20r$$

$$0 = r^2 - 35r - 450$$

$$0 = (r-45)(r+10)$$

$$r - 45 = 0 \quad or \quad r + 10 = 0$$

$$r = 45 \quad or \qquad r = -10$$

Check. Since the speed cannot be negative, we only check 45. If $r = 45$, then the time for the first part is $\frac{30}{45}$, or $\frac{2}{3}$ hr. If $r = 45$, then $r + 15 = 60$ and the time for the second part is $\frac{20}{60}$, or $\frac{1}{3}$ hr. The total time is $\frac{2}{3} + \frac{1}{3}$, or 1 hour. The value checks.

State. The speed for the first 30 miles was 45 mph.

68. $66\frac{2}{3}$ ft

69. Familiarize. Let $t =$ the number of minutes after 5:00 at which the hands will first be together. When the minute hand moves through t minutes, the hour hand moves through $t/12$ minutes. At 5:00 the hour hand is on the 25-minute mark, so at t minutes after 5:00 it is at $25 + t/12$.

Translate. We equate the positions of the minute hand and the hour hand.

$$t = 25 + \frac{t}{12}$$

Carry out. We solve the equation.

$$t = 25 + \frac{t}{12}$$

$$12 \cdot t = 12\left(25 + \frac{t}{12}\right)$$

$$12t = 300 + t$$

$$11t = 300$$

$$t = \frac{300}{11}, \text{ or } 27\frac{3}{11}$$

Check. When the minute hand is at $27\frac{3}{11}$ minutes after 5:00, the hour hand is at $25 + \dfrac{\frac{300}{11}}{12} =$

$25 + \frac{300}{11} \cdot \frac{1}{12} = 25 + \frac{25}{11} = 25 + 2\frac{3}{11} = 27\frac{3}{11}$ minutes after 5:00 also. The answer checks.

State. After 5:00 the hands on a clock will first be together in $27\frac{3}{11}$ minutes or at $27\frac{3}{11}$ minutes after 5:00.

70. $\dfrac{D}{B} = \dfrac{C}{A}; \dfrac{A}{C} = \dfrac{B}{D}; \dfrac{D}{C} = \dfrac{B}{A}$

71. ◈

72. ◈

Chapter 7

Systems and More Graphing

1. We check by substituting alphabetically 3 for x and 2 for y.

$$\frac{2x+3y=12}{\begin{array}{c|c} 2\cdot 3+3\cdot 2 \ ?\ 12 \\ 6+6 \\ 12 \end{array}}$$

$$\frac{x-4y=-5}{\begin{array}{c|c} 3-4\cdot 2 \ ?\ -5 \\ 3-8 \\ -5 \end{array}}$$

| 12 | 12 TRUE | | −5 | −5 TRUE |

The ordered pair $(3, 2)$ is a solution of each equation. Therefore it is a solution of the system of equations.

2. Yes

3. We check by substituting alphabetically 3 for a and 2 for b.

$$\frac{3b-2a=0}{\begin{array}{c|c} 3\cdot 2-2\cdot 3 \ ?\ 0 \\ 6-6 \\ 0 \end{array}}$$

$$\frac{b+2a=15}{\begin{array}{c|c} 2+2\cdot 3 \ ?\ 15 \\ 2+6 \\ 8 \end{array}}$$

| 0 | 0 TRUE | | 8 | 15 FALSE |

The ordered pair $(3, 2)$ is not a solution of $b+2a=15$. Therefore it is not a solution of the system of equations.

4. Yes

5. We check by substituting alphabetically 15 for x and 20 for y.

$$\frac{3x-2y=5}{\begin{array}{c|c} 3\cdot 15-2\cdot 20 \ ?\ 5 \\ 45-40 \\ 5 \end{array}}$$

| 5 | 5 TRUE |

$$\frac{6x-5y=-10}{\begin{array}{c|c} 6\cdot 15-5\cdot 20 \ ?\ -10 \\ 90-100 \\ -10 \end{array}}$$

| −10 | −10 TRUE |

The ordered pair $(15, 20)$ is a solution of each equation. Therefore it is a solution of the system of equations.

6. Yes

7. We graph the equations.

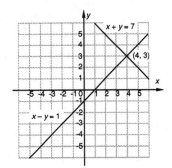

The "apparent" solution of the system, $(4, 3)$, should be checked in both equations.

Check:

$$\frac{x+y=7}{\begin{array}{c|c} 4+3 \ ?\ 7 \\ 7 \end{array}}$$

$$\frac{x-y=1}{\begin{array}{c|c} 4-3 \ ?\ 1 \\ 1 \end{array}}$$

| 7 | 7 TRUE | | 1 | 1 TRUE |

The solution is $(4, 3)$.

8. $(2, 1)$

9. We graph the equations.

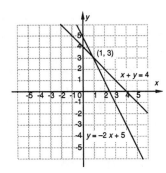

The "apparent" solution of the system, $(1, 3)$, should be checked in both equations.

Check:

$$\frac{y=-2x+5}{\begin{array}{c|c} 3 \ ?\ -2\cdot 1+5 \\ -2+5 \\ 3 \end{array}}$$

$$\frac{x+y=4}{\begin{array}{c|c} 1+3 \ ?\ 4 \\ 4 \end{array}}$$

| 3 | 3 TRUE | | 4 | 4 TRUE |

The solution is $(1, 3)$.

10. $(3, 1)$

11. We graph the equations.

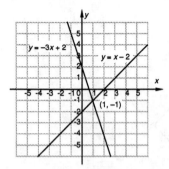

The "apparent" solution of the system, $(1, -1)$, should be checked in both equations.

Check:

$$\begin{array}{c|c} y = x - 2 \\ \hline -1 \ ? \ 1 - 2 \\ -1 \ \big| \ -1 \quad \text{TRUE} \end{array}$$

$$\begin{array}{c|c} y = -3x + 2 \\ \hline -1 \ ? \ -3 \cdot 1 + 2 \\ \ \big| \ -3 + 2 \\ -1 \ \big| \ -1 \quad \textit{TRUE} \end{array}$$

The solution is $(1, -1)$.

12. $(2, -1)$

13. We graph the equations.

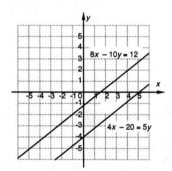

The lines are parallel. There is no solution.

14. Infinite number of solutions

15. We graph the equations.

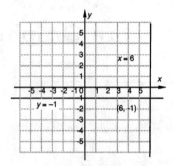

The "apparent" solution of the system, $(6, -1)$, should be checked in both equations.

Check:

$$\begin{array}{c|c} x = 6 \\ \hline 6 \ ? \ 6 \quad \text{TRUE} \end{array} \qquad \begin{array}{c|c} y = -1 \\ \hline -1 \ ? \ -1 \quad \text{TRUE} \end{array}$$

The solution is $(6, -1)$.

16. $(-2, 5)$

17. We graph the equations.

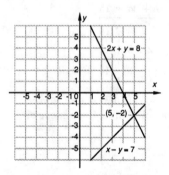

The "apparent" solution of the system, $(5, -2)$, should be checked in both equations.

Check:

$$\begin{array}{c|c} 2x + y = 8 \\ \hline 2 \cdot 5 + (-2) \ ? \ 8 \\ 10 - 2 \ \big| \\ 8 \ \big| \ 8 \quad \text{TRUE} \end{array}$$

$$\begin{array}{c|c} x - y = 7 \\ \hline 5 - (-2) \ ? \ 7 \\ 5 + 2 \ \big| \\ 7 \ \big| \ 7 \quad \text{TRUE} \end{array}$$

The solution is $(5, -2)$.

18. $(2, -2)$

19. We graph the equations.

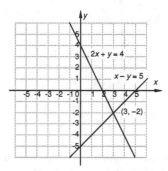

The "apparent" solution of the system, $(3, -2)$, should be checked in both equations.

Check:

$$x - y = 5$$

$$3 - (-2) \ ? \ 5$$

$$3 + 2$$

$$5 \ \big| \ 5 \quad \text{TRUE}$$

$$2x + y = 4$$

$$2 \cdot 3 + (-2) \ ? \ 4$$

$$6 - 2$$

$$4 \ \big| \ 4 \quad \text{TRUE}$$

The solution is $(3, -2)$.

20. $(5, -3)$

21. We graph the equations.

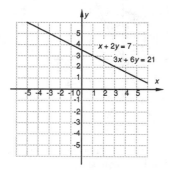

We see that the equations represent the same line. This means that any solution of one equation is a solution of the other equation as well. Thus, there is an infinite number of solutions.

22. Infinite number of solutions

23. We graph the equations.

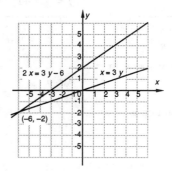

The "apparent" solution of the system, $(-6, -2)$, should be checked in both equations.

Check:

$$2x = 3y - 6$$

$$2(-6) \ ? \ 3(-2) - 6$$

$$-12 \ \big| \ -6 - 6$$

$$-12 \ \big| \ -12 \quad \text{TRUE}$$

$$x = 3y$$

$$-6 \ ? \ 3(-2)$$

$$-6 \ \big| \ -6 \quad \text{TRUE}$$

The solution is $(-6, -2)$.

24. $(-3, -3)$

25. $y = \dfrac{1}{5}x + 4,$

$2y = \dfrac{2}{5}x + 8$

Observe that we can obtain the second equation by multiplying both sides of the first equation by 2. Thus, the equations are dependent and there is an infinite number of solutions.

26. No solution

27. We graph the equations.

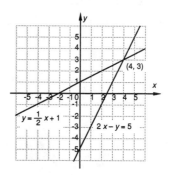

The "apparent" solution of the system, $(4, 3)$, should be checked in both equations.

Check:

$$\begin{array}{c|c} 2x - y = 5 \\ \hline 2\cdot 4 - 3 \;?\; 5 \\ 8 - 3 \\ \hspace{2em} 5 & 5 \;\; \text{TRUE} \end{array}$$

$$\begin{array}{c|c} y = \dfrac{1}{2}x + 1 \\ \hline 3 \;?\; \dfrac{1}{2}\cdot 4 + 1 \\ 2 + 1 \\ 3 & 3 \hspace{2em} \text{TRUE} \end{array}$$

The solution is $(4, 3)$.

28. $(1, -3)$

29. We graph the equations.

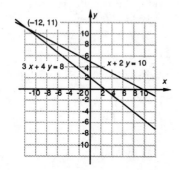

The "apparent" solution of the system, $(-12, 11)$, should be checked in both equations.

Check:

$$\begin{array}{c|c} 3x + 4y = 8 \\ \hline 3(-12) + 4\cdot 11 \;?\; 8 \\ -36 + 44 \\ 8 & 8 \;\; \text{TRUE} \end{array}$$

$$\begin{array}{c|c} x + 2y = 10 \\ \hline -12 + 2\cdot 11 \;?\; 10 \\ -12 + 22 \\ 10 & 10 \;\; \text{TRUE} \end{array}$$

The solution is $(-12, 11)$.

30. $(-8, -7)$

31. We graph the equations.

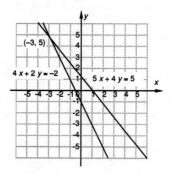

The "apparent" solution of the system, $(-3, 5)$, should be checked in both equations.

Check:

$$\begin{array}{c|c} 4x + 2y = -2 \\ \hline 4(-3) + 2\cdot 5 \;?\; -2 \\ -12 + 10 \\ -2 & -2 \;\; \text{TRUE} \end{array}$$

$$\begin{array}{c|c} 5x + 4y = 5 \\ \hline 5(-3) + 4\cdot 5 \;?\; 5 \\ -15 + 20 \\ 5 & 5 \;\; \text{TRUE} \end{array}$$

The solution is $(-3, 5)$.

32. $(3, -4)$

33. We graph the equations.

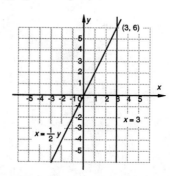

The "apparent" solution of the system, $(3, 6)$, should be checked in both equations.

Check:

$$\begin{array}{c|c} x = \dfrac{1}{2}y \\ \hline 3 \;?\; \dfrac{1}{2}\cdot 6 \\ 3 & 3 \hspace{2em} \text{TRUE} \end{array} \qquad \begin{array}{c|c} x = 3 \\ \hline 3 \;?\; 3 \;\; \text{TRUE} \\ \end{array}$$

The solution is $(3, 6)$.

34. $(2, 6)$

35. ◈

36. ◈

37. $4x - 5(9 - 2x) = 7$
$4x - 45 + 10x = 7$
$14x - 45 = 7$
$14x = 52$
$x = \dfrac{52}{14} = \dfrac{26}{7}$
The solution is $\dfrac{26}{7}$.

38. $\dfrac{15}{2}$

39. $3(4 - 2y) - 5y = 6$
$12 - 6y - 5y = 6$
$12 - 11y = 6$
$-11y = -6$
$y = \dfrac{6}{11}$
The solution is $\dfrac{6}{11}$.

40. $\dfrac{9}{10}$

41. $5(2x + 3y) - 3(7x + 5y)$
$= 10x + 15y - 21x - 15y$
$= 10x - 21x + 15y - 15y$
$= -11x$

42. $-11y$

43. ◈

44. ◈

45. Systems in which the graphs of the equations coincide contain dependent equations. This is the case in Exercises 21, 22, and 25.

46. All but Exercises 13 and 26

47. Systems in which the graphs of the equations are parallel are inconsistent. This is the case in Exercises 13 and 26.

48. All but Exercises 14, 21, 22, and 25

49. Answers may vary. Any equation with $(-1, 4)$ as a solution that is independent of $5x + 2y = 3$ will do. One such equation is $x + y = 3$.

50. Answers may vary. $x + y = 1$

51. $(2, -3)$ is a solution of $Ax - 3y = 13$. Substitute 2 for x and -3 for y and solve for A.
$$Ax - 3y = 13$$
$$A \cdot 2 - 3(-3) = 13$$
$$2A + 9 = 13$$
$$2A = 4$$
$$A = 2$$

$(2, -3)$ is a solution of $x - By = 8$. Substitute 2 for x and -3 for y and solve for B.
$$x - By = 8$$
$$2 - B(-3) = 8$$
$$2 + 3B = 8$$
$$3B = 6$$
$$B = 2$$

52. $(1.25, 1.5)$

53. a) Let $x =$ the number of copies, up to 500, and $y =$ the cost. Then the cost equation for the copy card method of payment is $y = 20$. The cost equation for the method of paying per page is $y = 0.06x$.

b)
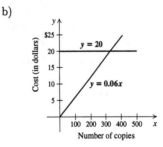

c) We see that the graphs intersect at approximately $(333, 20)$ and that for x-values greater than 333 the graph of $y = 20$ lies below the graph of $y = 0.06x$. Thus, Shelby must make more than 333 copies if the card is to be more economical.

54. No solution

55. Graph $y_1 = 1.2x - 32.7$ and $y_2 = -0.7x + 46.15$ and use the INTERSECT feature from the CALC menu to find the point of intersection of the graphs. It is $(41.5, 17.1)$.

Exercise Set 7.2

1. $x + y = 9,$ (1)

$x = y + 1$ (2)

Substitute $y + 1$ for x in Equation (1) and solve for y.

$$x + y = 9 \quad (1)$$
$$(y + 1) + y = 9 \quad \text{Substituting}$$
$$2y + 1 = 9$$
$$2y = 8$$
$$y = 4$$

Next we substitute 4 for y in either equation of the original system and solve for x.

$$x = y + 1 \quad (2)$$
$$x = 4 + 1 \quad \text{Substituting}$$
$$x = 5$$

We check the ordered pair $(5, 4)$.

$x + y = 9$			$x = y + 1$		
$5 + 4$? 9			5 ? $4 + 1$		
9	9	TRUE	5	5	TRUE

Since $(5, 4)$ checks in both equations, it is the solution.

2. $(2, 5)$

3. $y = x - 3,$ (1)

$3x + y = 5$ (2)

Substitute $x - 3$ for y in Equation (2) and solve for x.

$$3x + y = 5 \quad (1)$$
$$3x + (x - 3) = 5 \quad \text{Substituting}$$
$$4x - 3 = 5$$
$$4x = 8$$
$$x = 2$$

Next we substitute 2 for x in either equation of the original system and solve for y.

$$y = x + 3 \quad (1)$$
$$y = 2 - 3 \quad \text{Substituting}$$
$$y = -1$$

We check the ordered pair $(2, -1)$.

$y = x - 3$			$3x + y = 5$		
-1 ? $2 - 3$			$3 \cdot 2 - 1$? 5		
-1	-1	TRUE	$6 - 1$		
			5	5	TRUE

Since $(2, -1)$ checks in both equations, it is the solution.

4. $(2, 1)$

5. $y = 2x + 1,$ (1)

$x + y = 4$ (2)

Substitute $2x + 1$ for y in Equation (2) and solve for x.

$$x + y = 4 \quad (2)$$
$$x + (2x + 1) = 4 \quad \text{Substituting}$$
$$3x + 1 = 4$$
$$3x = 3$$
$$x = 1$$

Next we substitute 1 for x in either equation of the original system and solve for y.

$$y = 2x + 1 \quad (1)$$
$$y = 2 \cdot 1 + 1 \quad \text{Substituting}$$
$$y = 2 + 1$$
$$y = 3$$

We check the ordered pair $(1, 3)$.

$y = 2x + 1$			$x + y = 4$		
3 ? $2 \cdot 1 + 1$			$1 + 3$? 4		
	$2 + 1$		4	4	TRUE
3	3	TRUE			

Since $(1, 3)$ checks in both equations, it is the solution.

6. $(4, 3)$

7. $r = -3s,$ (1)

$r + 4s = 10$ (2)

Substitute $-3s$ for r in Equation (2) and solve for s.

$$r + 4s = 10 \quad (2)$$
$$-3s + 4s = 10 \quad \text{Substituting}$$
$$s = 10$$

Next we substitute 10 for s in either equation of the original system and solve for r.

$$r = -3s \quad (1)$$
$$r = -3 \cdot 10 \quad \text{Substituting}$$
$$r = -30$$

We check the ordered pair $(-30, 10)$.

$r = -3s$			$r + 4s = 10$		
-30 ? $-3 \cdot 10$			$-30 + 4 \cdot 10$? 10		
-30	-30	TRUE	$-30 + 40$		
			10	10	TRUE

Since $(-30, 10)$ checks in both equations, it is the solution.

8. $(2, -4)$

9.
$$x = y - 8, \quad (1)$$
$$3x + 2y = 1 \qquad (2)$$

Substitute $y - 8$ for x in Equation (2) and solve for y.
$$3x + 2y = 1 \qquad (2)$$
$$3(y - 8) + 2y = 1 \qquad \text{Substituting}$$
$$3y - 24 + 2y = 1$$
$$5y = 25$$
$$y = 5$$

Next we substitute 5 for y in either equation of the original system and solve for x.
$$x = y - 8 \quad (1)$$
$$x = 5 - 8 \quad \text{Substituting}$$
$$x = -3$$

We check the ordered pair $(-3, 5)$.

$x = y - 8$		$3x + 2y = 1$	
-3 ? $5 - 8$		$3(-3) + 2 \cdot 5$? 1	
-3	-3 TRUE	$-9 + 10$	
		1	1 TRUE

Since $(-3, 5)$ checks in both equations, it is the solution.

10. $(-2, 4)$

11.
$$y = 3x - 1, \quad (1)$$
$$6x - 2y = 2 \qquad (2)$$

Substitute $3x - 1$ for y in Equation (2) and solve for x.
$$6x - 2y = 2 \quad (2)$$
$$6x - 2(3x - 1) = 2$$
$$6x - 6x + 2 = 2$$
$$2 = 2$$

The last equation is true for any choice of x, so there is an infinite number of solutions.

12. No solution

13.
$$x - y = 6, \quad (1)$$
$$x + y = -2 \quad (2)$$

We solve Equation (1) for x.
$$x - y = 6 \qquad (1)$$
$$x = y + 6 \quad (3)$$

We substitute $y + 6$ for x in Equation (2) and solve for y.
$$x + y = -2 \quad (2)$$
$$(y + 6) + y = -2 \quad \text{Substituting}$$
$$2y + 6 = -2$$
$$2y = -8$$
$$y = -4$$

Now we substitute -4 for y in either of the original equations or in Equation (3) and solve for x. It is easiest to use (3).
$$x = y + 6 = -4 + 6 = 2$$

We check the ordered pair $(2, -4)$.

$x - y = 6$		$x + y = -2$	
$2 - (-4)$? 6		$2 + (-4)$? -2	
$2 + 4$			-2 -2 TRUE
6	6 TRUE		

Since $(2, -4)$ checks in both equations, it is the solution.

14. $(-1, -3)$

15.
$$y - 2x = -6, \quad (1)$$
$$2y - x = 5 \qquad (2)$$

We solve Equation (1) for y.
$$y - 2x = -6, \qquad (1)$$
$$y = 2x - 6 \quad (3)$$

We substitute $2x - 6$ for y in Equation (2) and solve for x.
$$2y - x = 5 \qquad (2)$$
$$2(2x - 6) - x = 5 \quad \text{Substituting}$$
$$4x - 12 - x = 5$$
$$3x - 12 = 5$$
$$3x = 17$$
$$x = \frac{17}{3}$$

Now we substitute $\dfrac{17}{3}$ for x in (3) in either of the original equations or in Equation (3) and solve for x. It is easiest to use (3).
$$y = 2x - 6 = 2\left(\frac{17}{3}\right) - 6 = \frac{34}{3} - \frac{18}{3} = \frac{16}{3}$$

The ordered pair $\left(\dfrac{17}{3}, \dfrac{16}{3}\right)$ checks in both equations. It is the solution.

16. $\left(\dfrac{17}{3}, \dfrac{2}{3}\right)$

17. $x - 4y = 3,$ (1)

     $2x - 6 = 8y$ (2)

We solve Equation (1) for x.

$$x - 4y = 3$$
$$x = 4y + 3$$

We substitute $4y + 3$ for x in Equation (2) and solve for y.

$$2x - 6 = 8y \quad (2)$$
$$2(4y + 3) - 6 = 8y \quad \text{Substituting}$$
$$8y + 6 - 6 = 8y$$
$$8y = 8y$$

The last equation is true for any choice of y, so there is an infinite number of solutions.

18. Infinite number of solutions

19. $y = -2x + 3,$ (1)

     $3y = -6x + 9$ (2)

Substitute $-2x + 3$ for y in Equation (2) and solve for x.

$$3y = -6x + 9 \quad (2)$$
$$3(-2x + 3) = -6x + 9 \quad \text{Substituting}$$
$$-6x + 9 = -6x + 9$$

The last equation is true for any choice of x, so there is an infinite number of solutions.

20. Infinite number of solutions

21. $x + 2y = 10,$ (1)

     $3x + 4y = 8$ (2)

Solve Equation (1) for x.

$$x + 2y = 10 \quad (1)$$
$$x = -2y + 10 \quad (3)$$

Substitute $-2y + 10$ for x in Equation (2) and solve for y.

$$3x + 4y = 8 \quad (2)$$
$$3(-2y + 10) + 4y = 8 \quad \text{Substituting}$$
$$-6y + 30 + 4y = 8$$
$$-2y + 30 = 8$$
$$-2y = -22$$
$$y = 11$$

Substitute 11 for y in Equation (3) and compute x.

$$x = -2y + 10 = -2 \cdot 11 + 10 = -22 + 10 = -12$$

The ordered pair $(-12, 11)$ checks in both equations. It is the solution.

22. $\left(\dfrac{25}{8}, -\dfrac{11}{4} \right)$

23. $3a + 2b = 2,$ (1)

     $-2a + b = 8$ (2)

Solve Equation (2) for b.

$$-2a + b = 8 \quad (2)$$
$$b = 2a + 8 \quad (3)$$

Substitute $2a + 8$ for b in Equation (1) and solve for a.

$$3a + 2b = 2 \quad (1)$$
$$3a + 2(2a + 8) = 2$$
$$3a + 4a + 16 = 2$$
$$7a + 16 = 2$$
$$7a = -14$$
$$a = -2$$

Substitute -2 for a in Equation (3) and compute b.

$$b = 2(-2) + 8 = -4 + 8 = 4$$

The ordered pair $(-2, 4)$ checks in both equations. It is the solution.

24. $(-3, 0)$

25. $y - 2x = 0,$ (1)

     $3x + 7y = 17$ (2)

Solve Equation (1) for y.

$$y - 2x = 0 \quad (1)$$
$$y = 2x \quad (3)$$

Substitute $2x$ for y in Equation (2) and solve for x.

$$3x + 7y = 17 \quad (2)$$
$$3x + 7(2x) = 17 \quad \text{Substituting}$$
$$3x + 14x = 17$$
$$17x = 17$$
$$x = 1$$

Substitute 1 for x in Equation (3) and compute y.

$$y = 2x = 2 \cdot 1 = 2$$

The ordered pair $(1, 2)$ checks in both equations. It is the solution.

26. $(6, 3)$

27. $8x + 2y = 6,$ (1)

     $y = 3 - 4x$ (2)

Substitute $3 - 4x$ for y in Equation (1) and solve for x.

$$8x + 2y = 6 \quad (1)$$
$$8x + 2(3 - 4x) = 6 \quad \text{Substituting}$$
$$8x + 6 - 8x = 6$$
$$6 = 6$$

The last equation is true for any choice of x, so there is an infinite number of solutions.

28. No solution

29. $x - 3y = -1$,   (1)

  $5y - 2x = 4$ (2)

Solve Equation (1) for x.

$$x - 3y = -1 \qquad (1)$$
$$x = 3y - 1 \qquad (3)$$

Substitute $3y - 1$ for x in Equation (2) and solve for y.

$$5y - 2x = 4 \qquad (2)$$
$$5y - 2(3y - 1) = 4 \qquad \text{Substituting}$$
$$5y - 6y + 2 = 4$$
$$-y + 2 = 4$$
$$-y = 2$$
$$y = -2$$

Now substitute -2 for y in Equation (3) and compute x.

$$x = 3(-2) - 1 = -6 - 1 = -7$$

The ordered pair $(-7, -2)$ checks in both equations. It is the solution.

30. $(-3, -4)$

31. $2x - y = 0$,   (1)

  $2x - y = -2$ (2)

Solve Equation (1) for y.

$$2x - y = 0 \quad (1)$$
$$2x = y \quad (3)$$

Substitute $2x$ for y in Equation (2) and solve for x.

$$2x - y = -2 \quad (2)$$
$$2x - 2x = -2 \quad \text{Substituting}$$
$$0 = -2$$

We obtain a false equation, so the system has no solution.

32. No solution

33. *Familiarize.* We let x = the larger number and y = the smaller number.

Translate.

The sum of two numbers is 87.

$$\downarrow \qquad\qquad\qquad \downarrow\ \downarrow$$
$$x + y \qquad\qquad\quad = 87$$

One number is 3 more than the other.

$$\downarrow \qquad \downarrow\ \downarrow \qquad \downarrow \qquad\quad \downarrow$$
$$x \qquad = 3 \qquad + \qquad\quad y$$

The resulting system is

$$x + y = 87, \quad (1)$$
$$x = 3 + y. \quad (2)$$

Carry out. We solve the system of equations. We substitute $3 + y$ for x in Equation (1) and solve for y.

$$x + y = 87 \quad (1)$$
$$(3 + y) + y = 87 \quad \text{Substituting}$$
$$3 + 2y = 87$$
$$2y = 84$$
$$y = 42$$

Next we substitute 42 for y in either equation of the original system and solve for x.

$$x + y = 87 \quad (1)$$
$$x + 42 = 87 \quad \text{Substituting}$$
$$x = 45$$

Check. The sum of 42 and 45 is 87. The number 45 is 3 more than 42. These numbers check.

State. The numbers are 45 and 42.

34. 37, 39

35. *Familiarize.* Let x = the larger number and y = the smaller number.

Translate.

The sum of two numbers is 58.

$$\downarrow \qquad\qquad\qquad \downarrow\ \downarrow$$
$$x + y \qquad\qquad\quad = 58$$

The difference of two numbers is 14.

$$\downarrow \qquad\qquad\qquad\quad \downarrow\ \downarrow$$
$$x - y \qquad\qquad\qquad = 14$$

The resulting system is

$$x + y = 58, \quad (1)$$
$$x - y = 14. \quad (2)$$

Carry out. We solve the system.

We solve Equation (2) for x.

$$x - y = 14 \qquad (2)$$
$$x = y + 14 \quad (3)$$

We substitute $y + 14$ for x in Equation (1) and solve for y.

$$x + y = 58 \quad (1)$$
$$(y + 14) + y = 58 \quad \text{Substituting}$$
$$2y + 14 = 58$$
$$2y = 44$$
$$y = 22$$

Now we substitute 22 for y in Equation (3) and compute x.

$$x = y + 14 = 22 + 14 = 36$$

Check. The sum of 36 and 22 is 58. The difference between 36 and 22, $36 - 22$, is 14. The numbers check.

State. The numbers are 36 and 22.

36. 32, 44

37. Familiarize. Let $x =$ the larger number and $y =$ the smaller number.

Translate.

The difference between two numbers is 16.

$$x - y = 16$$

Three times the larger number is seven times the smaller number.

$$3x = 7y$$

The resulting system is

$$x - y = 16, \quad (1)$$
$$3x = 7y. \quad (2)$$

Carry out. We solve the system.

We solve Equation (1) for x.

$$x - y = 16 \quad (1)$$
$$x = y + 16 \quad (3)$$

We substitute $y + 16$ for x in Equation (2) and solve for y.

$$3x = 7y \quad (2)$$
$$3(y + 16) = 7y \quad \text{Substituting}$$
$$3y + 48 = 7y$$
$$48 = 4y$$
$$12 = y$$

Next we substitute 12 for y in Equation (3) and compute x.

$$x = y + 16 = 12 + 16 = 28$$

Check. The difference between 28 and 12, $28 - 12$, is 16. Three times the larger, $3 \cdot 28$ or 84, is seven times the smaller, $7 \cdot 12 = 84$. The numbers check.

State. The numbers are 28 and 12.

38. 4, 22

39. Familiarize. Let $x =$ one angle and $y =$ the other angle.

Translate. Since the angles are supplementary, we have one equation.

$$x + y = 180$$

The second sentence can be translated as follows:

One angle is 60° less than twice the other.

$$x = 2y - 60$$

The resulting system is

$$x + y = 180, \quad (1)$$
$$x = 2y - 60. \quad (2)$$

Carry out. We solve the system.

We substitute $2y - 60$ for x in Equation (1) and solve for y.

$$x + y = 180 \quad (1)$$
$$2y - 60 + y = 180$$
$$3y - 60 = 180$$
$$3y = 240$$
$$y = 80$$

Next we substitute 80 for y in Equation (2) and find x.

$$x = 2y - 60 = 2 \cdot 80 - 60 = 160 - 60 = 100$$

Check. The sum of the angles is $80° + 100°$, or 180°, so the angles are supplementary. If 60° is subtracted from twice 80°, we have $2 \cdot 80° - 60°$, or 100°, which is the other angle. The answer checks.

State. One angle is 80°, and the other is 100°.

40. 47°, 133°

41. Familiarize. We let $x =$ the larger angle and $y =$ the smaller angle.

Translate. Since the angles are complementary, we have one equation.

$$x + y = 90$$

We reword and translate the second statement.

The difference of two angles is 42°.

$$x - y = 42$$

The resulting system is

$$x + y = 90, \quad (1)$$
$$x - y = 42. \quad (2)$$

Carry out. We solve the system.

We first solve Equation (2) for x.

$$x - y = 42 \quad (2)$$
$$x = y + 42 \quad (3)$$

Substitute $y + 42$ for x in Equation (1) and solve for y.

$$x + y = 90 \quad (1)$$
$$y + 42 + y = 90$$
$$2y + 42 = 90$$
$$2y = 48$$
$$y = 24$$

Next we substitute 24 for y in Equation (3) and solve for x.

$$x = y + 42 = 24 + 42 = 66$$

Check. The sum of the angles is $66° + 24°$, or $90°$, so the angles are complementary. The difference of the angles is $66° - 24°$, or $42°$. These numbers check.

State. The angles are $66°$ and $24°$.

42. $32°$, $58°$

43. Familiarize. Recall that the perimeter of a rectangle with length l and width w is given by $2l + 2w$.

Translate.

The perimeter is 1300 mi.
$$2l + 2w = 1300$$

The length is 110 mi more than the width.
$$l = w + 110$$

The resulting system is

$$2l + 2w = 1300, \quad (1)$$
$$l = w + 110. \quad (2)$$

Carry out. We solve the system.

Substitute $w + 110$ for l in Equation (1) and solve for w.

$$2l + 2w = 1300 \quad (1)$$
$$2(w + 110) + 2w = 1300 \quad \text{Substituting}$$
$$2w + 220 + 2w = 1300$$
$$4w + 220 = 1300$$
$$4w = 1080$$
$$w = 270$$

Now substitute 270 for w in Equation (2).

$$l = w + 110 \quad (2)$$
$$l = 270 + 110 \quad \text{Substituting}$$
$$l = 380$$

Check. If the length is 380 mi and the width is 270 mi, the perimeter would be $2 \cdot 380 + 2 \cdot 270$, or $760 + 540$, or 1300 mi. Also, the length is 110 mi more than the width. These numbers check.

State. The length is 380 mi, and the width is 270 mi.

44. Length: 365 mi; width: 275 mi

45. Familiarize. Recall that the perimeter of a rectangle with length l and width w is given by $2l + 2w$.

Translate.

The perimeter is 120 ft.
$$2l + 2w = 120$$

The length is twice the width.
$$l = 2w$$

The resulting system is

$$2l + 2w = 120, \quad (1)$$
$$l = 2w. \quad (2)$$

Carry out. We solve the system.

Substitute $2w$ for l in Equation (1) and solve for w.

$$2 \cdot 2w + 2w = 120 \quad (1)$$
$$4w + 2w = 120$$
$$6w = 120$$
$$w = 20$$

Now substitute 20 for w in Equation (2).

$$l = 2w \quad (2)$$
$$l = 2 \cdot 20 \quad \text{Substituting}$$
$$l = 40$$

Check. If the length is 40 ft and the width is 20 ft, the perimeter would be $2 \cdot 40 + 2 \cdot 20$, or $80 + 40$, or 120 ft. Also, the length is twice the width. These numbers check.

State. The length is 40 ft, and the width is 20 ft.

46. Height: 20 ft; width: 5 ft

47. Familiarize. Let $l =$ the length and $w =$ the width, in yards. The perimeter is $l + l + w + w$, or $2l + 2w$.

Translate.

The perimeter is 340 yd.
$$2l + 2w = 340$$

The length is 10 yd less than twice the width.
$$l = 2w - 10$$

The resulting system is

$$2l + 2w = 340, \quad (1)$$
$$l = 2w - 10. \quad (2)$$

Carry out. We solve the system. We substitute $2w - 10$ for l in Equation (1) and solve for w.

$$2l + 2w = 340 \quad (1)$$
$$2(w - 10) + 2w = 340$$
$$4w - 20 + 2w = 340$$
$$6w - 20 = 340$$
$$6w = 360$$
$$w = 60$$

Next we substitute 60 for w in Equation (2) and solve for l.

$$l = 2w - 10 = 2 \cdot 60 - 10 = 120 - 10 = 110$$

Check. The perimeter is $2 \cdot 110 + 2 \cdot 60$, or 340 yd. Also 10 yd less than twice the width is $2 \cdot 60 - 10 = 120 - 10 = 110$. The answer checks.

State. The length is 110 yd, and the width is 60 yd.

48. Length: 90 yd; width: 50 yd

49. ◈

50. ◈

51.
$$2(5x + 3y) - 3(5x + 3y)$$
$$= 10x + 6y - 15x - 9y$$
$$= -5x - 3y$$

We could also simplify this expression as follows:
$$2(5x + 3y) - 3(5x + 3y)$$
$$= -1(5x + 3y)$$
$$= -5x - 3y$$

52. $-15y$

53.
$$3(8x + 2y) - 2(7x + 3y)$$
$$= 24x + 6y - 14x - 6y$$
$$= 10x$$

54. $-6y - 25$

55.
$$2(5x - 3y) - 5(2x + y)$$
$$= 10x - 6y - 10x - 5y$$
$$= -11y$$

56. $23x$

57. ◈

58. ◈

59. $\dfrac{1}{6}(a + b) = 1, \quad (1)$

$\dfrac{1}{4}(a - b) = 2 \quad (2)$

Observe that $\dfrac{1}{6}(a + b) = 1$, so $a + b = 6$. Also, $\dfrac{1}{4}(a - b) = 2$, so $a - b = 8$. We need to find two numbers whose sum is 6 and whose difference is 8. The numbers are 7 and -1, so the solution of the system of equations is $(7, -1)$.

We could also solve this system of equations using the substitution method. We first clear the fractions.

$$a + b = 6 \quad (1a)$$
$$a - b = 8 \quad (2a)$$

We solve Equation (2a) for a.

$$a - b = 8 \quad (2a)$$
$$a = b + 8$$

We substitute $b + 8$ for a in Equation (1a) and solve for b.

$$(b + 8) + b = 6$$
$$2b + 8 = 6$$
$$2b = -2$$
$$b = -1$$

Next we substitute -1 for b in Equation (2a) and solve for a.

$$a - b = 8$$
$$a - (-1) = 8$$
$$a + 1 = 8$$
$$a = 7$$

Since $(7, -1)$ checks in both equations, it is the solution.

60. $(10, -2)$

61. Graph the equations and use the INTERSECT feature from the CALC menu to find the coordinates of the point of intersection. (It might be necessary to solve each equation for y before entering them on a grapher.) The solution is approximately $(4.38, 4.33)$.

62. $(5.36, 4.7)$

63. $x + y + z = 4, \quad (1)$
$\quad\quad x - 2y - z = 1, \quad (2)$
$\quad\quad y = -1 \quad\quad\quad (3)$

Substitute -1 for y in Equations (1) and (2).

$$x + y + z = 4 \quad (1) \quad\quad x - 2y - z = 1 \quad (2)$$
$$x + (-1) + z = 4 \quad\quad x - 2(-1) - z = 1$$
$$x + z = 5 \quad\quad\quad x + 2 - z = 1$$
$$\quad\quad\quad\quad\quad\quad\quad\quad x - z = -1$$

We now have a system of two equations in two variables.

$$x + z = 5, \quad (4)$$
$$x - z = -1 \quad (5)$$

We solve Equation (5) for x.

$$x - z = -1 \quad (5)$$
$$x = z - 1 \quad (6)$$

We substitute $z - 1$ for x in Equation (4) and solve for z.

$$x + z = 5 \quad (4)$$
$$(z - 1) + z = 5 \quad \text{Substituting}$$
$$2z - 1 = 5$$
$$2z = 6$$
$$z = 3$$

Next we substitute 3 for z in Equation (6) and compute x.

$$x = z - 1 = 3 - 1 = 2$$

We check the ordered triple $(2, -1, 3)$.

$x + y + z = 4$	$x - 2y - z = 1$
$2 + (-1) + 3 \ ? \ 4$	$2 - 2(-1) - 3 \ ? \ 1$
$4 \ \mid \ 4$ TRUE	$2 + 2 - 3$
	$1 \ \mid \ 1$ TRUE

$y = -1$
$-1 \ ? \ -1$ TRUE

Since $(2, -1, 3)$ checks in all three equations, it is the solution.

64. $(30, 50, 100)$

65. *Familiarize.* Let $s = $ the perimeter of a softball diamond, in yards, and $b = $ the perimeter of a baseball diamond, in yards.

Translate.

$$\underbrace{\text{Perimeter of a}}_{s} \text{ is } \underbrace{\frac{2}{3}}_{=\frac{2}{3}} \text{ of } \underbrace{\text{perimeter of a}}_{b}$$
softball diamond   baseball diamond.

$$\underbrace{\text{The sum of the perimeters}}_{s + b} \text{ is } \underbrace{200 \text{ yd.}}_{= \quad 200}$$

The resulting system is

$$s = \frac{2}{3}b, \quad (1)$$
$$s + b = 200. \quad (2)$$

Carry out. We solve the system of equations. We substitute $\frac{2}{3}b$ for s in Equation (2) and solve for b.

$$s + b = 200$$
$$\frac{2}{3}b + b = 200$$
$$\frac{5}{3}b = 200$$
$$\frac{3}{5} \cdot \frac{5}{3}b = \frac{3}{5} \cdot 200$$
$$b = 120$$

Next we substitute 120 for b in Equation (1) and solve for s.

$$s = \frac{2}{3}b = \frac{2}{3} \cdot 120 = 80$$

Each diamond has four sides of equal length, so we divide each perimeter by 4 to find the distance between bases in each sport. For the softball diamond the distance is 80/4, or 20 yd. For the baseball diamond it is 120/4, or 30 yd.

Check. The perimeter of the softball diamond, 80 yd, is $\frac{2}{3}$ of 120 yd, the perimeter of the baseball diamond. The sum of the perimeters is $80 + 120$, or 200 yd. We can also recheck the calculations of the distances between the bases. The answer checks.

State. The distance between bases on a softball diamond is 20 yd and the distance between bases on a baseball diamond is 30 yd.

66. ◈

67. Answers may vary.

$$2x + 3y = 5,$$
$$5x + 4y = 2$$

Exercise Set 7.3

1. $x + y = 3 \quad (1)$
$$\underline{x - y = 7 \quad (2)}$$
$$2x \quad = 10 \quad \text{Adding}$$
$$x = 5$$

Substitute 5 for x in one of the original equations and solve for y.

$$x + y = 3 \quad (1)$$
$$5 + y = 3 \quad \text{Substituting}$$
$$y = -2$$

Check:

$x + y = 3$	$x - y = 7$
$5 + (-2) \ ? \ 3$	$5 - (-2) \ ? \ 7$
$3 \ \mid \ 3$ TRUE	$7 \ \mid \ 7$ TRUE

Since $(5, -2)$ checks, it is the solution.

2. $(9, 3)$

3.
$$x + y = 6 \quad (1)$$
$$\underline{-x + 2y = 15} \quad (2)$$
$$3y = 21 \quad \text{Adding}$$
$$y = 7$$

Substitute 7 for y in one of the original equations and solve for x.
$$x + y = 6 \quad (1)$$
$$x + 7 = 6 \quad \text{Substituting}$$
$$x = -1$$

Check:
$$\frac{x + y = 6}{-1 + 7 \ ? \ 6}$$
$$6 \ | \ 6 \quad \text{TRUE}$$

$$\frac{-x + 2y = 15}{-(-1) + 2 \cdot 7 \ ? \ 15}$$
$$1 + 14 \ |$$
$$15 \ | \ 15 \quad \text{TRUE}$$

Since $(-1, 7)$ checks, it is the solution.

4. $(5,1)$

5.
$$3x - y = 9 \quad (1)$$
$$\underline{2x + y = 6} \quad (2)$$
$$5x = 15 \quad \text{Adding}$$
$$x = 3$$

Substitute 3 for x in one of the original equations and solve for y.
$$2x + y = 6 \quad (2)$$
$$2 \cdot 3 + y = 6 \quad \text{Substituting}$$
$$6 + y = 6$$
$$y = 0$$

Check:
$$\frac{3x - y = 9}{3 \cdot 3 - 0 \ ? \ 9} \qquad \frac{2x + y = 6}{2 \cdot 3 + 0 \ ? \ 6}$$
$$9 - 0 \ | \qquad\qquad 6 + 0 \ |$$
$$9 \ | \ 9 \ \text{TRUE} \qquad 6 \ | \ 6 \ \text{TRUE}$$

Since $(3, 0)$ checks, it is the solution.

6. $(2,7)$

7.
$$5a + 4b = 7 \quad (1)$$
$$\underline{-5a + b = 8} \quad (2)$$
$$5b = 15$$
$$b = 3$$

Substitute 3 for b in one of the original equations and solve for a.
$$5a + 4b = 7 \quad (1)$$
$$5a + 4 \cdot 3 = 7$$
$$5a + 12 = 7$$
$$5a = -5$$
$$a = -1$$

Check:
$$\frac{5a + 4b = 7}{5(-1) + 4 \cdot 3 \ ? \ 7}$$
$$-5 + 12 \ |$$
$$7 \ | \ 7 \quad \text{TRUE}$$

$$\frac{-5a + b = 8}{-5(-1) + 3 \ ? \ 8}$$
$$5 + 3 \ |$$
$$8 \ | \ 8 \quad \text{TRUE}$$

Since $(-1, 3)$ checks, it is the solution.

8. $\left(\dfrac{3}{2}, \dfrac{11}{8}\right)$

9.
$$8x - 5y = -9 \quad (1)$$
$$\underline{3x + 5y = -2} \quad (2)$$
$$11x = -11 \quad \text{Adding}$$
$$x = -1$$

Substitute -1 for x in either of the original equations and solve for y.
$$3x + 5y = -2 \quad \text{Equation (2)}$$
$$3(-1) + 5y = -2 \quad \text{Substituting}$$
$$-3 + 5y = -2$$
$$5y = 1$$
$$y = \frac{1}{5}$$

Check:
$$\frac{8x - 5y = -9}{8(-1) - 5\left(\frac{1}{5}\right) \ ? \ -9}$$
$$-8 - 1 \ |$$
$$-9 \ | \ -9 \quad \text{TRUE}$$

$$3x + 5y = -2$$

$$3(-1) + 5\left(\frac{1}{5}\right) \ ? \ -2$$
$$-3 + 1$$
$$-2 \ \big| \ -2 \quad \text{TRUE}$$

Since $\left(-1, \frac{1}{5}\right)$ checks, it is the solution.

10. $(-2, 3)$

11. $3a - 6b = 8,$
$$\underline{-3a + 6b = -8}$$
$$0 = 0 \quad \text{Adding}$$

The equation $0 = 0$ is always true, so the system has an infinite number of solutions.

12. Infinite number of solutions

13. $-x - y = 8, \quad (1)$
$$2x - y = -1 \quad (2)$$

We multiply by -1 on both sides of Equation (1) and then add.
$$x + y = -8 \quad \text{Multiplying by } -1$$
$$\underline{2x - y = -1}$$
$$3x \quad\ \ = -9 \quad \text{Adding}$$
$$x = -3$$

Substitute -3 for x in one of the original equations and solve for y.
$$2x - y = -1 \quad (2)$$
$$2(-3) - y = -1 \quad \text{Substituting}$$
$$-6 - y = -1$$
$$-y = 5$$
$$y = -5$$

Check:

$$-x - y = 8$$
$$-(-3) - (-5) \ ? \ 8$$
$$3 + 5$$
$$8 \ \big| \ 8 \quad \text{TRUE}$$

$$2x - y = -1$$
$$2(-3) - (-5) \ ? \ -1$$
$$-6 + 5$$
$$-1 \ \big| \ -1 \quad \text{TRUE}$$

Since $(-3, -5)$ checks, it is the solution.

14. $(-1, -6)$

15. $x + 3y = 19,$
$$x - y = -1$$

We multiply by -1 on both sides of Equation (2) and then add.
$$x + 3y = 19$$
$$\underline{-x + y = 1} \quad \text{Multiplying by } -1$$
$$4y = 20 \quad \text{Adding}$$
$$y = 5$$

Substitute 5 for y in one of the original equations and solve for x.
$$x - y = -1 \quad (2)$$
$$x - 5 = -1 \quad \text{Substituting}$$
$$x = 4$$

Check:

$$x + 3y = 19$$
$$4 + 3 \cdot 5 \ ? \ 19$$
$$4 + 15$$
$$19 \ \big| \ 19 \quad \text{TRUE}$$

$$x - y = -1$$
$$4 - 5 \ ? \ -1$$
$$-1 \ \big| \ -1 \quad \text{TRUE}$$

Since $(4, 5)$ checks, it is the solution.

16. $(3, 1)$

17. $x + y = 5, \quad (1)$
$$4x - 3y = 13 \quad (2)$$

We multiply by 3 on both sides of Equation (1) and then add.
$$3x + 3y = 15 \quad \text{Multiplying by } 3$$
$$\underline{4x - 3y = 13}$$
$$7x \quad\ \ = 28$$
$$x = 4$$

Substitute 4 for x in one of the original equations and solve for y.
$$x + y = 5 \quad (1)$$
$$4 + y = 5 \quad \text{Substituting}$$
$$y = 1$$

Check:

$$x + y = 5$$
$$4 + 1 \ ? \ 5$$
$$5 \ \big| \ 5 \quad \text{TRUE}$$

$$4x - 3y = 13$$
$$4 \cdot 4 - 3 \cdot 1 \ ? \ 13$$
$$16 - 3$$
$$13 \ \big| \ 13 \quad \text{TRUE}$$

Since $(4, 1)$ checks, it is the solution.

18. $(10, 3)$

19. $2w - 3z = -1,$ (1)
$3w + 4z = 24$ (2)

We use the multiplication principle with both equations and then add.

$8w - 12z = -4$ Multiplying (1) by 4
$\underline{9w + 12z = 72}$ Multiplying (2) by 3
$17w \quad\quad\; = 68$ Adding
$\quad\quad\; w = 4$

Substitute 4 for w in one of the original equations and solve for z.

$3w + 4z = 24$ Equation (2)
$3 \cdot 4 + 4z = 24$ Substituting
$12 + 4z = 24$
$4z = 12$
$z = 3$

Check:

$$\begin{array}{c|c}
\multicolumn{2}{c}{2w - 3z = -1} \\
\hline
2 \cdot 4 - 3 \cdot 3 \ ? \ -1 & \\
8 - 9 & \\
\hline
-1 & -1 \quad \text{TRUE}
\end{array}$$

$$\begin{array}{c|c}
\multicolumn{2}{c}{3w + 4z = 24} \\
\hline
3 \cdot 4 + 4 \cdot 3 \ ? \ 24 & \\
12 + 12 & \\
\hline
24 & 24 \quad \text{TRUE}
\end{array}$$

Since $(4, 3)$ checks, it is the solution.

20. $(1, -1)$

21. $2a + 3b = -1,$ (1)
$3a + 5b = -2$ (2)

We use the multiplication principle with both equations and then add.

$-10a - 15b = 5$ Multiplying (1) by -5
$\underline{9a + 15b = -6}$ Multiplying (2) by 3
$-a \quad\quad\; = -1$ Adding
$\quad\; a = 1$

Substitute 1 for a in one of the original equations and solve for b.

$2a + 3b = -1$ Equation (1)
$2 \cdot 1 + 3b = -1$ Substituting
$2 + 3b = -1$
$3b = -3$
$b = -1$

Check:

$$\begin{array}{c|c}
\multicolumn{2}{c}{2a + 3b = -1} \\
\hline
2 \cdot 1 + 3(-1) \ ? \ -1 & \\
2 - 3 & \\
\hline
-1 & -1 \quad \text{TRUE}
\end{array}$$

$$\begin{array}{c|c}
\multicolumn{2}{c}{3a + 5b = -2} \\
\hline
3 \cdot 1 + 5(-1) \ ? \ -2 & \\
3 - 5 & \\
\hline
-2 & -2 \quad \text{TRUE}
\end{array}$$

Since $(1, -1)$ checks, it is the solution.

22. $(4, -1)$

23. $3y = x,$ (1)
$5x + 14 = y$ (2)

We first get each equation in the form $Ax + By = C$.

$x - 3y = 0,$ (1a)   Adding $-3y$
$5x - y = -14$ (2a)   Adding $-y - 14$

We multiply by -5 on both sides of Equation (1a) and add.

$-5x + 15y = 0$ Multiplying by -5
$\underline{5x - y = -14}$
$14y = -14$ Adding
$y = -1$

Substitute -1 for y in Equation (1a) and solve for x.

$x - 3y = 0$
$x - 3(-1) = 0$ Substituting
$x + 3 = 0$
$x = -3$

Check:

$$\begin{array}{c|c}
\multicolumn{2}{c}{x = 3y} \\
\hline
-3 \ ? \ 3(-1) & \\
\hline
-3 & -3 \quad\quad \text{TRUE}
\end{array}$$

$$\begin{array}{c|c}
\multicolumn{2}{c}{5x + 14 = y} \\
\hline
5(-3) + 14 \ ? \ -1 & \\
-15 + 14 & \\
\hline
-1 & -1 \quad \text{TRUE}
\end{array}$$

Since $(-3, -1)$ checks, it is the solution.

24. $(2, 5)$

25. $4x - 10y = 13,$ (1)
$-2x + 5y = 8$ (2)

We multiply by 2 on both sides of Equation (2) and then add.

$$4x - 10y = 13$$
$$\underline{-4x + 10y = 16} \quad \text{Multiplying by 2}$$
$$0 = 29$$

The equation $0 = 29$ is false for any pair (x, y), so there is no solution.

26. $(2, 1)$

27. $8n + 6 - 3m = 0,$
$32 = m - n$

We first get each equation in the form $Am + Bn = C$.
$$-3m + 8n = -6, \quad (1) \quad \text{Subtracting 6}$$
$$m - n = 32 \quad (2)$$

We multiply by 3 on both sides of Equation (2) and add.
$$-3m + 8n = -6$$
$$\underline{3m - 3n = 96}$$
$$5n = 90$$
$$n = 18$$

Substitute 18 for n in Equation (2) and solve for m.
$$m - n = 32$$
$$m - 18 = 32$$
$$m = 50$$

Check:
$$8n + 6 - 3m = 0$$
$$\overline{8 \cdot 18 + 6 - 3 \cdot 50 \ ? \ 0}$$
$$144 + 6 - 150 \quad \bigg| $$
$$0 \ \bigg| \ 0 \quad \text{TRUE}$$

$$32 = m - n$$
$$\overline{32 \ ? \ 50 - 18}$$
$$32 \ \bigg| \ 32 \qquad \text{TRUE}$$

Since $(50, 18)$ checks, it is the solution.

28. $\left(\dfrac{1}{2}, 5\right)$

29. $3x + 5y = 4, \quad (1)$
$-2x + 3y = 10 \quad (2)$

We use the multiplication principle with both equations and then add.
$$6x + 10y = 8 \quad \text{Multiplying (1) by 2}$$
$$\underline{-6x + 9y = 30} \quad \text{Multiplying (2) by 3}$$
$$19y = 38 \quad \text{Adding}$$
$$y = 2$$

Substitute 2 for y in one of the original equations and solve for x.
$$3x + 5y = 4 \quad (1)$$
$$3x + 5 \cdot 2 = 4$$
$$3x + 10 = 4$$
$$3x = -6$$
$$x = -2$$

Check:
$$3x + 5y = 4$$
$$\overline{3(-2) + 5 \cdot 2 \ ? \ 4}$$
$$-6 + 10 \quad \bigg|$$
$$4 \ \bigg| \ 4 \quad \text{TRUE}$$

$$-2x + 3y = 10$$
$$\overline{-2(-2) + 3 \cdot 2 \ ? \ 10}$$
$$4 + 6 \quad \bigg|$$
$$10 \ \bigg| \ 10 \quad \text{TRUE}$$

Since $(-2, 2)$ checks, it is the solution.

30. No solution

31. $0.06x + 0.05y = 0.07,$
$0.04x - 0.03y = 0.11$

We first multiply each equation by 100 to clear the decimals.
$$6x + 5y = 7, \quad (1)$$
$$4x - 3y = 11 \quad (2)$$

We use the multiplication principle with both equations of the resulting system.
$$18x + 15y = 21 \quad \text{Multiplying (1) by 3}$$
$$\underline{20x - 15y = 55} \quad \text{Multiplying (2) by 5}$$
$$38x = 76 \quad \text{Adding}$$
$$x = 2$$

Substitute 2 for x in Equation (1) and solve for y.
$$6x + 5y = 7$$
$$6 \cdot 2 + 5y = 7$$
$$12 + 5y = 7$$
$$5y = -5$$
$$y = -1$$

Check:
$$0.06x + 0.05y = 0.07$$
$$\overline{0.06(2) + 0.05(-1) \ ? \ 0.07}$$
$$0.12 - 0.05 \quad \bigg|$$
$$0.07 \ \bigg| \ 0.07 \quad \text{TRUE}$$

$$0.04x - 0.03y = 0.11$$
$$\overline{0.04(2) - 0.03(-1) \ ? \ 0.11}$$
$$0.08 + 0.03 \quad \bigg|$$
$$0.11 \ \bigg| \ 0.11 \quad \text{TRUE}$$

Since $(2, -1)$ checks, it is the solution.

32. $(10, -2)$

33. $x + \dfrac{9}{2}y = \dfrac{15}{4},$

$\dfrac{9}{10}x - y = \dfrac{9}{20}$

First we clear fractions. We multiply both sides of the first equation by 4 and both sides of the second equation by 20.

$$4\left(x + \frac{9}{2}y\right) = 4 \cdot \frac{15}{4}$$

$$4x + 4 \cdot \frac{9}{2}y = 15$$

$$4x + 18 = 15$$

$$20\left(\frac{9}{10}x - y\right) = 20 \cdot \frac{9}{20}$$

$$20 \cdot \frac{9}{10}x - 20y = 9$$

$$18x - 20y = 9$$

The resulting system is

$$4x + 18y = 15, \quad (1)$$
$$18x - 20y = 9. \quad (2)$$

We use the multiplication principle with both equations.

$72x + 324y = 270$ Multiplying (1) by 18

$-72x + 80y = -36$ Multiplying (2) by -4

$$404y = 234$$

$$y = \frac{234}{404}, \text{ or } \frac{117}{202}$$

Substitute $\dfrac{117}{202}$ for y in (1) and solve for x.

$$4x + 18\left(\frac{117}{202}\right) = 15$$

$$4x + \frac{1053}{101} = 15$$

$$4x = \frac{462}{101}$$

$$x = \frac{1}{4} \cdot \frac{462}{101}$$

$$x = \frac{231}{202}$$

The ordered pair $\left(\dfrac{231}{202}, \dfrac{117}{202}\right)$ checks in both equations. It is the solution.

34. $\left(\dfrac{231}{202}, \dfrac{117}{202}\right)$

35. *Familiarize*. We let $m =$ the number of miles driven and $c =$ the total cost of the truck rental.

Translate. We reword and translate the first statement, using \$0.39 for 39¢.

$\$49$ plus 39¢ times the number of miles driven is cost.

$\downarrow \quad \downarrow \quad \downarrow \quad \downarrow \qquad \downarrow \qquad \downarrow \quad \downarrow$

$49 \ + \ 0.39 \ \cdot \qquad m \qquad = \ c$

We reword and translate the second statement using \$0.49 for 49¢.

$\$29.95$ plus 49¢ times the number of miles driven is cost.

$\downarrow \quad \downarrow \quad \downarrow \quad \downarrow \qquad \downarrow \qquad \downarrow \quad \downarrow$

$29.95 \ + \ 0.49 \ \cdot \qquad m \qquad = \ c$

We have a system of equations:

$$49 + 0.39m = c,$$
$$29.95 + 0.49m = c$$

Carry out. To solve the system of equations, we multiply the second equation by -1 and add to eliminate c.

$$49 + 0.39m = c$$
$$\underline{-29.95 - 0.49m = -c}$$
$$19.05 - \quad 0.1m = 0$$
$$19.05 = 0.1m$$
$$190.5 = m$$

Check. For 190.5 mi, the cost of the Budget truck is $49 + 0.39(190.5)$, or $49 + 74.295$, or about \$123.30. For 190.5 mi, the cost of the U-Haul rental is $29.95 + 0.49(190.5)$, or $29.95 + 93.345$, or about \$123.30. The cost is the same for 190.5 mi.

State. When the trucks are driven 190.5 mi, the cost is the same.

36. 211.7 mi

37. *Familiarize*. We let $x =$ the larger angle and $y =$ the smaller angle.

Translate. We reword and translate the first statement.

The sum of two angles is $90°$.

$\downarrow \qquad\qquad \downarrow \quad \downarrow$

$x + y \qquad = \ 90$

We reword and translate the second statement.

The larger angle is $12°$ more than twice the smaller angle.

$\downarrow \qquad \downarrow\ \downarrow \qquad \downarrow \qquad \downarrow \qquad \downarrow$

$x \qquad = 12 \ + \ 2\cdot \qquad y$

We have a system of equations:

$$x + y = 90,$$
$$x = 12 + 2y$$

Carry out. We solve the system. We will use the elimination method, although we could also easily use the substitution method. First we get the second equation in the form $Ax + By = C$.

$$x + y = 90 \quad (1)$$
$$x - 2y = 12 \quad (2) \quad \text{Adding } -2y$$

Now we multiply Equation (2) by 2 and add.

$$2x + 2y = 180$$
$$\underline{x - 2y = 12}$$
$$3x = 192$$
$$x = 64$$

Then we substitute 64 for x in Equation (1) and solve for y.

$$x + y = 90 \quad (1)$$
$$64 + y = 90 \quad \text{Substituting}$$
$$y = 26$$

Check. The sum of the angles is $64° + 26°$, or $90°$, so the angles are complementary. The larger angle, $64°$, is $12°$ more than twice the smaller angle, $26°$. These numbers check.

State. The angles are $64°$ and $26°$.

38. $32°$, $58°$

39. *Familiarize*. Let $m =$ the number of long distance minutes used in a month and $c =$ the cost of the calls, in cents.

Translate. We reword the problem and translate. We will express \$3.95 as 395¢.

The One-Plus' cost is \$3.95 plus 7¢ times the length of the call, in minutes.

$$c = 395 + 7 \cdot m$$

Other plan's cost is 9¢ times the length of the call, in minutes.

$$c = 9 \cdot m$$

We have a system of equations:

$$c = 395 + 7m,$$
$$c = 9m$$

Carry out. To solve the system, we multiply the second equation by -1 and add to eliminate c.

$$c = 395 + 7m$$
$$\underline{-c = - 9m}$$
$$0 = 395 - 2m$$
$$2m = 395$$
$$m = 197.75$$

Check. For 197.5 min, the cost of the One-Plus plan is $395 + 7(197.5)$, or $395 + 1382.5$, or 1777.5¢ and the cost of the other plan is $9(197.5)$, or 1777.5¢. Thus the costs are the same for 197.5 long distance minutes.

State. For 197.5 minutes of long distance calls, the monthly costs are the same.

40. About 22.2 minutes

41. *Familiarize*. Let $x =$ the measure of one angle and $y =$ the measure of the other angle.

Translate. We reword the problem.

The measure of one angle plus the measure of the other angle is 180°.

$$x + y = 180$$

One angle is 4 times the other angle minus 5°.

$$x = 4 \cdot y - 5$$

The resulting system is

$$x + y = 180,$$
$$x = 4y - 5.$$

Carry out. We solve the system. We will use the elimination method although we could also easily use the substitution method. First we get the second equation in the form $Ax + By = C$.

$$x + y = 180 \quad (1)$$
$$x - 4y = -5 \quad (2) \text{ Adding } -4y$$

Now we multiply Equation (2) by -1 and add.

$$x + y = 180$$
$$\underline{-x + 4y = 5}$$
$$5y = 185$$
$$y = 37$$

Then we substitute 37 for y in Equation (1) and solve for x.

$$x + y = 180$$
$$x + 37 = 180$$
$$x = 143$$

Check. The sum of the angle measures is $37° + 143°$, or $180°$, so the angles are supplementary. Also, $5°$ less than four times the $37°$ angle is $4 \cdot 37° - 5°$, or $148° - 5°$, or $143°$, the measure of the other angle. These numbers check.

State. The measures of the angles are $37°$ and $143°$.

42. $45°$, $135°$

43. *Familiarize*. We let c = the number of acres of Chardonnay grapes that should be planted and r = the number of acres of Riesling grapes that should be planted.

***Translate*.** We reword and translate the first statement.

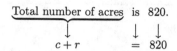

$$c + r \qquad = \quad 820$$

Now we reword and translate the second statement.

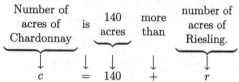

$$c \quad = \quad 140 \quad + \quad r$$

The resulting system is

$$c + r = 820,$$
$$c = 140 + r$$

***Carry out*.** We solve the system. We will use the elimination method, although we could also easily use the substitution method. First we get the second equation in the form $Ax + By = C$. Then we add the equations.

$$\begin{array}{rl} c + r = 820 & (1) \\ \underline{c - r = 140} & (2) \quad \text{Adding} -r \\ 2c \quad\;\; = 960 & \text{Adding} \\ c = 480 & \end{array}$$

Now we substitute 480 for c in Equation (1) and solve for r.

$$\begin{array}{rl} c + r = 820 & (1) \\ 480 + r = 820 & \text{Substituting} \\ r = 340 & \end{array}$$

***Check*.** The total number of acres is $480 + 340$, or 820. Also, the number of acres of Chardonnay grapes is 140 more than the number of acres of Riesling grapes. These numbers check.

***State*.** The vintner should plant 480 acres of Chardonnay grapes and 340 acres of Riesling grapes.

44. Oats: 20 acres, hay: 11 acres

45. *Familiarize*. Let l = the length of the mirror and w = the width, in feet.

***Translate*.**

The perimeter is 18 ft.

$$2l + 2w \quad = \quad 18$$

The length is twice the width.

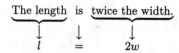

$$l \quad = \quad 2w$$

The resulting system is

$$2l + 2w = 18,$$
$$l = 2w.$$

***Carry out*.** We solve the system. We will use the elimination method, although we could also easily use the substitution method. First we get the second equation in the form $Al + Bw = C$. Then we add the equations.

$$\begin{array}{rl} 2l + 2w = 18 & (1) \\ \underline{l - 2w = 0} & (2) \\ 3l \quad\;\;\; = 18 & \\ l = 6 & \end{array}$$

Substitute 6 for l in Equation (2) and solve for w.

$$\begin{array}{rl} l - 2w = 0 \\ 6 - 2w = 0 \\ 6 = 2w \\ 3 = w \end{array}$$

***Check*.** The perimeter is $2 \cdot 6 + 2 \cdot 3$, or 18 ft. Twice the width is $2 \cdot 3$, or 6 ft, which is the length. These numbers check.

***State*.** The length of the mirror is 6 ft, and the width is 3 ft.

46. 32.4 ft by 21.6 ft

47. ◈

48. ◈

49. $8\% = 8 \times 0.01$ Replacing % by $\times 0.01$
$\quad\;\; = 0.08$

This is equivalent to moving the decimal point two places to the left and dropping the percent symbol.

50. 0.073

51. $0.4\% = 0.4 \times 0.01$ Replacing % by $\times 0.01$
$\quad\quad\;\; = 0.004$

This is equivalent to moving the decimal point two places to the left and dropping the percent symbol.

52. 40%

53. *Translate*.

What is 9% of 350?

$$x \quad = \quad 9\% \quad \cdot \quad 350$$

We solve the equation.

$$x = 0.09(350) \quad (9\% = 0.09)$$
$$x = 31.5 \qquad \text{Multiplying}$$

The answer is 31.5.

54. 16%

55. ◈

56. ◈

57. $x + y = 7, \qquad (1)$
$\quad\ 3(y - x) = 9 \quad (2)$

Multiply Equation (1) by 3 and remove parentheses in Equation (2) and then rewrite this equation in the form $Ax + By = C$. Then add.

$$\begin{array}{r} 3x + 3y = 21 \\ \underline{-3x + 3y = \ 9} \\ 6y = 30 \\ y = 5 \end{array}$$

Substitute 5 for y in Equation (1).

$$x + 5 = 7$$
$$x = 2$$

The ordered pair $(2, 5)$ checks, so it is the solution.

58. $(-1, 1)$

59. $2(5a - 5b) = 10, \quad (1)$
$\quad -5(2a + 6b) = 10 \quad (2)$

Remove parentheses and add.

$$\begin{array}{r} 10a - 10b = 10 \\ \underline{-10a - 30b = 10} \\ -40b = 20 \\ b = -\dfrac{1}{2} \end{array}$$

Substitute $-\dfrac{1}{2}$ for b in Equation (2).

$$-5\left(2a + 6\left(-\frac{1}{2}\right)\right) = 10$$
$$-5(2a - 3) = 10$$
$$-10 + 15 = 10$$
$$-10a = -5$$
$$a = \frac{1}{2}$$

The ordered pair $\left(\dfrac{1}{2}, -\dfrac{1}{2}\right)$ checks, so it is the solution.

60. $(0, 4)$

61. $y = -\dfrac{2}{7}x + 3,$

$\quad\ y = \dfrac{4}{5}x + 3$

Observe that these equations represent lines with different slopes and the same y-intercept. Thus, their point of intersection is the y-intercept, $(0, 3)$ and this is the solution of the system of equations.

62. No solution

63. $y = ax + b, \quad (1)$
$\quad\ y = x + c \qquad (2)$

Substitute $x + c$ for y in Equation (1) and solve for x.

$$y = ax + b$$
$$x + c = ax + b \quad \text{Substituting}$$
$$x - ax = b - c$$
$$(1 - a)x = b - c$$
$$x = \frac{b - c}{1 - a}$$

Substitute $\dfrac{b - c}{1 - a}$ for x in Equation (2) and simplify to find y.

$$y = x + c$$
$$y = \frac{b - c}{1 - a} + c$$
$$y = \frac{b - c}{1 - a} + c \cdot \frac{1 - a}{1 - a}$$
$$y = \frac{b - c + c - ac}{1 - a}$$
$$y = \frac{b - ac}{1 - a}$$

The ordered pair $\left(\dfrac{b - c}{1 - a}, \dfrac{b - ac}{1 - a}\right)$ checks and is the solution. This ordered pair could also be expressed as $\left(\dfrac{c - b}{a - 1}, \dfrac{ac - b}{a - 1}\right)$.

64. $\left(\dfrac{-b - c}{a}, 1\right)$

65. *Familiarize.* Let x represent the number of rabbits and y the number of pheasants in the cage. Each rabbit has one head and four feet. Thus, there are x rabbit heads and $4x$ rabbit feet in the cage. Each pheasant has one head and two feet. Thus, there y pheasant heads and $2y$ pheasant feet in the cage.

Translate. We reword the problem.

$$\underbrace{\text{Rabbit heads}}_{x} \underset{+}{\text{ plus }} \underbrace{\text{pheasant heads}}_{y} \underset{=}{\text{ is }} \underset{35}{35.}$$

$\underbrace{\text{Rabbit feet}}_{4x} \;\; \underbrace{\text{plus}}_{+} \;\; \underbrace{\text{pheasant feet}}_{2y} \;\; \underbrace{\text{is}}_{=} \;\; \underbrace{94.}_{94}$

The resulting system is

$$x + y = 35, \quad (1)$$
$$4x + 2y = 94. \quad (2)$$

Carry out. We solve the system of equations. We multiply Equation (1) by -2 and then add.

$$
\begin{aligned}
-2x - 2y &= -70 \\
4x + 2y &= 94 \\
\hline
2x \phantom{{}+ 2y} &= 24 \quad \text{Adding} \\
x &= 12
\end{aligned}
$$

Substitute 12 for x in one of the original equations and solve for y.

$$x + y = 35 \quad (1)$$
$$12 + y = 35 \quad \text{Substituting}$$
$$y = 23$$

Check. If there are 12 rabbits and 23 pheasants, the total number of heads in the cage is $12 + 23$, or 35. The total number of feet in the cage is $4 \cdot 12 + 2 \cdot 23$, or $48 + 46$, or 94. The numbers check.

State. There are 12 rabbits and 23 pheasants.

66. Patrick: 6 yr; mother: 30

67. Familiarize. Let $x = $ the man's age and $y = $ his daughter's age. Five years ago their ages were $x - 5$ and $y - 5$.

Translate.

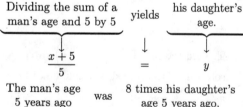

$$\underbrace{\frac{x+5}{5}}_{} \;\; \underbrace{\text{yields}}_{=} \;\; \underbrace{y}_{}$$

$$\underbrace{x - 5}_{} \;\; \underbrace{\text{was}}_{=} \;\; \underbrace{8(y-5)}_{}$$

We have a system of equations:

$$\frac{x+5}{5} = y$$
$$x - 5 = 8(y - 5)$$

Carry out. Solve the system.

Multiply the first Equation by 5 to clear the fraction.

$$x + 5 = 5y$$
$$x - 5y = -5$$

Simplify the second equation.

$$x - 5 = 8(y - 5)$$
$$x - 5 = 8y - 40$$
$$x - 8y = -35$$

The resulting system is

$$x - 5y = -5, \quad (1)$$
$$x - 8y = -35. \quad (2)$$

Multiply Equation (2) by -1 and add.

$$
\begin{aligned}
x - 5y &= -5 \\
-x + 8y &= 35 \quad \text{Multiplying by } -1 \\
\hline
3y &= 30 \quad \text{Adding} \\
y &= 10
\end{aligned}
$$

Substitute 10 for y in Equation (1) and solve for x.

$$x - 5y = -5$$
$$x - 5 \cdot 10 = -5 \quad \text{Substituting}$$
$$x - 50 = -5$$
$$x = 45$$

Possible solution: Man is 45, daughter is 10.

Check. If 5 is added to the man's age, $5 + 45$, the result is 50. If 50 is divided by 5, the result is 10, the daughter's age. Five years ago the father and daughter were 40 and 5, respectively, and $40 = 8 \cdot 5$. The numbers check.

State. The man is 45 years old; his daughter is 10 years old.

68. Base: 9 ft; height: 6 ft

Exercise Set 7.4

1. Familiarize. Let $x = $ the number of two-point shots that were made and $y = $ the number of three-pointers made. Then Terry scored $2x$ points on two-point shots and $3y$ points on three-pointers.

Translate. We reword the problem and translate.

$$\underbrace{\text{Total number of shots}}_{x + y} \;\; \underbrace{\text{is}}_{=} \;\; \underbrace{9.}_{9}$$

$$\underbrace{\text{Total number of points}}_{2x + 3y} \;\; \underbrace{\text{is}}_{=} \;\; \underbrace{20.}_{20}$$

The resulting system is

$$x + y = 9, \quad (1)$$
$$2x + 3y = 20. \quad (2)$$

Carry out. We solve using the elimination method.

$$
\begin{aligned}
-2x - 2y &= -18 \quad \text{Multiplying (1) by } -2 \\
2x + 3y &= 20 \quad (2) \\
\hline
y &= 2 \quad \text{Adding}
\end{aligned}
$$

Substitute 2 for y in Equation (1) and solve for x.

$$x + y = 9 \quad (1)$$
$$x + 2 = 9$$
$$x = 7$$

Check. If Terry made 7 two-pointers and 2 three-pointers, then he made $7 + 2$, or 9 shots for a total of $2 \cdot 7 + 3 \cdot 2$, or $14 + 6$, or 20 points. The numbers check.

State. Terry made 7 two-point shots and 2 three-point shots.

2. 5 two-point shots; 2 three-point shots

3. Familiarize. Let $x =$ the number of two-point shots that were made and $y =$ the number of three-pointers made. Then the Pacers scored $2x$ points on two-point shots and $3y$ points on three-pointers.

Translate. We reword the problem and translate.

Total number of shots is 36.
$$x + y \qquad = 36$$

Total number of points is 84.
$$2x + 3y \qquad = 84$$

The resulting system is
$$x + y = 36, \quad (1)$$
$$2x + 3y = 84. \quad (2)$$

Carry out. We solve using the elimination method.
$$-2x - 2y = -72 \quad \text{Multiplying (1) by } -2$$
$$\underline{2x + 3y = 84} \quad (2)$$
$$y = 12 \quad \text{Adding}$$

Substitute 12 for y in Equation (1) and solve for x.
$$x + y = 36 \quad (1)$$
$$x + 12 = 36$$
$$x = 24$$

Check. If the Pacers made 24 two-pointers and 12 three-pointers, then they made $24 + 12$, or 36 shots and scored $2 \cdot 24 + 3 \cdot 12$, or $48 + 36$, or 84 points. The numbers check.

State. The Pacers made 24 two-point shots and 12 three-point shots.

4. 24 two-point shots; 7 three-point shots

5. Familiarize. Let $x =$ the number of rolls of 24-exposure film purchased and $y =$ the number of rolls of 36-exposure film purchased.

Translate. We can present the information in a table.

	24-exposure	36-exposure	Total
Cost per Roll	$1.50	$2.50	
Number of Rolls	x	y	17
Money Paid	$1.50x$	$2.50y$	34.50

The last two rows of the table give us a system of equations.
$$x + y = 17,$$
$$1.5x + 2.5y = 34.5$$

Carry out. First we multiply both sides of the second equation by 10 to clear the decimals.
$$x + y = 17, \quad (1)$$
$$15x + 25y = 345 \quad (2)$$

Now multiply Equation (1) by -15 and then add.
$$-15x - 15y = -255$$
$$\underline{15x + 25y = 345}$$
$$10y = 90$$
$$y = 9$$

Substitute 9 for y in Equation (1) and solve for x.
$$x + y = 17$$
$$x + 9 = 17$$
$$x = 8$$

Check. If $x = 8$ and $y = 9$, a total of $8 + 9$, or 17 rolls of film was purchased. The amount paid was $1.50(8)$, or $12 for the 24-exposure rolls and $2.50(9)$, or $22.50 for the 36-exposure rolls. Then the total amount paid was $12 + $22.50, or $34.50. The numbers check.

State. Linda bought 8 rolls of 24-exposure film and 9 rolls of 36-exposure film.

6. 24-exposure: 7; 36-exposure: 12

7. Familiarize. Let $x =$ the number of 5-cent bottles or cans collected and $y =$ the number of 10-cent bottles or cans collected.

Translate. We organize the given information in a table.

	$0.05	$0.10	Total
Number	x	y	430
Total Value	$0.05x$	$0.10y$	26.20

A system of two equations can be formed using the rows of the table.
$$x + y = 430,$$
$$0.05x + 0.10y = 26.20$$

Carry out. First we multiply on both sides of the second equation by 100 to clear the decimals.

$$x + y = 430, \quad (1)$$
$$5x + 10y = 2620 \quad (2)$$

Now multiply Equation (1) by -5 and add.

$$-5x - 5y = -2150$$
$$\underline{5x + 10y = 2620}$$
$$5y = 470$$
$$y = 94$$

Substitute 94 for y in Equation (1) and solve for x.

$$x + y = 430$$
$$x + 94 = 430$$
$$x = 336$$

Check. If 336 5-cent bottles and cans and 94 10-cent bottles and cans were collected, then a total of $336 + 94$, or 430 bottles and cans were collected. Their total value is $0.05(336) + \$0.10(94)$, or $16.80 + \$9.40$, or 26.20. These numbers check.

State. 336 5-cent bottles and cans and 94 10-cent bottles and cans were collected.

8. 24 soft-serve; 16 hard-pack

9. *Familiarize*. Let $x =$ the number of adult admissions and $y =$ the number of children and senior admissions.

Translate. We organize the information in a table.

	Adult	Children and Senior	Total
Admission	$9	$5	
Number	x	y	960
Money Taken In	$9x$	$5y$	6320

We use the last two rows of the table to form a system of equations.

$$x + y = 960, \quad (1)$$
$$9x + 5y = 6320 \quad (2)$$

Carry out. We solve using the elimination method. We first multiply Equation (1) by -5 and add.

$$-5x - 5y = -4800$$
$$\underline{9x + 5y = 6320}$$
$$4x = 1520$$
$$x = 380$$

Since the problem asks only for the number of adult admissions, this is the number that we needed to find. We will find y also, however, in order to be able to check the solution.

Substitute 380 for x in Equation (1) and solve for y.

$$x + y = 960$$
$$380 + y = 960$$
$$y = 580$$

Check. If $x = 380$ and $y = 580$, then there were $380 + 580$, or 960 admissions sold. The amount collected for adult admissions was $9 \cdot 380$, or 3420, and the amount collected for children and senior admissions was $5 \cdot 580$, or 2900. The total amount collected was $3420 + \$2900$, or 6320. The numbers check.

State. There were 380 adult admissions.

10. 124

11. *Familiarize*. Let $a =$ the number of adults in the group and $c =$ the number of children.

Translate. We present the information in a table.

	Adults	Children	Total
Price	$86	$65	
Number	a	c	23
Money Paid	$86a$	$65c$	1684

The last two rows of the table give us a system of equations.

$$a + c = 23, \quad (1)$$
$$86a + 65c = 1684 \quad (2)$$

Carry out. We solve using the elimination method. First we multiply Equation (1) by -65 and add.

$$-65a - 65c = -1495$$
$$\underline{86a + 65c = 1684}$$
$$21a = 189$$
$$a = 9$$

Substitute 9 for a in Equation (1) and solve for c.

$$a + c = 23$$
$$9 + c = 23$$
$$c = 14$$

Check. If $a = 9$ and $c = 14$, then there are $9 + 14$, or 23 members in the club. The cost of the adults' passports is $86 \cdot 9$, or 774, and the cost of the children's passports is $65 \cdot 14$, or 910. Then the total cost of the passports is $774 + 910$, or 1684. The numbers check.

State. There are 14 children and 9 adults in the club.

12. 2225

13. *Familiarize*. Let $x =$ the number of students receiving private lessons and $y =$ the number of students receiving group lessons.

Translate. We present the information in a table.

	Private	Group	Total
Price	$25	$18	
Number	x	y	12
Money Earned	$25x$	$18y$	265

The last two rows of the table give us a system of equations.

$$x + y = 12, \quad (1)$$
$$25x + 18y = 265 \quad (2)$$

Carry out. We solve using the elimination method. First we multiply Equation (1) by -18 and add.

$$
\begin{array}{r}
-18x - 18y = -216 \\
25x + 18y = 265 \\
\hline
7x = 49 \\
x = 7
\end{array}
$$

Substitute 7 for x in Equation (1) and solve for y.

$$x + y = 12$$
$$7 + y = 12$$
$$y = 5$$

Check. If $x = 7$ and $y = 5$, then a total of $7 + 5$, or 12 students received lessons. Alice earned $\$25 \cdot 7$, or $175 teaching private lessons and $\$18 \cdot 5$, or $90 teaching group lessons. Thus, she earned a total of $175 + \$90$, or $265. The numbers check.

State. Alice gave private lessons to 7 students and group lessons to 5 students.

14. Private lessons: 6 students; group lessons: 8 students

15. *Familiarize*. Let $x =$ the number of kg of Brazilian coffee to be used and $y =$ the number of kg of Turkish coffee to be used.

Translate. Organize the given information in a table.

Type of coffee	Brazilian	Turkish	Mixture
Cost of coffee	$19	$22	$20
Amount (in kg)	x	y	300
Value	$19x$	$22y$	$20(300)$; or $6000

The last two rows of the table give us two equations. Since the total amount of the mixture is 300 lb, we have

$$x + y = 300.$$

The value of the Brazilian coffee is $19x$ (x lb at $19 per pound), the value of the Turkish coffee is $22y$

(y lb at $22 per pound), and the value of the mixture is $20(300)$ or $6000. Thus we have

$$19x + 22y = 6000.$$

The resulting system is

$$x + y = 300, \quad (1)$$
$$19x + 22y = 6000. \quad (2)$$

Carry out. We use the elimination method. We multiply on both sides of Equation (1) by -19 and then add.

$$
\begin{array}{r}
-19x - 19y = -5700 \\
19x + 22y = 6000 \\
\hline
3y = 300 \\
y = 100
\end{array}
$$

Now substitute 100 for y in Equation (1) and solve for x.

$$x + y = 300$$
$$x + 100 = 300$$
$$x = 200$$

Check. The sum of 100 and 200 is 300. The value of the mixture is $\$19(200) + \$22(100)$, or $3800 + \$2200$, or $6000. These numbers check.

State. 200 kg of Brazilian coffee and 100 kg of Turkish coffee should be used.

16. 30 lb of sunflower seed; 20 lb of rolled oats

17. *Familiarize*. Let x and y represent the number of pounds of peanuts and Brazil nuts to be used, respectively.

Translate. We organize the given information in a table.

Type of nut	Peanuts	Brazil nuts	Mixture
Cost per pound	$2.52	$3.80	$3.44
Amount	x	y	480
Value	$2.52x$	$3.80y$	$3.44(480)$; or 1651.20

The last two rows of the table form a system of equations.

$$x + y = 480,$$
$$2.52x + 3.80y = 1651.20$$

Carry out. First we multiply the second equation by 100 to clear decimals.

$$x + y = 480, \quad (1)$$
$$252x + 380y = 165,120 \quad (2)$$

Now multiply Equation (1) by -252 and add.

$$-252x - 252y = -120,960$$
$$\underline{252x + 380y = 165,120}$$
$$128y = 44,160$$
$$y = 345$$

Substitute 345 for y in Equation (1) and solve for x.

$$x + y = 480$$
$$x + 345 = 480$$
$$x = 135$$

Check. The sum of 135 and 345 is 480. The value of the mixture is $2.52(135) + \$3.80(345)$, or $340.20 + \$1311$, or 1651.20. These numbers check.

State. 135 lb of peanuts and 345 lb of Brazil nuts should be used.

18. 6 kg of cashews; 4 kg of pecans

19. Familiarize. From the table in the text, note that x represents the number of milliliters of solution A to be used and y represents the number of milliliters of solution B.

Translate. We complete the table in the text.

Type of solution	50%-acid	80%-acid	68%-acid
Amount of solution	x	y	200
Percent acid	50%	80%	68%
Amount of acid in solution	$0.5x$	$0.8y$	0.68×200, or 136

Since the total amount of solution is 200 mL, we have

$$x + y = 200.$$

The amount of acid in the mixture is to be 68% of 200 mL, or 136 mL. The amounts of acid from the two solutions are $50\%x$ and $80\%y$. Thus

$$50\%x + 80\%y = 136,$$
or $\quad 0.5x + 0.8y = 136,$
or $\qquad 5x + 8y = 1360$ Clearing decimals

Carry out. We use the elimination method.

$$x + y = 200, \qquad (1)$$
$$5x + 8y = 1360 \qquad (2)$$

We multiply Equation (1) by -5 and then add.

$$-5x - 5y = -1000$$
$$\underline{5x + 8y = 1360}$$
$$3y = 360$$
$$y = 120$$

Next we substitute 120 for y in one of the original equations and solve for x.

$$x + y = 200 \quad (1)$$
$$x + 120 = 200 \quad \text{Substituting}$$
$$x = 80$$

Check. The sum of 80 and 120 is 200. Now 50% of 80 is 40 and 80% of 120 is 96. These add up to 136. The numbers check.

State. 80 mL of the 50%-acid solution and 120 mL of the 80%-acid solution should be used.

20. 50 oz of Clear Shine; 40 oz of Sunstream

21. Familiarize. Let x and y represent the number of liters of 28%-fungicide solution and 40%-fungicide solution to be used in the mixture, respectively.

Translate. We organize the given information in a table.

Type of solution	28%	40%	36%
Amount of solution	x	y	300
Percent fungicide	28%	40%	36%
Amount of fungicide in solution	$0.28x$	$0.4y$	0.36(300), or 108

We get a system of equations from the first and third rows of the table.

$$x + y = 300,$$
$$0.28x + 0.4y = 108$$

Clearing decimals we have

$$x + y = 300, \qquad (1)$$
$$28x + 40y = 10,800 \qquad (2)$$

Carry out. We use the elimination method. Multiply Equation (1) by -28 and add.

$$-28x - 28y = -8400$$
$$\underline{28x + 40y = 10,800}$$
$$12y = 2400$$
$$y = 200$$

Now substitute 200 for y in Equation (1) and solve for x.

$$x + y = 300$$
$$x + 200 = 300$$
$$x = 100$$

Check. The sum of 100 and 200 is 300. The amount of fungicide in the mixture is $0.28(100) + 0.4(200)$, or $28 + 80$, or 108 L. These numbers check.

State. 100 L of the 28%-fungicide solution and 200 L of the 40%-fungicide solution should be used in the mixture.

22. 128 L of 80%-base; 72 L of 30%-base

23. *Familiarize.* Let x and y represent the number of gallons of 87-octane gas and 95-octane gas to be blended, respectively. We organize the given information in a table.

Type of gasoline	87-octane	95-octane	93-octane
Amount of gas	x	y	10
Octane rating	87	95	93
Mixture	$87x$	$95y$	$93 \cdot 10$, or 930

Translate. We get a system of equations from the first and third rows of the table.
$$x + y = 10, \quad (1)$$
$$87x + 95y = 930 \quad (2)$$

Carry out. We use the elimination method. First we multiply Equation (1) by -87 and add.

$$
\begin{array}{r}
-87x - 87y = -870 \\
87x + 95y = 930 \\
\hline
8y = 60 \\
y = 7.5
\end{array}
$$

Now substitute 7.5 for y in Equation (1) and solve for x.
$$x + y = 10$$
$$x + 7.5 = 10$$
$$x = 2.5$$

Check. The sum of 2.5 and 7.5 is 10. The mixture is $87(2.5) + 95(7.5)$, or $217.5 + 712.5$, or 930. These numbers check.

State. 2.5 gal of 87-octane gas and 7.5 gal of 95-octane gas should be blended.

24. 4 gal of 87-octane; 8 gal of 93-octane

25. *Familiarize.* Let $x =$ the number of 1300-word pages that were filled and $y =$ the number of 1850-word pages that were filled. Then the number of words on the x pages that hold 1300 words each is $1300x$ and on the y pages that hold 1850 words each is $1850y$.

Translate. We reword the problem and translate.

Total number of pages is 12.
$$x + y \qquad = 12$$

Total number of words is $18,350$.
$$1300x + 1850y \qquad = 18,350$$

The resulting system of equations is
$$x + y = 12, \quad (1)$$
$$1300x + 1850y = 18,350. \quad (2)$$

Carry out. We solve using the elimination method. First we multiply Equation (1) by -1300 and add.

$$
\begin{array}{r}
-1300x - 1300y = -15,600 \\
1300x + 1850y = 18,350 \\
\hline
550y = 2750 \\
y = 5
\end{array}
$$

Substitute 5 for y in Equation (1) and solve for x.
$$x + y = 12$$
$$x + 5 = 12$$
$$x = 7$$

Check. If $x = 7$ and $y = 5$, then $7 + 5$, or 12 pages are filled. The 1300-word pages contain $1300 \cdot 7$, or 9100 words and the 1850-word pages contain $1850 \cdot 5$ or 9250 words. The total number of words is $9100 + 9250$, or $18,350$. The numbers check.

State. The typesetter used 7 1300-word pages and 5 1850-word pages.

26. 10 nickels; 3 quarters

27. *Familiarize.* Let $f =$ the number of foul shots made and $t =$ the number of two point shots made. Then Chamberlain scored f points from foul shots and $2t$ points from two-pointers.

Translate. We reword the problem and translate.

Total number of shots is 64.
$$f + t \qquad = 64$$

Total number of points is 100.
$$f + 2t \qquad = 100$$

The resulting system of equations is
$$f + t = 64, \quad (1)$$
$$f + 2t = 100. \quad (2)$$

Carry out. We solve using the elimination method. First we multiply Equation (1) by -1 and add.

$$
\begin{array}{r}
-f - t = -64 \\
f + 2t = 100 \\
\hline
t = 36
\end{array}
$$

Substitute 36 for t in Equation (1) and solve for f.

$$f + t = 64$$
$$f + 36 = 64$$
$$f = 28$$

Check. If Chamberlain made 28 foul shots and 36 two point shots, he made $28 + 36$, or 64 shots for a total of $28 + 2 \cdot 36$, or $28 + 72$, or 100 points. The numbers check.

State. Chamberlain made 28 foul shots and 36 two point shots.

28. 8 foul shots; 13 two point shots

29. **Familiarize**. Let $x =$ the number of fluid ounces of Kinney's suntan lotion that should be used and $y =$ the number of fluid ounces of Coppertone that should be used.

Translate. We present the information in a table.

	Kinney's	Coppertone	Mixture
spf Rating	15	30	20
Amount	x	y	50
spf Value	$15x$	$30y$	$20 \cdot 50$, or 1000

The last two rows of the table give us a system of equations.

$$x + y = 50, \quad (1)$$
$$15x + 30y = 1000. \quad (2)$$

Carry out. We solve using the elimination method. First we multiply Equation (1) by -15 and add.

$$-15x - 15y = -750$$
$$\underline{15x + 30y = 1000}$$
$$15y = 250$$
$$y = 16\frac{2}{3}$$

Substitute $16\frac{2}{3}$ for y in Equation (1) and solve for x.

$$x + y = 50$$
$$x + 16\frac{2}{3} = 50$$
$$x = 33\frac{1}{3}$$

Check. If $x = 33\frac{1}{3}$ and $y = 16\frac{2}{3}$, then the total amount of suntan lotion is $33\frac{1}{3} + 16\frac{1}{3}$, or 50 fluid ounces. The spf value of the mixture is $15\left(33\frac{1}{3}\right) + 30\left(16\frac{2}{3}\right)$, or $500 + 500$, or 1000. The numbers check.

State. The mixture should contain $33\frac{1}{3}$ fluid ounces of Kinney's and $16\frac{2}{3}$ fluid ounces of Coppertone.

30. $53\frac{1}{3}$ oz of Dr. Zeke's; $26\frac{2}{3}$ oz of Vitabrite

31. Observe that the number 27 is midway between the number 20 and 34. Thus, the mixture would contain equal parts of the 20% and 34% goose down insulation. Since 50 lb of 27% goose down insulation is to be made, the mixture should contain 25 lb of 20% and 25 lb of 34% goose down insulation.

32. 15 lb of muesli; 30 lb of granola

33. ◈

34. ◈

35. $7 - 3x < 22$
$$-3x < 15$$
$$x > -5$$
The solution set is $\{x | x > -5\}$.

36. $\{x | x \le -7\}$

37. $x + 2 \ge 6$
$$x \ge 4$$

38.

39. $6 < -\frac{1}{2}x + 1$
$$5 < -\frac{1}{2}x$$
$$-10 > x, \text{ or } x < -10$$

40.

41. ◈

42. ◈

43. Familiarize. Let k = the number of pounds of pure Kona beans that should be added to the Columbian beans and m = the total weight of the final mixture, in pounds.

Translate. We present the information in a table.

	Kona	Columbian	Total
Amount	k	45	m
Percent of Kona	100%	0%	30%
Amount of Kona	$1 \cdot k$, or k	$0 \cdot 45$, or 0	$0.3m$

The table gives us two equations.

$$k + 45 = m, \quad (1)$$
$$k = 0.3m \quad (2)$$

Carry out. We use the substitution method. First substitute $k + 45$ for m in Equation (2) and solve for k.

$$k = 0.3(x + 45)$$
$$k = 0.3x + 13.5$$
$$0.7k = 13.5$$
$$k = \frac{13.5}{0.7} = \frac{135}{7} = 19\frac{2}{7}$$

This is the number the problem asks for. We will also find m so that we can check the answer. Substitute $19\frac{2}{7}$ for k in Equation (1) and compute m.

$$k + 45 = m$$
$$19\frac{2}{7} + 45 = m$$
$$64\frac{2}{7} = m$$

Check. If a coffee mixture that weighs $64\frac{2}{7}$ lb contains $19\frac{2}{7}$ lb of Kona coffee, then the percent of Kona coffee in the mixture is $\frac{19\frac{2}{7}}{64\frac{2}{7}}$, or 0.3, or 30%. The answer checks.

State. $19\frac{2}{7}$ lb of Kona coffee should be added to 45 lb of Columbian coffee to obtain the desired mixture.

44. $2666\frac{2}{3}$ L

45. Familiarize. In a table we arrange the information regarding the solution *after* some of the 30% solution is drained and replaced with pure antifreeze. We let x represent the amount of the original (30%) solution remaining, and we let y represent the amount of the 30% mixture that is drained and replaced with pure antifreeze.

Type of solution	Original (30%)	Pure anti-freeze	Mixture
Amount of solution	x	y	6.3
Percent of antifreeze	30%	100%	50%
Amount of antifreeze in solution	$0.3x$	$1 \cdot y$, or y	0.5(6.3), or 3.15

Translate. The table gives us two equations.

Amount of solution: $x + y = 6.3$

Amount of antifreeze in solution: $0.3x + y = 3.15$, or $30x + 100y = 315$

The resulting system is

$$x + y = 63, \quad (1)$$
$$30x + 100y = 315. \quad (2)$$

Carry out. We multiply Equation (1) by -30 and then add.

$$
\begin{array}{r}
-30x - 30y = -189 \\
30x + 100y = 315 \\
\hline
70y = 126
\end{array}
$$

$$y = \frac{126}{70} = 1.8$$

Then we substitute 1.8 for y in Equation (1) and solve for x.

$$x + y = 6.3$$
$$x + 1.8 = 6.3$$
$$x = 4.5$$

Check. When $x = 4.5$ L and $y = 1.8$ L, the total is 6.3 L. The amount of antifreeze in the mixture is $0.3(4.5) + 1.8$, or $1.35 + 1.8$, or 3.15 L. This is 50% of 6.3 L, so the numbers check.

State. 1.8 L of the original mixture should be drained and replaced with pure antifreeze.

46. 9%; 10.5%

47. Familiarize. Let x represent the number of gallons of 91-octane gas to be added to the tank and let y represent the total number of gallons in the tank after the 91-octane gas is added. We organize the given information in a table.

Type of gasoline	85-octane	91-octane	Mixture
Amount of gas	5	x	y
Octane rating	85	91	87
Mixture	$85 \cdot 5$, or 425	$91x$	$87y$

Translate. We get a system of equations from the first and third rows of the table.

$$5 + x = y, \qquad (1)$$
$$425 + 91x = 87y \quad (2)$$

Carry out. Substitute $5 + x$ for y in Equation (2) and solve for x.

$$425 + 91x = 87y$$
$$425 + 91x = 87(5 + x)$$
$$425 + 91x = 435 + 87x$$
$$425 + 4x = 435$$
$$4x = 10$$
$$x = 2.5$$

Although the original problem asks us to find only x, we will find y also in order to check the answer. Substitute 2.5 for x in Equation (1) and compute y.

$$y = 5 + 2.5 = 7.5$$

Check. The mixture is $425 + 91(2.5)$, or 652.5. This is equal to $87(7.5)$, so the answer checks.

State. Kim should add 2.5 gal of 91-octane gas to her tank.

48. Bat: $14.50;, ball: $4.55; glove: $79.95

49. *Familiarize*. Let x = the price per gallon of the inexpensive paint and y = the price per gallon of the expensive paint.

Translate. We present the information in a table.

	Inexpensive paint	Expensive paint	Mixture
Price per gallon	x	y	
First mixture amount	9 gal	7 gal	16 gal
First mixture cost	$9x$	$7y$	$19.70(16)$, or $315.20
Second mixture amount	3 gal	5 gal	8 gal
Second mixture cost	$3x$	$5y$	$19.825(8)$, or $158.60

The third and fifth rows of the table give us two equations. The total cost of the first mixture is:

$$9x + 7y = 315.20, \text{ or}$$
$$90x + 70y = 3152 \qquad \text{Clearing the decimal}$$

The total cost of the second mixture is:

$$3x + 5y = 158.60, \text{ or}$$
$$30x + 50y = 1586 \qquad \text{Clearing the decimal}$$

Carry out. We use the elimination method.

$$90x + 70y = 3152, \quad (1)$$
$$30x + 50y = 1586. \quad (2)$$

We multiply Equation (2) by -3 and add.

$$90x + 70y = 3152$$
$$\underline{-90x - 150y = -4758}$$
$$-80y = -1606$$
$$y = 20.075$$

Now we substitute 20.075 for y in Equation (2) and solve for x.

$$30x + 50y = 1586$$
$$30x + 50(20.075) = 1586$$
$$30x + 1003.75 = 1586$$
$$30x = 582.25$$
$$x \approx 19.408$$

Check. We check $x \approx \$19.408$ and $y = \$20.075$. The cost of the first mixture is $\$19.408(9) + \$20.075(7)$, or about $315.20. The cost of the second mixture is $\$19.408(3) + \$20.075(5)$, or about $158.60. These values check.

State. The inexpensive paint costs about $19.41 per gallon, and the expensive paint costs about $20.08 per gallon.

50. $25,000 at 6%; $29,000 at 6.5%

51. *Familiarize*. Let x = the number of liters of skim milk and y = the number of liters of 2% milk.

Translate. We present the information in a table.

Type of milk	4.6%	Skim	2% (Mixture)
Amount of milk	100 L	x	y
Percent of butterfat	4.6%	0%	2%
Amount of butterfat in milk	$4.6\% \times 100$, or 4.6 L	$0\% \cdot x$, or 0 L	$2\%y$

The first and third rows of the table give us two equations.

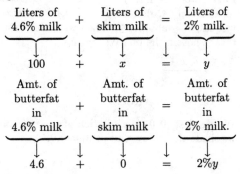

The resulting system is

$$100 + x = y,$$
$$4.6 = 2\%y, \text{ or}$$

$$100 + x = y,$$
$$4.6 = 0.02y.$$

Carry out. We solve the second equation for y.

$$4.6 = 0.02y$$
$$\frac{4.6}{0.02} = y$$
$$230 = y$$

We substitute 230 for y in the first equation and solve for x.

$$100 + x = y$$
$$100 + x = 230$$
$$x = 130$$

Check. We consider $x = 130$ L and $y = 230$ L. The difference between 130 L and 230 L is 100 L. There is no butterfat in the skim milk. There are 4.6 liters of butterfat in the 100 liters of the 4.6% milk. Thus there are 4.6 liters of butterfat in the mixture. This checks because 2% of 230 is 4.6.

State. 130 L of skim milk should be added.

52. 10 at \$20 per hour; 5 at \$25 per hour

53. Familiarize. Let x represent the ten's digit and y the one's digit. Then the number is $10x + y$.

Translate.

We simplify the first equation.

$$10x + y = 6(x + y)$$
$$10x + y = 6x + 6y$$
$$4x - 5y = 0$$

The system of equations is

$$4x - 5y = 0, \quad (1)$$
$$x = 1 + y. \quad (2)$$

Carry out. We use the substitution method. We substitute $1 + y$ for x in Equation (1) and solve for y.

$$4(1 + y) - 5y = 0$$
$$4 + 4y - 5y = 0$$
$$4 - y = 0$$
$$4 = y$$

Then we substitute 4 for y in Equation (2) and compute x.

$$x = 1 + y = 1 + 4 = 5$$

Check. We consider the number 54. The number is 6 times the sum of the digits, 9. The ten's digit is 1 more than the one's digit. This number checks.

State. The number is 54.

54. 75

55. Familiarize. Let $x =$ Tweedledum's weight and $y =$ Tweedledee's weight, in pounds.

Translate.

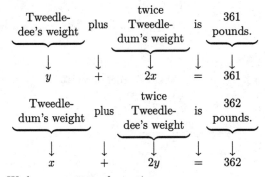

We have a system of equations.

$$y + 2x = 361,$$
$$x + 2y = 362, \text{ or}$$

$$2x + y = 361, \quad (1)$$
$$x + 2y = 362 \quad (2)$$

Carry out. We use elimination. First multiply Equation (2) by -2 and add.

$$2x + \quad y = 361$$
$$\underline{-2x - \quad 4y = -724}$$
$$-3y = -363$$
$$y = 121$$

Now substitute 121 for y in Equation (2) and solve for x.

$$x + 2y = 362 \quad (2)$$
$$x + 2 \cdot 121 = 362 \quad \text{Substituting}$$
$$x + 242 = 362$$
$$x = 120$$

Check. If Tweedledum weighs 120 lb and Tweedledee weighs 121 lb, then the sum of Tweedledee's weight and twice Tweedledum's is $121 + 2 \cdot 120$, or $121 + 240$, or 361 lb. The sum of Tweedledum's weight and twice Tweedledee's is $120 + 2 \cdot 121$, or $120 + 242$, or 362 lb. The answer checks.

State. Tweedledum weighs 120 lb, and Tweedledee weighs 121 lb.

Exercise Set 7.5

1. We use alphabetical order of variables. We replace x by -3 and y by -5.

$$\frac{x + 3y < -18}{}$$
$$-3 + 3(-5) \ ? \ -18$$
$$-3 - 15 \ \Big|$$
$$-18 \ \Big| \ -18 \ \text{FALSE}$$

Since $-18 < -18$ is false, $(-3, -5)$ is not a solution.

2. Yes

3. We use alphabetical order of variables. We replace x by $\frac{7}{8}$ and y by $\frac{1}{2}$.

$$\frac{6y + 5x \geq -3}{}$$
$$6 \cdot \frac{1}{2} + 5\left(\frac{7}{8}\right) \ ? \ -3$$
$$3 + \frac{35}{8} \ \Big|$$
$$\frac{59}{8} \ \Big| \ -3 \ \text{TRUE}$$

Since $\frac{59}{8} \geq 7$ is true, $\left(\frac{7}{8}, \frac{1}{2}\right)$ is a solution.

4. Yes

5. Graph $y \leq x + 4$.

First graph the line $y = x + 4$. The intercepts are $(0, 4)$ and $(-4, 0)$. We draw a solid line since the inequality symbol is $\leq$. Then we pick a test point that is not on the line We try $(0, 0)$.

$$\frac{y \leq x + 4}{}$$
$$0 \ ? \ 0 + 4$$
$$0 \ \Big| \ 4 \qquad \text{TRUE}$$

We see that $(0, 0)$ is a solution of the inequality, so we shade the region that contains $(0, 0)$.

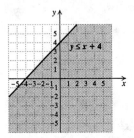

6.

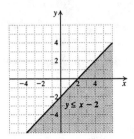

7. Graph $y < x - 1$.

First graph the line $y = x - 1$. The intercepts are $(0, -1)$ and $(1, 0)$. We draw a dashed line since the inequality symbol is $<$. Then we pick a test point that is not on the line. We try $(0, 0)$.

$$\frac{y < x - 1}{}$$
$$0 \ ? \ 0 - 1$$
$$0 \ \Big| \ -1 \qquad \text{FALSE}$$

Since $(0, 0)$ is not a solution of the inequality, we shade the region that does not contain $(0, 0)$.

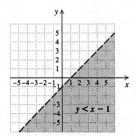

8.

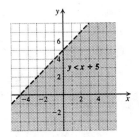

9. Graph $y \geq x - 3$.

First graph the line $y = x - 3$. The intercepts are $(0, -3)$ and $(3, 0)$. We draw a solid line since the inequality symbol is $\geq$. Then we test the point $(0, 0)$.

$$\frac{y \geq x - 3}{0\ ?\ 0 - 3}$$
$$0\ \Big|\ -3 \quad \text{TRUE}$$

Since $(0, 0)$ is a solution of the inequality, we shade the region containing $(0, 0)$.

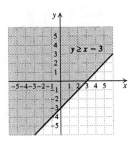

10.

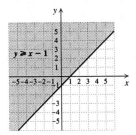

11. Graph $y \leq 2x - 1$.

First graph the line $y = 2x - 1$. The intercepts are $(0, -1)$ and $\left(\dfrac{1}{2}, 0\right)$. We draw a solid line since the inequality symbol is $\leq$. Then we test the point $(0, 0)$.

$$\frac{y \leq 2x - 1}{0\ ?\ 2 \cdot 0 - 1}$$
$$0\ \Big|\ -1 \quad \text{FALSE}$$

Since $(0, 0)$ is not a solution of the inequality, we shade the region that does not contain $(0, 0)$.

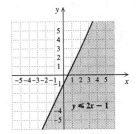

12.

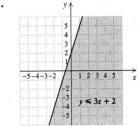

13. Graph $x + y \leq 4$.

First graph the line $x + y = 4$. The intercepts are $(0, 4)$ and $(4, 0)$. We draw a solid line since the inequality symbol is $\leq$. Then we test the point $(0, 0)$.

$$\frac{x + y \leq 4}{0 + 0\ ?\ 4}$$
$$0\ \Big|\ 4 \quad \text{TRUE}$$

Since $(0, 0)$ is a solution of the inequality, we shade the region that contains $(0, 0)$.

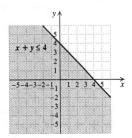

14.

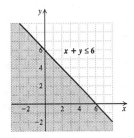

15. Graph $x - y > 7$.

First graph the line $x - y = 7$. The intercepts are

$(0, -7)$ and $(7, 0)$. We draw a dashed line since the inequality symbol is $>$. Then we test the point $(0, 0)$.

$$\frac{x - y > 7}{\begin{array}{c|c} 0 - 0 \ ? \ 7 \\ \hline 0 & 7 \quad \text{FALSE} \end{array}}$$

Since $(0, 0)$ is not a solution of the inequality, we shade the region that does not contain $(0, 0)$.

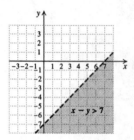

16.

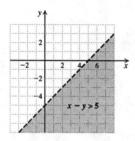

17. Graph $y \geq 1 - 2x$.

First graph the line $y = 1 - 2x$. The intercepts are $(0, 1)$ and $\left(\frac{1}{2}, 0\right)$. We draw a solid line since the inequality symbol is $\geq$. Then we test the point $(0, 0)$.

$$\frac{y \geq 1 - 2x}{\begin{array}{c|c} 0 \ ? \ 1 - 2 \cdot 0 \\ \hline 0 & 1 \quad\quad \text{FALSE} \end{array}}$$

Since $(0, 0)$ is not a solution of the inequality, we shade the region that does not contain $(0, 0)$.

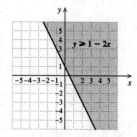

18.

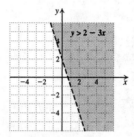

19. Graph $y - 3x > 0$.

First graph the line $y - 3x = 0$, or $y = 3x$. Two points on the line are $(0, 0)$ and $(1, 3)$. We draw a dashed line, since the inequality symbol is $>$. Then we test the point $(2, 1)$, which is not a point on the line.

$$\frac{y - 3x > 0}{\begin{array}{c|c} 1 - 3 \cdot 2 \ ? \ 0 \\ 1 - 6 & \\ \hline -5 & 0 \quad \text{FALSE} \end{array}}$$

Since $(2, 1)$ is not a solution of the inequality, we shade the region that does not contain $(2, 1)$.

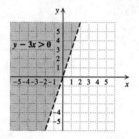

20.

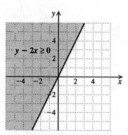

21. Graph $x \geq 3$.

First graph the line $x = 3$ using a solid line since the inequality symbol is $\geq$. Then use $(4, -3)$ as a test point. We can write the inequality as $x + 0y \geq 3$.

$$\frac{x + 0y \geq 3}{\begin{array}{c|c} 4 + 0(-3) \ ? \ 3 \\ \hline 4 & 3 \quad \text{TRUE} \end{array}}$$

Since $(4, -3)$ is a solution of the inequality, we shade the region containing $(4, -3)$.

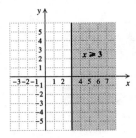

25. Graph $y \geq -5$.

Graph the line $y = -5$ using a solid line since the inequality symbol is $\geq$. Then use $(2,3)$ as a test point. We can write the inequality as $0x + y \geq -5$.

$$\begin{array}{c} 0x + y \geq -5 \\ \hline 0 \cdot 2 + 3 \ ? \ -5 \\ 3 \ \bigg| \ -5 \ \text{TRUE} \end{array}$$

Since $(2,3)$ is a solution of the inequality, we shade the region containing $(2,3)$.

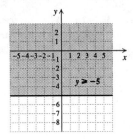

22.

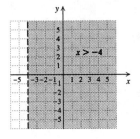

26.

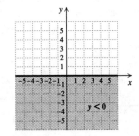

23. Graph $y \leq 3$.

Graph the line $y = 3$ using a solid line since the inequality symbol is $\leq$. Then use $(1,-2)$ as a test point. We can write the inequality as $0x + y \leq 3$.

$$\begin{array}{c} x + 0y \leq 3 \\ \hline 0 \cdot 1 + (-2) \ ? \ 3 \\ -2 \ \bigg| \ 3 \ \text{TRUE} \end{array}$$

Since $(1,-2)$ is a solution of the inequality, we shade the region containing $(1,-2)$.

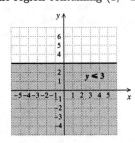

27. Graph $x < 4$.

Graph the line $x = 4$ using a dashed line since the inequality symbol is $<$. Then use $(-1,2)$ as a test point. We can write the inequality as $x + 0y < 4$.

$$\begin{array}{c} x + 0y < 4 \\ \hline -1 + 0 \cdot 2 \ ? \ 4 \\ -1 \ \bigg| \ 4 \ \text{TRUE} \end{array}$$

Since $(-1,2)$ is a solution of the inequality, we shade the region containing $(-1,2)$.

24.

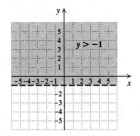

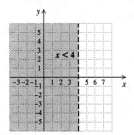

28.

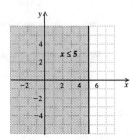

29. Graph $x - y < -10$.

Graph the line $x - y = -10$. The intercepts are $(0, 10)$ and $(-10, 0)$. We draw a dashed line since the inequality symbol is $<$. Then we test the point $(0,0)$.

$$\frac{x - y < -10}{\begin{array}{c|c} 0 - 0 \ ? \ -10 \\ \hline 0 & -10 \ \text{FALSE} \end{array}}$$

Since $(0, 0)$ is not a solution of the inequality, we shade the region that does not contain $(0, 0)$.

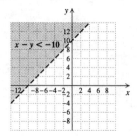

30.

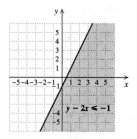

31. Graph $2x + 3y \leq 12$.

First graph the line $2x + 3y = 12$. The intercepts are $(0, 4)$ and $(6, 0)$. We draw a solid line since the inequality symbol is $\leq$. Then we test the point $(0, 0)$.

$$\frac{2x + 3y \leq 12}{\begin{array}{c|c} 2 \cdot 0 + 3 \cdot 0 \ ? \ 12 \\ \hline 0 & 12 \ \text{TRUE} \end{array}}$$

Since $(0, 0)$ is a solution of the inequality, we shade the region containing $(0, 0)$.

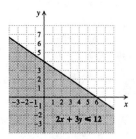

32.

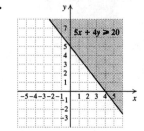

33. ◈

34. ◈

35. $3x + 5y = 3(-2) + 5(4)$
$\qquad\quad = -6 + 20$
$\qquad\quad = 14$

36. 2

37. We find June, 1999, on the horizontal axis, go up to the line representing corrugated cardboard, then go across to the vertical axis, and read the price there. We see that recyclers were paying about $60 per ton for corrugated cardboard in June, 1999.

38. Between June, 1998, and December, 1998

39. Find the highest point on the graph representing newspaper and then go directly down to the horizontal axis and read the date there. We see that the value of newspaper peaked in June, 2000.

40. December, 1998

41. The only periods during which the price paid for corrugated cardboard dropped are from June, 1998, to December, 1998, and from June, 2000, to December, 2000. During the first period the price dropped from about $60 to $20, or $60 − $20, or $40. During the second period the price dropped from about $120 to $35, or $120 − $35, or $85. Thus, the price paid for corrugated cardboard dropped the most during the period from June, 2000, to December, 2000.

42. December, 1998, to June, 1999

43.

44.

45. The c children weigh $75c$ lb, and the a adults weigh $150a$ lb. Together, the children and adults weigh $75c + 150a$ lb. When this total is more than 1000 lb the elevator is overloaded, so we have $75c + 150a > 1000$. (Of course, c and a would also have to be non-negative, so we show only the portion of the graph that is in the first quadrant.)

To graph $75c + 150a > 1000$, we first graph $75c + 150a = 1000$ using a dashed line. (Remember to use alphabetical order of variables.) Then we test the point $(0,0)$.

$$\begin{array}{c} 75c + 150a > 1000 \\ \hline 75 \cdot 0 + 150 \cdot 0 \; ? \; 1000 \\ 0 \; \mid \; 1000 \quad \textbf{FALSE} \end{array}$$

Since $(0,0)$ is not a solution of the inequality, we shade the region that does not contain $(0,0)$.

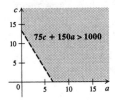

46. $2w + t \geq 60$

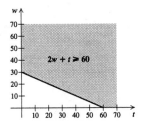

47. The charge for developing w rolls of 24-exposure film is $10w$, and the charge for developing r rolls of 36-exposure film is $14r$. Then the total cost of developing the film is $10w + 14r$ and when this total exceeds \$140 we have $10w + 14r > 140$.

To graph $10w + 14r > 140$, we first graph $10w + 14r = 140$. (Remember to use alphabetical order of variables.)

Then we test the point $(0,0)$.

$$\begin{array}{c} 10w + 14r > 140 \\ \hline 10 \cdot 0 + 14 \cdot 0 \; ? \; 140 \\ 0 \; \mid \; 140 \quad \textbf{FALSE} \end{array}$$

Since $(0,0)$ is not a solution of the inequality, we shade the region that does not contain $(0,0)$.

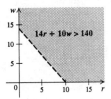

48. $r + t \geq 17$

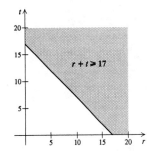

49. First find the equation of the line containing the points $(2,0)$ and $(0,-2)$. The slope is

$$\frac{-2 - 0}{0 - 2} = \frac{-2}{-2} = 1.$$

We know that the y-intercept is $(0, -2)$, so we write the equation using the slope-intercept form.

$$y = mx + b$$
$$y = 1 \cdot x + (-2)$$
$$y = x - 2$$

Since the line is dashed, the inequality symbol will be $<$ or $>$. To determine which, we substitute the coordinates of a point in the shaded region. We will use $(0,0)$.

$$\begin{array}{c|c} y & x - 2 \\ \hline 0 \; ? & 0 - 2 \\ 0 \; \mid & -2 \end{array}$$

Since $0 > -2$ is true, the correct symbol is $>$. The inequality is $y > x - 2$.

50. $x \geq -2$

51. Graph $xy \leq 0$.

From the principle of zero products, we know that $xy = 0$ when $x = 0$ or $y = 0$. Therefore, the graph contains the lines $x = 0$ and $y = 0$, or the y- and x-axes. Also, $xy < 0$ when x and y have different signs. This is the case for all points in the second quadrant (x is negative and y is positive) and in the fourth quadrant (x is positive and y is negative). Thus, we shade the second and fourth quadrants.

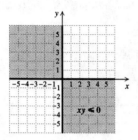

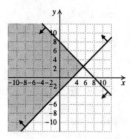

52.

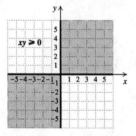

2.

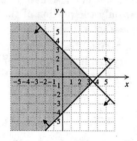

53. First solve the equation for y.

$$y + 3x \leq 4.9$$
$$y \leq -3x + 4.9$$

Then graph the line $y = -3x + 4.9$ and shade below the line.

3. $y - 2x > 1,$
      $y - 2x < 3$

We graph the lines $y - 2x = 1$ and $y - 2x = 3$ using dashed lines. We indicate the region for each inequality by the arrows at the ends of the lines. We shade the area where the regions overlap.

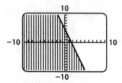

54.

$y \geq 0.7x - 2.3$

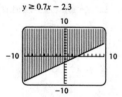

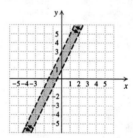

4.

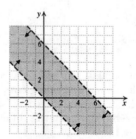

Exercise Set 7.6

1. $x + y \leq 8,$
      $x - y \leq 3$

We graph the lines $x + y = 8$ and $x - y = 3$ using solid lines. We indicate the region for each inequality by the arrows at the ends of the lines. We shade the area where the regions overlap.

5. $y \geq -3$,

$x > 2 + y$

We graph the line $y = -3$ using a solid line and the line $x = 2 + y$ using a dashed line. We indicate the region for each inequality by the arrows at the ends of the lines. We shade the area where the regions overlap.

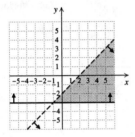

6.

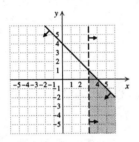

7. $y > 3x - 2$,

$y < -x + 4$

We graph the lines $y = 3x - 2$ and $y = -x + 4$ using dashed lines. We indicate the region for each inequality by the arrows at the ends of the lines. We shade the area where the regions overlap.

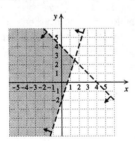

8.

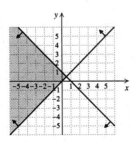

9. $x \leq 4$,

$y \leq 5$

We graph the lines $x = 4$ and $y = 5$ using sold lines. We indicate the region for each inequality by the arrows at the ends of the lines. We shade the area where the regions overlap.

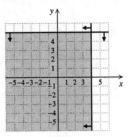

10.

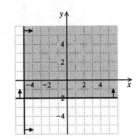

11. $x \leq 0$,

$y \leq 0$

We graph the lines $x = 0$ and $y = 0$ using solid lines. We indicate the region for each inequality by the arrows at the ends of the lines. We shade the area where the regions overlap.

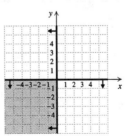

12.

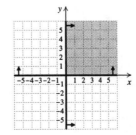

13. $2x - 3y \geq 9,$

$2y + x > 6$

We graph the line $2x - 3y = 9$ using a solid line and the line $2y + x = 6$ using a dashed line. We indicate the region for each inequality by the arrows at the ends of the lines. We shade the area where the regions overlap.

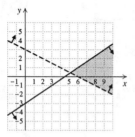

14.

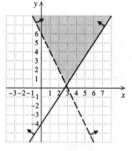

15. $y > 5x + 2,$

$y \leq 1 - x$

We graph the line $y = 5x + 2$ using a dashed line and the line $y = 1 - x$ using a solid line. We indicate the region for each inequality by the arrows at the ends of the lines. We shade the area where the regions overlap.

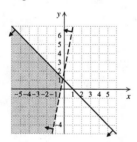

16.

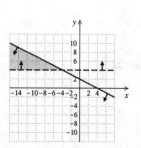

17. $x + y \leq 5,$

$x \geq 0,$

$y \geq 0,$

$y \leq 3$

We graph the lines $x + y = 5$, $x = 0$, $y = 0$, and $y = 3$ using solid lines. We indicate the region for each inequality by the arrows at the ends of the lines. We shade the area where the regions overlap.

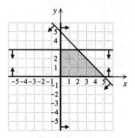

18.

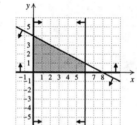

19. $y - x \geq 1,$

$y - x \leq 3,$

$x \leq 5,$

$x \geq 2$

We graph the lines $y - x = 1$, $y - x = 3$, $x = 5$, and $x = 2$ using solid lines. We indicate the region for each inequality by the arrows at the ends of the lines. We shade the area where the regions overlap.

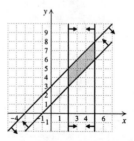

20.

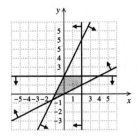

21. $y \leq x,$

$x \geq -2,$

$x \leq -y$

We graph the lines $y = x$, $x = -2$ and $x = -y$ using solid lines. We indicate the region for each inequality by the arrows at the ends of the lines. We shade the area where the regions overlap.

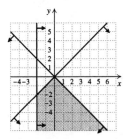

22.

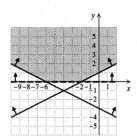

23. ◈

24. ◈

25. $7 = \dfrac{k}{5}$

$5 \cdot 7 = 5 \cdot \dfrac{k}{5}$

$35 = k$

The solution is 35.

26. 24

27. $18 = k \cdot 3$

$\dfrac{18}{3} = \dfrac{k \cdot 3}{3}$

$6 = k$

The solution is 6.

28. 5

29. $5 = k \cdot 45$

$\dfrac{5}{45} = \dfrac{k \cdot 45}{45}$

$\dfrac{1}{9} = k$

The solution is $\dfrac{1}{9}$.

30. $\dfrac{1}{3}$

31. ◈

32. ◈

33. $2x + 5y \geq 18,$

$4x + 3y \geq 22,$

$2x + y \geq 8,$

$x \geq 0,$

$y \geq 0$

We graph the related equations, find the region for each inequality, and shade the area where the regions overlap.

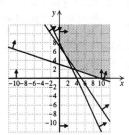

34.

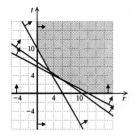

35.　　$2x + 5y \geq 10,$

　　　　$x - 3y \leq 6,$

　　$4x + 10y \leq 20$

We graph the related equations and find the region for each inequality. We see that the solution set consists of a single point, $\left(\dfrac{60}{11}, -\dfrac{2}{11}\right)$, or $\left(5\dfrac{5}{11}, -\dfrac{2}{11}\right)$.

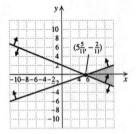

36.

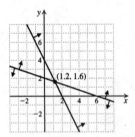

37. Note that the graphs of the related equations for the first two inequalities are parallel lines since $4x + 6y = 2(2x + 3y)$ and $9 \neq 2 \cdot 1$. Also observe that we would shade below the first line when graphing the solution set of the first inequality and we would shade above the second line when graphing the solution set of the second inequality. Thus there is no region of overlap for these two inequalities and hence the system of inequalities has no solution.

38. No solution

39.

40.

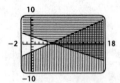

41.

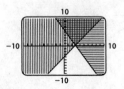

Exercise Set 7.7

1. We substitute to find k.

　　$y = kx$　　y varies directly as x.

　　$28 = k \cdot 2$　Substituting 28 for y and 2 for x

　　$\dfrac{28}{2} = k$

　　$14 = k$　　k is the constant of variation.

The equation of variation is $y = 14x$.

2. $y = \dfrac{15}{4}x$

3. We substitute to find k.

　　$y = kx$　　　y varies directly as x.

　　$0.7 = k \cdot 0.4$　Substituting 0.7 for y and
　　　　　　　　　　0.4 for x

　　$\dfrac{0.7}{0.4} = k$

　　$\dfrac{7}{4} = k$, or $k = 1.75$

The equation of variation is $y = \dfrac{7}{4}x$, or $1.75x$.

4. $y = \dfrac{8}{5}x$

5. We substitute to find k.

　　$y = kx$

　　$400 = k \cdot 75$　Substituting 400 for y
　　　　　　　　　and 75 for x

　　$\dfrac{400}{75} = k$

　　$\dfrac{16}{3} = k$

The equation of variation is $y = \dfrac{16}{3}x$.

6. $y = \dfrac{26}{7}$

7. We substitute to find k.

　　$y = kx$

　　$200 = k \cdot 300$　　Substituting 200 for y and
　　　　　　　　　　　300 for x

　　$\dfrac{200}{300} = k$

　　$\dfrac{2}{3} = k$

The equation of variation is $y = \dfrac{2}{3}x$.

8. $y = \dfrac{25}{3}x$

9. We substitute to find k.

$$y = \frac{k}{x}$$

$$45 = \frac{k}{2} \quad \text{Substituting 45 for } y \text{ and} \atop \text{2 for } x$$

$$90 = k$$

The equation of variation is $y = \dfrac{90}{x}$.

10. $y = \dfrac{24}{x}$

11. We substitute to find k.

$$y = \frac{k}{x}$$

$$7 = \frac{k}{10} \quad \text{Substituting 7 for } y \text{ and 10 for } x$$

$$70 = k$$

The equation of variation is $y = \dfrac{70}{x}$.

12. $y = \dfrac{1}{x}$

13. We substitute to find k.

$$y = \frac{k}{x}$$

$$6.25 = \frac{k}{25} \quad \text{Substituting 6.25 for } y \text{ and} \atop \text{25 for } x$$

$$156.25 = k$$

The equation of variation is $y = \dfrac{156.25}{x}$.

14. $y = \dfrac{2100}{x}$

15. We substitute to find k.

$$y = \frac{k}{x}$$

$$42 = \frac{k}{5} \quad \text{Substituting 42 for } y \text{ and} \atop \text{5 for } x$$

$$210 = k$$

The equation of variation is $y = \dfrac{210}{x}$.

16. $y = \dfrac{2}{x}$

17. *Familiarize and Translate.* The problem states that we have direct variation between the variables P and H. Thus, an equation $P = kH$ applies.

Carry out. First find an equation of variation.

$$P = kH$$

$$135 = k \cdot 15 \quad \text{Substituting 135 for } P \atop \text{and 15 for } H$$

$$\frac{135}{15} = k$$

$$9 = k$$

The equation of variation is $P = 9H$. When $H = 23$, we have:

$$P = 9H$$

$$P = 9(23) \quad \text{Substituting 23 for } H$$

$$P = 207$$

Check. This check might be done by repeating the computations. We might also do some reasoning about the answer. The paycheck increased from \$135 to \$207. Similarly, the hours increased from 15 to 23. The ratios 15/135 and 23/207 are the same value: $0.\overline{1}$.

State. For 23 hours work, the paycheck is \$207.

18. 16,445

19. *Familiarize and Translate.* The problem states that we have direct variation between S and W. Thus, an equation $S = kW$ applies.

Carry out. First find an equation of variation.

$$S = kW$$

$$40 = k \cdot 15 \quad \text{Substituting 40 for } S \text{ and} \atop \text{15 for } W$$

$$\frac{40}{15} = k$$

$$\frac{8}{3} = k$$

The equation of variation is $S = \dfrac{8}{3}W$. When $W = 9$, we have:

$$S = \frac{8}{3}W$$

$$S = \frac{8}{3} \cdot 9 \quad \text{Substituting 9 for } W$$

$$S = 24$$

Check. A check might be done by repeating the computation. We can also do some reasoning about the answer. The number of servings decreased from 40 to 24. Similarly, the weight decreased from 15 kg to 9 kg. The ratios $\dfrac{15}{40}$ and $\dfrac{9}{24}$ are the same value: 0.375.

State. 24 servings can be obtained from a 9-kg turkey.

20. 320 cm^3

21. *Familiarize and Translate*. The problem states that we have direct variation between the variables M and E. Thus, an equation $M = kE$ applies.

Carry out. First find an equation of variation.

$$M = kE$$

$$32 = k \cdot 192 \quad \text{Substituting 32 for } M \text{ and } 192 \text{ for } E$$

$$\frac{32}{192} = k$$

$$\frac{1}{6} = k$$

The equation of variation is $M = \frac{1}{6}E$.

When $E = 185$,

$$M = \frac{1}{6}E$$

$$M = \frac{1}{6} \cdot 185 \quad \text{Substituting 185 for } E$$

$$M \approx 30.8$$

Check. In addition to repeating the computations we can do some reasoning. The weight on the earth decreased from 192 lb to 185 lb. Similarly, the weight on the moon decreased from 32 lb to about 30.8 lb. The ratios 192/32 and 185/30.8 are about the same value: 6.

State. David Ellenbogen would weigh about 30.8 lb on the moon.

22. $5\frac{1}{3}$ amperes

23. *Familiarize and Translate*. The problem states that we have inverse variation between the variables P and W. Thus, an equation $P = \frac{k}{W}$ applies.

Carry out. First find an equation of variation.

$$P = \frac{k}{W}$$

$$440 = \frac{k}{2.4} \quad \text{Substituting 440 for } P \text{ and } 2.4 \text{ for } W$$

$$1056 = k$$

The equation of variation is $P = \frac{1056}{W}$. When $P = 660$, we have

$$P = \frac{1056}{W}$$

$$660 = \frac{1056}{W} \quad \text{Substituting 660 for } P$$

$$660W = 1056$$

$$W = \frac{1056}{660} = 1.6$$

Check. A check might be done by repeating the computations. We can also do some reasoning about

the answer. Note that as the pitch increases, the wavelength decreases as expected. Also note that 2.4(440) and 1.6(660) are both 1056.

State. The wavelength of a trumpet's E above concert A is 1.6 ft.

24. 54 min

25. *Familiarize and Translate*. This problem states that we have direct variation between the variables c and n. Thus, an equation $c = kn$, applies.

Carry out. First we find an equation of variation.

$$c = kn$$

$$14 = k \cdot 720 \quad \text{Substituting 14 for } c \text{ and 720 (the number of hours in 30 days) for } n$$

$$\frac{14}{720} = k$$

$$\frac{7}{360} = k$$

The equation of variation is $c = \frac{7}{360}n$.

Use the equation of variation to find the operating cost for 1 day and for 1 hr.

For 1 day: 1 day is equivalent to 24 hr, so we substitute 24 for n.

$$c = \frac{7}{360} \cdot 24$$

$$c = \frac{168}{360}, \text{ or } c \approx 0.47$$

For 1 hr:

$$c = \frac{7}{360}n$$

$$c = \frac{7}{360} \cdot 1 \quad \text{Substituting 1 for } n$$

$$c \approx 0.02$$

Check. In addition to repeating the computations we can do some reasoning. The hours decreased from 720 to 24 and from 720 to 1. Similarly, the cost decreased from \$14 to \$0.47 and from \$14 to \$0.02.

State. It cost \$0.47 to operate a television for 1 day. It cost \$0.02 to operate a television for 1 hr.

26. 10

27. Observe that \$31.50 is twice \$15.75. Since we have inverse variation and the cost doubles, the number of people would be divided by 2, so there would be 10/2, or 5 people going fishing.

28. 48.8 lb

29. ◈

30.

31.

32.

33. $(-7)^2 = (-7)(-7) = 49$

34. 25

35. $13^2 = 13 \cdot 13 = 169$

36. 225

37. $(3x)^2 = 3^2 \cdot x^2 = 9x^2$

38. $36a^2$

39. $(a^2b)^2 = (a^2)^2(b)^2 = a^4b^2$

40. s^6t^2

41.

42.

43. $S = kv^6$

44. $P^2 = kt$

45. $I = \dfrac{k}{d^2}$

46. $B = kN$

47. $P = kv^3$

48. $P = 8S$

49. $C = kr$

From the formula for the circumference of a circle we know that $k = 2\pi$, and we have $C = 2\pi r$.

50. $A = \pi r^2$

51. $V = kr^3$

From the formula for the volume of a sphere we know that $k = \dfrac{4}{3}\pi$, and we have $V = \dfrac{4}{3}\pi r^3$.

Chapter 8

Radical Expressions and Equations

1. The square roots of 4 are 2 and -2, since $2^2 = 4$ and $(-2)^2 = 4$.

2. $3, -3$

3. The square roots of 16 are 4 and -4, because $4^2 = 16$ and $(-4)^2 = 16$.

4. $1, -1$

5. The square roots of 49 are 7 and -7, because $7^2 = 49$ and $(-7)^2 = 49$.

6. $11, -11$

7. The square roots of 144 are 12 and -12, because $12^2 = 144$ and $(-12)^2 = 144$.

8. $13, -13$

9. $\sqrt{9} = 3$, taking the principal square root.

10. 2

11. $\sqrt{1} = 1$, so $-\sqrt{1} = -1$.

12. -5

13. $\sqrt{0} = 0$

14. -9

15. $\sqrt{121} = 11$, so $-\sqrt{121} = -11$.

16. 19

17. $\sqrt{900} = 30$

18. 21

19. $\sqrt{169} = 13$

20. 12

21. $\sqrt{625} = 25$, so $-\sqrt{625} = -25$.

22. -20

23. The radicand is the expression under the radical, $a - 7$.

24. $t - 5$

25. The radicand is the expression under the radical, $t^2 + 1$.

26. $x^2 + 2$

27. The radicand is the expression under the radical, $\dfrac{3}{x+2}$.

28. $\dfrac{a}{a-b}$

29. $\sqrt{100}$ is rational, since 100 is a perfect square.

30. Irrational

31. $\sqrt{8}$ is irrational, since 8 is not a perfect square.

32. Irrational

33. $\sqrt{32}$ is irrational, since 32 is not a perfect square.

34. Rational

35. $\sqrt{98}$ is irrational, since 98 is not a perfect square.

36. Irrational

37. $-\sqrt{4}$ is rational, since 4 is a perfect square.

38. Rational

39. Since the radicand is expressed as a number squared, we know that it is a perfect square and thus the number in the exercise is rational.

40. Irrational

41. 2.236

42. 2.449

43. 4.123

44. 4.359

45. 9.644

46. 6.557

47. $\sqrt{t^2} = t$ Since t is assumed to be nonnegative

48. x

49. $\sqrt{9x^2} = \sqrt{(3x)^2} = 3x$ Since x is assumed
to be nonnegative

50. $5a$

51. $\sqrt{(7a)^2} = 7a$ Since a is assumed to be
nonnegative

52. $4x$

53. $\sqrt{(17x)^2} = 17x$ Since x is assumed to be
nonnegative

54. $8ab$

55. a) We substitute 36 in the formula.
$$N = 2.5\sqrt{36} = 2.5(6) = 15$$
For an average of 36 arrivals, 15 spaces are
needed.

    b) We substitute 29 in the formula. We use a calculator or Table 2 to find an approximation.
$$N = 2.5\sqrt{29} \approx 2.5(5.385) \approx 13.463 \approx 14$$
For an average of 29 arrivals, 14 spaces are
needed.

56. a) 18

    b) 19

57. Substitute 36 in the formula.
$$T = 0.144\sqrt{36} = 0.144(6) = 0.864 \text{ sec}$$

58. 0.72 sec

59. ◈

60. ◈

61. $(7x)^2 = 7^2 x^2 = 49x^2$

62. $9a^2$

63. $(4t^7)^2 = 4^2(t^7)^2 = 16t^{7\cdot 2} = 16t^{14}$

64. $45x^7$

65. $3a \cdot 16a^{10} = 3 \cdot 16 \cdot a \cdot a^{10} = 48a^{1+10} = 48a^{11}$

66. $8t^9 u^6$

67. ◈

68. ◈

69. We find the inner square root first.
$$\sqrt{\sqrt{81}} = \sqrt{9} = 3$$

70. 5

71. $-\sqrt{36} < -\sqrt{33} < -\sqrt{25}$, or $-6 < -\sqrt{33} < -5$
$-\sqrt{33}$ is between -6 and -5.

72. 64; answers may vary

73. If $\sqrt{t^2} = 7$, then $t^2 = 7^2$, or 49. Thus, $t = 7$ or
$t = -7$. The solutions are 7 and -7.

74. No real solution

75. If $-\sqrt{x^2} = -3$, then $\sqrt{x^2} = 3$ and $x^2 = 3^2$, or 9.
Thus, $x = 3$ or $x = -3$. The solutions are 3 and -3.

76. -6, 6

77. $\sqrt{(9a^3 b^4)^2} = 9a^3 b^4$ Since all variables represent
positive numbers

78. $3a$

79. $\sqrt{\dfrac{144x^8}{36y^6}} = \sqrt{\dfrac{4x^8}{y^6}} = \sqrt{\left(\dfrac{2x^4}{y^3}\right)^2} = \dfrac{2x^4}{y^3}$

80. $\dfrac{y^6}{90}$

81. $\sqrt{\dfrac{400}{m^{16}}} = \sqrt{\left(\dfrac{20}{m^8}\right)^2} = \dfrac{20}{m^8}$

82. $\dfrac{p}{60}$

83. a) Locate 3 on the x-axis, move up vertically to the
graph, and then move left horizontally to the y-axis to read the approximation.
$$\sqrt{3} \approx 1.7 \quad \text{(Answers may vary.)}$$

    b) Locate 5 on the x-axis, move up vertically to the
graph, and then move left horizontally to the y-axis to read the approximation.
$$\sqrt{5} \approx 2.2 \quad \text{(Answers may vary.)}$$

    c) Locate 7 on the x-axis, move up vertically to the
graph, and then move left horizontally to the y-axis to read the approximation.
$$\sqrt{7} \approx 2.6 \quad \text{(Answers may vary.)}$$

84. 1141.6 ft/sec

85. We substitute 5 in the formula.
$$V = \frac{1087\sqrt{273 + 5}}{16.52}$$
$$V = \frac{1087\sqrt{278}}{16.52}$$
$$V \approx \frac{1087(16.673)}{16.52}$$
$$V \approx 1097.1 \text{ ft/sec}$$

86. 1067.1 ft/sec

87. We substitute 100 in the formula.

$$V = \frac{1087\sqrt{273 + 100}}{16.52}$$

$$V = \frac{1087\sqrt{373}}{16.52}$$

$$V \approx \frac{1087(19.313)}{16.52}$$

$$V \approx 1270.8 \text{ ft/sec}$$

88. $y_1 = \sqrt{x-2};\ y_2 = \sqrt{x+7};$ $y_3 = 5 + \sqrt{x};\ y_4 = -4 + \sqrt{x}$

89. ◈

Exercise Set 8.2

1. $\sqrt{5}\sqrt{3} = \sqrt{5 \cdot 3} = \sqrt{15}$

2. $\sqrt{35}$

3. $\sqrt{4}\sqrt{3} = \sqrt{12}$, or
  $\sqrt{4}\sqrt{3} = 2\sqrt{3}$ Taking the square root of 4

4. $\sqrt{18}$, or $3\sqrt{2}$

5. $\sqrt{\frac{2}{5}}\sqrt{\frac{3}{4}} = \sqrt{\frac{2 \cdot 3}{5 \cdot 4}} = \sqrt{\frac{3}{10}}$

6. $\sqrt{\frac{3}{40}}$, or $\frac{1}{2}\sqrt{\frac{3}{10}}$

7. $\sqrt{13}\sqrt{13} = \sqrt{13 \cdot 13} = \sqrt{169} = 13$

8. 17

9. $\sqrt{25}\sqrt{3} = \sqrt{75}$, or
  $\sqrt{25}\sqrt{3} = 5\sqrt{3}$ Taking the square root of 25

10. $\sqrt{72}$, or $6\sqrt{2}$

11. $\sqrt{2}\sqrt{x} = \sqrt{2 \cdot x} = \sqrt{2x}$

12. $\sqrt{3a}$

13. $\sqrt{7}\sqrt{2a} = \sqrt{7 \cdot 2a} = \sqrt{14a}$

14. $\sqrt{35t}$

15. $\sqrt{5x} \cdot 7 = \sqrt{5x \cdot 7} = \sqrt{35x}$

16. $\sqrt{10mn}$

17. $\sqrt{3a}\sqrt{2c} = \sqrt{3a \cdot 2c} = \sqrt{6ac}$

18. $\sqrt{3xyz}$

19. $\sqrt{20} = \sqrt{4 \cdot 5}$ 4 is a perfect square.
  $= \sqrt{4}\sqrt{5}$ Factoring into a product of radicals
  $= 2\sqrt{5}$

20. $2\sqrt{7}$

21. $\sqrt{50} = \sqrt{25 \cdot 2}$ 25 is a perfect square.
  $= \sqrt{25}\sqrt{2}$ Factoring into a product of radicals
  $= 5\sqrt{2}$

22. $3\sqrt{5}$

23. $\sqrt{700} = \sqrt{100 \cdot 7} = \sqrt{100}\sqrt{7} = 10\sqrt{7}$

24. $10\sqrt{2}$

25. $\sqrt{9x} = \sqrt{9 \cdot x} = \sqrt{9}\sqrt{x} = 3\sqrt{x}$

26. $2\sqrt{y}$

27. $\sqrt{75a} = \sqrt{25 \cdot 3a} = \sqrt{25}\sqrt{3a} = 5\sqrt{3a}$

28. $2\sqrt{10m}$

29. $\sqrt{16a} = \sqrt{16 \cdot a} = \sqrt{16}\sqrt{a} = 4\sqrt{a}$

30. $7\sqrt{b}$

31. $\sqrt{64y^2} = \sqrt{64}\sqrt{y^2} = 8y$, or
  $\sqrt{64y^2} = \sqrt{(8y)^2} = 8y$

32. $3x$

33. $\sqrt{13x^2} = \sqrt{13}\sqrt{x^2} = \sqrt{13} \cdot x$, or $x\sqrt{13}$

34. $t\sqrt{29}$

35. $\sqrt{28t^2} = \sqrt{4 \cdot t^2 \cdot 7} = \sqrt{4}\sqrt{t^2}\sqrt{7} = 2t\sqrt{7}$

36. $5a\sqrt{5}$

37. $\sqrt{80} = \sqrt{2 \cdot 2 \cdot 2 \cdot 2 \cdot 5}$ Writing the prime factorization
  $= \sqrt{2^2}\sqrt{2^2}\sqrt{5}$
  $= 2 \cdot 2 \cdot \sqrt{5}$
  $= 4\sqrt{5}$

38. $7\sqrt{2}$

39. $\sqrt{288y} = \sqrt{144 \cdot 2y} = \sqrt{144}\sqrt{2y} = 12\sqrt{2y}$

40. $11\sqrt{3p}$

41. $\sqrt{a^{14}} = \sqrt{(a^7)^2} = a^7$

42. t^{10}

43. $\sqrt{x^{12}} = \sqrt{(x^6)^2} = x^6$

44. x^8

45. $\sqrt{r^7} = \sqrt{r^6 \cdot r}$ One factor is a perfect square
$$= \sqrt{r^6}\,\sqrt{r}$$
$$= \sqrt{(r^3)^2}\,\sqrt{r}$$
$$= r^3\sqrt{r}$$

46. $t^2\sqrt{t}$

47. $\sqrt{t^{19}} = \sqrt{t^{18} \cdot t} = \sqrt{t^{18}}\,\sqrt{t} = \sqrt{(t^9)^2}\,\sqrt{t} = t^9\sqrt{t}$

48. $p^8\sqrt{p}$

49. $\sqrt{90a^3} = \sqrt{9 \cdot a^2 \cdot 10 \cdot a} = \sqrt{9}\sqrt{a^2}\sqrt{10a} = 3a\sqrt{10a}$

50. $5y\sqrt{10y}$

51. $\sqrt{8a^5} = \sqrt{4a^4(2a)} = \sqrt{4(a^2)^2(2a)} =$
$$\sqrt{4}\sqrt{(a^2)^2}\sqrt{2a} = 2a^2\sqrt{2a}$$

52. $2b^3\sqrt{3b}$

53. $\sqrt{104p^{17}} = \sqrt{4p^{16}(26p)} = \sqrt{4(p^8)^2(26p)} =$
$$\sqrt{4}\sqrt{(p^8)^2}\sqrt{26p} = 2p^8\sqrt{26p}$$

54. $3m^{11}\sqrt{10m}$

55. $\sqrt{7} \cdot \sqrt{14} = \sqrt{7 \cdot 14}$ Multiplying
$$= \sqrt{7 \cdot 2 \cdot 7}$$ Writing the prime factorization
$$= \sqrt{2}\sqrt{7^2}$$
$$= \sqrt{2} \cdot 7, \text{ or}$$
$$7\sqrt{2}$$

56. $3\sqrt{2}$

57. $\sqrt{3} \cdot \sqrt{27} = \sqrt{3 \cdot 27}$ Multiplying
$$= \sqrt{3 \cdot 3 \cdot 3 \cdot 3}$$ Writing the prime factorization
$$= \sqrt{3^4}$$
$$= 3^2$$
$$= 9$$

58. 4

59. $\sqrt{3x}\sqrt{12y} = \sqrt{3x \cdot 12y}$
$$= \sqrt{3 \cdot x \cdot 2 \cdot 2 \cdot 3 \cdot y}$$
$$= \sqrt{2^2}\sqrt{3^2}\sqrt{xy}$$
$$= 2 \cdot 3\sqrt{xy}$$
$$= 6\sqrt{xy}$$

60. $10\sqrt{xy}$

61. $\sqrt{13}\sqrt{13x} = \sqrt{13 \cdot 13x} = \sqrt{13 \cdot 13 \cdot x} = \sqrt{13^2}\sqrt{x} = 13\sqrt{x}$

62. $11\sqrt{x}$

63. $\sqrt{10b}\sqrt{50b} = \sqrt{10b \cdot 50b}$
$$= \sqrt{10 \cdot b \cdot 5 \cdot 10 \cdot b}$$
$$= \sqrt{10^2}\sqrt{b^2}\sqrt{5}$$
$$= 10b\sqrt{5}$$

64. $6a\sqrt{3}$

65. Since the radicands are identical, the product will be the radicand, $7x$. We could also do this problem as follows.
$$\sqrt{7x} \cdot \sqrt{7x} = \sqrt{7x \cdot 7x} = \sqrt{(7x)^2} = 7x$$

66. $3a$

67. $\sqrt{ab}\sqrt{ac} = \sqrt{a^2bc} = \sqrt{a^2}\sqrt{bc} = a\sqrt{bc}$

68. $x\sqrt{yz}$

69. $\sqrt{2x}\sqrt{14x^5} = \sqrt{2x \cdot 14x^5}$
$$= \sqrt{2 \cdot 2 \cdot 7 \cdot x^6}$$
$$= \sqrt{2^2}\sqrt{7}\sqrt{x^6}$$
$$= 2 \cdot \sqrt{7} \cdot x^3, \text{ or}$$
$$2x^3\sqrt{7}$$

70. $5m^4\sqrt{3}$

71. $\sqrt{x^2y^3}\sqrt{xy^4}$
$$= \sqrt{x^2y^3}\sqrt{x(y^2)^2}\quad x^2 \text{ and } (y^2)^2 \text{ are}$$
$$\qquad\qquad\qquad\qquad\quad \text{perfect squares}$$
$$= \sqrt{x^2} \cdot \sqrt{y^3} \cdot \sqrt{x} \cdot \sqrt{(y^2)^2}$$
$$= x \cdot \sqrt{y^3} \cdot \sqrt{x} \cdot y^2$$
$$= xy^2\sqrt{y^3 \cdot x}$$
$$= xy^2\sqrt{x \cdot y^2 \cdot y}$$
$$= xy^2\sqrt{y^2}\sqrt{xy}$$
$$= xy^2 \cdot y \cdot \sqrt{xy}$$
$$= xy^3\sqrt{xy}$$

72. $x^2 y \sqrt{y}$

73. $\sqrt{50ab}\sqrt{10a^2b^7} = \sqrt{50ab} \cdot \sqrt{10} \cdot \sqrt{a^2} \cdot \sqrt{(b^3)^2 \cdot b}$
$= ab^3 \sqrt{50ab}\sqrt{10}\sqrt{b}$
$= ab^3 \sqrt{50ab \cdot 10b}$
$= ab^3 \sqrt{2 \cdot 5 \cdot 5 \cdot a \cdot b \cdot 2 \cdot 5 \cdot b}$
$= ab^3 \sqrt{2^2 \cdot 5^2 \cdot b^2 \cdot 5 \cdot a}$
$= ab^3 \sqrt{2^2}\sqrt{5^2}\sqrt{b^2}\sqrt{5a}$
$= ab^3 \cdot 2 \cdot 5 \cdot b\sqrt{5a}$
$= 10ab^4\sqrt{5a}$

74. $5xy^2\sqrt{2xy}$

75. First we substitute 20 for L in the formula:
$r = 2\sqrt{5L} = 2\sqrt{5 \cdot 20} = 2\sqrt{100} = 2 \cdot 10 = 20$ mph
Then we substitute 150 for L:
$r = 2\sqrt{5 \cdot 150} = 2\sqrt{750} = 2\sqrt{25 \cdot 30} = 2\sqrt{25}\,\sqrt{30} = 2 \cdot 5\sqrt{30} = 10\sqrt{30} \approx 10(5.477) \approx 54.77$ mph, or 54.8 mph (rounded to the nearest tenth)

76. 24.5 mph; 37.4 mph

77. ◈

78. ◈

79. $\dfrac{a^7 b^3}{a^2 b} = a^{7-2} b^{3-1} = a^5 b^2$

80. $x^3 y^8$

81. $\dfrac{3x}{5y} \cdot \dfrac{2x}{7y} = \dfrac{3x \cdot 2x}{5y \cdot 7y} = \dfrac{6x^2}{35y^2}$

82. $\dfrac{35a^2}{6b^2}$

83. $\dfrac{2r^3}{7t} \cdot \dfrac{rt}{rt} = \dfrac{2r^3}{7t} \cdot 1 = \dfrac{2r^3}{7t}$

84. $\dfrac{5x^7}{11y}$

85. ◈

86. ◈

87. $\sqrt{0.01} = \sqrt{(0.1)^2} = 0.1$

88. 0.5

89. $\sqrt{0.0625} = \sqrt{(0.25)^2} = 0.25$

90. 0.001

91. $\sqrt{450} = \sqrt{225 \cdot 2} = 15\sqrt{2}$, so $15\sqrt{2} = \sqrt{450}$.

92. $15 > 4\sqrt{14}$

93. $3\sqrt{11} = \sqrt{9}\sqrt{11} = \sqrt{99}$ and
$7\sqrt{2} = \sqrt{49}\sqrt{2} = \sqrt{98}$, so
$3\sqrt{11} > 7\sqrt{2}$.

94. $16 > \sqrt{15}\sqrt{17}$

95. $8^2 = 64$
$(\sqrt{15} + \sqrt{17})^2 = 15 + 2\sqrt{255} + 17 = 32 + 2\sqrt{255}$
Now $\sqrt{255} < \sqrt{256}$, or $\sqrt{255} < 16$, so
$2\sqrt{255} < 2 \cdot 16 = 32$. Then $32 + 2\sqrt{255} < 32 + 32$, or
$(\sqrt{15} + \sqrt{17})^2 < 64$. Thus, $8 > \sqrt{15} + \sqrt{17}$.

96. $5\sqrt{7} < 4\sqrt{11}$

97. $\sqrt{27(x+1)}\,\sqrt{12y(x+1)^2} =$
$\sqrt{27(x+1) \cdot 12y(x+1)^2} =$
$\sqrt{9 \cdot 3 \cdot (x+1) \cdot 4 \cdot 3 \cdot y(x+1)^2} =$
$\sqrt{9 \cdot 3 \cdot 3 \cdot 4 \cdot (x+1)^2 \cdot (x+1)y} =$
$\sqrt{9}\,\sqrt{3 \cdot 3}\,\sqrt{4}\,\sqrt{(x+1)^2}\,\sqrt{(x+1)y} =$
$3 \cdot 3 \cdot 2(x+1)\sqrt{(x+1)y} = 18(x+1)\sqrt{(x+1)y}$

98. $6(x-2)^2\sqrt{10}$

99. $\sqrt{x^9}\,\sqrt{2x}\,\sqrt{10x^5} = \sqrt{x^9 \cdot 2x \cdot 10x^5} =$
$\sqrt{x^8 \cdot x \cdot 2 \cdot x \cdot 2 \cdot 5 \cdot x^4 \cdot x} =$
$\sqrt{x^8 \cdot x \cdot x \cdot 2 \cdot 2 \cdot x^4 \cdot 5 \cdot x} =$
$\sqrt{x^8}\,\sqrt{x \cdot x}\,\sqrt{2 \cdot 2}\,\sqrt{x^4}\,\sqrt{5x} =$
$x^4 \cdot x \cdot 2 \cdot x^2\sqrt{5x} = 2x^7\sqrt{5x}$

100. $2^{54}x^{158}\sqrt{2x}$

101. $7x^{14}\sqrt{6x^7} = \sqrt{(7x^{14})^2} \cdot \sqrt{6x^7} = \sqrt{49x^{28}} \cdot \sqrt{6x^7} =$
$\sqrt{49x^{28} \cdot 6x^7} = \sqrt{294x^{35}} = \sqrt{21x^9 \cdot 14x^{26}} =$
$\sqrt{21x^9} \cdot \sqrt{14x^{26}}$

102. $\sqrt{10x^6}$

103. $\sqrt{x^{16n}} = \sqrt{(x^{8n})^2} = x^{8n}$

104. $0.2x^{2n}$

105. If n is an odd whole number greater than or equal to 3, then $n = 2k + 1$, where k is a natural number.
$\sqrt{y^n} = \sqrt{y^{2k+1}} = \sqrt{y^{2k} \cdot y} = y^k\sqrt{y}$, where
$k = \dfrac{n-1}{2}$.

Exercise Set 8.3

1. $\dfrac{\sqrt{12}}{\sqrt{3}} = \sqrt{\dfrac{12}{3}} = \sqrt{4} = 2$

2. 2

3. $\dfrac{\sqrt{75}}{\sqrt{3}} = \sqrt{\dfrac{75}{3}} = \sqrt{25} = 5$

4. 6

5. $\dfrac{\sqrt{35}}{\sqrt{5}} = \sqrt{\dfrac{35}{5}} = \sqrt{7}$

6. $\sqrt{6}$

7. $\dfrac{\sqrt{7}}{\sqrt{63}} = \sqrt{\dfrac{7}{63}} = \sqrt{\dfrac{1}{9}} = \dfrac{1}{3}$

8. $\dfrac{1}{4}$

9. $\dfrac{\sqrt{18}}{\sqrt{32}} = \sqrt{\dfrac{18}{32}} = \sqrt{\dfrac{9}{16}} = \dfrac{3}{4}$

10. $\dfrac{2}{5}$

11. $\dfrac{\sqrt{8x}}{\sqrt{2x}} = \sqrt{\dfrac{8x}{2x}} = \sqrt{4} = 2$

12. 3

13. $\dfrac{\sqrt{48y^3}}{\sqrt{3y}} = \sqrt{\dfrac{48y^3}{3y}} = \sqrt{16y^2} = 4y$

14. $3x$

15. $\dfrac{\sqrt{27x^5}}{\sqrt{3x}} = \sqrt{\dfrac{27x^5}{3x}} = \sqrt{9x^4} = 3x^2$

16. $2a^3$

17. $\dfrac{\sqrt{21a^9}}{\sqrt{7a^3}} = \sqrt{\dfrac{21a^9}{7a^3}} = \sqrt{3a^6} = a^3\sqrt{3}$

18. $t^3\sqrt{7}$

19. $\sqrt{\dfrac{36}{25}} = \dfrac{\sqrt{36}}{\sqrt{25}} = \dfrac{6}{5}$

20. $\dfrac{3}{7}$

21. $\sqrt{\dfrac{49}{16}} = \dfrac{\sqrt{49}}{\sqrt{16}} = \dfrac{7}{4}$

22. $\dfrac{10}{7}$

23. $-\sqrt{\dfrac{25}{81}} = -\dfrac{\sqrt{25}}{\sqrt{81}} = -\dfrac{5}{9}$

24. $-\dfrac{5}{8}$

25. $\sqrt{\dfrac{2a^3}{50a}} = \sqrt{\dfrac{a^2}{25}} = \dfrac{\sqrt{a^2}}{\sqrt{25}} = \dfrac{a}{5}$

26. $\dfrac{a^2}{2}$

27. $\sqrt{\dfrac{6x^7}{32x}} = \sqrt{\dfrac{3x^6}{16}} = \dfrac{\sqrt{3x^6}}{\sqrt{16}} = \dfrac{x^3\sqrt{3}}{4}$

28. $\dfrac{x\sqrt{2}}{5}$

29. $\sqrt{\dfrac{21t^9}{28t^3}} = \sqrt{\dfrac{3t^6}{4}} = \dfrac{\sqrt{3t^6}}{\sqrt{4}} = \dfrac{t^3\sqrt{3}}{2}$

30. $\dfrac{t^2\sqrt{5}}{3}$

31. $\dfrac{5}{\sqrt{3}} = \dfrac{5}{\sqrt{3}} \cdot \dfrac{\sqrt{3}}{\sqrt{3}} = \dfrac{5\sqrt{3}}{3}$

32. $\dfrac{7\sqrt{2}}{2}$

33. $\dfrac{\sqrt{3}}{\sqrt{7}} = \dfrac{\sqrt{3}}{\sqrt{7}} \cdot \dfrac{\sqrt{7}}{\sqrt{7}} = \dfrac{\sqrt{21}}{7}$

34. $\dfrac{\sqrt{77}}{11}$

35. $\dfrac{\sqrt{4}}{\sqrt{27}} = \dfrac{\sqrt{4}}{\sqrt{9}\sqrt{3}} = \dfrac{2}{3\sqrt{3}} = \dfrac{2}{3\sqrt{3}} \cdot \dfrac{\sqrt{3}}{\sqrt{3}} = \dfrac{2\sqrt{3}}{3 \cdot 3} = \dfrac{2\sqrt{3}}{9}$

36. $\dfrac{3\sqrt{2}}{4}$

37. $\dfrac{\sqrt{2}}{\sqrt{3}} = \dfrac{\sqrt{2}}{\sqrt{3}} \cdot \dfrac{\sqrt{3}}{\sqrt{3}} = \dfrac{\sqrt{6}}{3}$

38. $\dfrac{\sqrt{35}}{7}$

39. $\dfrac{\sqrt{3}}{\sqrt{50}} = \dfrac{\sqrt{3}}{\sqrt{25}\sqrt{2}} = \dfrac{\sqrt{3}}{5\sqrt{2}} = \dfrac{\sqrt{3}}{5\sqrt{2}} \cdot \dfrac{\sqrt{2}}{\sqrt{2}} =$
$\dfrac{\sqrt{6}}{5 \cdot 2} = \dfrac{\sqrt{6}}{10}$

40. $\dfrac{\sqrt{10}}{6}$

41. $\dfrac{\sqrt{2a}}{\sqrt{45}} = \dfrac{\sqrt{2a}}{\sqrt{9}\sqrt{5}} = \dfrac{\sqrt{2a}}{3\sqrt{5}} = \dfrac{\sqrt{2a}}{3\sqrt{5}} \cdot \dfrac{\sqrt{5}}{\sqrt{5}} =$

$\dfrac{\sqrt{10a}}{3 \cdot 5} = \dfrac{\sqrt{10a}}{15}$

42. $\dfrac{\sqrt{6a}}{8}$

43. $\sqrt{\dfrac{12}{5}} = \dfrac{\sqrt{4}\sqrt{3}}{\sqrt{5}} = \dfrac{2\sqrt{3}}{\sqrt{5}} = \dfrac{2\sqrt{3}}{\sqrt{5}} \cdot \dfrac{\sqrt{5}}{\sqrt{5}} = \dfrac{2\sqrt{15}}{5}$

44. $\dfrac{2\sqrt{6}}{3}$

45. $\sqrt{\dfrac{2}{x}} = \dfrac{\sqrt{2}}{\sqrt{x}} = \dfrac{\sqrt{2}}{\sqrt{x}} \cdot \dfrac{\sqrt{x}}{\sqrt{x}} = \dfrac{\sqrt{2x}}{x}$

46. $\dfrac{\sqrt{3x}}{x}$

47. $\sqrt{\dfrac{t}{32}} = \dfrac{\sqrt{t}}{\sqrt{32}} = \dfrac{\sqrt{t}}{\sqrt{16}\sqrt{2}} = \dfrac{\sqrt{t}}{4\sqrt{2}} =$

$\dfrac{\sqrt{t}}{4\sqrt{2}} \cdot \dfrac{\sqrt{2}}{\sqrt{2}} = \dfrac{\sqrt{2t}}{4 \cdot 2} = \dfrac{\sqrt{2t}}{8}$

48. $\dfrac{\sqrt{3a}}{6}$

49. $\sqrt{\dfrac{x}{40}} = \dfrac{\sqrt{x}}{\sqrt{40}} = \dfrac{\sqrt{x}}{\sqrt{4}\sqrt{10}} = \dfrac{\sqrt{x}}{2\sqrt{10}} =$

$\dfrac{\sqrt{x}}{2\sqrt{10}} \cdot \dfrac{\sqrt{10}}{\sqrt{10}} = \dfrac{\sqrt{10x}}{2 \cdot 10} = \dfrac{\sqrt{10x}}{20}$

50. $\dfrac{\sqrt{10y}}{30}$

51. Since the denominator, 25, is a perfect square we need only to simplify the expression in order to rationalize the denominator.

$\sqrt{\dfrac{3a}{25}} = \dfrac{\sqrt{3a}}{\sqrt{25}} = \dfrac{\sqrt{3a}}{5}$

52. $\dfrac{\sqrt{7t}}{4}$

53. $\sqrt{\dfrac{5x^3}{12x}} = \sqrt{\dfrac{5x^2}{12}} = \dfrac{\sqrt{5x^2}}{\sqrt{12}} = \dfrac{\sqrt{5x^2}}{\sqrt{4}\sqrt{3}} = \dfrac{x\sqrt{5}}{2\sqrt{3}} =$

$\dfrac{x\sqrt{5}}{2\sqrt{3}} \cdot \dfrac{\sqrt{3}}{\sqrt{3}} = \dfrac{x\sqrt{15}}{2 \cdot 3} = \dfrac{x\sqrt{15}}{6}$

54. $\dfrac{t\sqrt{14}}{8}$

55. 32 ft: $T \approx 2(3.14)\sqrt{\dfrac{32}{32}} \approx 6.28\sqrt{1} \approx 6.28(1) \approx$

6.28 sec

50 ft: $T \approx 2(3.14)\sqrt{\dfrac{50}{32}} \approx 6.28\sqrt{\dfrac{25}{16}} \approx 6.28\dfrac{\sqrt{25}}{\sqrt{16}} \approx$

$6.28\left(\dfrac{5}{4}\right) \approx 7.85$ sec

56. 3.14 sec; 1.57 sec

57. Substitute $\dfrac{2}{\pi^2}$ for L in the formula.

$T = 2\pi\sqrt{\dfrac{L}{32}} = 2\pi\sqrt{\dfrac{\frac{2}{\pi^2}}{32}} = 2\pi\sqrt{\dfrac{2}{\pi^2} \cdot \dfrac{1}{32}} =$

$2\pi\sqrt{\dfrac{2}{32\pi^2}} = 2\pi\sqrt{\dfrac{1}{16\pi^2}} = 2\pi \cdot \dfrac{\sqrt{1}}{\sqrt{16\pi^2}} = 2\pi \cdot \dfrac{1}{4\pi} =$

$\dfrac{2\pi}{4\pi} = 0.5$ sec

It takes 0.5 sec to move from one side to the other and back.

58. 3.5 sec

59. Substitute 72 for L in the formula.

$T \approx 2(3.14)\sqrt{\dfrac{72}{32}} \approx 6.28\sqrt{2.25} \approx 9.42$ sec

60. 4.71 sec

61. ◈

62. ◈

63. $5x + 9 + 7x + 4 = 5x + 7x + 9 + 4 = 12x + 13$

64. $10a + 6$

65. $2a^3 - a^2 - 3a^3 - 7a^2 = 2a^3 - 3a^3 - a^2 - 7a^2 = -a^3 - 8a^2$

66. $-11t^2 + 3t$

67. $9x(2x - 7) = 9x \cdot 2x - 9x \cdot 7 = 18x^2 - 63x$

68. $6x^2 - 10x$

69. $(3 + 4x)(2 + 5x) = 3 \cdot 2 + 3 \cdot 5x + 4x \cdot 2 + 4x \cdot 5x = 6 + 15x + 8x + 20x^2 = 6 + 23x + 20x^2$

70. $14 - 13x + 3x^2$

71. ◈

72. ◈

73. $\sqrt{\dfrac{7}{1000}} = \dfrac{\sqrt{7}}{\sqrt{1000}} = \dfrac{\sqrt{7}}{\sqrt{100}\sqrt{10}} = \dfrac{\sqrt{7}}{10\sqrt{10}} =$

$\dfrac{\sqrt{7}}{10\sqrt{10}} \cdot \dfrac{\sqrt{10}}{\sqrt{10}} = \dfrac{\sqrt{70}}{10 \cdot 10} = \dfrac{\sqrt{70}}{100}$

74. $\dfrac{\sqrt{6}}{40}$

75. $\sqrt{\dfrac{5x^2}{8x^7y^3}} = \sqrt{\dfrac{5}{8x^5y^3}} = \dfrac{\sqrt{5}}{\sqrt{8x^5y^3}} =$

$\dfrac{\sqrt{5}}{\sqrt{4x^4y^2}\sqrt{2xy}} = \dfrac{\sqrt{5}}{2x^2y\sqrt{2xy}} =$

$\dfrac{\sqrt{5}}{2x^2y\sqrt{2xy}} \cdot \dfrac{\sqrt{2xy}}{\sqrt{2xy}} = \dfrac{\sqrt{10xy}}{2x^2y \cdot 2xy} = \dfrac{\sqrt{10xy}}{4x^3y^2}$

76. $\dfrac{\sqrt{3xy}}{ax^2}$

77. $\sqrt{\dfrac{2a}{5b^3c^9}} = \dfrac{\sqrt{2a}}{\sqrt{5b^3c^9}} = \dfrac{\sqrt{2a}}{\sqrt{b^2c^8}\sqrt{5bc}} = \dfrac{\sqrt{2a}}{bc^4\sqrt{5bc}} =$

$\dfrac{\sqrt{2a}}{bc^4\sqrt{5bc}} \cdot \dfrac{\sqrt{5bc}}{\sqrt{5bc}} = \dfrac{\sqrt{10abc}}{bc^4 \cdot 5bc} = \dfrac{\sqrt{10abc}}{5b^2c^5}$

78. $\dfrac{\sqrt{5z}}{5zw}$

79. $\dfrac{3}{\sqrt{\sqrt{7}}} = \dfrac{3}{\sqrt{\sqrt{7}}} \cdot \dfrac{\sqrt{7\sqrt{7}}}{\sqrt{7\sqrt{7}}} = \dfrac{3\sqrt{7\sqrt{7}}}{\sqrt{\sqrt{7}\cdot 7\sqrt{7}}} = \dfrac{3\sqrt{7\sqrt{7}}}{\sqrt{7\cdot 7}} =$

$\dfrac{3\sqrt{7\sqrt{7}}}{7}$

80. $\dfrac{2\sqrt{5\sqrt{5}}}{5}$

81. $\sqrt{\dfrac{1}{x^2} - \dfrac{2}{xy} + \dfrac{1}{y^2}}$, LCD is x^2y^2

$= \sqrt{\dfrac{1}{x^2}\cdot\dfrac{y^2}{y^2} - \dfrac{2}{xy}\cdot\dfrac{xy}{xy} + \dfrac{1}{y^2}\cdot\dfrac{x^2}{x^2}}$

$= \sqrt{\dfrac{y^2 - 2xy + x^2}{x^2y^2}}$

$= \sqrt{\dfrac{(y-x)^2}{x^2y^2}}$

$= \dfrac{\sqrt{(y-x)^2}}{\sqrt{x^2y^2}}$

$= \dfrac{y-x}{xy}$

An alternate method of simplifying this expression is shown below.

$\sqrt{\dfrac{1}{x^2} - \dfrac{2}{xy} + \dfrac{1}{y^2}} = \sqrt{\left(\dfrac{1}{x} - \dfrac{1}{y}\right)^2}$

$= \dfrac{1}{x} - \dfrac{1}{y}$

The two answers are equivalent.

82. $\left(\dfrac{1}{z^2} - 1\right)\sqrt{2}$, or $\dfrac{\sqrt{2}(1-z^2)}{z^2}$

Exercise Set 8.4

1. $\quad 4\sqrt{3} + 8\sqrt{3}$

$= (4+8)\sqrt{3}$ Using the distributive law

$= 12\sqrt{3}$

2. $11\sqrt{2}$

3. $\quad 7\sqrt{2} - 5\sqrt{2}$

$= (7-5)\sqrt{2}$ Using the distributive law

$= 2\sqrt{2}$

4. $3\sqrt{5}$

5. $9\sqrt{y} + 3\sqrt{y} = (9+3)\sqrt{y} = 12\sqrt{y}$

6. $13\sqrt{x}$

7. $6\sqrt{a} - 12\sqrt{a} = (6-12)\sqrt{a} = -6\sqrt{a}$

8. $2\sqrt{x}$

9. $5\sqrt{6x} + 2\sqrt{6x} = (5+2)\sqrt{6x} = 7\sqrt{6x}$

10. $8\sqrt{2a}$

11. $12\sqrt{14y} - \sqrt{14y} = (12-1)\sqrt{14y} = 11\sqrt{14y}$

12. $8\sqrt{10y}$

13. $2\sqrt{5} + 7\sqrt{5} + 5\sqrt{5} = (2+7+5)\sqrt{5} = 14\sqrt{5}$

14. $11\sqrt{7}$

15. $3\sqrt{6} - 7\sqrt{6} + 2\sqrt{6} = (3-7+2)\sqrt{6} = -2\sqrt{6}$

16. $2\sqrt{2}$

17. $\quad 2\sqrt{5} + \sqrt{45} = 2\sqrt{5} + \sqrt{9\cdot 5}$ Factoring 45

$= 2\sqrt{5} + \sqrt{9}\sqrt{5}$

$= 2\sqrt{5} + 3\sqrt{5}$

$= (2+3)\sqrt{5}$

$= 5\sqrt{5}$

18. $5\sqrt{3} + 2\sqrt{2}$

19. $\sqrt{25a} - \sqrt{a} = \sqrt{25}\sqrt{a} - \sqrt{a}$
$= 5\sqrt{a} - \sqrt{a}$
$= (5-1)\sqrt{a}$
$= 4\sqrt{a}$

20. $-2\sqrt{x}$

21. $7\sqrt{50} - 3\sqrt{2} = 7\sqrt{25\cdot 2} - 3\sqrt{2}$
$= 7\sqrt{25}\sqrt{2} - 3\sqrt{2}$
$= 7\cdot 5\sqrt{2} - 3\sqrt{2}$
$= 35\sqrt{2} - 3\sqrt{2}$
$= 32\sqrt{2}$

22. $-18\sqrt{3}$

23. $3\sqrt{12} + 2\sqrt{300} = 3\sqrt{4\cdot 3} + 2\sqrt{100\cdot 3}$
$= 3\sqrt{4}\sqrt{3} + 2\sqrt{100}\sqrt{3}$
$= 3\cdot 2\sqrt{3} + 2\cdot 10\sqrt{3}$
$= 6\sqrt{3} + 20\sqrt{3}$
$= 26\sqrt{3}$

24. $28\sqrt{2}$

25. $\sqrt{45} + \sqrt{80} = \sqrt{9\cdot 5} + \sqrt{16\cdot 5}$
$= 3\sqrt{5} + 4\sqrt{5}$
$= 7\sqrt{5}$

26. $13\sqrt{2}$

27. $9\sqrt{8} + \sqrt{72} - 9\sqrt{8}$
Observe that $9\sqrt{8} - 9\sqrt{8} = 0$, so we need only to simplify $\sqrt{72}$.
$\sqrt{72} = \sqrt{36\cdot 2} = \sqrt{36}\sqrt{2} = 6\sqrt{2}$

28. $9\sqrt{3}$

29. $7\sqrt{12} - 2\sqrt{27} + \sqrt{75}$
$= 7\sqrt{4\cdot 3} - 2\sqrt{9\cdot 3} + \sqrt{25\cdot 3}$
$= 7\cdot 2\sqrt{3} - 2\cdot 3\sqrt{3} + 5\sqrt{3}$
$= 14\sqrt{3} - 6\sqrt{3} + 5\sqrt{3}$
$= 13\sqrt{3}$

30. $2\sqrt{2}$

31. $\sqrt{9x} + \sqrt{49x} - 9\sqrt{x} = 3\sqrt{x} + 7\sqrt{x} - 9\sqrt{x}$
$= (3+7-9)\sqrt{x}$
$= 1\sqrt{x}$
$= \sqrt{x}$

32. $5\sqrt{a}$

33. $\sqrt{2}(\sqrt{5} + \sqrt{7})$
$= \sqrt{2}\sqrt{5} + \sqrt{2}\sqrt{7}$ Using the distributive law
$= \sqrt{10} + \sqrt{14}$

34. $\sqrt{10} + \sqrt{55}$

35. $\sqrt{5}(\sqrt{6} - \sqrt{10}) = \sqrt{5}\sqrt{6} - \sqrt{5}\sqrt{10}$
$= \sqrt{30} - \sqrt{50}$
$= \sqrt{30} - \sqrt{25\cdot 2}$
$= \sqrt{30} - 5\sqrt{2}$

36. $3\sqrt{10} - \sqrt{42}$

37. $(3+\sqrt{2})(4+\sqrt{2})$
$= 3\cdot 4 + 3\cdot\sqrt{2} + \sqrt{2}\cdot 4 + \sqrt{2}\cdot\sqrt{2}$ Using FOIL
$= 12 + 3\sqrt{2} + 4\sqrt{2} + 2$
$= 14 + 7\sqrt{2}$

38. $26 + 8\sqrt{11}$

39. $(\sqrt{7}-2)(\sqrt{7}-5)$
$= \sqrt{7}\cdot\sqrt{7} - \sqrt{7}\cdot 5 - 2\cdot\sqrt{7} + 2\cdot 5$ Using FOIL
$= 7 - 5\sqrt{7} - 2\sqrt{7} + 10$
$= 17 - 7\sqrt{7}$

40. $-18 - 3\sqrt{10}$

41. $(\sqrt{5}+4)(\sqrt{5}-4)$
$= (\sqrt{5})^2 - 4^2$ Using $(A+B)(A-B) = A^2 - B^2$
$= 5 - 16$
$= -11$

42. -4

43. $(\sqrt{6}-\sqrt{3})(\sqrt{6}+\sqrt{3})$
$= (\sqrt{6})^2 - (\sqrt{3})^2$ Using $(A-B)(A+B) = A^2 - B^2$
$= 6 - 3$
$= 3$

44. -4

45. $(4+3\sqrt{2})(1-\sqrt{2})$
$= 4\cdot 1 - 4\cdot\sqrt{2} + 3\sqrt{2}\cdot 1 - 3\sqrt{2}\cdot\sqrt{2}$ Using FOIL
$= 4 - 4\sqrt{2} + 3\sqrt{2} - 3\cdot 2$
$= 4 - \sqrt{2} - 6$
$= -2 - \sqrt{2}$

46. $10 + 13\sqrt{7}$

47. $(7 + \sqrt{3})^2$
$= 7^2 + 2 \cdot 7 \cdot \sqrt{3} + (\sqrt{3})^2$ Using $(A + B)^2 =$
$\qquad\qquad\qquad\qquad\qquad A^2 + 2AB + B^2$
$= 49 + 14\sqrt{3} + 3$
$= 52 + 14\sqrt{3}$

48. $9 + 4\sqrt{5}$

49. $(1 - 2\sqrt{3})^2$
$= 1^2 - 2 \cdot 1 \cdot 2\sqrt{3} + (2\sqrt{3})^2$ Using $(A - B)^2 =$
$\qquad\qquad\qquad\qquad\qquad A^2 - 2AB + B^2$
$= 1 - 4\sqrt{3} + 4 \cdot 3$
$= 1 - 4\sqrt{3} + 12$
$= 13 - 4\sqrt{3}$

50. $81 - 36\sqrt{5}$

51. $(\sqrt{x} - \sqrt{10})^2 = (\sqrt{x})^2 - 2\sqrt{x}\sqrt{10} + (\sqrt{10})^2$
$\qquad\qquad\qquad = x - 2\sqrt{10x} + 10$

52. $a - 2\sqrt{6a} + 6$

53. $\dfrac{4}{7 + \sqrt{2}}$
$= \dfrac{4}{7 + \sqrt{2}} \cdot \dfrac{7 - \sqrt{2}}{7 - \sqrt{2}}$ Multiplying by 1
$= \dfrac{4(7 - \sqrt{2})}{(7 + \sqrt{2})(7 - \sqrt{2})}$
$= \dfrac{28 - 4\sqrt{2}}{7^2 - (\sqrt{2})^2}$
$= \dfrac{28 - 4\sqrt{2}}{49 - 2}$
$= \dfrac{28 - 4\sqrt{2}}{47}$

54. $\dfrac{3 - \sqrt{5}}{2}$

55. $\dfrac{6}{2 - \sqrt{7}}$
$= \dfrac{6}{2 - \sqrt{7}} \cdot \dfrac{2 + \sqrt{7}}{2 + \sqrt{7}}$
$= \dfrac{6(2 + \sqrt{7})}{(2 - \sqrt{7})(2 + \sqrt{7})}$
$= \dfrac{12 + 6\sqrt{7}}{2^2 - (\sqrt{7})^2}$
$= \dfrac{12 + 6\sqrt{7}}{4 - 7}$
$= \dfrac{12 + 6\sqrt{7}}{-3}$ Since 3 is a common factor, we simplify.
$= \dfrac{\cancel{3}(4 + 2\sqrt{7})}{\cancel{3}(-1)}$ Factoring and removing a factor equal to 1
$= \dfrac{4 + 2\sqrt{7}}{-1}$
$= -4 - 2\sqrt{7}$

56. $\dfrac{12 + 3\sqrt{2}}{14}$

57. $\dfrac{2}{\sqrt{7} + 5}$
$= \dfrac{2}{\sqrt{7} + 5} \cdot \dfrac{\sqrt{7} - 5}{\sqrt{7} - 5}$
$= \dfrac{2(\sqrt{7} - 5)}{(\sqrt{7})^2 - 5^2}$
$= \dfrac{2\sqrt{7} - 10}{7 - 25}$
$= \dfrac{2\sqrt{7} - 10}{-18}$ Since 2 is a common factor, we simplify.
$= \dfrac{\cancel{2}(\sqrt{7} - 5)}{\cancel{2}(-9)}$ Factoring and removing a factor equal to 1
$= \dfrac{\sqrt{7} - 5}{-9}$
$= \dfrac{-(\sqrt{7} - 5)}{9}$
$= \dfrac{5 - \sqrt{7}}{9}$

58. $6\sqrt{10} - 18$

59. $\dfrac{\sqrt{6}}{\sqrt{6} - 5} = \dfrac{\sqrt{6}}{\sqrt{6} - 5} \cdot \dfrac{\sqrt{6} + 5}{\sqrt{6} + 5} = \dfrac{\sqrt{6}\sqrt{6} + \sqrt{6} \cdot 5}{(\sqrt{6})^2 - 5^2} =$
$\dfrac{6 + 5\sqrt{6}}{6 - 25} = \dfrac{6 + 5\sqrt{6}}{-19} = -\dfrac{6 + 5\sqrt{6}}{19}, \text{ or } \dfrac{-6 - 5\sqrt{6}}{19}$

60. $-\dfrac{10 + 7\sqrt{10}}{39}$, or $\dfrac{-10 - 7\sqrt{10}}{39}$

61. $\dfrac{\sqrt{5}}{\sqrt{5} + \sqrt{3}} = \dfrac{\sqrt{5}}{\sqrt{5} + \sqrt{3}} \cdot \dfrac{\sqrt{5} - \sqrt{3}}{\sqrt{5} - \sqrt{3}} =$

$\dfrac{\sqrt{5}\sqrt{5} - \sqrt{5}\sqrt{3}}{(\sqrt{5})^2 - (\sqrt{3})^2} = \dfrac{5 - \sqrt{15}}{5 - 3} = \dfrac{5 - \sqrt{15}}{2}$

62. $\dfrac{7 + \sqrt{35}}{2}$

63. $\dfrac{\sqrt{3}}{\sqrt{5} - \sqrt{3}} = \dfrac{\sqrt{3}}{\sqrt{5} - \sqrt{3}} \cdot \dfrac{\sqrt{5} + \sqrt{3}}{\sqrt{5} + \sqrt{3}} =$

$\dfrac{\sqrt{3}\sqrt{5} + \sqrt{3}\sqrt{3}}{(\sqrt{5})^2 - (\sqrt{3})^2} = \dfrac{\sqrt{15} + 3}{5 - 3} = \dfrac{\sqrt{15} + 3}{2}$

64. $\sqrt{42} - 6$

65. $\dfrac{2}{\sqrt{7} + \sqrt{2}} = \dfrac{2}{\sqrt{7} + \sqrt{2}} \cdot \dfrac{\sqrt{7} - \sqrt{2}}{\sqrt{7} - \sqrt{2}} =$

$\dfrac{2\sqrt{7} - 2\sqrt{2}}{(\sqrt{7})^2 - (\sqrt{2})^2} = \dfrac{2\sqrt{7} - 2\sqrt{2}}{7 - 2} = \dfrac{2\sqrt{7} - 2\sqrt{2}}{5}$

66. $3\sqrt{5} + 3\sqrt{3}$

67. $\dfrac{\sqrt{7} + \sqrt{5}}{\sqrt{7} - \sqrt{5}} = \dfrac{\sqrt{7} + \sqrt{5}}{\sqrt{7} - \sqrt{5}} \cdot \dfrac{\sqrt{7} + \sqrt{5}}{\sqrt{7} + \sqrt{5}}$

$= \dfrac{(\sqrt{7} + \sqrt{5})^2}{(\sqrt{7} - \sqrt{5})(\sqrt{7} + \sqrt{5})}$

$= \dfrac{(\sqrt{7})^2 + 2\sqrt{7}\sqrt{5} + (\sqrt{5})^2}{(\sqrt{7})^2 - (\sqrt{5})^2}$

$= \dfrac{7 + 2\sqrt{35} + 5}{7 - 5} = \dfrac{12 + 2\sqrt{35}}{2}$

$= \dfrac{\cancel{2}(6 + \sqrt{35})}{\cancel{2} \cdot 1} = 6 + \sqrt{35}$

68. $4 - \sqrt{15}$

69. ◈

70. ◈

71. $3x + 5 + 2(x - 3) = 4 - 6x$

$3x + 5 + 2x - 6 = 4 - 6x$

$5x - 1 = 4 - 6x$

$11x - 1 = 4$

$11x = 5$

$x = \dfrac{5}{11}$

The solution is $\dfrac{5}{11}$.

72. $-\dfrac{38}{13}$

73. $x^2 - 5x = 6$

$x^2 - 5x - 6 = 0$

$(x + 1)(x - 6) = 0$

$x + 1 = 0 \quad or \quad x - 6 = 0$

$x = -1 \quad or \qquad x = 6$

The solutions are -1 and 6.

74. $2, 5$

75. *Familiarize.* Let x = the number of liters of Jolly Juice and y = the number of liters of Real Squeeze in the mixture. We organize the given information in a table.

	Jolly Juice	Real Squeeze	Mixture
Amount	x	y	8
Percent real fruit juice	3%	6%	5.4%
Amount of real fruit juice	$0.03x$	$0.06y$	$0.054(8)$, or 0.432

Translate. We get two equation from the first and third rows of the table.

$x + \quad y = \quad 8,$

$0.03x + 0.06y = 0.432$

Clearing decimals gives

$x + \quad y = \quad 8, \quad (1)$

$30x + 60y = 432. \quad (2)$

Carry out. We use elimination. Multiply Equation (1) by -30 and add.

$-30x - 30y = -240$

$\underline{30x + 60y = \quad 432}$

$30y = \quad 192$

$y = \quad 6.4$

Now substitute 6.4 for y in Equation (1) and solve for x.

$x + y = 8$

$x + 6.4 = 8$

$x = 1.6$

Check. The sum of 1.6 and 6.4 is 8. The amount of real fruit juice in this mixture is $0.03(1.6) + 0.06(6.4)$, or $0.048 + 0.384$, or 0.432 L. The answer checks.

State. 1.6 L of Jolly Juice and 6.4 L of Real Squeeze should be used.

76. At least 3 hr

77.

78.

79. $5\sqrt{\dfrac{1}{2}} + \dfrac{7}{2}\sqrt{18} - 4\sqrt{98}$

$= 5 \cdot \dfrac{1}{\sqrt{2}} + \dfrac{7}{2}\sqrt{9 \cdot 2} - 4\sqrt{49 \cdot 2}$

$= \dfrac{5}{\sqrt{2}} \cdot \dfrac{\sqrt{2}}{\sqrt{2}} + \dfrac{7}{2} \cdot 3\sqrt{2} - 4 \cdot 7\sqrt{2}$

$= \dfrac{5\sqrt{2}}{2} + \dfrac{21\sqrt{2}}{2} - 28\sqrt{2}$

$= \left(\dfrac{5}{2} + \dfrac{21}{2} - 28\right)\sqrt{2}$

$= -15\sqrt{2}$

80. $\dfrac{11\sqrt{x} - 5x\sqrt{2}}{2x}$

81. $a\sqrt{a^{17}b^9} - b\sqrt{a^{13}b^{11}} + a\sqrt{a^9b^{15}}$

$= a\sqrt{a^{16} \cdot a \cdot b^8 \cdot b} - b\sqrt{a^{12} \cdot a \cdot b^{10} \cdot b} +$
$\qquad\qquad\qquad\qquad a\sqrt{a^8 \cdot a \cdot b^{14} \cdot b}$

$= a \cdot a^8 b^4\sqrt{ab} - b \cdot a^6 b^5\sqrt{ab} + a \cdot a^4 b^7\sqrt{ab}$

$= a^9 b^4\sqrt{ab} - a^6 b^6\sqrt{ab} + a^5 b^7\sqrt{ab}$

$= a^5 b^4\sqrt{ab}(a^4 - ab^2 + b^3)$

82. $xy\sqrt{2y}(2x^2 - y^2 - xy^3)$

83. $7x\sqrt{12xy^2} - 9y\sqrt{27x^3} + 5\sqrt{300x^3y^2}$

$= 7x\sqrt{4y^2 \cdot 3x} - 9y\sqrt{9x^2 \cdot 3x} + 5\sqrt{100x^2y^2 \cdot 3x}$

$= 7x \cdot 2y\sqrt{3x} - 9y \cdot 3x\sqrt{3x} + 5 \cdot 10xy\sqrt{3x}$

$= 14xy\sqrt{3x} - 27xy\sqrt{3x} + 50xy\sqrt{3x}$

$= (14xy - 27xy + 50xy)\sqrt{3x}$

$= 37xy\sqrt{3x}$

84. Any pair of numbers a, b such that $a = 0$ or $b = 0$ will do.

85. $\sqrt{10} + \sqrt{50} = \sqrt{10} + \sqrt{10}\,\sqrt{5} = \sqrt{10}(1 + \sqrt{5})$

$\sqrt{10} + \sqrt{50} = \sqrt{10} + \sqrt{25 \cdot 2} = \sqrt{10} + 5\sqrt{2}$

$\sqrt{10} + \sqrt{50} = \sqrt{2}\,\sqrt{5} + \sqrt{2}\,\sqrt{25} =$

$\sqrt{2}(\sqrt{5} + \sqrt{25}) = \sqrt{2}(\sqrt{5} + 5)$, or $\sqrt{2}(5 + \sqrt{5})$

All three are correct.

Exercise Set 8.5

1. $\sqrt{x} = 5$

$(\sqrt{x})^2 = 5^2$ Squaring both sides

$x = 25$ Simplifying

Check: $\dfrac{\sqrt{x} = 5}{}$

$\sqrt{25} \ ? \ 5$

$5 \ \Big| \ 5$ TRUE

The solution is 25.

2. 64

3. $\sqrt{x} + 4 = 12$

$\sqrt{x} = 8$ Subtracting 4

$(\sqrt{x})^2 = 8^2$

$x = 64$

Check: $\dfrac{\sqrt{x} + 4 = 12}{}$

$\sqrt{64} + 4 \ ? \ 12$

$8 + 4 \ \Big|$

$12 \ \Big| \ 12$ TRUE

The solution is 64.

4. 144

5. $\sqrt{2x + 1} = 13$

$(\sqrt{2x + 1})^2 = 13^2$ Squaring both sides

$2x + 1 = 169$

$2x = 168$

$x = 84$

Check: $\dfrac{\sqrt{2x + 1} = 13}{}$

$\sqrt{2 \cdot 84 + 1} \ ? \ 13$

$\sqrt{169} \ \Big|$

$13 \ \Big| \ 13$ TRUE

The solution is 84.

6. $\dfrac{77}{2}$

7. $2 + \sqrt{3 - y} = 9$

$\sqrt{3 - y} = 7$ Subtracting 2

$(\sqrt{3 - y})^2 = 7^2$ Squaring both sides

$3 - y = 49$

$-y = 46$

$y = -46$

Check:
$$2 + \sqrt{3 - y} = 9$$
$$\begin{array}{c|c} 2 + \sqrt{3 - (-46)} & ? \ 9 \\ 2 + \sqrt{49} \\ 2 + 7 \\ \hline 9 & 9 \quad \text{TRUE} \end{array}$$

The solution is -46.

8. -3

9. $8 - 4\sqrt{5n} = 0$
$$8 = 4\sqrt{5n} \quad \text{Adding } 4\sqrt{5n}$$
$$2 = \sqrt{5n} \quad \text{Dividing by 4}$$
$$2^2 = (\sqrt{5n})^2 \quad \text{Squaring both sides}$$
$$4 = 5n$$
$$\frac{4}{5} = n$$

Check:
$$8 - 4\sqrt{5n} = 0$$
$$\begin{array}{c|c} 8 - 4\sqrt{5 \cdot \frac{4}{5}} & ? \ 0 \\ 8 - 4\sqrt{4} \\ 8 - 4 \cdot 2 \\ 8 - 8 \\ \hline 0 & 0 \quad \text{TRUE} \end{array}$$

The solution is $\frac{4}{5}$.

10. 3

11. $\sqrt{4x - 6} = \sqrt{x + 9}$
$$(\sqrt{4x - 6})^2 = (\sqrt{x + 9})^2$$
$$4x - 6 = x + 9$$
$$3x - 6 = 9$$
$$3x = 15$$
$$x = 5$$

Check:
$$\sqrt{4x - 6} = \sqrt{x + 9}$$
$$\begin{array}{c|c} \sqrt{4 \cdot 5 - 6} & ? \ \sqrt{5 + 9} \\ \sqrt{20 - 6} & \sqrt{14} \\ \sqrt{14} & \sqrt{14} \quad \text{TRUE} \end{array}$$

The solution is 5.

12. 3

13. $\sqrt{x} = -7$

Since the principal square root of a number cannot be negative, we see that this equation has no solution.

14. No solution

15. $\sqrt{2t + 5} = \sqrt{3t + 7}$
$$(\sqrt{2t+5})^2 = (\sqrt{3t+7})^2 \quad \text{Squaring both sides}$$
$$2t + 5 = 3t + 7$$
$$5 = t + 7$$
$$-2 = t$$

Check:
$$\sqrt{2t + 5} = \sqrt{3t + 7}$$
$$\begin{array}{c|c} \sqrt{2(-2) + 5} & ? \ \sqrt{3(-2) + 7} \\ \sqrt{-4 + 5} & \sqrt{-6 + 7} \\ \sqrt{1} & \sqrt{1} \\ 1 & 1 \quad \text{TRUE} \end{array}$$

The solution is -2.

16. -1

17. $\sqrt{3x - 2} = x - 4$
$$(\sqrt{3x - 2})^2 = (x - 4)^2$$
$$3x - 2 = x^2 - 8x + 16$$
$$0 = x^2 - 11x + 18$$
$$0 = (x - 2)(x - 9)$$
$$x - 2 = 0 \quad or \quad x - 9 = 0$$
$$x = 2 \quad or \quad x = 9$$

Check:
$$\sqrt{3x - 2} = x - 4$$
$$\begin{array}{c|c} \sqrt{3 \cdot 2 - 2} & ? \ 2 - 4 \\ \sqrt{4} & -2 \\ 2 & -2 \quad \text{FALSE} \end{array}$$
$$\sqrt{3x - 2} = x - 4$$
$$\begin{array}{c|c} \sqrt{3 \cdot 9 - 2} & ? \ 9 - 4 \\ \sqrt{25} & 5 \\ 5 & 5 \quad \text{TRUE} \end{array}$$

The number 2 does not check, but 9 does. The solution is 9.

18. 9

19. $a - 9 = \sqrt{a - 3}$
$$(a - 9)^2 = (\sqrt{a - 3})^2$$
$$a^2 - 18a + 81 = a - 3$$
$$a^2 - 19a + 84 = 0$$
$$(a - 12)(a - 7) = 0$$
$$a - 12 = 0 \quad or \quad a - 7 = 0$$
$$a = 12 \quad or \quad a = 7$$

Check:

$$\begin{array}{c|c} a - 9 = \sqrt{a - 3} \\ \hline 12 - 9 \ ? \ \sqrt{12 - 3} \\ 3 \ \big| \ \sqrt{9} \\ 3 \ \big| \ 3 \qquad \text{TRUE} \end{array}$$

$$\begin{array}{c|c} a - 9 = \sqrt{a - 3} \\ \hline 7 - 9 \ ? \ \sqrt{7 - 3} \\ -2 \ \big| \ \sqrt{4} \\ -2 \ \big| \ 2 \qquad \text{FALSE} \end{array}$$

The number 12 checks, but 7 does not. The solution is 12.

20. 7

21.
$$x - 1 = 6\sqrt{x - 9}$$
$$(x - 1)^2 = (6\sqrt{x - 9})^2$$
$$x^2 - 2x + 1 = 36(x - 9)$$
$$x^2 - 2x + 1 = 36x - 324$$
$$x^2 - 38x + 325 = 0$$
$$(x - 13)(x - 25) = 0$$
$$x - 13 = 0 \quad or \quad x - 25 = 0$$
$$x = 13 \quad or \qquad x = 25$$

Check:
$$\begin{array}{c|c} x - 1 = 6\sqrt{x - 9} \\ \hline 13 - 1 \ ? \ 6\sqrt{13 - 9} \\ 12 \ \big| \ 6\sqrt{4} \\ \big| \ 6 \cdot 2 \\ 12 \ \big| \ 12 \qquad \text{TRUE} \end{array}$$

$$\begin{array}{c|c} x - 1 = 6\sqrt{x - 9} \\ \hline 25 - 1 \ ? \ 6\sqrt{25 - 9} \\ 24 \ \big| \ 6\sqrt{16} \\ \big| \ 6 \cdot 4 \\ 24 \ \big| \ 24 \qquad \text{TRUE} \end{array}$$

The solutions are 13 and 25.

22. 5

23.
$$\sqrt{5x + 21} = x + 3$$
$$(\sqrt{5x + 21})^2 = (x + 3)^2$$
$$5x + 21 = x^2 + 6x + 9$$
$$0 = x^2 + x - 12$$
$$0 = (x + 4)(x - 3)$$
$$x + 4 = 0 \quad or \quad x - 3 = 0$$
$$x = -4 \quad or \qquad x = 3$$

Check:

$$\begin{array}{c|c} \sqrt{5x + 21} = x + 3 \\ \hline \sqrt{5(-4) + 21} \ ? \ -4 + 3 \\ \sqrt{1} \ \big| \ -1 \\ 1 \ \big| \ -1 \qquad \text{FALSE} \end{array}$$

$$\begin{array}{c|c} \sqrt{5x + 21} = x + 3 \\ \hline \sqrt{5 \cdot 3 + 21} \ \big| \ 3 + 3 \\ \sqrt{36} \ \big| \ 6 \\ 6 \ \big| \ 6 \qquad \text{TRUE} \end{array}$$

The number 3 checks, but -4 does not. The solution is 3.

24. 6

25.
$$t + 4 = 4\sqrt{t + 1}$$
$$(t + 4)^2 = (4\sqrt{t + 1})^2$$
$$t^2 + 8t + 16 = 16(t + 1)$$
$$t^2 + 8t + 16 = 16t + 16$$
$$t^2 - 8t = 0$$
$$t(t - 8) = 0$$
$$t = 0 \quad or \quad t - 8 = 0$$
$$t = 0 \quad or \qquad t = 8$$

Check:
$$\begin{array}{c|c} t + 4 = 4\sqrt{t + 1} \\ \hline 0 + 4 \ ? \ 4\sqrt{0 + 1} \\ 4 \ \big| \ 4\sqrt{1} \\ \big| \ 4 \cdot 1 \\ 4 \ \big| \ 4 \qquad \text{TRUE} \end{array}$$

$$\begin{array}{c|c} t + 4 = 4\sqrt{t + 1} \\ \hline 8 + 4 \ ? \ 4\sqrt{8 + 1} \\ 12 \ \big| \ 4\sqrt{9} \\ \big| \ 4 \cdot 3 \\ 12 \ \big| \ 12 \qquad \text{TRUE} \end{array}$$

The solutions are 0 and 8.

26. 1, 5

27.
$$\sqrt{x^2 + 5} - x + 2 = 0$$
$$\sqrt{x^2 + 5} = x - 2 \quad \text{Isolating the radical}$$
$$(\sqrt{x^2 + 5})^2 = (x - 2)^2$$
$$x^2 + 5 = x^2 - 4x + 4$$
$$1 = -4x \quad \text{Adding } -x^2 \text{ and } -4$$
$$-\frac{1}{4} = x$$

Check: $\dfrac{\sqrt{x^2+5}-x+2=0}{}$

$$\sqrt{\left(-\frac{1}{4}\right)^2+5}-\left(-\frac{1}{4}\right)+2 \;\Big|\; 0$$

$$\sqrt{\frac{81}{16}}+\frac{1}{4}+2$$

$$\frac{9}{4}+\frac{1}{4}+2$$

$$\frac{18}{4}\;\Big|\;0 \text{ FALSE}$$

The number $-\dfrac{1}{4}$ does not check. There is no solution.

28. No solution

29. $\sqrt{(p+6)(p+1)}-2=p+1$

$\quad\sqrt{(p+6)(p+1)}=p+3 \quad$ Isolating the radical

$\quad\left(\sqrt{(p+6)(p+1)}\right)^2=(p+3)^2$

$\quad(p+6)(p+1)=p^2+6p+9$

$\quad p^2+7p+6=p^2+6p+9$

$\quad p=3$

The number 3 checks. It is the solution.

30. 5

31. $\sqrt{2-7x}=\sqrt{5-2x}$

$\quad(\sqrt{2-7x})^2=(\sqrt{5-2x})^2$

$\quad 2-7x=5-2x$

$\quad 2=5+5x$

$\quad -3=5x$

$\quad -\dfrac{3}{5}=x$

The number $-\dfrac{3}{5}$ checks. It is the solution.

32. $-\dfrac{5}{4}$

33. $x-1=\sqrt{(x+1)(x-2)}$

$\quad(x-1)^2=(\sqrt{(x+1)(x-2)})^2$

$\quad x^2-2x+1=(x+1)(x-2)$

$\quad x^2-2x+1=x^2-x-2$

$\quad -2x+1=-x-2 \quad$ Adding $-x^2$ on both sides

$\quad -x+1=-2$

$\quad -x=-3$

$\quad x=3$

The number 3 checks. It is the solution.

34. 1

35. $r=2\sqrt{5L}$

$\quad 48=2\sqrt{5L} \quad$ Substituting 48 for r

$\quad 24=\sqrt{5L}$

$\quad 24^2=(\sqrt{5L})^2$

$\quad 576=5L$

$\quad 115.2=L$

The car will skid 115.2 ft at 48 mph.

$\quad 80=2\sqrt{5L} \quad$ Substituting 80 for r

$\quad 40=\sqrt{5L}$

$\quad 40^2=(\sqrt{5L})^2$

$\quad 1600=5L$

$\quad 320=L$

The car will skid 320 ft at 80 mph.

36. 80 ft; 180 ft

37. *Familiarize.* We will use the formula $s=21.9\sqrt{5t+2457}$, where t is in degrees Fahrenheit and s is in feet per second.

Translate. We substitute 1113 for s in the formula.

$\quad 1113=21.9\sqrt{5t+2457}$

Carry out. We solve for t.

$\quad 1113=21.9\sqrt{5t+2457}$

$\quad \dfrac{1113}{21.9}=\sqrt{5t+2457}$

$\quad \left(\dfrac{1113}{21.9}\right)^2=(\sqrt{5t+2457})^2$

$\quad 2582.9\approx 5t+2457$

$\quad 125.9\approx 5t$

$\quad 25.2\approx t$

Check. We can substitute 25.2 for t in the formula.

$\quad 21.9\sqrt{5(25.2)+2457}=21.9\sqrt{2583}\approx 1113$

The answer checks.

State. The temperature was about 25.2°F.

38. 85.3°F

39. *Familiarize and Translate.* We substitute 99.4 for V in equation $V=3.5\sqrt{h}$.

$\quad 99.4=3.5\sqrt{h}$

Carry out. We solve the equation.

$\quad 99.4=3.5\sqrt{h}$

$\quad 28.4=\sqrt{h} \quad$ Dividing by 3.5

$\quad (28.4)^2=(\sqrt{h})^2$

$\quad 806.56=h$

Check. We go over the computation.

State. The mast is 806.56 m high.

40. 36 m

41. Familiarize and Translate. We substitute 84 for V in the equation $V = 3.5\sqrt{h}$.

$$84 = 3.5\sqrt{h}$$

Carry out.

$$84 = 3.5\sqrt{h}$$
$$24 = \sqrt{h} \quad \text{Dividing by 3.5}$$
$$24^2 = (\sqrt{h})^2$$
$$576 = h$$

Check. We go over the computation.

State. The altitude of the scout's eyes is 576 m.

42. 11,664 m

43.
$$T = 2\pi\sqrt{\frac{L}{32}}$$

$$1.0 = 2(3.14)\sqrt{\frac{L}{32}} \quad \begin{array}{l}\text{Substituting 1.0 for } T \\ \text{and 3.14 for } \pi\end{array}$$

$$1.0 = 6.28\sqrt{\frac{L}{32}}$$

$$\frac{1.0}{6.28} = \sqrt{\frac{L}{32}}$$

$$\left(\frac{1.0}{6.28}\right)^2 = \left(\sqrt{\frac{L}{32}}\right)^2$$

$$0.0254 \approx \frac{L}{32}$$

$$0.81 \approx L$$

The pendulum is about 0.81 ft long.

44. About 3.25 ft

45.

46.

47. Familiarize. Let a and b represent the number of questions of type A and type B answered correctly, respectively. Then Amy scores $10a$ points on the type A items and $15b$ points on the type B items.

Translate. Sixteen questions were answered correctly, so we can write one equation:

$$a + b = 16$$

Amy scored a total of 180 points, so we can write a second equation:

$$10a + 15b = 180$$

The resulting system is

$$a + b = 16, \quad (1)$$
$$10a + 15b = 180. \quad (2)$$

Carry out. We first multiply Equation (1) by -10 and add.

$$-10a - 10b = -160$$
$$\underline{10a + 15b = 180}$$
$$5b = 20$$
$$b = 4$$

Now substitute 4 for b in Equation (1) and solve for a.

$$a + b = 16$$
$$a + 4 = 16$$
$$a = 12$$

Check. If Amy correctly answers 12 questions of type A and 4 questions of type B, she answers a total of $12 + 4$, or 16 questions. She scores $10 \cdot 12 + 15 \cdot 4$, or $120 + 60$, 180 points. The answer checks.

State. Amy answered 12 questions of type A and 4 questions of type B correctly.

48. 45 oz of the three-fourths pure; 15 oz of the five-twelfths pure

49. Familiarize. We present the information in a table.

	d	$=$	r	$\cdot$	t
	Distance		Speed		Time
First car	d		56		t
Second car	d		84		$t - 1$

Translate. From the rows of the table we get two equations:

$$d = 56t,$$
$$d = 84(t - 1).$$

Carry out. We use the substitution method.

$$56t = 84(t - 1) \quad \text{Substituting } 56t \text{ for } d$$
$$56t = 84t - 84$$
$$-28t = -84$$
$$t = 3$$

The problem asks how far from Hereford the second car will overtake the first, so we need to find d. Substitute 3 for t in the first equation.

$$d = 56t$$
$$d = 56 \cdot 3$$
$$d = 168$$

Check. If $t = 3$, then the first car travels $56 \cdot 3$, or 168 km, and the second car travels $84(3 - 1)$, or

84 · 2, or 168 km. Since the distances are the same, the answer checks.

State. The second car overtakes the first 168 km from Hereford.

50. 70 dimes; 33 quarters

51.

52.

53. Familiarize. Let x represent the number. Then the square root of twice the number is $\sqrt{2x}$.

Translate. We reword the problem.

$$\underbrace{\text{The square root of twice the number}}_{\downarrow} \quad \text{less} \quad 1 \quad \text{is} \quad 7.$$
$$\sqrt{2x} \qquad\qquad - \quad 1 \;=\; 7$$

Carry out. We solve the equation.

$$\sqrt{2x} - 1 = 7$$
$$\sqrt{2x} = 8$$
$$(\sqrt{2x})^2 = 8^2$$
$$2x = 64$$
$$x = 32$$

Check. The square root of two times 32 is $\sqrt{2 \cdot 32}$, or $\sqrt{64}$, or 8, and 1 less than this number is $8 - 1$, or 7. The answer checks.

State. The number is 32.

54. 121

55.
$$5 - \sqrt{x} = \sqrt{x - 5}$$
$$(5 - \sqrt{x})^2 = (\sqrt{x - 5})^2$$
$$25 - 10\sqrt{x} + x = x - 5$$
$$25 - 10\sqrt{x} = -5 \qquad \text{Adding } -x$$
$$-10\sqrt{x} = -30$$
$$\sqrt{x} = 3 \qquad \text{Dividing by } -10$$
$$(\sqrt{x})^2 = 3^2$$
$$x = 9$$

The number 9 checks. It is the solution.

56. 16

57.
$$\sqrt{t + 4} = 1 - \sqrt{3t + 1}$$
$$(\sqrt{t + 4})^2 = (1 - \sqrt{3t + 1})^2$$
$$t + 4 = 1 - 2\sqrt{3t + 1} + 3t + 1$$
$$t + 4 = 2 - 2\sqrt{3t + 1} + 3t$$
$$-2t + 2 = -2\sqrt{3t + 1}$$
$$\qquad\qquad\qquad \text{Isolating the radical}$$
$$t - 1 = \sqrt{3t + 1}$$
$$\qquad\qquad\qquad \text{Multiplying by } -\frac{1}{2}$$
$$(t - 1)^2 = (\sqrt{3t + 1})^2$$
$$t^2 - 2t + 1 = 3t + 1$$
$$t^2 - 5t = 0$$
$$t(t - 5) = 0$$
$$t = 0 \quad or \quad t - 5 = 0$$
$$t = 0 \quad or \qquad t = 5$$

Check:

$$\begin{array}{c}
\sqrt{t + 4} = 1 - \sqrt{3t + 1} \\
\hline
\sqrt{0 + 4} \;?\; 1 - \sqrt{3 \cdot 0 + 1} \\
\sqrt{4} \;\Big|\; 1 - \sqrt{1} \\
2 \;\Big|\; 1 - 1 \\
2 \;\Big|\; 0 \qquad\qquad \text{FALSE}
\end{array}$$

$$\begin{array}{c}
\sqrt{t + 4} = 1 - \sqrt{3t + 1} \\
\hline
\sqrt{5 + 4} \;?\; 1 - \sqrt{3 \cdot 5 + 1} \\
\sqrt{9} \;\Big|\; 1 - \sqrt{16} \\
3 \;\Big|\; 1 - 4 \\
3 \;\Big|\; -3 \qquad\qquad \text{FALSE}
\end{array}$$

Neither number checks. There is no solution.

58. 1

59.
$$3 + \sqrt{19 - x} = 5 + \sqrt{4 - x}$$
$$\sqrt{19 - x} = 2 + \sqrt{4 - x} \quad \text{Isolating one radical}$$
$$(\sqrt{19 - x})^2 = (2 + \sqrt{4 - x})^2$$
$$19 - x = 4 + 4\sqrt{4 - x} + (4 - x)$$
$$19 - x = 4\sqrt{4 - x} + 8 - x$$
$$11 = 4\sqrt{4 - x}$$
$$11^2 = (4\sqrt{4 - x})^2$$
$$121 = 16(4 - x)$$
$$121 = 64 - 16x$$
$$57 = -16x$$
$$-\frac{57}{16} = x$$

$-\frac{57}{16}$ checks, so it is the solution.

60. 3

61.
$$2\sqrt{x-1} - \sqrt{x-9} = \sqrt{3x-5}$$
$$(2\sqrt{x-1} - \sqrt{x-9})^2 = (\sqrt{3x-5})^2$$
$$4(x-1) - 4\sqrt{(x-1)(x-9)} + x - 9 = 3x - 5$$
$$4x - 4 - 4\sqrt{x^2 - 10x + 9} + x - 9 = 3x - 5$$
$$5x - 13 - 4\sqrt{x^2 - 10x + 9} = 3x - 5$$
$$-4\sqrt{x^2 - 10x + 9} = -2x + 8$$
$$2\sqrt{x^2 - 10x + 9} = x - 4$$
$$\text{Multiplying by } -\frac{1}{2}$$
$$(2\sqrt{x^2 - 10x + 9})^2 = (x-4)^2$$
$$4(x^2 - 10x + 9) = x^2 - 8x + 16$$
$$4x^2 - 40x + 36 = x^2 - 8x + 16$$
$$3x^2 - 32x + 20 = 0$$
$$(3x - 2)(x - 10) = 0$$

$$3x - 2 = 0 \quad or \quad x - 10 = 0$$
$$3x = 2 \quad or \qquad x = 10$$
$$x = \frac{2}{3} \quad or \qquad x = 10$$

The number 10 checks, but $\frac{2}{3}$ does not. The solution is 10.

62. 0, 4

63. Familiarize. We will use the formula $V = 3.5\sqrt{h}$. We present the information in a table.

	Height	Distance to the horizon
First sighting	h	V
Second sighting	$h+100$	$V+20$

Translate. The rows of the table give us two equations.
$$V = 3.5\sqrt{h}, \qquad (1)$$
$$V + 20 = 3.5\sqrt{h+100} \quad (2)$$

Carry out. We substitute $3.5\sqrt{h}$ for V in Equation (2) and solve for h.

$$3.5\sqrt{h} + 20 = 3.5\sqrt{h+100}$$
$$(3.5\sqrt{h} + 20)^2 = (3.5\sqrt{h+100})^2$$
$$12.25h + 140\sqrt{h} + 400 = 12.25(h+100)$$
$$12.25h + 140\sqrt{h} + 400 = 12.25h + 1225$$
$$140\sqrt{h} = 825$$
$$28\sqrt{h} = 165 \quad \text{Multiplying by } \frac{1}{5}$$
$$(28\sqrt{h})^2 = (165)^2$$
$$784h = 27,225$$
$$h = 34\frac{569}{784}, \text{ or}$$
$$h \approx 34.726$$

Check. When $h \approx 34.726$, then $V \approx 3.5\sqrt{34.726} \approx 20.625$ km. When $h \approx 100 + 34.726$, or 134.726, then $V \approx 3.5\sqrt{134.726} \approx 40.625$ km. This is 20 km more than 20.625. The answer checks.

State. The climber was at a height of $34\frac{569}{784}$ m, or about 34.726 m when the first computation was made.

64. $b = \dfrac{a}{A^4 - 2A^2 + 1}$

65. Graph $y = \sqrt{x}$.

We make a table of values. Note that we must choose nonnegative values of x in order to have a nonnegative radicand.

x	y
0	0
1	1
2	1.414
4	2
5	2.236

We plot these points and connect them with a smooth curve.

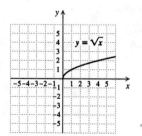

66.

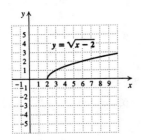

x	y
5	0
6	1
7	1.414
8	1.732
9	2

We plot these points and connect them with a smooth curve.

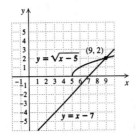

The graphs intersect at $(9, 2)$, so the solution of the equation $x - 7 = \sqrt{x - 5}$ is 9.

70.

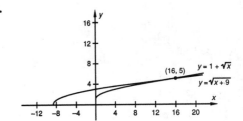

The solution is 16.

71. Graph $y_1 = \sqrt{x + 3}$ and $y_2 = 2x - 1$ and then find the first coordinate(s) of the point(s) of intersection. The solution is about 1.57.

72. -0.32

73. ◈

67. Graph $y = \sqrt{x - 3}$.

We make a table of values. Note that we must choose values for x that are greater than or equal to 3 in order to have a nonnegative radicand.

x	y
3	0
4	1
5	1.414
6	1.732
7	2

We plot these points and connect them with a smooth curve.

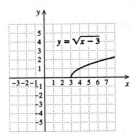

68.

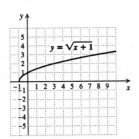

69. We can graph $y = x - 7$ using the intercepts, $(0, -7)$ and $(7, 0)$.

We make a table of values for $y = \sqrt{x - 5}$.

Exercise Set 8.6

1. $a^2 + b^2 = c^2$

$\quad 8^2 + 15^2 = x^2$ Substituting

$\quad 64 + 225 = x^2$

$\quad\quad\quad 289 = x^2$

$\quad\quad \sqrt{289} = x$

$\quad\quad\quad\quad 17 = x$

2. $\sqrt{34} \approx 5.831$

3. $a^2 + b^2 = c^2$

$\quad 6^2 + 6^2 = x^2$ Substituting

$\quad 36 + 36 = x^2$

$\qquad 72 = x^2$

$\qquad \sqrt{72} = x$ Exact answer

$\quad 8.485 \approx x$ Approximation

4. $\sqrt{98} \approx 9.899$

5. $a^2 + b^2 = c^2$

$\quad 5^2 + x^2 = 13^2$

$\quad 25 + x^2 = 169$

$\qquad x^2 = 144$

$\qquad x = 12$

6. 5

7. $a^2 + b^2 = c^2$

$\quad (6\sqrt{3})^2 + x^2 = 12^2$

$\quad 36 \cdot 3 + x^2 = 144$

$\quad 108 + x^2 = 144$

$\qquad x^2 = 36$

$\qquad x = 6$

8. $\sqrt{31} \approx 5.568$

9. $a^2 + b^2 = c^2$

$\quad 12^2 + 5^2 = c^2$

$\quad 144 + 25 = c^2$

$\qquad 169 = c^2$

$\qquad 13 = c$

10. 26

11. $a^2 + b^2 = c^2$

$\quad 18^2 + b^2 = 30^2$

$\quad 324 + b^2 = 900$

$\qquad b^2 = 576$

$\qquad b = 24$

12. 12

13. $a^2 + b^2 = c^2$

$\quad a^2 + 1^2 = (\sqrt{5})^2$

$\quad a^2 + 1 = 5$

$\quad a^2 = 4$

$\qquad a = 2$

14. 1

15. $a^2 + b^2 = c^2$

$\quad 1^2 + b^2 = (\sqrt{3})^2$

$\quad 1 + b^2 = 3$

$\qquad b^2 = 2$

$\qquad b = \sqrt{2}$ Exact answer

$\qquad b \approx 1.414$ Approximation

16. $\sqrt{8} \approx 2.828$

17. $a^2 + b^2 = c^2$

$\quad a^2 + (5\sqrt{3})^2 = 10^2$

$\quad a^2 + 25 \cdot 3 = 100$

$\quad a^2 + 75 = 100$

$\qquad a^2 = 25$

$\qquad a = 5$

18. $\sqrt{50} \approx 7.071$

19. *Familiarize*. Let h = the height of the back of the jump, in inches.

Translate. We use the Pythagorean theorem, substituting 30 for a, h for b, and 33 for c.

$$30^2 + h^2 = 33^2$$

Carry out. We solve the equation.

$\quad 30^2 + h^2 = 33^2$

$\quad 900 + h^2 = 1089$

$\qquad h^2 = 189$

$\qquad h = \sqrt{189}$ Exact answer

$\qquad h \approx 13.748$

Check. We check by substituting 30, $\sqrt{189}$, and 33 in the Pythagorean equation.

$$\frac{a^2 + b^2 = c^2}{}$$

$30^2 + (\sqrt{189})^2$? 33^2	
$900 + 189$	1089
1089	1089 TRUE

State. The back of the jump should be $\sqrt{189}$ in., or about 13.748 in. high.

20. $\sqrt{32}$ cm ≈ 5.657 cm

21. *Familiarize*. We first make a drawing. We label the unknown length w.

Translate. We use the Pythagorean theorem, substituting 8 for a, 12 for b, and w for c.

$$8^2 + 12^2 = w^2$$

Carry out. We solve the equation.

$$8^2 + 12^2 = w^2$$
$$64 + 144 = w^2$$
$$208 = w^2$$
$$\sqrt{208} = w \qquad \text{Exact answer}$$
$$14.422 \approx w \qquad \text{Approximation}$$

Check. We check by substituting 8, 12, and $\sqrt{208}$ into the Pythagorean equation:

$$a^2 + b^2 = c^2$$

$8^2 + 12^2$? $(\sqrt{208})^2$	
$64 + 144$	208
208	208 TRUE

State. The pipe should be $\sqrt{208}$ feet or about 14.422 feet long.

22. $\sqrt{250}$ m ≈ 15.811 m

23. *Familiarize*. We first make a drawing. We label the diagonal d.

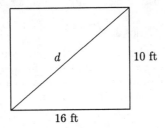

Translate.We use the Pythagorean theorem, substituting 16 for a, 10 for b, and d for c.

$$a^2 + b^2 = c^2$$
$$16^2 + 10^2 = d^2$$

Carry out. We solve the equation.

$$16^2 + 10^2 = d^2$$
$$256 + 100 = d^2$$
$$356 = d^2$$
$$\sqrt{356} = d$$
$$18.868 \approx d$$

Check. We check by substituting 16, 10, and $\sqrt{356}$ in the Pythagorean equation.

$$a^2 + b^2 = c^2$$

$16^2 + 10^2$? $(\sqrt{356})^2$	
$256 + 100$	356
356	356 TRUE

State. The wire needs to be $\sqrt{356}$ ft, or about 18.868 ft long.

24. $\sqrt{12,500}$ yd ≈ 111.803 yd

25. *Familiarize*. We make a drawing. we label the diagonal d.

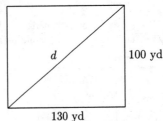

Translate. We use the Pythagorean theorem, substituting 130 for a, 100 for b, and d for c.

$$130^2 + 100^2 = d^2$$

Carry out. We solve the equation.

$$130^2 + 100^2 = d^2$$
$$16,900 + 10,000 = d^2$$
$$26,900 = d^2$$
$$\sqrt{26,900} = d$$
$$164.012 \approx d$$

Check. We check by substituting 130 for a, 100 for b, and $\sqrt{26,900}$ for c in the Pythagorean equation.

$$a^2 + b^2 = c^2$$

$130^2 + 100^2$? $(\sqrt{26,900})^2$	
$16,900 + 10,000$	$26,900$
$26,900$	$26,900$ TRUE

State. The length of a diagonal is $\sqrt{26,900}$ yd, or about 164.012 yd.

26. $\sqrt{16,200}$ ft ≈ 127.279 ft

27. *Familiarize*. We make a drawing. we label the diagonal d.

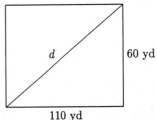

Translate. We use the Pythagorean theorem, substituting 110 for a, 60 for b, and d for c.

$$110^2 + 60^2 = d^2$$

Carry out. We solve the equation.

$$110^2 + 60^2 = d^2$$
$$12,100 + 3600 = d^2$$
$$15,700 = d^2$$
$$\sqrt{15,700} = d$$
$$125.300 \approx d$$

Check. We check by substituting 110 for a, 60 for b, and $\sqrt{15,700}$ for c in the Pythagorean equation.

$$
\begin{array}{c|c}
\multicolumn{2}{c}{a^2 + b^2 = c^2} \\
\hline
110^2 + 60^2 \; ? \; (\sqrt{15,700})^2 & \\
12,100 + 3600 & 15,700 \\
15,700 & 15,700 \qquad \text{TRUE}
\end{array}
$$

State. The length of a diagonal is $\sqrt{15,700}$ yd, or about 125.300 yd.

28. $\sqrt{74}$ km ≈ 8.602 km

29.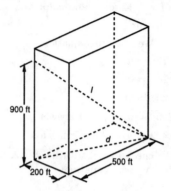

30.

31. Rational numbers can be expressed in the form $\dfrac{a}{b}$, where a and b are integers and $b \neq 0$. Decimal notation for rational numbers either terminates or repeats. The rational numbers in the given list are -45, -9.7, 0, $\dfrac{2}{7}$, 5.09, and 19.

32. $-\sqrt{5}, \sqrt{7}, \pi$

33. Each of the numbers in the list is either rational or irrational, so all of them are real numbers.

34. $-45, 0, 19$

35. $(-2)^5 = (-2)(-2)(-2)(-2)(-2) = -32$

36. $\dfrac{25}{9}$

37. $(2x)^3 = 2^3 x^3 = 8x^3$

38. $-8x^3$

39.

40.

41. *Familiarize*. We make a drawing.

First we will find d, the diagonal distance, in feet, across the building. This is the length of the hypotenuse of a right triangle with legs of 200 ft and 500 ft. Then we will find l, the distance from Vance's office to the restaurant. This is the length, in feet, of the hypotenuse of a right triangle with legs of 900 ft and d.

Note that $\dfrac{1}{4}$ mi $= \dfrac{1}{4} \cdot 5280$ ft $= 1320$ ft. Then if $l \leq 1320$ ft, Vance can use the handset at the restaurant.

Translate. To find d we use the Pythagorean theorem, substituting 200 for a, 500 for b, and d for c.

$$200^2 + 500^2 = d^2$$

After we find d we will use the Pythagorean theorem again, this time substituting 900 for a, d for b, and l for c. (Since we will use d^2 in this equation we need only solve the equation above for d^2.)

$$900^2 + d^2 = l^2$$

Carry out. First find d^2.

$$200^2 + 500^2 = d^2$$
$$40,000 + 250,000 = d^2$$
$$290,000 = d^2$$

Now we substitute 290,000 for d^2 in the second equation in the Translate step.

$$900^2 + d^2 = l^2$$
$$900^2 + 290,000 = l^2$$
$$810,000 + 290,000 = l^2$$
$$1,100,000 = l^2$$
$$\sqrt{1,100,000} = l$$
$$1049 \approx l$$

Check. We can check by substituting in the Pythagorean equation as before. This is left to the student.

State. The distance from Vance's office to the restaurant is about 1049 ft. Since this is less than

1320 ft, or 1/4 mi, Vance can use the handset at the restaurant.

42. Yes

43. *Familiarize*. Referring to the drawing in the text, we let d represent the distance the plane will travel. The vertical length of the descent is $32,000 - 21,000$ or $11,000$ ft.

We convert 5 miles to feet

$$5 \text{ mi} = 5 \cancel{\text{mi}} \times \frac{5280 \text{ ft}}{1 \cancel{\text{mi}}} = 26,400 \text{ ft}$$

Translate. We use the Pythagorean theorem, substituting $11,000$ for a, $26,400$ for b, and d for c.

$$11,000^2 + 26,400^2 = d^2$$

Carry out. We solve the equation.

$$11,000^2 + 26,400^2 = d^2$$
$$121,000,000 + 696,960,000 = d^2$$
$$817,960,000 = d^2$$
$$28,600 = d$$

Check. We check by substituting $11,000$ for a, $26,400$ for b, and $28,600$ for c in the Pythagorean equation.

$$\frac{a^2 + b^2 = c^2}{}$$

$$11,000^2 + 26,400^2 \ ? \ 28,600^2$$
$$817,960,000 \ \big| \ 817,960,000 \quad \text{TRUE}$$

State. The plane will travel 28,600 ft during the descent.

44. 8 ft

45. *Familiarize*. Let s = the length of a side of the square. Recall that the area A of a square with side s is given by $A = s^2$.

Translate. We substitute 7 for A in the formula.

$$7 = s^2$$

Carry out. We solve the equation.

$$7 = s^2$$
$$\sqrt{7} = s$$
$$2.646 \approx s$$

Check. If the length of a side of a square is $\sqrt{7}$ m, then the area of the square is $(\sqrt{7} \text{ m})^2$, or 7 m^2. The result checks.

State. The length of a side of the square is $\sqrt{7}$ m or about 2.646 m.

46. 6, 8, 10

47.

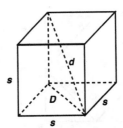

From the drawing we see that the diagonal d of the cube is the hypotenuse of a right triangle with one leg of length s, where s is the length of a side of the cube, and the other leg of length D, where D is the length of the diagonal of the base of the cube. First we find D:

Using the Pythagorean theorem we have:

$$s^2 + s^2 = D^2$$
$$2s^2 = D^2$$
$$\sqrt{2s^2} = D$$
$$s\sqrt{2} = D$$

Then we find d:

Using the Pythagorean theorem again we have:

$$s^2 + (s\sqrt{2})^2 = d^2$$
$$s^2 + 2s^2 = d^2$$
$$3s^2 = d^2$$
$$\sqrt{3s^2} = d$$
$$s\sqrt{3} = d$$

48. $\dfrac{2}{3}$

49. If the area of square $PQRS$ is 100 ft^2, then each side measures 10 ft. If A, B, C, and D are midpoints, then each of the segments PB, BQ, QC, CR, RD, DS, SA, and AP measures 5 ft. We can label the figure with additional information.

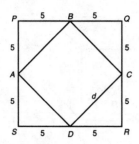

We label a side of the square $ABCD$ with d. Then we use the Pythagorean theorem.

$$5^2 + 5^2 = d^2$$
$$25 + 25 = d^2$$
$$50 = d^2$$
$$\sqrt{50} = d$$

If a side of square $ABCD$ is $\sqrt{50}$, then its area is $\sqrt{50} \cdot \sqrt{50}$, or 50 ft^2.

50. $\dfrac{a\sqrt{3}}{2}$

51.

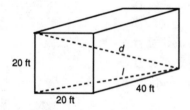

First find l^2:
$$20^2 + 40^2 = l^2$$
$$2000 = l^2$$

Then find d:
$$20^2 + l^2 = d^2$$
$$400 + 2000 = d^2$$
$$2400 = d^2$$
$$\sqrt{2400} = d$$
$$48.990 \approx d$$

The longest straight-line distance that can be measured is $\sqrt{2400}$ ft ≈ 48.990 ft.

52. $\sqrt{1525}$ mi ≈ 39.051 mi

53.

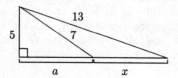

$$a^2 + 5^2 = 7^2$$
$$a^2 + 25 = 49$$
$$a^2 = 24$$
$$a = \sqrt{24}, \text{ or } 2\sqrt{6}$$

$$(a+x)^2 + 5^2 = 13^2$$
$$(2\sqrt{6}+x)^2 + 5^2 = 13^2 \quad \text{Substituting } 2\sqrt{6} \text{ for } a$$
$$(2\sqrt{6}+x)^2 + 25 = 169$$
$$(2\sqrt{6}+x)^2 = 144$$
$$2\sqrt{6} + x = 12 \quad \text{Taking the principal} \atop \text{square root}$$
$$x = 12 - 2\sqrt{6}$$
$$x \approx 7.101$$

54. 640 acres

Exercise Set 8.7

1. $\sqrt[3]{-64} = -4 \qquad (-4)^3 = (-4)(-4)(-4) = -64$

2. -2

3. $\sqrt[3]{-125} = -5 \qquad (-5)^3 = (-5)(-5)(-5) = -125$

4. -3

5. $\sqrt[3]{1000} = 10 \qquad 10^3 = 10 \cdot 10 \cdot 10 = 1000$

6. 2

7. $-\sqrt[3]{216} = -6 \qquad \sqrt[3]{216} = 6$, so $-\sqrt[3]{216} = -6$.

8. 7

9. $\sqrt[4]{81} = 3 \qquad 3^4 = 3 \cdot 3 \cdot 3 \cdot 3 = 81$

10. 5

11. $\sqrt[5]{0} = 0 \qquad 0^5 = 0 \cdot 0 \cdot 0 \cdot 0 \cdot 0 = 0$

12. 1

13. $\sqrt[5]{-1} = -1 \quad (-1)^5 = (-1)(-1)(-1)(-1)(-1) = -1$

14. -3

15. $\sqrt[4]{-81}$ is not a real number, because it is an even root of a negative number.

16. Not a real number

17. $\sqrt[4]{10,000} = 10 \quad 10^4 = 10 \cdot 10 \cdot 10 \cdot 10 = 10,000$

18. 10

19. $\sqrt[3]{7^3} = 7 \qquad 7^3 = 7 \cdot 7 \cdot 7$

20. 5

21. $\sqrt[6]{64} = 2 \qquad 2^6 = 2 \cdot 2 \cdot 2 \cdot 2 \cdot 2 \cdot 2 = 64$

22. 1

23. $\sqrt[9]{a^9} = a \qquad a^9 = a \cdot a \cdot a \cdot a \cdot a \cdot a \cdot a \cdot a \cdot a$

24. n

25. $\sqrt[3]{54} = \sqrt[3]{27 \cdot 2} = \sqrt[3]{27}\sqrt[3]{2} = 3\sqrt[3]{2}$

26. $2\sqrt[3]{4}$

27. $\sqrt[4]{48} = \sqrt[4]{16 \cdot 3} = \sqrt[4]{16}\sqrt[4]{3} = 2\sqrt[4]{3}$

28. $2\sqrt[5]{5}$

29. $\sqrt[3]{\dfrac{64}{125}} = \dfrac{\sqrt[3]{64}}{\sqrt[3]{125}} = \dfrac{4}{5}$

30. $\dfrac{5}{3}$

31. $\sqrt[5]{\dfrac{32}{243}} = \dfrac{\sqrt[5]{32}}{\sqrt[5]{243}} = \dfrac{2}{3}$

32. $\dfrac{5}{4}$

33. $\sqrt[3]{\dfrac{7}{8}} = \dfrac{\sqrt[3]{7}}{\sqrt[3]{8}} = \dfrac{\sqrt[3]{7}}{2}$

34. $\dfrac{\sqrt[5]{17}}{2}$

35. $\sqrt[4]{\dfrac{14}{81}} = \dfrac{\sqrt[4]{14}}{\sqrt[4]{81}} = \dfrac{\sqrt[4]{14}}{3}$

36. $\dfrac{\sqrt[3]{10}}{3}$

37. $25^{1/2} = \sqrt{25} = 5$

38. 3

39. $125^{1/3} = \sqrt[3]{125} = 5$

40. 10

41. $32^{1/5} = \sqrt[5]{32} = 2$

42. 2

43. $16^{3/4} = (16^{1/4})^3 = (\sqrt[4]{16})^3 = 2^3 = 8$

44. 16

45. $9^{5/2} = (9^{1/2})^5 = (\sqrt{9})^5 = 3^5 = 243$

46. 8

47. $64^{2/3} = (64^{1/3})^2 = (\sqrt[3]{64})^2 = 4^2 = 16$

48. 4

49. $8^{5/3} = (8^{1/3})^5 = (\sqrt[3]{8})^5 = 2^5 = 32$

50. 32

51. $4^{5/2} = (4^{1/2})^5 = (\sqrt{4})^5 = 2^5 = 32$

52. $\dfrac{1}{6}$

53. $25^{-1/2} = \dfrac{1}{25^{1/2}} = \dfrac{1}{\sqrt{25}} = \dfrac{1}{5}$

54. $\dfrac{1}{2}$

55. $256^{-1/4} = \dfrac{1}{256^{1/4}} = \dfrac{1}{\sqrt[4]{256}} = \dfrac{1}{4}$

56. $\dfrac{1}{1000}$

57. $16^{-3/4} = \dfrac{1}{16^{3/4}} = \dfrac{1}{(\sqrt[4]{16})^3} = \dfrac{1}{2^3} = \dfrac{1}{8}$

58. $\dfrac{1}{27}$

59. $81^{-5/4} = \dfrac{1}{81^{5/4}} = \dfrac{1}{(\sqrt[4]{81})^5} = \dfrac{1}{3^5} = \dfrac{1}{243}$

60. $\dfrac{1}{4}$

61. $8^{-2/3} = \dfrac{1}{8^{2/3}} = \dfrac{1}{(\sqrt[3]{8})^2} = \dfrac{1}{2^2} = \dfrac{1}{4}$

62. $\dfrac{1}{125}$

63. ❖

64. ❖

65. $\quad x^2 - 5x - 6 = 0$

$\quad (x - 6)(x + 1) = 0$

$\quad x - 6 = 0 \;\; or \;\; x + 1 = 0$

$\qquad x = 6 \;\; or \qquad x = -1$

The solutions are 6 and -1.

66. $-5, \; 1$

67.
$$4t^2 - 9 = 0$$
$$(2t + 3)(2t - 3) = 0$$
$$2t + 3 = 0 \quad or \quad 2t - 3 = 0$$
$$2t = -3 \quad or \quad 2t = 3$$
$$t = -\frac{3}{2} \quad or \quad t = \frac{3}{2}$$

The solutions are $-\dfrac{3}{2}$ and $\dfrac{3}{2}$.

68. $-\dfrac{2}{5}, \dfrac{2}{5}$

69.
$$3x^2 + 8x + 4 = 0$$
$$(3x + 2)(x + 2) = 0$$
$$3x + 2 = 0 \quad or \quad x + 2 = 0$$
$$3x = -2 \quad or \quad x = -2$$
$$x = -\frac{2}{3} \quad or \quad x = -2$$

The solutions are $-\dfrac{2}{3}$ and -2.

70. $-3, \dfrac{2}{5}$

71. ◈

72. ◈

73. Enter 10, press the power key, enter 0.8 (or $(4 \div 5)$), and then press $\boxed{=}$.

$$10^{4/5} \approx 6.310$$

(Some calculators have a 10^x key which might have to be accessed using the $\boxed{\text{SHIFT}}$ key.)

74. 2.213

75. Enter 36, press the power key, enter 0.375 (or $(3 \div 8)$), and then press $\boxed{=}$.

$$36^{3/8} \approx 3.834$$

76. 31.623

77. $\left(x^{2/3}\right)^{7/3} = x^{\frac{2}{3} \cdot \frac{7}{3}} = x^{14/9}$

78. $a^{7/4}$

79. $\dfrac{p^{5/6}}{p^{2/3}} = p^{5/6 - 2/3} = p^{5/6 - 4/6} = p^{1/6}$

80. $m^{13/12}$

81. Graph $y = \sqrt[3]{x}$

We make a table of values.

x	y
-8	-2
-1	-1
0	0
1	1
8	2

We plot these points and connect them with a smooth curve.

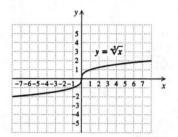

82.

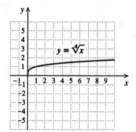

83.

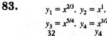
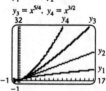

Chapter 9

Quadratic Equations

Exercise Set 9.1

1.
$$x^2 = 49$$
$x = \sqrt{49}$ or $x = -\sqrt{49}$ Using the principle
$x = 7$ or $x = -7$ of square roots

We can check mentally that $7^2 = 49$ and $(-7)^2 = 49$.
The solutions are 7 and -7.

2. $-10,\ 10$

3.
$$a^2 = 25$$
$a = \sqrt{25}$ or $a = -\sqrt{25}$ Using the principle
$a = 5$ or $a = -5$ of square roots

We can check mentally that $5^2 = 25$ and $(-5)^2 = 25$.
The solutions are 5 and -5.

4. $-6,\ 6$

5.
$$t^2 = 17$$
$t = \sqrt{17}$ or $t = -\sqrt{17}$ Using the principle
of square roots

Check: For $\sqrt{17}$:   For $-\sqrt{17}$:

$t^2 = 17$	$t^2 = 17$
$(\sqrt{17})^2$? 17	$(-\sqrt{17})^2$? 17
$17 \mid 17$ TRUE	$17 \mid 17$ TRUE

The solutions are $\sqrt{17}$ and $-\sqrt{17}$.

6. $-\sqrt{13},\ \sqrt{13}$

7.
$$3x^2 = 27$$
$$x^2 = 9$$ Dividing by 3
$x = \sqrt{9}$ or $x = -\sqrt{9}$ Using the principle
of square roots
$x = 3$ or $x = -3$

Both numbers check. The solutions are $\sqrt{3}$ and $-\sqrt{3}$.

8. $-2,\ 2$

9. $9t^2 = 0$

Observe that t^2 must be 0, so $t = 0$. The solution
is 0.

10. $-\sqrt{3},\ \sqrt{3}$

11.
$$4 - 9x^2 = 0$$
$$4 = 9x^2$$
$$\frac{4}{9} = x^2$$
$x = \sqrt{\dfrac{4}{9}}$ or $x = -\sqrt{\dfrac{4}{9}}$
$x = \dfrac{2}{3}$ or $x = -\dfrac{2}{3}$

Both numbers check. The solutions are $\dfrac{2}{3}$ and $-\dfrac{2}{3}$.

12. $-\dfrac{5}{2},\ \dfrac{5}{2}$

13.
$$49y^2 - 5 = 15$$
$$49y^2 = 20$$
$$y^2 = \frac{20}{49}$$
$y = \sqrt{\dfrac{20}{49}}$ or $y = -\sqrt{\dfrac{20}{49}}$
$y = \dfrac{\sqrt{20}}{7}$ or $y = -\dfrac{\sqrt{20}}{7}$
$y = \dfrac{\sqrt{4 \cdot 5}}{7}$ or $y = -\dfrac{\sqrt{4 \cdot 5}}{7}$
$y = \dfrac{2\sqrt{5}}{7}$ or $y = -\dfrac{2\sqrt{5}}{7}$

The solutions are $\dfrac{2\sqrt{5}}{7}$ and $-\dfrac{2\sqrt{5}}{7}$.

14. $-\sqrt{3},\ \sqrt{3}$

15.
$$8x^2 - 28 = 0$$
$$8x^2 = 28$$
$$x^2 = \frac{7}{2}$$
$x = \sqrt{\dfrac{7}{2}}$ or $x = -\sqrt{\dfrac{7}{2}}$ Using the principle
of square roots
$x = \dfrac{\sqrt{7}}{\sqrt{2}}$ or $x = -\dfrac{\sqrt{7}}{\sqrt{2}}$
$x = \dfrac{\sqrt{7}}{\sqrt{2}} \cdot \dfrac{\sqrt{2}}{\sqrt{2}}$ or $x = -\dfrac{\sqrt{7}}{\sqrt{2}} \cdot \dfrac{\sqrt{2}}{\sqrt{2}}$
$x = \dfrac{\sqrt{14}}{2}$ or $x = -\dfrac{\sqrt{14}}{2}$

The solutions are $\dfrac{\sqrt{14}}{2}$ and $-\dfrac{\sqrt{14}}{2}$.

16. $-\dfrac{\sqrt{7}}{5}$, $\dfrac{\sqrt{7}}{5}$, or $-\dfrac{\sqrt{35}}{5}$, $-\dfrac{\sqrt{35}}{5}$

17.
$$(x-1)^2 = 25$$
$x - 1 = 5 \ \ or \ \ x - 1 = -5$ Using the principle
                               of square roots
$\ \ \ \ x = 6 \ \ or \ \ \ \ x \ \ = -4$
The solutions are 6 and -4.

18. -5, 9

19.
$$(x+4)^2 = 81$$
$x + 4 = 9 \ \ or \ \ x + 4 = -9$ Using the principle
                                 of square roots
$\ \ \ \ x = 5 \ \ or \ \ \ \ \ \ x = -13$
The solutions are 5 and -13.

20. -9, 3

21.
$$(m+3)^2 = 6$$
$m + 3 = \sqrt{6} \ \ \ \ or \ \ m + 3 = -\sqrt{6}$
$\ \ \ m = -3 + \sqrt{6} \ \ or \ \ \ \ \ \ m = -3 - \sqrt{6}$
The solutions are $-3 + \sqrt{6}$ and $-3 - \sqrt{6}$, or $-3 \pm \sqrt{6}$.

22. $4 + \sqrt{21}$, $4 - \sqrt{21}$, or $4 \pm \sqrt{21}$

23. $(a-7)^2 = 0$

Observe that $a - 7$ must be 0, so $a - 7 = 0$, or $a = 7$.
The solution is 7.

24. -21, -3

25.
$$(x-5)^2 = 14$$
$x - 5 = \sqrt{14} \ \ \ \ or \ \ x - 5 = -\sqrt{14}$
$\ \ x = 5 + \sqrt{14} \ \ or \ \ \ \ \ \ x = 5 - \sqrt{14}$
The solutions are $5 + \sqrt{14}$ and $5 - \sqrt{14}$, or $5 \pm \sqrt{14}$.

26. $7 + 2\sqrt{3}$, $7 - 2\sqrt{3}$, or $7 \pm 2\sqrt{3}$

27.
$$(t+3)^2 = 25$$
$t + 3 = 5 \ \ or \ \ t + 3 = -5$
$\ \ t = 2 \ \ or \ \ \ \ \ \ t = -8$
The solutions are 2 and -8.

28. -12, 2

29.
$$\left(y - \frac{3}{4}\right)^2 = \frac{17}{16}$$
$y - \dfrac{3}{4} = \sqrt{\dfrac{17}{16}} \ \ \ \ or \ \ y - \dfrac{3}{4} = -\sqrt{\dfrac{17}{16}}$
$y - \dfrac{3}{4} = \dfrac{\sqrt{17}}{4} \ \ \ \ or \ \ y - \dfrac{3}{4} = -\dfrac{\sqrt{17}}{4}$
$y = \dfrac{3}{4} + \dfrac{\sqrt{17}}{4} \ \ or \ \ \ \ \ \ \ y = \dfrac{3}{4} - \dfrac{\sqrt{17}}{4}$
$y = \dfrac{3 + \sqrt{17}}{4} \ \ or \ \ \ \ \ \ \ y = \dfrac{3 - \sqrt{17}}{4}$
The solutions are $\dfrac{3 + \sqrt{17}}{4}$ and $\dfrac{3 - \sqrt{17}}{4}$, or
$\dfrac{3 \pm \sqrt{17}}{4}$.

30. $\dfrac{-3 + \sqrt{13}}{2}$, $\dfrac{-3 - \sqrt{13}}{2}$, $\dfrac{-3 \pm \sqrt{13}}{2}$

31.
$$x^2 - 10x + 25 = 100$$
$$(x-5)^2 = 100$$
$x - 5 = 10 \ \ or \ \ x - 5 = -10$
$\ \ x = 15 \ \ or \ \ \ \ \ \ x = -5$
The solutions are 15 and -5.

32. -5, 11

33.
$$p^2 + 8p + 16 = 1$$
$$(p+4)^2 = 1$$
$p + 4 = 1 \ \ or \ \ p + 4 = -1$
$\ \ p = -3 \ \ or \ \ \ \ \ \ p = -5$
The solutions are -3 and -5.

34. -9, -5

35.
$$t^2 - 6t + 9 = 13$$
$$(t-3)^2 = 13$$
$t - 3 = \sqrt{13} \ \ \ \ or \ \ t - 3 = -\sqrt{13}$
$\ \ t = 3 + \sqrt{13} \ \ or \ \ \ \ \ \ t = 3 - \sqrt{13}$
The solutions are $3 + \sqrt{13}$ and $3 - \sqrt{13}$, or
$3 \pm \sqrt{13}$.

36. $1 + \sqrt{5}$, $1 - \sqrt{5}$, or $1 \pm \sqrt{5}$

37.
$$x^2 + 12x + 36 = 18$$
$$(x+6)^2 = 18$$
$x + 6 = \sqrt{18} \ \ \ \ \ \ or \ \ x + 6 = -\sqrt{18}$
$x + 6 = 3\sqrt{2} \ \ \ \ \ \ or \ \ x + 6 = -3\sqrt{2}$
$\ \ \ \ x = -6 + 3\sqrt{2} \ \ or \ \ \ \ \ \ x = -6 - 3\sqrt{2}$
The solutions are $-6 + 3\sqrt{2}$ and $-6 - 3\sqrt{2}$, or
$-6 \pm 3\sqrt{2}$.

38. $-2 + 2\sqrt{3}, \ -2 - 2\sqrt{3}$, or $-2 \pm 2\sqrt{3}$

39. ◈

40. ◈

41. $3x^2 + 12x + 3 = 3(x^2 + 4x + 1)$

42. $5(t^2 + 4t + 5)$

43. $t^2 + 16t + 64 = t^2 + 2 \cdot t \cdot 8 + 8^2 = (t + 8)^2$

44. $(x + 7)^2$

45. $x^2 - 10x + 25 = x^2 - 2 \cdot x \cdot 5 + 5^2 = (x - 5)^2$

46. $(t - 4)^2$

47. ◈

48. ◈

49. $x^2 + \dfrac{7}{3}x + \dfrac{49}{36} = \dfrac{7}{36}$

$\left(x + \dfrac{7}{6}\right)^2 = \dfrac{7}{36}$

$x + \dfrac{7}{6} = \dfrac{\sqrt{7}}{6} \quad or \quad x + \dfrac{7}{6} = -\dfrac{\sqrt{7}}{6}$

$x = -\dfrac{7}{6} + \dfrac{\sqrt{7}}{6} \quad or \quad x = -\dfrac{7}{6} - \dfrac{\sqrt{7}}{6}$

The solutions are $\dfrac{-7 + \sqrt{7}}{6}$ and $\dfrac{-7 - \sqrt{7}}{6}$, or $\dfrac{-7 \pm \sqrt{7}}{6}$.

50. $\dfrac{5}{2} + \dfrac{\sqrt{13}}{2}, \ \dfrac{5}{2} - \dfrac{\sqrt{13}}{2}$, or $\dfrac{5 \pm \sqrt{13}}{2}$

51. $m^2 - \dfrac{3}{2}m + \dfrac{9}{16} = \dfrac{17}{16}$

$\left(m - \dfrac{3}{4}\right)^2 = \dfrac{17}{16}$

$m - \dfrac{3}{4} = \dfrac{\sqrt{17}}{4} \quad or \quad m - \dfrac{3}{4} = -\dfrac{\sqrt{17}}{4}$

$m = \dfrac{3 + \sqrt{17}}{4} \quad or \quad m = \dfrac{3 - \sqrt{17}}{4}$

The solutions are $\dfrac{3 \pm \sqrt{17}}{4}$.

52. $-5, \ 2$

53. $x^2 + 2.5x + 1.5625 = 9.61$

$(x + 1.25)^2 = 9.61$

$x + 1.25 = 3.1 \quad or \quad x + 1.25 = -3.1$

$x = 1.85 \quad or \qquad x = -4.35$

The solutions are 1.85 and -4.35.

54. $-3.3, \ 7.1$

55. From the graph we see that when $y = 1$, then $x = -4$ or $x = -2$. Thus, the solutions of $(x + 3)^2 = 1$ are -4 and -2.

56. $-5, \ -1$

57. From the graph we see that when $y = 9$, then $x = -6$ or $x = 0$. Thus, the solutions of $(x + 3)^2 = 9$ are -6 and 0.

58. -3

59.
$$f = \frac{kMm}{d^2}$$

$d^2 f = kMm \qquad$ Multiplying by d^2

$d^2 = \dfrac{kMm}{f} \qquad$ Dividing by f

$d = \sqrt{\dfrac{kMm}{f}} \qquad$ Taking the principal square root

$d = \dfrac{\sqrt{kMmf}}{f} \qquad$ Rationalizing the denominator

Exercise Set 9.2

1. To complete the square for $x^2 + 8x$, we take half the coefficient of x and square it:

$$\left(\frac{8}{2}\right)^2 = 4^2 = 16$$

The trinomial $x^2 + 8x + 16$ is the square of $x + 4$. That is, $x^2 + 8x + 16 = (x + 4)^2$.

2. $x^2 + 4x + 4$

3. To complete the square for $x^2 - 12x$, we take half the coefficient of x and square it:

$$\left(\frac{-12}{2}\right)^2 = (-6)^2 = 36$$

The trinomial $x^2 - 12x + 36$ is the square of $x - 6$. That is, $x^2 - 12x + 36 = (x - 6)^2$.

4. $x^2 - 14x + 49$

5. To complete the square for $x^2 - 3x$, we take half the coefficient of x and square it:
$$\left(\frac{-3}{2}\right)^2 = \frac{9}{4}$$
The trinomial $x^2 - 3x + \frac{9}{4}$ is the square of $x - \frac{3}{2}$.
That is, $x^2 - 3x + \frac{9}{4} = \left(x - \frac{3}{2}\right)^2$.

6. $x^2 - 9x + \frac{81}{4}$

7. To complete the square for $t^2 + t$, we take half the coefficient of t and square it:
$$\left(\frac{1}{2}\right)^2 = \frac{1}{4}$$
The trinomial $t^2 + t + \frac{1}{4}$ is the square of $t + \frac{1}{2}$. That is, $t^2 + t + \frac{1}{4} = \left(t + \frac{1}{2}\right)^2$.

8. $y^2 - y + \frac{1}{4}$

9. To complete the square for $x^2 + \frac{5}{4}x$, we take half the coefficient of x and square it:
$$\left(\frac{1}{2} \cdot \frac{5}{4}\right)^2 = \left(\frac{5}{8}\right)^2 = \frac{25}{64}$$
The trinomial $x^2 + \frac{5}{4}x + \frac{25}{64}$ is the square of $x + \frac{5}{8}$.
That is, $x^2 + \frac{5}{4}x + \frac{25}{64} = \left(x + \frac{5}{8}\right)^2$.

10. $x^2 + \frac{4}{3}x + \frac{4}{9}$

11. To complete the square for $m^2 - \frac{9}{2}m$, we take half the coefficient of m and square it:
$$\left[\frac{1}{2}\left(-\frac{9}{2}\right)\right]^2 = \left(-\frac{9}{4}\right)^2 = \frac{81}{16}$$
The trinomial $m^2 - \frac{9}{2}m + \frac{81}{16}$ is the square of $m - \frac{9}{4}$.
That is, $m^2 - \frac{9}{2}m + \frac{81}{16} = \left(x - \frac{9}{4}\right)^2$.

12. $r^2 - \frac{2}{5}r + \frac{1}{25}$

13. $x^2 + 8x + 12 = 0$
$x^2 + 8x \qquad = -12$ \qquad Subtracting 12
$x^2 + 8x + 16 = -12 + 16$ \quad Adding 16:
$$\left(\frac{8}{2}\right)^2 = 4^2 = 16$$
$(x + 4)^2 = 4$

$x + 4 = 2 \quad or \quad x + 4 = -2$ \quad Principle of square roots
$x = -2 \quad or \qquad x = -6$
The solutions are -2 and -6.

14. $-7, 1$

15. $x^2 - 24x + 21 = 0$
$x^2 - 24x \qquad = -21$ \qquad Subtracting 21
$x^2 - 24x + 144 = -21 + 144$ \quad Adding 144:
$$\left(\frac{-24}{2}\right)^2 = (-12)^2 = 144$$
$(x - 12)^2 = 123$
$x - 12 = \sqrt{123} \qquad or \quad x - 12 = -\sqrt{123}$
\hfill Principle of square roots
$x = 12 + \sqrt{123} \quad or \qquad x = 12 - \sqrt{123}$
The solutions are $12 + \sqrt{123}$ and $12 - \sqrt{123}$, or $12 \pm \sqrt{123}$.

16. No real-number solution

17. $\qquad 3x^2 - 6x - 15 = 0$
$\frac{1}{3}(3x^2 - 6x - 15) = \frac{1}{3} \cdot 0$
$x^2 - 2x - 5 = 0$
$x^2 - 2x \qquad = 5$
$x^2 - 2x + 1 = 5 + 1$ \quad Adding 1: $\left(\frac{-2}{2}\right)^2 =$
\hfill $(-1)^2 = 1$
$(x - 1)^2 = 6$
$x - 1 = \sqrt{6} \quad or \quad x - 1 = -\sqrt{6}$
$x = 1 + \sqrt{6} \quad or \qquad x = 1 - \sqrt{6}$
The solutions are $1 \pm \sqrt{6}$.

18. $2 \pm \sqrt{15}$

19. $\quad x^2 - 22x + 102 = 0$
$x^2 - 22x \qquad = -102$
$x^2 - 22x + 121 = -102 + 121$ \quad Adding 121:
$$\left(\frac{-22}{2}\right)^2 = (-11)^2 = 121$$
$(x - 11)^2 = 19$
$x - 11 = \sqrt{19} \qquad or \quad x - 11 = -\sqrt{19}$
$x = 11 + \sqrt{19} \quad or \qquad x = 11 - \sqrt{19}$
The solutions are $11 \pm \sqrt{19}$.

20. $9 \pm \sqrt{7}$

21. $t^2 + 8t - 5 = 0$

$t^2 + 8t = 5$

$t^2 + 8t + 16 = 5 + 16$ Adding 16:

$$\left(\frac{8}{2}\right)^2 = 4^2 = 16$$

$(t+4)^2 = 21$

$t + 4 = \sqrt{21}$ or $t + 4 = -\sqrt{21}$
$t = -4 + \sqrt{21}$ or $t = -4 - \sqrt{21}$

The solutions are $t = -4 \pm \sqrt{21}$.

22. $\dfrac{7 \pm \sqrt{57}}{2}$

23. $2x^2 + 10x - 12 = 0$

$2(x^2 + 5x - 6) = 0$

$x^2 + 5x - 6 = 0$ Dividing by 2
$x^2 + 5x = 6$

$x^2 + 5x + \dfrac{25}{4} = 6 + \dfrac{25}{4}$ Adding $\dfrac{25}{4}$:

$$\left(\frac{5}{2}\right)^2 = \frac{25}{4}$$

$\left(x + \dfrac{5}{2}\right)^2 = \dfrac{24}{4} + \dfrac{25}{4} = \dfrac{49}{4}$

$x + \dfrac{5}{2} = \dfrac{7}{2}$ or $x + \dfrac{5}{2} = -\dfrac{7}{2}$

$x = -\dfrac{5}{2} + \dfrac{7}{2}$ or $x = -\dfrac{5}{2} - \dfrac{7}{2}$

$x = \dfrac{2}{2}$ or $x = -\dfrac{12}{2}$

$x = 1$ or $x = -6$

The solutions are 1 and -6.

24. $-7,\ 4$

25. $x^2 + \dfrac{3}{2}x - 2 = 0$

$x^2 + \dfrac{3}{2}x = 2$

$x^2 + \dfrac{3}{2}x + \dfrac{9}{16} = 2 + \dfrac{9}{16}$ Adding $\dfrac{9}{16}$:

$$\left(\frac{1}{2} \cdot \frac{3}{2}\right)^2 = \left(\frac{3}{4}\right)^2 = \frac{9}{16}$$

$\left(x + \dfrac{3}{4}\right)^2 = \dfrac{32}{16} + \dfrac{9}{16} = \dfrac{41}{16}$

$x + \dfrac{3}{4} = \dfrac{\sqrt{41}}{4}$ or $x + \dfrac{3}{4} = -\dfrac{\sqrt{41}}{4}$

$x = -\dfrac{3}{4} + \dfrac{\sqrt{41}}{4}$ or $x = -\dfrac{3}{4} - \dfrac{\sqrt{41}}{4}$

$x = \dfrac{-3 + \sqrt{41}}{4}$ or $x = \dfrac{-3 - \sqrt{41}}{4}$

The solutions are $\dfrac{-3 \pm \sqrt{41}}{4}$.

26. $\dfrac{3 \pm \sqrt{41}}{4}$

27. $2x^2 + 3x - 16 = 0$

$\dfrac{1}{2}(2x^2 + 3x - 16) = \dfrac{1}{2} \cdot 0$

$x^2 + \dfrac{3}{2}x - 8 = 0$

$x^2 + \dfrac{3}{2}x = 8$

$x^2 + \dfrac{3}{2}x + \dfrac{9}{16} = 8 + \dfrac{9}{16}$ Adding $\dfrac{9}{16}$:

$$\left(\frac{1}{2} \cdot \frac{3}{2}\right)^2 = \left(\frac{3}{4}\right)^2 = \frac{9}{16}$$

$\left(x + \dfrac{3}{4}\right)^2 = \dfrac{128}{16} + \dfrac{9}{16} = \dfrac{137}{16}$

$x + \dfrac{3}{4} = \dfrac{\sqrt{137}}{4}$ or $x + \dfrac{3}{4} = -\dfrac{\sqrt{137}}{4}$

$x = -\dfrac{3}{4} + \dfrac{\sqrt{137}}{4}$ or $x = -\dfrac{3}{4} - \dfrac{\sqrt{137}}{4}$

The solutions are $\dfrac{-3 \pm \sqrt{137}}{4}$.

28. $\dfrac{3 \pm \sqrt{73}}{4}$

29. $3t^2 + 6t - 1 = 0$

$\dfrac{1}{3}(3t^2 + 6t - 1) = \dfrac{1}{3} \cdot 0$

$t^2 + 2t - \dfrac{1}{3} = 0$

$t^2 + 2t = \dfrac{1}{3}$

$t^2 + 2t + 1 = \dfrac{1}{3} + 1$

$(t+1)^2 = \dfrac{4}{3}$

$t + 1 = \dfrac{2}{\sqrt{3}}$ or $t + 1 = -\dfrac{2}{\sqrt{3}}$

$t + 1 = \dfrac{2}{\sqrt{3}} \cdot \dfrac{\sqrt{3}}{\sqrt{3}}$ or $t + 1 = -\dfrac{2}{\sqrt{3}} \cdot \dfrac{\sqrt{3}}{\sqrt{3}}$

$t + 1 = \dfrac{2\sqrt{3}}{3}$ or $t + 1 = -\dfrac{2\sqrt{3}}{3}$

$t = -1 + \dfrac{2\sqrt{3}}{3}$ or $t = -1 - \dfrac{2\sqrt{3}}{3}$

$t = \dfrac{-3 + 2\sqrt{3}}{3}$ or $t = \dfrac{-3 - 2\sqrt{3}}{3}$

The solutions are $\dfrac{-3 \pm 2\sqrt{3}}{3}$.

30. $\dfrac{2 \pm \sqrt{13}}{3}$

31.
$$2x^2 = 9 + 5x$$
$$2x^2 - 5x - 9 = 0$$
$$\frac{1}{2}(2x^2 - 5x - 9) = \frac{1}{2} \cdot 0$$
$$x^2 - \frac{5}{2}x - \frac{9}{2} = 0$$
$$x^2 - \frac{5}{2}x = \frac{9}{2}$$
$$x^2 - \frac{5}{2}x + \frac{25}{16} = \frac{9}{2} + \frac{25}{16}$$
$$\left(x - \frac{5}{4}\right)^2 = \frac{72 + 25}{16} = \frac{97}{16}$$
$$x - \frac{5}{4} = \frac{\sqrt{97}}{4} \quad or \quad x - \frac{5}{4} = -\frac{\sqrt{97}}{4}$$
$$x = \frac{5}{4} + \frac{\sqrt{97}}{4} \quad or \quad x = \frac{5}{4} - \frac{\sqrt{97}}{4}$$

The solutions are $\dfrac{5 \pm \sqrt{97}}{4}$.

32. $-\dfrac{1}{2}, \ 5$

33.
$$4x^2 + 12x = 7$$
$$x^2 + 3x = \frac{7}{4}$$
$$x^2 + 3x + \frac{9}{4} = \frac{7}{4} + \frac{9}{4}$$
$$\left(x + \frac{3}{2}\right)^2 = 4$$
$$x + \frac{3}{2} = 2 \quad or \quad x + \frac{3}{2} = -2$$
$$x = \frac{1}{2} \quad or \quad x = -\frac{7}{2}$$

The solutions are $\dfrac{1}{2}$ and $-\dfrac{7}{2}$.

34. $-\dfrac{5}{2}, \ \dfrac{2}{3}$

35. ◈

36. ◈

37. $\dfrac{3 + 6x}{3} = \dfrac{\not{3}(1 + 2x)}{\not{3} \cdot 1} = 1 + 2x$

38. $2 + x$

39. $\dfrac{15 - 10x}{5} = \dfrac{\not{5}(3 - 2x)}{\not{5} \cdot 1} = 3 - 2x$

40. $6 - 5x$

41. $\dfrac{24 - 3\sqrt{5}}{9} = \dfrac{\not{3}(8 - \sqrt{5})}{\not{3} \cdot 3} = \dfrac{8 - \sqrt{5}}{3}$

42. $\dfrac{5 - \sqrt{6}}{2}$

43. ◈

44. ◈

45. $x^2 + bx + 49$

The trinomial is a square if the square of one-half the x-coefficient is equal to 49. Thus, we have:
$$\left(\frac{b}{2}\right)^2 = 49$$
$$\frac{b^2}{4} = 49$$
$$b^2 = 196$$
$$b = 14 \quad or \quad b = -14$$

46. $-12, \ 12$

47. $x^2 + bx + 50$

The trinomial is a square if the square of one-half the x-coefficient is equal to 50. Thus, we have:
$$\left(\frac{b}{2}\right)^2 = 50$$
$$\frac{b^2}{4} = 50$$
$$b^2 = 200$$
$$b = \sqrt{200} \quad or \quad b = -\sqrt{200}$$
$$b = \sqrt{100 \cdot 2} \quad or \quad b = -\sqrt{100 \cdot 2}$$
$$b = 10\sqrt{2} \quad or \quad b = -10\sqrt{2}$$

48. $-6\sqrt{5}, \ 6\sqrt{5}$

49. $x^2 - bx + 48$

The trinomial is a square if the square of one-half the x-coefficient is equal to 48. Thus, we have:
$$\left(\frac{-b}{2}\right)^2 = 48$$
$$\frac{b^2}{4} = 48$$
$$b^2 = 192$$
$$b = \sqrt{192} \quad or \quad b = -\sqrt{192}$$
$$b = \sqrt{64 \cdot 3} \quad or \quad b = -\sqrt{64 \cdot 3}$$
$$b = 8\sqrt{3} \quad or \quad b = -8\sqrt{3}$$

50. $-16, \ 16$

51. $-0.39, \ -7.61$

52. $-4.59, -7.41$

53. $23.09, 0.91$

54. $7.27, -0.27$

55. $3.71, -1.21$

56. $5, -0.5$

57. ◈

Exercise Set 9.3

1.
$$x^2 - 7x = 18$$
$$x^2 - 7x - 18 = 0 \qquad \text{Standard form}$$
We can factor.
$$x^2 - 7x - 18 = 0$$
$$(x - 9)(x + 2) = 0$$
$$x - 9 = 0 \ \ or \ \ x + 2 = 0$$
$$x = 9 \ \ or \ \ \quad x = -2$$
The solutions are 9 and -2.

2. $-7, 3$

3.
$$x^2 = 8x - 16$$
$$x^2 - 8x + 16 = 0 \qquad \text{Standard form}$$
We can factor.
$$x^2 - 8x + 16 = 0$$
$$(x - 4)(x - 4) = 0$$
$$x - 4 = 0 \ \ or \ \ x - 4 = 0$$
$$x = 4 \ \ or \ \ \quad x = 4$$
The solution is 4.

4. 3

5. $3y^2 + 7y + 4 = 0$

We can factor.
$$3y^2 + 7y + 4 = 0$$
$$(3y + 4)(y + 1) = 0$$
$$3y + 4 = 0 \ \ or \ \ y + 1 = 0$$
$$3y = -4 \ \ or \ \ \quad y = -1$$
$$y = -\frac{4}{3} \ \ or \ \ \quad y = -1$$
The solutions are $-\frac{4}{3}$ and -1.

6. $-2, \dfrac{4}{3}$

7.
$$4x^2 - 12x = 7$$
$$4x^2 - 12x - 7 = 0$$
We can factor.
$$4x^2 - 12x - 7 = 0$$
$$(2x + 1)(2x - 7) = 0$$
$$2x + 1 = 0 \quad or \ \ 2x - 7 = 0$$
$$2x = -1 \quad or \ \ \quad 2x = 7$$
$$x = -\frac{1}{2} \quad or \ \ \quad x = \frac{7}{2}$$
The solutions are $-\dfrac{1}{2}$ and $\dfrac{7}{2}$.

8. $-\dfrac{5}{2}, \dfrac{3}{2}$

9.
$$t^2 = 64$$
$$t = 8 \ \ or \ \ t = -8 \quad \text{Principle of square roots}$$
The solutions are 8 and -8.

10. $-9, 9$

11. $x^2 + 4x - 7 = 0$

We use the quadratic formula.
$$a = 1, \ b = 4, \ c = -7$$
$$x = \frac{-b \pm \sqrt{b^2 - 4ac}}{2a}$$
$$x = \frac{-4 \pm \sqrt{4^2 - 4 \cdot 1 \cdot (-7)}}{2 \cdot 1}$$
$$x = \frac{-4 \pm \sqrt{16 + 28}}{2}$$
$$x = \frac{-4 \pm \sqrt{44}}{2} = \frac{-4 \pm \sqrt{4 \cdot 11}}{2}$$
$$x = \frac{-4 \pm 2\sqrt{11}}{2} = \frac{2(-2 \pm \sqrt{11})}{2 \cdot 1}$$
$$x = -2 \pm \sqrt{11}$$
The solutions are $-2 + \sqrt{11}$ and $-2 - \sqrt{11}$, or $-2 \pm \sqrt{11}$.

12. $-1 \pm \sqrt{3}$

13. $y^2 - 10y + 19 = 0$

We use the quadratic formula.
$$a = 1, \ b = -10, \ c = 19$$

$$y = \frac{-b \pm \sqrt{b^2 - 4ac}}{2a}$$

$$y = \frac{-(-10) \pm \sqrt{(-10)^2 - 4 \cdot 1 \cdot 19}}{2 \cdot 1}$$

$$y = \frac{10 \pm \sqrt{100 - 76}}{2}$$

$$y = \frac{10 \pm \sqrt{24}}{2} = \frac{10 \pm \sqrt{4 \cdot 6}}{2}$$

$$y = \frac{10 \pm 2\sqrt{6}}{2} = \frac{2(5 \pm \sqrt{6})}{2 \cdot 1}$$

$$y = 5 \pm \sqrt{6}$$

The solutions are $5 + \sqrt{6}$ and $5 - \sqrt{6}$, or $5 \pm \sqrt{6}$.

14. $-3 \pm \sqrt{11}$

15. $x^2 + 2x + 1 = 7$

Observe that $x^2 + 2x + 1$ is a perfect-square trinomial. Then we can use the principle of square roots.

$$x^2 + 2x + 1 = 7$$
$$(x + 1)^2 = 7$$
$$x + 1 = \sqrt{7} \quad or \quad x + 1 = -\sqrt{7}$$
$$x = -1 + \sqrt{7} \quad or \quad x = -1 - \sqrt{7}$$

The solutions are $-1 + \sqrt{7}$ and $-1 - \sqrt{7}$, or $-1 \pm \sqrt{7}$.

16. $2 \pm \sqrt{5}$

17. $3t^2 + 8t + 2 = 0$

We use the quadratic formula.

$a = 3, b = 8, c = 2$

$$t = \frac{-b \pm \sqrt{b^2 - 4ac}}{2a}$$

$$t = \frac{-8 \pm \sqrt{8^2 - 2 \cdot 3 \cdot 2}}{2 \cdot 3}$$

$$t = \frac{-8 \pm \sqrt{64 - 24}}{6} = \frac{-8 \pm \sqrt{40}}{6}$$

$$t = \frac{-8 \pm \sqrt{4 \cdot 10}}{6} = \frac{-8 \pm 2\sqrt{10}}{6}$$

$$t = \frac{2(-4 \pm \sqrt{10})}{2 \cdot 3} = \frac{-4 \pm \sqrt{10}}{3}$$

The solutions are $\dfrac{-4 + \sqrt{10}}{3}$ and $\dfrac{-4 - \sqrt{10}}{3}$, or $\dfrac{-4 \pm \sqrt{10}}{3}$.

18. $\dfrac{2 \pm \sqrt{10}}{3}$

19. $2x^2 - 5x = 1$

$2x^2 - 5x - 1 = 0$ Standard form

We use the quadratic formula.

$a = 2, b = -5, c = -1$

$$x = \frac{-b \pm \sqrt{b^2 - 4ac}}{2a}$$

$$x = \frac{-(-5) \pm \sqrt{(-5)^2 - 4 \cdot 2 \cdot (-1)}}{2 \cdot 2}$$

$$x = \frac{5 \pm \sqrt{25 + 8}}{4} = \frac{5 \pm \sqrt{33}}{4}$$

The solutions are $\dfrac{5 + \sqrt{33}}{4}$ and $\dfrac{5 - \sqrt{33}}{4}$, or $\dfrac{5 \pm \sqrt{33}}{4}$.

20. $\dfrac{-1 \pm \sqrt{7}}{2}$

21. $4y^2 + 2y - 3 = 0$

We use the quadratic formula.

$a = 4, b = 2, c = -3$

$$y = \frac{-b \pm \sqrt{b^2 - 4ac}}{2a}$$

$$y = \frac{-2 \pm \sqrt{2^2 - 4 \cdot 4 \cdot (-3)}}{2 \cdot 4}$$

$$y = \frac{-2 \pm \sqrt{4 + 48}}{8} = \frac{-2 \pm \sqrt{52}}{8}$$

$$y = \frac{-2 \pm \sqrt{4 \cdot 13}}{8} = \frac{-2 \pm 2\sqrt{13}}{8}$$

$$y = \frac{2(-1 \pm \sqrt{13})}{2 \cdot 4} = \frac{-1 \pm \sqrt{13}}{4}$$

The solutions are $\dfrac{-1 + \sqrt{13}}{4}$ and $\dfrac{-1 - \sqrt{13}}{4}$, or $\dfrac{-1 \pm \sqrt{13}}{4}$.

22. $-\dfrac{1}{2}, \dfrac{3}{2}$

23. $2t^2 - 3t + 2 = 0$

We use the quadratic formula.

$a = 2, b = -3, c = 2$

$$t = \frac{-b \pm \sqrt{b^2 - 4ac}}{2a}$$

$$t = \frac{-(-3) \pm \sqrt{(-3)^2 - 4 \cdot 2 \cdot 2}}{2 \cdot 2}$$

$$t = \frac{3 \pm \sqrt{9 - 16}}{4} = \frac{3 \pm \sqrt{-7}}{4}$$

Since the radicand, -7, is negative, there are no real-number solutions.

24. No real-number solution

25. $3x^2 - 5x = 4$

$3x^2 - 5x - 4 = 0$

We use the quadratic formula.

$a = 3,\ b = -5,\ c = -4$

$$x = \frac{-b \pm \sqrt{b^2 - 4ac}}{2a}$$

$$x = \frac{-(-5) \pm \sqrt{(-5)^2 - 4 \cdot 3 \cdot (-4)}}{2 \cdot 3}$$

$$x = \frac{5 \pm \sqrt{25 + 48}}{6} = \frac{5 \pm \sqrt{73}}{6}$$

The solutions are $\dfrac{5 + \sqrt{73}}{6}$ and $\dfrac{5 - \sqrt{73}}{6}$, or $\dfrac{5 \pm \sqrt{73}}{6}$.

26. $\dfrac{-3 \pm \sqrt{17}}{4}$

27. $2y^2 - 6y = 10$

$2y^2 - 6y - 10 = 0$

$y^2 - 3y - 5 = 0$ Multiplying by $\dfrac{1}{2}$

We use the quadratic formula.

$a = 1,\ b = -3,\ c = -5$

$$y = \frac{-b \pm \sqrt{b^2 - 4ac}}{2a}$$

$$y = \frac{-(-3) \pm \sqrt{(-3)^2 - 4 \cdot 1 \cdot (-5)}}{2 \cdot 1}$$

$$y = \frac{3 \pm \sqrt{9 + 20}}{2} = \frac{3 \pm \sqrt{29}}{2}$$

The solutions are $\dfrac{3 + \sqrt{29}}{2}$ and $\dfrac{3 - \sqrt{29}}{2}$, or $\dfrac{3 \pm \sqrt{29}}{2}$.

28. $\dfrac{11 \pm \sqrt{181}}{10}$

29. $10x^2 - 15x = 0$

We can factor.

$10x^2 - 15x = 0$

$5x(2x - 3) = 0$

$5x = 0 \quad or \quad 2x - 3 = 0$

$x = 0 \quad or \qquad 2x = 3$

$x = 0 \quad or \qquad\quad x = \dfrac{3}{2}$

The solutions are 0 and $\dfrac{3}{2}$.

30. No real-number solution

31. $5t^2 - 7t = -4$

$5t^2 - 7t + 4 = 0$ Standard form

We use the quadratic formula.

$a = 5,\ b = -7,\ c = 4$

$$t = \frac{-b \pm \sqrt{b^2 - 4ac}}{2a}$$

$$t = \frac{-(-7) \pm \sqrt{(-7)^2 - 4 \cdot 5 \cdot 4}}{2 \cdot 5}$$

$$t = \frac{7 \pm \sqrt{49 - 80}}{10} = \frac{7 \pm \sqrt{-31}}{10}$$

Since the radicand, -31, is negative, there are no real-number solutions.

32. $-\dfrac{2}{3},\ 0$

33. $9y^2 = 162$

$y^2 = 18$ Dividing by 9

$y = \sqrt{18} \quad or \quad y = -\sqrt{18}$ Principle of square roots

$y = 3\sqrt{2} \quad or \quad y = -3\sqrt{2}$

The solutions are $3\sqrt{2}$ and $-3\sqrt{2}$, or $\pm 3\sqrt{2}$.

34. $\pm 2\sqrt{5}$

35. $x^2 - 4x - 7 = 0$

$a = 1,\ b = -4,\ c = -7$

$$x = \frac{-(-4) \pm \sqrt{(-4)^2 - 4 \cdot 1 \cdot (-7)}}{2 \cdot 1}$$

$$x = \frac{4 \pm \sqrt{16 + 28}}{2} = \frac{4 \pm \sqrt{44}}{2}$$

$$x = \frac{4 \pm \sqrt{4 \cdot 11}}{2} = \frac{4 \pm 2\sqrt{11}}{2}$$

$$x = \frac{2(2 \pm \sqrt{11})}{2} = 2 \pm \sqrt{11}$$

Using a calculator or Table 2, we see that $\sqrt{11} \approx 3.317$:

$2 + \sqrt{11} \approx 2 + 3.317 \quad or \quad 2 - \sqrt{11} \approx 2 - 3.317$

$\approx 5.317 \qquad or \qquad\qquad \approx -1.317$

The approximate solutions, to the nearest thousandth, are 5.317 and -1.317.

36. $-2.732,\ 0.732$

37. $y^2 - 5y - 1 = 0$

$a = 1,\ b = -5,\ c = -1$

$$y = \frac{-b \pm \sqrt{b^2 - 4ac}}{2a}$$

$$y = \frac{-(-5) \pm \sqrt{(-5)^2 - 4 \cdot 1 \cdot (-1)}}{2 \cdot 1}$$

$$y = \frac{5 \pm \sqrt{25 + 4}}{2} = \frac{5 \pm \sqrt{29}}{2}$$

Using a calculator or Table 2, we see that $\sqrt{29} \approx 5.385$:

$$\frac{5 + \sqrt{29}}{2} \approx \frac{5 + 5.385}{2} \quad or \quad \frac{5 - \sqrt{29}}{2} \approx \frac{5 - 5.385}{2}$$

$$\approx 5.193 \quad or \quad \approx -0.193$$

The approximate solutions, to the nearest thousandth, are 5.193 and −0.193.

38. −6.541, −0.459

39.
$$4x^2 + 4x = 1$$

$$4x^2 + 4x - 1 = 0 \qquad \text{Standard form}$$
$$a = 4, \ b = 4, \ c = -1$$
$$x = \frac{-4 \pm \sqrt{4^2 - 4 \cdot 4 \cdot (-1)}}{2 \cdot 4}$$
$$x = \frac{-4 \pm \sqrt{16 + 16}}{8} = \frac{-4 \pm \sqrt{32}}{8}$$
$$x = \frac{-4 \pm \sqrt{16 \cdot 2}}{8} = \frac{-4 \pm 4\sqrt{2}}{8}$$
$$x = \frac{4(-1 \pm \sqrt{2})}{4 \cdot 2} = \frac{-1 \pm \sqrt{2}}{2}$$

Using a calculator or Table 2, we see that $\sqrt{2} \approx 1.414$:

$$\frac{-1 + \sqrt{2}}{2} \approx \frac{-1 + 1.414}{2} \quad or \quad \frac{-1 - \sqrt{2}}{2} \approx \frac{-1 - 1.414}{2}$$

$$\approx \frac{0.414}{2} \quad or \quad \approx \frac{-2.414}{2}$$

$$\approx 0.207 \quad or \quad \approx -1.207$$

The approximate solutions, to the nearest thousandth, are 0.207 and −1.207.

40. −0.207, 1.207

41. *Familiarize*. We will use the formula
$$d = \frac{n^2 - 3n}{2},$$
where d is the number of diagonals and n is the number of sides.

***Translate*.** We substitute 35 for d.
$$35 = \frac{n^2 - 3n}{2}$$

***Carry out*.** We solve the equation.
$$\frac{n^2 - 3n}{2} = 35$$
$$n^2 - 3n = 70 \qquad \text{Multiplying by 2}$$
$$n^2 - 3n - 70 = 0$$
$$(n - 10)(n + 7) = 0$$
$$n - 10 = 0 \quad or \quad n + 7 = 0$$
$$n = 10 \quad or \quad n = -7$$

***Check*.** Since the number of sides cannot be negative, −7 cannot be a solution. To check 10, we substitute 10 for n in the original formula and determine if this yields $d = 35$. This is left to the student.

***State*.** The polygon has 10 sides.

42. 8

43. *Familiarize*. We will use the formula $s = 16t^2$.

***Translate*.** We substitute 1490 for s.
$$1490 = 16t^2$$

***Carry out*.** We solve the equation.
$$1490 = 16t^2$$
$$\frac{1490}{16} = t^2$$
$$\sqrt{\frac{1490}{16}} = t \quad or \quad -\sqrt{\frac{1490}{16}} = t \quad \begin{array}{l}\text{Principle of} \\ \text{square roots}\end{array}$$
$$9.65 \approx t \quad or \quad -9.65 \approx t$$

***Check*.** The number −9.65 cannot be a solution, because time cannot be negative in this situation. We substitute 9.65 in the original equation:
$$s = 16(9.65)^2 = 16(93.1225) = 1489.96.$$

This is close to 1490. Remember that we approximated the solution. Thus, we have a check.

***State*.** It would take about 9.65 sec for an object to fall to the ground from the top of the Universal Financial Center.

44. 9.53 sec

45. *Familiarize*. We will use the formula $s = 16t^2$.

***Translate*.** We substitute 700 for s.
$$700 = 16t^2$$

***Carry out*.** We solve the equation.
$$700 = 16t^2$$
$$\frac{700}{16} = t^2$$
$$\sqrt{\frac{700}{16}} = t \quad or \quad -\sqrt{\frac{700}{16}} = t \quad \begin{array}{l}\text{Principle of} \\ \text{square roots}\end{array}$$
$$6.61 \approx t \quad or \quad -6.61 \approx t$$

***Check*.** The number −6.61 cannot be a solution, because time cannot be negative in this situation. We substitute 6.61 in the original equation:
$$s = 16(6.61)^2 = 16(43.6921) = 699.0736.$$

This is close to 700. Remember that we approximated the solution. Thus, we have a check.

***State*.** The free-fall portion of the jump lasted about 6.61 sec.

46. 3.31 sec

47. *Familiarize.* From the drawing in the text we have s = the length of the shorter leg and $s + 17$ = the length of the longer leg, in feet.

Translate. We use the Pythagorean theorem.
$$x^2 + (x + 17)^2 = 25^2$$

Carry out. We solve the equation.
$$x^2 + x^2 + 34x + 289 = 625$$
$$2x^2 + 34x - 336 = 0$$
$$x^2 + 17x - 168 = 0 \qquad \text{Multiplying by } \frac{1}{2}$$
$$(x - 7)(x + 24) = 0$$
$$x - 7 = 0 \ \ or \ \ x + 24 = 0$$
$$x = 7 \ \ or \ \ \qquad x = -24$$

Check. Since the length of a leg cannot be negative, -24 does not check. But 7 does check. If the smaller leg is 7, the other leg is $7+17$, or 24. Then, $7^2 + 24^2 = 49 + 576 = 625$, and $\sqrt{625} = 25$, the length of the hypotenuse.

State. The legs measure 7 ft and 24 ft.

48. 10 yd, 24 yd

49. *Familiarize.* From the drawing in the text, we see that w represents the width of the rectangle and $w+4$ represents the length, in centimeters.

Translate. The area is length $\times$ width. Thus, we have two expressions for the area of the rectangle: $(w + 4)w$ and 60. This gives us a translation.
$$(w + 4)w = 60$$

Carry out. We solve the equation.
$$w^2 + 4w = 60$$
$$w^2 + 4w - 60 = 0$$
$$(w + 10)(w - 6) = 0$$
$$w + 10 = 0 \quad or \quad w - 6 = 0$$
$$w = -10 \ \ or \qquad w = 6$$

Check. Since the length of a side cannot be negative, -10 does not check. But 6 does check. If the width is 6, then the length is $6 + 4$, or 10. The area is 10×6, or 60. This checks.

State. The length is 10 cm, and the width is 6 cm.

50. Length: 10 m; width: 7 m

51. *Familiarize.* We make a drawing. We let w = the width of the yard. Then $w + 6$ = the length, in meters.

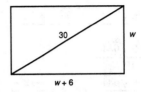

Translate. We use the Pythagorean theorem.
$$w^2 + (w + 6)^2 = 30^2$$

Carry out. We solve the equation.
$$w^2 + w^2 + 12w + 36 = 900$$
$$2w^2 + 12w - 864 = 0$$
$$w^2 + 6w - 432 = 0$$
$$(w + 24)(w - 18) = 0$$
$$w + 24 = 0 \qquad or \ \ w - 18 = 0$$
$$w = -24 \ \ or \qquad w = 18$$

Check. Since the width cannot be negative, -24 does not check. But 18 does check. If the width is 18, then the length is $18 + 6$, or 24, and $18^2 + 24^2 = 324 + 576 = 900 = 30^2$.

State. The yard is 18 m by 24 m.

52. 24 ft

53. *Familiarize.* We make a drawing. Let x = the length of the shorter leg of the right triangle. Then $x + 2.4$ = the length of the longer leg, in meters.

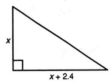

Translate. Using the formula $A = \frac{1}{2}bh$, we substitute 31 for A, $x + 2.4$ for b, and x for h.
$$31 = \frac{1}{2}(x + 2.4)(x)$$

Carry out. We solve the equation.
$$31 = \frac{1}{2}(x + 2.4)(x)$$
$$62 = (x + 2.4)(x) \qquad \text{Multiplying by 2}$$
$$62 = x^2 + 2.4x$$
$$0 = x^2 + 2.4x - 62$$
$$0 = 10x^2 + 24x - 620 \qquad \begin{array}{l}\text{Multiplying by 10 to} \\ \text{clear the decimal}\end{array}$$
$$0 = 2(5x^2 + 12x - 310)$$

We use the quadratic formula.
$$a = 5, \ b = 12, \ b = -310$$

$$x = \frac{-12 \pm \sqrt{12^2 - 4 \cdot 5 \cdot (-310)}}{2 \cdot 5}$$

$$x = \frac{-12 \pm \sqrt{6344}}{10}$$

$$x = \frac{-12 + \sqrt{6344}}{10} \quad or \quad x = \frac{-12 - \sqrt{6344}}{10}$$

$$x \approx 6.76 \qquad\qquad or \quad x \approx -9.16$$

Check. Since the length of a leg cannot be negative, -9.16 does not check. But 6.76 does. If the shorter leg is 6.76 m, then the longer leg is $6.76 + 2.4$, or 9.16 m, and $A = \frac{1}{2}(9.16)(6.76) \approx 31 \approx 30$.

State. The lengths of the legs are about 6.76 m and 9.16 m.

54. 5.13 cm, 10.13 cm

55. Familiarize. We first make a drawing. We let x represent the width and $x + 3$ the length, in inches.

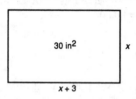

Translate. The area is length $\times$ width. We have two expressions for the area of the rectangle: $(x+3)x$ and 30. This gives us a translation.

$$(x + 3)x = 30$$

Carry out. We solve the equation.

$$x^2 + 3x = 30$$
$$x^2 + 3x - 30 = 0$$
$$a = 1, \, b = 3, \, c = -30$$
$$x = \frac{-3 \pm \sqrt{3^2 - 4 \cdot 1 \cdot (-30)}}{2 \cdot 1}$$
$$x = \frac{-3 \pm \sqrt{129}}{2}$$
$$x = \frac{-3 + \sqrt{129}}{2} \, or \, x = \frac{-3 - \sqrt{129}}{2}$$
$$x \approx 4.18 \qquad or \, x \approx -7.18$$

Check. Since the width cannot be negative, -7.18 does not check. But 4.18 does check. If the width is 4.18 in., then the length is $4.18 + 3$, or 7.18 in., and the area is $7.18(4.18) \approx 30$ in^2.

State. The length is about 7.18 in., and the width is about 4.18 in.

56. Length: 8.09 ft; width: 3.09 ft

57. Familiarize. We first make a drawing. We let x represent the width and $2x$ the length, in meters.

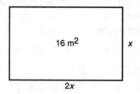

Translate. The area is length $\times$ width. We have two expressions for the area of the rectangle: $2x \cdot x$ and 16. This gives us a translation.

$$2x \cdot x = 16$$

Carry out. We solve the equation.

$$2x^2 = 16$$
$$x^2 = 8$$
$$x = \sqrt{8} \quad or \quad x = -\sqrt{8}$$
$$x = 2.83 \quad or \quad x \approx -2.83 \quad \text{Using a calculator}$$
$$\text{or Table 2}$$

Check. Since the length cannot be negative, -2.83 does not check. But 2.83 does check. If the width is $\sqrt{8}$ m, then the length is $(2\sqrt{8})$ or 5.66 m. The area is $(5.66)(2.83)$, or $16.0178 \approx 16$.

State. The length is about 5.66 m, and the width is about 2.83 m.

58. Length: 6.32 cm; width: 3.16 cm

59. Familiarize. We will use the formula $A = P(1+r)^t$.

Translate. We substitute 2000 for P, 2880 for A, and 2 for t.

$$2880 = 2000(1 + r)^2$$

Carry out. We solve the equation.

$$2880 = 2000(1 + r)^2$$
$$1.44 = (1 + r)^2$$
$$\sqrt{1.44} = 1 + r \, or \, -\sqrt{1.44} = 1 + r \quad \text{Principle}$$
$$\text{of square roots}$$
$$1.2 = 1 + r \, or \qquad -1.2 = 1 + r$$
$$0.2 = r \qquad or \qquad -2.2 = r$$

Check. Since the interest rate cannot be negative, we check only 0.2, or 20%. We substitute in the formula:

$$2000(1 + 0.2)^2 = 2000(1.2)^2 = 2000(1.44) = 2880.$$

The answer checks.

State. The interest rate is 20%.

60. 18.75%

61. Familiarize. We will use the formula $A = P(1+r)^t$.

Translate. We substitute 6000 for P, 6615 for A, and 2 for t.

$$6615 = 6000(1 + r)^2$$

Carry out. We solve the equation.

$$6615 = 6000(1 + r)^2$$
$$1.1025 = (1 + r)^2$$
$$\sqrt{1.1025} = 1 + r \quad or \quad -\sqrt{1.1025} = 1 + r$$

Principle of square roots

$$1.05 = 1 + r \quad or \quad -1.05 = 1 + r$$
$$0.05 = r \quad or \quad -2.05 = r$$

Check. Since the interest rate cannot be negative, we check only 0.05, or 5%. We substitute in the formula:

$$6000(1 + 0.05)^2 = 6000(1.05)^2 = 6615.$$

The answer checks.

State. The interest rate is 5%.

62. 8%

63. *Familiarize*. Let d = the diameter (or width) of the oil slick, in km. Then $d/2$ = the radius. We will use the formula for the area of the circle, $A = \pi r^2$.

Translate. We substitute 20 for A, 3.14 for π, and $d/2$ for r in the formula.

$$20 = 3.14\left(\frac{d}{2}\right)^2$$
$$20 = 3.14\left(\frac{d^2}{4}\right)$$

Carry out. We solve the equation.

$$20 = 3.14\left(\frac{d^2}{4}\right)$$
$$20 = 0.785d^2 \quad \left(\frac{3.14}{4} = 0.785\right)$$
$$\frac{20}{0.785} = d^2$$
$$d = \sqrt{\frac{20}{0.785}} \quad or \quad d = -\sqrt{\frac{20}{0.785}}$$
$$d \approx 5.05 \quad or \quad d \approx -5.05$$

Check. Since the diameter cannot be negative, -5.05 cannot be a solution. If $d = 5.05$, then $A = 3.14\left(\frac{5.05}{2}\right)^2 \approx 20$. The answer checks.

State. The oil slick was about 5.05 km wide.

64. 17.85 ft (If we had used the π key on a calculator, the result would have been 17.84 ft.)

65. ◈

66. ◈

67. $2x - 7 = 43$
$$2x = 50 \quad \text{Adding 7}$$
$$x = 25 \quad \text{Dividing by 2}$$
The solution is 25.

68. 4

69. $\dfrac{3}{5}t + 6 = 15$
$$\frac{3}{5}t = 9 \quad \text{Subtracting 9}$$
$$\frac{5}{3} \cdot \frac{3}{5}t = \frac{5}{3} \cdot 9$$
$$t = 15$$
The solution is 15.

70. 9

71. $\sqrt{4x} - 3 = 5$
$$\sqrt{4x} = 8 \quad \text{Adding 3}$$
$$(\sqrt{4x})^2 = 8^2 \quad \text{Squaring both sides}$$
$$4x = 64$$
$$x = 16 \quad \text{Dividing by 4}$$
The solution is 16.

72. 50

73. ◈

74. ◈

75. $$5x = -x(x - 7)$$
$$5x = -x^2 + 7x$$
$$5x + x^2 - 7x = 0$$
$$x^2 - 2x = 0$$
$$x(x - 2) = 0$$
$$x = 0 \quad or \quad x - 2 = 0$$
$$x = 0 \quad or \quad x = 2$$
The solutions are 0 and 2.

76. $-\dfrac{4}{3}$, 0

77. $$3 - x(x - 3) = 4$$
$$3 - x^2 + 3x = 4$$
$$0 = x^2 - 3x + 1$$
$$a = 1, \, b = -3, \, c = 1$$

$$x = \frac{-(-3) \pm \sqrt{(-3)^2 - 4 \cdot 1 \cdot 1}}{2 \cdot 1}$$

$$x = \frac{3 \pm \sqrt{5}}{2}$$

The solutions are $\dfrac{3 + \sqrt{5}}{2}$ and $\dfrac{3 - \sqrt{5}}{2}$, or $\dfrac{3 \pm \sqrt{5}}{2}$.

78. $\dfrac{7 \pm \sqrt{69}}{10}$

79. $(y + 4)(y + 3) = 15$

$$y^2 + 7y + 12 = 15$$
$$y^2 + 7y - 3 = 0$$
$$a = 1,\, b = 7,\, c = -3$$
$$y = \frac{-7 \pm \sqrt{7^2 - 4 \cdot 1 \cdot (-3)}}{2 \cdot 1}$$
$$y = \frac{-7 \pm \sqrt{61}}{2}$$

The solutions are $\dfrac{-7 + \sqrt{61}}{2}$ and $\dfrac{-7 - \sqrt{61}}{2}$, or $\dfrac{-7 \pm \sqrt{61}}{2}$.

80. $\dfrac{-2 \pm \sqrt{10}}{2}$

81.
$$\frac{x^2}{x + 3} - \frac{5}{x + 3} = 0, \qquad \text{LCM is } x + 3$$
$$(x + 3)\left(\frac{x^2}{x + 3} - \frac{5}{x + 3}\right) = (x + 3) \cdot 0$$
$$x^2 - 5 = 0$$
$$x^2 = 5$$

$x = \sqrt{5}$ or $x = -\sqrt{5}$ Principle of square roots

Both numbers check. The solutions are $\sqrt{5}$ and $-\sqrt{5}$, or $\pm\sqrt{5}$.

82. $\pm\sqrt{7}$

83.
$$\frac{1}{x} + \frac{1}{x + 1} = \frac{1}{3}, \qquad \text{LCM is } 3x(x + 1)$$
$$3x(x + 1)\left(\frac{1}{x} + \frac{1}{x + 1}\right) = 3x(x + 1) \cdot \frac{1}{3}$$
$$3(x + 1) + 3x = x(x + 1)$$
$$3x + 3 + 3x = x^2 + x$$
$$6x + 3 = x^2 + x$$
$$0 = x^2 - 5x - 3$$
$$x = \frac{-(-5) \pm \sqrt{(-5)^2 - 4 \cdot 1 \cdot (-3)}}{2 \cdot 1}$$
$$x = \frac{5 \pm \sqrt{37}}{2}$$

The solutions are $\dfrac{5 + \sqrt{37}}{2}$ and $\dfrac{5 - \sqrt{37}}{2}$, or $\dfrac{5 \pm \sqrt{37}}{2}$.

84. $2 \pm \sqrt{34}$

85. *Familiarize.* From the drawing in the text, we see that we have a right triangle where $r =$ the length of each leg and $r + 2 =$ the length of the hypotenuse, in centimeters.

Translate. We use the Pythagorean theorem.
$$r^2 + r^2 = (r + 2)^2.$$

Carry out. We solve the equation.
$$2r^2 = r^2 + 4r + 4$$
$$r^2 - 4r - 4 = 0$$
$$a = 1,\, b = -4,\, c = -4$$
$$r = \frac{-(-4) \pm \sqrt{(-4)^2 - 4 \cdot 1 \cdot (-4)}}{2 \cdot 1}$$
$$r = \frac{4 \pm \sqrt{16 + 16}}{2} = \frac{4 \pm \sqrt{32}}{2}$$
$$r = \frac{4 \pm \sqrt{16 \cdot 2}}{2} = \frac{4 \pm 4\sqrt{2}}{2}$$
$$r = \frac{2(2 \pm 2\sqrt{2})}{2 \cdot 1} = 2 \pm 2\sqrt{2}$$

$x = 2 - 2\sqrt{2}$ *or* $x = 2 + 2\sqrt{2}$

$x \approx 2 - 2.828$ *or* $x \approx 2 + 2.828$

$x \approx -0.828$ *or* $x \approx 4.828$

$x \approx -0.83$ *or* $x \approx 4.83$ Rounding to the nearest hundredth

Check. Since the length of a leg cannot be negative, -0.83 cannot be a solution of the original equation. When $x \approx 4.83$, then $x + 2 \approx 6.83$ and $(4.83)^2 + (4.83)^2 = 23.3289 + 23.3289 = 46.6578 \approx (5.83)^2$. This checks.

State. In the figure, $r = 2 + 2\sqrt{2}$ cm ≈ 4.83 cm.

86. 52.5 square units

87. *Familiarize.* Let $w =$ the width of the rectangle, in meters. Then $1.6w =$ the length. Recall that the formula for the area of a rectangle is $A = l \times w$.

Translate. We substitute 9000 for A, $1.6w$ for w, and w for w in the formula.
$$9000 = 1.6w(w)$$
$$9000 = 1.6w^2$$

Carry out. We solve the equation.
$$9000 = 1.6w^2$$
$$\frac{9000}{1.6} = w^2$$

$$w = \sqrt{\frac{9000}{1.6}} \quad or \quad w = -\sqrt{\frac{9000}{1.6}}$$
$$w = 75 \quad or \quad w = -75$$

Check. Since the width of the rectangle cannot be negative we will not check -75. If $w = 75$, then the length is $1.6(75)$, or 120, and the area is $120(75) = 9000 \text{ m}^2$. The answer checks.

State. The length of the rectangle is 120 m, and the width is 75 m.

88. 7.5 ft

89. Familiarize. We will use the formula $A = P(1 + r)^2$ twice. The amount in the account for the $4000 invested for 2 yr is given by $4000(1+r)^2$. The amount for the $2000 invested for 1 yr is $2000(1 + r)$.

Translate. The total amount in the account at the end of 2 yr is the sum of the amounts above.

$$6510 = 4000(1 + r)^2 + 2000(1 + r)$$

Carry out. We solve the equation. Begin by letting $x = 1 + r$.

$$6510 = 4000x^2 + 2000x$$
$$0 = 4000x^2 + 2000x - 6510$$
$$0 = 400x^2 + 200x - 651 \quad \text{Dividing by 10}$$

We will use the quadratic formula.

$$a = 400, \; b = 200, \; c = -651$$
$$x = \frac{-200 \pm \sqrt{200^2 - 4 \cdot 400(-651)}}{2 \cdot 400}$$
$$x = \frac{-200 \pm \sqrt{1,081,600}}{800} = \frac{-200 \pm 1040}{800}$$
$$x = \frac{-200 + 1040}{800} \quad or \quad x = \frac{-200 - 1040}{800}$$
$$x = 1.05 \quad or \quad x = -1.55$$
$$1 + r = 1.05 \quad or \quad 1 + r = -1.55$$
$$r = 0.05 \quad or \quad r = -2.55$$

Check. Since the interest rate cannot be negative, we check only 0.05 or 5%. At the end of 2 yr, $4000 invested at 5% interest has grown to $4000(1+0.05)^2$, or $4410. At the end of 1 yr, $2000 invested at 5% interest has grown to $2000(1+0.05)$, or $2100. Then the total amount in the account is $4410 + $2100, or $6510. The answer checks.

State. The interest rate is 5%.

90. $2682.63

91. Familiarize. The area of the actual strike zone is $15(40)$, so the area of the enlarged zone is $15(40) + 0.4(15)(40)$, or $1.4(15)(40)$. From the drawing in the text we see that the dimensions of the enlarged strike zone are $15 + 2x$ by $40 + 2x$.

Translate. Using the formula $A = lw$, we write an equation for the area of the enlarged strike zone.

$$1.4(15)(40) = (15 + 2x)(40 + 2x)$$

Carry out. We solve the equation.

$$1.4(15)(40) = (15 + 2x)(40 + 2x)$$
$$840 = 600 + 110x + 4x^2 \quad \begin{array}{l}\text{Multiplying}\\\text{on both sides}\end{array}$$
$$0 = 4x^2 + 110x - 240$$
$$0 = 2x^2 + 55x - 120 \quad \text{Dividing by 2}$$
$$a = 2, \; b = 55, \; c = -120$$
$$x = \frac{-55 \pm \sqrt{55^2 - 4 \cdot 2 \cdot (-120)}}{2 \cdot 2}$$
$$x = \frac{-55 \pm \sqrt{3985}}{4}$$
$$x \approx 2.03 \quad or \quad x \approx -29.53$$

Check. Since the measurement cannot be negative, -29.53 cannot be a solution. If $x = 2.03$, then the dimensions of the enlarged strike zone are $15 + 2(2.03)$, or 19.06, by $40 + 2(2.03)$, or 44.06, and the area is $19.06(44.06) = 839.7836 \approx 840$. The answer checks.

State. The dimensions of the enlarged strike zone are 19.06 in. by 44.06 in.

92.

Exercise Set 9.4

---

1. $3x + 5 = 7x + 1$
$$4 = 4x \quad \text{Adding } -3x - 1$$
$$1 = x \quad \text{Dividing by 4}$$
The solution is 1.

2. $-\dfrac{13}{4}$

3. $$x^2 = 5x - 6$$
$$x^2 - 5x + 6 = 0 \quad \text{Adding } -5x + 6$$
$$(x - 2)(x - 3) = 0$$
$$x - 2 = 0 \quad or \quad x - 3 = 0$$
$$x = 2 \quad or \quad x = 3$$
The solutions are 2 and 3.

4. $-1, \; 4$

5. $4 + \sqrt{x} = 9$
$$\sqrt{x} = 5 \quad \text{Subtracting 4}$$
$$(\sqrt{x})^2 = 5^2 \quad \text{Squaring both sides}$$
$$x = 25$$
The number 25 checks. It is the solution.

6. 144

7.
$$\frac{5}{x} + \frac{3}{4} = 2 \qquad \text{Note that } x \neq 0;$$
$$\text{LCD} = 4x.$$

$$4x\left(\frac{5}{x} + \frac{3}{4}\right) = 4x \cdot 2 \quad \text{Multiplying by the LCD}$$

$$4x \cdot \frac{5}{x} + 4x \cdot \frac{3}{4} = 8x$$
$$20 + 3x = 8x$$
$$20 = 5x$$
$$4 = x$$

The number 4 checks. It is the solution.

8. $\frac{5}{6}$

9.
$$4(2x - 1) = 3x - 5$$
$$8x - 4 = 3x - 5$$
$$5x = -1 \quad \text{Adding } -3x + 4$$
$$x = -\frac{1}{5}$$

The solution is $-\frac{1}{5}$.

10. $-\frac{1}{6}$

11.
$$\frac{3}{10x} - \frac{4}{5x} = 6 \quad \text{Note that } x \neq 0;$$
$$\text{LCD} = 10x.$$

$$10x\left(\frac{3}{10x} - \frac{4}{5x}\right) = 10x \cdot 6$$
$$10x \cdot \frac{3}{10x} - 10x \cdot \frac{4}{5x} = 60x$$
$$3 - 8 = 60x$$
$$-5 = 60x$$
$$-\frac{1}{12} = x$$

This number checks. The solution is $-\frac{1}{12}$.

12. $\frac{4}{33}$

13.
$$2t^2 - 7t + 3 = 0$$
$$(2t - 1)(t - 3) = 0$$
$$2t - 1 = 0 \quad or \quad t - 3 = 0$$
$$2t = 1 \quad or \quad t = 3$$
$$t = \frac{1}{2} \quad or \quad t = 3$$

The solutions are $\frac{1}{2}$ and 3.

14. $\frac{1}{3}$, 2

15.
$$5\sqrt{2t} - 7 = 3$$
$$5\sqrt{2t} = 10$$
$$\sqrt{2t} = 2 \quad \text{Dividing by 5}$$
$$(\sqrt{2t})^2 = 2^2$$
$$2t = 4$$
$$t = 2$$

The number 2 checks. It is the solution.

16. $\frac{16}{3}$

17.
$$\frac{1}{4t} + \frac{t}{6} = 2 \qquad \text{Note that } t \neq 0;$$
$$\text{LCD} = 12t.$$
$$12t\left(\frac{1}{4t} + \frac{t}{6}\right) = 12t \cdot 2$$
$$12t \cdot \frac{1}{4t} + 12t \cdot \frac{t}{6} = 24t$$
$$3 + 2t^2 = 24t$$
$$2t^2 - 24t + 3 = 0$$
$$a = 2, b = -24, c = 3$$
$$t = \frac{-b \pm \sqrt{b^2 - 4ac}}{2a}$$
$$t = \frac{-(-24) \pm \sqrt{(-24)^2 - 4 \cdot 2 \cdot 3}}{2 \cdot 2}$$
$$t = \frac{24 \pm \sqrt{552}}{4} = \frac{24 \pm \sqrt{4 \cdot 138}}{4}$$
$$t = \frac{24 \pm 2\sqrt{138}}{4} = \frac{2(12 \pm \sqrt{138})}{2 \cdot 2}$$
$$t = \frac{12 \pm \sqrt{138}}{2}$$

Both numbers check. The solutions are $\frac{12 + \sqrt{138}}{2}$ and $\frac{12 - \sqrt{138}}{2}$, or $\frac{12 \pm \sqrt{138}}{2}$.

18. $\frac{5 \pm \sqrt{19}}{2}$

19.
$$7t - 1 = 2(3 - t) + 5$$
$$7t - 1 = 6 - 2t + 5$$
$$7t - 1 = 11 - 2t$$
$$9t = 12 \qquad \text{Adding } 2t + 1$$
$$t = \frac{4}{3}$$

The solution is $\frac{4}{3}$.

20. 1

21. $3\sqrt{2t-1} = 2\sqrt{5t+3}$

$(3\sqrt{2t-1})^2 = (2\sqrt{5t+3})^2$

$9(2t-1) = 4(5t+3)$

$18t - 9 = 20t + 12$

$-21 = 2t$ Adding $-18t - 12$

$-\dfrac{21}{2} = t$

Check:

$$3\sqrt{2t-1} = 2\sqrt{5t+3}$$

$$3\sqrt{2\left(-\dfrac{21}{2}\right)-1} \;?\; 2\sqrt{5\left(-\dfrac{21}{2}\right)+3}$$

$$3\sqrt{-21-1} \;\Big|\; 2\sqrt{-\dfrac{105}{2}+3}$$

$$3\sqrt{-22} \;\Big|\; 2\sqrt{-\dfrac{99}{2}} \quad \text{UNDEFINED}$$

The equation has no solution.

22. No solution

23. $2n - 1 = 3n^2$

$0 = 3n^2 - 2n + 1$

$a = 3, \; b = -2, \; c = 1$

$n = \dfrac{-b \pm \sqrt{b^2 - 4ac}}{2a}$

$n = \dfrac{-(-2) \pm \sqrt{(-2)^2 - 4 \cdot 3 \cdot 1}}{2 \cdot 3}$

$n = \dfrac{2 \pm \sqrt{-8}}{6}$

Since the radicand, -8, is negative, there are no real-number solutions.

24. $-3, \; \dfrac{1}{2}$

25. $1 - \dfrac{3}{7n} = \dfrac{5}{14}$ Note that $n \neq 0$;

                               LCD $= 14n$.

$14n\left(1 - \dfrac{3}{7n}\right) = 14n \cdot \dfrac{5}{14}$

$14n \cdot 1 - 14n \cdot \dfrac{3}{7n} = 5n$

$14n - 6 = 5n$

$-6 = -9n$

$\dfrac{2}{3} = n$

The solution is $\dfrac{2}{3}$.

26. $\dfrac{10}{47}$

27. $s = \dfrac{1}{2}gt^2$

$2s = gt^2$ Multiplying by 2

$\dfrac{2s}{t^2} = g$ Dividing by t^2

28. $h = \dfrac{2A}{b}$

29. $S = 2\pi rh$

$\dfrac{S}{2\pi r} = h$ Multiplying by $\dfrac{1}{2\pi r}$

30. $t = \dfrac{A - P}{Pr}$

31. $d = c\sqrt{h}$

$d^2 = (c\sqrt{h})^2$ Squaring both sides

$d^2 = c^2 \cdot h$

$\dfrac{d^2}{c^2} = h$ Dividing by c^2

32. $t = n^2 - 2nc + c^2$

33. $\dfrac{1}{R} = \dfrac{1}{r_1} + \dfrac{1}{r_2}$

$Rr_1r_2 \cdot \dfrac{1}{R} = Rr_1r_2\left(\dfrac{1}{r_1} + \dfrac{1}{r_2}\right)$ Multiplying by the LCD, Rr_1r_2, to clear fractions

$r_1r_2 = Rr_1r_2 \cdot \dfrac{1}{r_1} + Rr_1r_2 \cdot \dfrac{1}{r_2}$

$r_1r_2 = Rr_2 + Rr_1$

$r_1r_2 = R(r_2 + r_1)$

$\dfrac{r_1r_2}{r_2 + r_1} = R$

34. $f = \dfrac{pq}{q + p}$

35. $ax^2 + bx + c = 0$

Observe that this is standard form for a quadratic equation. The solution is given by the quadratic formula:

$$x = \dfrac{-b \pm \sqrt{b^2 - 4ac}}{2a}.$$

36. $x = d \pm \sqrt{k}$

37. $\dfrac{m}{n} = p - q$

$m = n(p - q)$ Multiplying by n

$\dfrac{m}{p - q} = n$ Dividing by $p - q$

38. $t = \dfrac{M - g}{r + s}$

39.
$$rl + rS = L$$
$$r(l + S) = L$$
$$r = \frac{L}{l + S}$$

40. $m = \dfrac{T}{g + f}$

41.
$$S = 2\pi r(r + h)$$
$$\frac{S}{2\pi r} = r + h \qquad \text{Dividing by } 2\pi r$$
$$\frac{S}{2\pi r} - r = h, \text{ or}$$
$$\frac{S - 2\pi r^2}{2\pi r} = h$$

42. $a = \dfrac{d}{b - c}$

43.
$$\frac{s}{h} = \frac{h}{t} \qquad \text{LCD} = ht$$
$$ht \cdot \frac{s}{h} = ht \cdot \frac{h}{t}$$
$$st = h^2$$
$$\pm\sqrt{st} = h$$

44. $x = \pm\sqrt{yz}$

45. $mt^2 + nt - p = 0$
$$a = m, \ b = n, \ c = -p$$
$$t = \frac{-n \pm \sqrt{n^2 - 4 \cdot m \cdot (-p)}}{2 \cdot m}$$
$$t = \frac{-n \pm \sqrt{n^2 + 4mp}}{2m}$$

46. $s = \dfrac{t \pm \sqrt{t^2 - 4rp}}{2r}$

47.
$$\frac{m}{n} = r$$
$$m = nr \quad \text{Multiplying by } n$$
$$\frac{m}{r} = n \quad \text{Dividing by } r$$

48. $v = \dfrac{r}{s + t}$

49.
$$m + t = \frac{n}{m}$$
$$m(m + t) = m \cdot \frac{n}{m}$$
$$m^2 + mt = n$$
$$m^2 + mt - n = 0$$
$$a = 1, \ b = t, \ c = -n$$

$$m = \frac{-t \pm \sqrt{t^2 - 4 \cdot 1 \cdot (-n)}}{2 \cdot 1}$$
$$m = \frac{-t \pm \sqrt{t^2 + 4n}}{2}$$

50. $y = \dfrac{x \pm \sqrt{x^2 - 4z}}{2}$

51.
$$n = p - 3\sqrt{t + c}$$
$$n - p = -3\sqrt{t + c}$$
$$(n - p)^2 = (-3\sqrt{t + c})^2$$
$$n^2 - 2np + p^2 = 9(t + c)$$
$$n^2 - 2np + p^2 = 9t + 9c$$
$$n^2 - 2np + p^2 - 9c = 9t$$
$$\frac{n^2 - 2np + p^2 - 9c}{9} = t$$

52. $t = \dfrac{r^2 - 2rM + M^2}{m^2 c}$

53.
$$\sqrt{m - n} = \sqrt{3t}$$
$$(\sqrt{m - n})^2 = (\sqrt{3t})^2$$
$$m - n = 3t$$
$$m = n + 3t$$
$$m - 3t = n$$

54. $t = 0$

55. ◈

56. ◈

57. $\sqrt{6} \cdot \sqrt{10} = \sqrt{6 \cdot 10} = \sqrt{2 \cdot 3 \cdot 2 \cdot 5} = \sqrt{2 \cdot 2}\sqrt{3 \cdot 5} = 2\sqrt{15}$

58. $3\sqrt{35}$

59. $\sqrt{150} = \sqrt{25 \cdot 6} = \sqrt{25}\sqrt{6} = 5\sqrt{6}$

60. $3\sqrt{7}$

61. $\sqrt{4a^7 b^4} = \sqrt{4 \cdot a^6 \cdot b^4 \cdot a} = \sqrt{4}\sqrt{a^6}\sqrt{b^4}\sqrt{a} = 4a^3 b^2 \sqrt{a}$

62. $3x^4 y^2 \sqrt{y}$

63. ◈

64. ◈

65. Substitute 8 for c and 224 for d and solve for a.

$$c = \frac{a}{a+12} \cdot d$$

$$8 = \frac{a}{a+12} \cdot 24$$

$$\frac{1}{3} = \frac{a}{a+12} \qquad \text{Dividing by 24}$$

$$3(a+12) \cdot \frac{1}{3} = 3(a+12) \cdot \frac{a}{a+12}$$

$$a + 12 = 3a$$

$$12 = 2a$$

$$6 = a$$

The child is 6 years old.

66. $r = \dfrac{Sl + l \pm \sqrt{S^2l^2 + 2Sl^2 + l^2 - 4Sa}}{2S}$

67.
$$fm = \frac{gm - t}{m}$$

$$fm^2 = gm - t \qquad \text{Multiplying by } m$$

$$fm^2 - gm + t = 0$$

$$a = f, \ b = -g, \ c = t$$

$$m = \frac{-(-g) \pm \sqrt{(-g)^2 - 4 \cdot f \cdot t}}{2 \cdot f}$$

$$m = \frac{g \pm \sqrt{g^2 - 4ft}}{2f}$$

68. $n_2 = \dfrac{n_1 p_2 R + p_1 p_2 n_1}{p_1 p_2 - p_1 R}$

69.
$$u = -F\left(E - \frac{P}{T}\right)$$

$$u = -EF + \frac{FP}{T}$$

$$T \cdot u = T\left(-EF + \frac{FP}{T}\right)$$

$$Tu = -EFT + FP$$

$$Tu + EFT = FP$$

$$T(u + EF) = FP$$

$$T = \frac{FP}{u + EF}$$

70. $v = \dfrac{Nbf_2 - bf_1 - df_1}{Nf_2 - 1}$

71. When $C = F$, we have

$$C = \frac{5}{9}(C - 32)$$

$$9C = 5(C - 32)$$

$$9C = 5C - 160$$

$$4C = -160$$

$$C = -40$$

At $-40°$ the Fahrenheit and Celsius readings are the same.

Exercise Set 9.5

1. $\sqrt{-1} = i$

2. $6i$

3. $\sqrt{-16} = \sqrt{-1 \cdot 16} = \sqrt{-1} \cdot \sqrt{16} = i \cdot 4 = 4i$

4. $9i$

5. $\sqrt{-50} = \sqrt{-1 \cdot 25 \cdot 2} = \sqrt{-1} \cdot \sqrt{25}\sqrt{2} = i \cdot 5\sqrt{2} = 5i\sqrt{2}$, or $5\sqrt{2}i$

6. $2i\sqrt{11}$, or $2\sqrt{11}i$

7. $-\sqrt{-45} = -\sqrt{-1 \cdot 9 \cdot 5} = -\sqrt{-1} \cdot \sqrt{9} \cdot \sqrt{5} = -i \cdot 3\sqrt{5} = -3i\sqrt{5}$, or $-3\sqrt{5}i$

8. $-2i\sqrt{5}$, or $-2\sqrt{5}i$

9. $-\sqrt{-18} = -\sqrt{-1 \cdot 9 \cdot 2} = -\sqrt{-1} \cdot \sqrt{9} \cdot \sqrt{2} = -i \cdot 3\sqrt{2} = -3i\sqrt{2}$, or $-3\sqrt{2}i$

10. $-2i\sqrt{7}$, or $-2\sqrt{7}i$

11. $3 + \sqrt{-49} = 3 + \sqrt{-1 \cdot 49} = 3 + \sqrt{-1} \cdot \sqrt{49} = 3 + i \cdot 7 = 3 + 7i$

12. $7 + 2i$

13. $5 + \sqrt{-9} = 5 + \sqrt{-1 \cdot 9} = 5 + \sqrt{-1} \cdot \sqrt{9} = 5 + i \cdot 3 = 5 + 3i$

14. $-8 - 6i$

15. $2 - \sqrt{-98} = 2 - \sqrt{-1 \cdot 98} = 2 - \sqrt{-1} \cdot \sqrt{98} = 2 - i \cdot 7\sqrt{2} = 2 - 7i\sqrt{2}$

16. $-2 + 5i\sqrt{5}$

17. $x^2 + 4 = 0$

$x^2 = -4$

$x = \sqrt{-4}$ or $x = -\sqrt{-4}$

$x = \sqrt{-1}\sqrt{4}$ or $x = -\sqrt{-1}\sqrt{4}$

$x = 2i$ or $x = -2i$ Principle of square roots

The solutions are $2i$ and $-2i$, or $\pm 2i$.

18. $\pm 3i$

19. $x^2 = -28$

$x = \sqrt{-28}$ or $x = -\sqrt{-28}$ Principle of square roots

$x = \sqrt{-1 \cdot 4 \cdot 7}$ or $x = -\sqrt{-1 \cdot 4 \cdot 7}$

$x = i \cdot 2\sqrt{7}$ or $x = -i \cdot 2\sqrt{7}$

$x = 2i\sqrt{7}$ or $x = -2i\sqrt{7}$

The solutions are $2i\sqrt{7}$ and $-2i\sqrt{7}$, or $\pm 2i\sqrt{7}$.

20. $\pm 4i\sqrt{3}$

21. $t^2 + 4t + 5 = 0$

$a = 1,\ b = 4,\ c = 5$

$t = \dfrac{-b \pm \sqrt{b^2 - 4ac}}{2a}$

$t = \dfrac{-4 \pm \sqrt{4^2 - 4 \cdot 1 \cdot 5}}{2 \cdot 1}$

$t = \dfrac{-4 \pm \sqrt{-4}}{2} = \dfrac{-4 \pm \sqrt{-1}\sqrt{4}}{2}$

$t = \dfrac{-4 \pm 2i}{2}$

$t = -2 \pm i$ Writing in the form $a + bi$

The solutions are $-2 \pm i$.

22. $2 \pm \sqrt{2}i$

23. $(t + 3)^2 = -16$

$t + 3 = \pm\sqrt{-16}$ Principle of square roots

$t + 3 = \pm\sqrt{-1}\sqrt{16}$

$t + 3 = \pm 4i$

$t = -3 \pm 4i$

The solutions are $-3 \pm 4i$.

24. $2 \pm 5i$

25. $x^2 + 5 = 2x$

$x^2 - 2x + 5 = 0$

$a = 1,\ b = -2,\ c = 5$

$x = \dfrac{-b \pm \sqrt{b^2 - 4ac}}{2a}$

$x = \dfrac{-(-2) \pm \sqrt{(-2)^2 - 4 \cdot 1 \cdot 5}}{2 \cdot 1}$

$x = \dfrac{2 \pm \sqrt{-16}}{2} = \dfrac{2 \pm i\sqrt{16}}{2}$

$x = \dfrac{2 \pm 4i}{2} = \dfrac{2}{2} \pm \dfrac{4i}{2}$

$x = 1 \pm 2i$

The solutions are $1 \pm 2i$.

26. $-1 \pm \sqrt{2}i$

27. $t^2 + 7 - 4t = 0$

$t^2 - 4t + 7 = 0$ Standard form

$a = 1,\ b = -4,\ c = 7$

$t = \dfrac{-b \pm \sqrt{b^2 - 4ac}}{2a}$

$t = \dfrac{-(-4) \pm \sqrt{(-4)^2 - 4 \cdot 1 \cdot 7}}{2 \cdot 1}$

$t = \dfrac{4 \pm \sqrt{-12}}{2} = \dfrac{4 \pm i\sqrt{12}}{2}$

$t = \dfrac{4 \pm 2i\sqrt{3}}{2} = \dfrac{4}{2} \pm \dfrac{2\sqrt{3}}{2}i = 2 \pm \sqrt{3}i$

The solutions are $2 \pm \sqrt{3}i$.

28. $-2 \pm 2i$

29. $2t^2 + 6t + 5 = 0$

$a = 2,\ b = 6,\ t = 5$

$t = \dfrac{-b \pm \sqrt{b^2 - 4ac}}{2a}$

$t = \dfrac{-6 \pm \sqrt{6^2 - 4 \cdot 2 \cdot 5}}{2 \cdot 2}$

$t = \dfrac{-6 \pm \sqrt{-4}}{4} = \dfrac{-6 \pm i\sqrt{4}}{4} = \dfrac{-6 \pm 2i}{4}$

$t = \dfrac{-6}{4} \pm \dfrac{2}{4}i = -\dfrac{3}{2} \pm \dfrac{1}{2}i$

The solutions are $-\dfrac{3}{2} \pm \dfrac{1}{2}i$.

30. $-\dfrac{3}{8} \pm \dfrac{\sqrt{23}}{8}i$

31. $1 + 2m + 3m^2 = 0$

$3m^2 + 2m + 1 = 0$ Standard form

$a = 3,\ b = 2,\ c = 1$

$$m = \frac{-b \pm \sqrt{b^2 - 4ac}}{2a}$$

$$m = \frac{-2 \pm \sqrt{2^2 - 4 \cdot 3 \cdot 1}}{2 \cdot 3}$$

$$m = \frac{-2 \pm \sqrt{-8}}{6} = \frac{-2 \pm i\sqrt{8}}{6} = \frac{-2 \pm 2i\sqrt{2}}{6}$$

$$m = \frac{-2}{6} \pm \frac{2\sqrt{2}}{6}i = -\frac{1}{3} \pm \frac{\sqrt{2}}{3}i$$

The solutions are $-\frac{1}{3} \pm \frac{\sqrt{2}}{3}i$.

32. $\frac{3}{4} \pm \frac{\sqrt{3}}{4}i$

33. ◈

34. ◈

35. Graph $y = \frac{3}{5}x$.

x	y	(x, y)
-5	-3	$(-5, -3)$
0	0	$(0, 0)$
5	3	$(5, 3)$

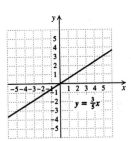

36.

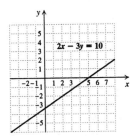

37. Graph $y = -4x$.

x	y	(x, y)
-1	4	$(-1, 4)$
0	0	$(0, 0)$
1	-4	$(1, -4)$

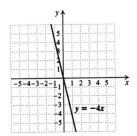

38.

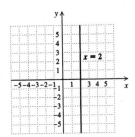

39. $(-17)^2 - (8 + 9)^2$

Observe that $8 + 9 = 17$, so we have $(-17)^2 - 17^2$. Since $(-17)^2 = 17^2$, then the difference is 0.

40. 0

41. ◈

42. ◈

43.
$$(x + 1)^2 + (x + 3)^2 = 0$$
$$x^2 + 2x + 1 + x^2 + 6x + 9 = 0$$
$$2x^2 + 8x + 10 = 0$$
$$x^2 + 4x + 5 = 0 \quad \text{Dividing by 2}$$

$a = 1, b = 4, c = 5$

$$x = \frac{-b \pm \sqrt{b^2 - 4ac}}{2a}$$

$$x = \frac{-4 \pm \sqrt{4^2 - 4 \cdot 1 \cdot 5}}{2 \cdot 1}$$

$$x = \frac{-4 \pm \sqrt{16 - 20}}{2} = \frac{-4 \pm \sqrt{-4}}{2}$$

$$x = \frac{-4 \pm 2i}{2} = \frac{2(-2 \pm i)}{2 \cdot 1}$$

$$x = -2 \pm i$$

The solutions are $-2 \pm i$.

44. $-3 \pm 2i$

45. $\frac{2x - 1}{5} - \frac{2}{x} = \frac{x}{2}$

We multiply by $10x$, the LCD.

$$10x\left(\frac{2x - 1}{5} - \frac{2}{x}\right) = 10x \cdot \frac{x}{2}$$

$$2x(2x - 1) - 10 \cdot 2 = 5x \cdot x$$

$$4x^2 - 2x - 20 = 5x^2$$

$$0 = x^2 + 2x + 20$$

$a = 1, b = 2, c = 20$

$$x = \frac{-b \pm \sqrt{b^2 - 4ac}}{2a}$$

$$x = \frac{-2 \pm \sqrt{2^2 - 4 \cdot 1 \cdot 20}}{2 \cdot 1}$$

$$x = \frac{-2 \pm \sqrt{-76}}{2} = \frac{-2 \pm i\sqrt{76}}{2} = \frac{-2 \pm 2i\sqrt{19}}{2}$$

$$x = \frac{-2}{2} \pm \frac{2\sqrt{19}}{2}i = -1 \pm \sqrt{19}i$$

The solutions are $-1 \pm \sqrt{19}i$.

46. $\dfrac{1}{2} \pm \dfrac{\sqrt{3}}{6}i$

47. Example 2(a):

Graph $y = x^2 + 3x + 4$.

There are no x-intercepts, so the equation $x^2 + 3x + 4 = 0$ has no real-number solutions.

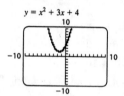

Example 2(b):

Graph $y_1 = x^2 + 2$ and $y_2 = 2x$.

The graphs do not intersect, so the equation $x^2 + 2 = 2x$ has no real-number solutions.

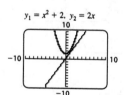

Exercise Set 9.6

1. $y = x^2 + 1$

We first find the vertex. The x-coordinate is

$$-\frac{b}{2a} = -\frac{0}{2 \cdot 1} = 0.$$

We substitute into the equation to find the second coordinate of the vertex.

$$x^2 + 1 = 0^2 + 1 = 1$$

The vertex is $(0, 1)$. The line of symmetry is $x = 0$, the y-axis.

We choose some x-values on both sides of the vertex and graph the parabola.

When $x = 1$, $y = 1^2 + 1 = 1 + 1 = 2$.
When $x = -1$, $y = (-1)^2 + 1 = 1 + 1 = 2$.
When $x = 2$, $y = 2^2 + 1 = 4 + 1 = 5$.
When $x = -2$, $y = (-2)^2 + 1 = 4 + 1 = 5$.

x	y
0	1
1	2
-1	2
2	5
-2	5

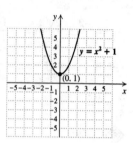

2.

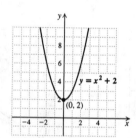

3. $y = -2x^2$

Find the vertex. The x-coordinate is

$$-\frac{b}{2a} = -\frac{0}{2(-2)} = 0.$$

The y-coordinate is

$$-2x^2 = -2 \cdot 0^2 = 0.$$

The vertex is $(0, 0)$. The line of symmetry is $x = 0$, the y-axis.

Choose some x-values on both sides of the vertex and graph the parabola.

When $x = -2$, $y = -2(-2)^2 = -2 \cdot 4 = -8$.
When $x = -1$, $y = -2(-1)^2 = -2 \cdot 1 = -2$.
When $x = 1$, $y = -2 \cdot 1^2 = -2 \cdot 1 = -2$.
When $x = 2$, $y = -2 \cdot 2^2 = -2 \cdot 4 = -8$.

x	y
0	0
-2	-8
-1	-2
1	-2
2	-8

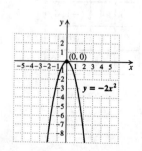

4.

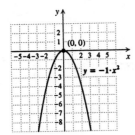

x	y
-4	5
-3	0
-1	-4
0	-3
2	5

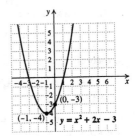

5. $y = -x^2 + 2x$

Find the vertex. The x-coordinate is

$$-\frac{b}{2a} = -\frac{2}{2(-1)} = -\frac{2}{-2} = 1.$$

The y-coordinate is

$$-x^2 + 2x = -(1)^2 + 2 \cdot 1 = -1 + 2 = 1.$$

The vertex is $(1, 1)$.

We choose some x-values on both sides of the vertex and graph the parabola. We make sure we find y when $x = 0$. This gives us the y-intercept.

x	y
1	1
0	0
-1	-3
2	0
3	-3

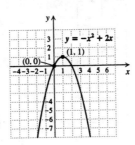

6.

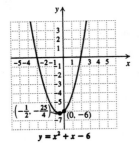

7. $y = x^2 + 2x - 3$

Find the vertex. The x-coordinate is

$$-\frac{b}{2a} = -\frac{2}{2 \cdot 1} = -1.$$

The y-coordinate is

$$x^2 + 2x - 3 = (-1)^2 + 2(-1) - 3 = 1 - 2 - 3 = -4.$$

The vertex is $(-1, -4)$.

We choose some x-values on both sides of the vertex and graph the parabola. We make sure we find y when $x = 0$. This gives us the y-intercept.

8.

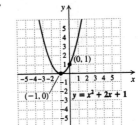

9. $y = 3x^2 - 12x + 11$

Find the vertex. The x-coordinate is

$$-\frac{b}{2a} = -\frac{-12}{2 \cdot 3} = 2.$$

The y-coordinate is

$$3 \cdot 2^2 - 12 \cdot 2 + 11 = 12 - 24 + 11 = -1.$$

The vertex is $(2, -1)$.

We choose some x-values on both sides of the vertex and graph the parabola.

x	y
0	11
1	2
3	2
4	11

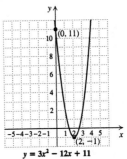

10.

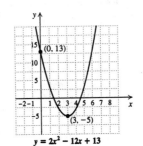

11. $y = -2x^2 - 4x + 1$

Find the vertex. The x-coordinate is
$$-\frac{b}{2a} = -\frac{-4}{2(-2)} = -1.$$
The y-coordinate is
$$-2(-1)^2 - 4(-1) + 1 = -2 + 4 + 1 = 3.$$
The vertex is $(-1, 3)$.

We choose some x-values on both sides of the vertex and graph the parabola.

x	y
-3	-5
-2	1
0	1
1	-5

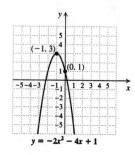

12.

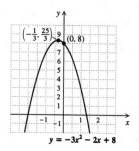

13. $y = \frac{1}{4}x^2$

Find the vertex. The x-coordinate is
$$-\frac{b}{2a} = -\frac{0}{2 \cdot \frac{1}{4}} = 0.$$
The y-coordinate is
$$\frac{1}{4} \cdot 0^2 = 0.$$
The vertex is $(0, 0)$.

We choose some x-values on both sides of the vertex and graph the parabola.

x	y
-4	4
-2	1
2	1
4	4

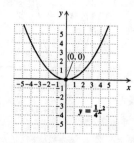

14.

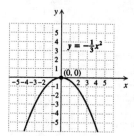

15. $y = -\frac{1}{2}x^2 + 5$

Find the vertex. The x-coordinate is
$$-\frac{b}{2a} = -\frac{0}{2\left(-\frac{1}{2}\right)} = 0.$$
The y-coordinate is
$$-\frac{1}{2} \cdot 0^2 + 5 = 0 + 5 = 5.$$
The vertex is $(0, 5)$.

We choose some x-values on both sides of the vertex and graph the parabola.

x	y
-4	-3
-2	3
2	3
4	-3

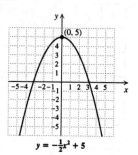

16.

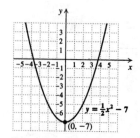

17. $y = x^2 - 3x$

Find the vertex. The x-coordinate is
$$-\frac{b}{2a} = -\frac{-3}{2 \cdot 1} = \frac{3}{2}.$$
The y-coordinate is
$$\left(\frac{3}{2}\right)^2 - 3 \cdot \frac{3}{2} = \frac{9}{4} - \frac{9}{2} = -\frac{9}{4}.$$
The vertex is $\left(\frac{3}{2}, -\frac{9}{4}\right)$.

We choose some x-values on both sides of the vertex and graph the parabola.

x	y
0	0
1	-2
2	-2
3	0

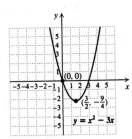

20.

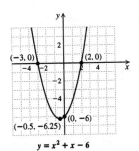

$y = x^2 + x - 6$

21. $y = 2x^2 - 5x$

Find the vertex. The x-coordinate is
$$-\frac{b}{2a} = -\frac{-5}{2 \cdot 2} = \frac{5}{4}.$$

The y-coordinate is
$$2\left(\frac{5}{4}\right)^2 - 5 \cdot \frac{5}{4} = \frac{25}{8} - \frac{25}{4} = -\frac{25}{8}.$$

The vertex is $\left(\frac{5}{4}, -\frac{25}{8}\right)$.

To find the y-intercept we replace x with 0 and compute y:
$$y = 2 \cdot 0^2 - 5 \cdot 0 = 0 - 0 = 0.$$

The y-intercept is $(0,0)$.

To find the x-intercepts we replace y with 0 and solve for x.
$$0 = 2x^2 - 5x$$
$$0 = x(2x - 5)$$
$$x = 0 \quad or \quad 2x - 5 = 0$$
$$x = 0 \quad or \quad 2x = 5$$
$$x = 0 \quad or \quad x = \frac{5}{2}$$

The x-intercepts are $(0,0)$ and $\left(\frac{5}{2}, 0\right)$.

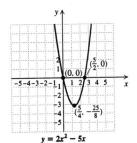

$y = 2x^2 - 5x$

18.

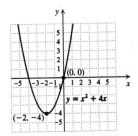

$y = x^2 + 4x$

19. $y = x^2 + 2x - 8$

Find the vertex. The x-coordinate is
$$-\frac{b}{2a} = -\frac{2}{2 \cdot 1} = -1.$$

The y-coordinate is
$$(-1)^2 + 2(-1) - 8 = 1 - 2 - 8 = -9.$$

To find the y-intercept we replace x with 0 and compute y:
$$y = 0^2 + 2 \cdot 0 - 8 = 0 + 0 - 8 = -8.$$

The y-intercept is $(0, -8)$.

To find the x-intercepts we replace y with 0 and solve for x.
$$0 = x^2 + 2x - 8$$
$$0 = (x + 4)(x - 2)$$
$$x + 4 = 0 \quad or \quad x - 2 = 0$$
$$x = -4 \quad or \quad x = 2$$

The x-intercepts are $(-4, 0)$ and $(2, 0)$.

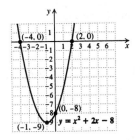

$y = x^2 + 2x - 8$

22.

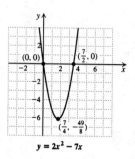

$$y = 2x^2 - 7x$$

23. $y = -x^2 - x + 12$

Find the vertex. The x-coordinate is
$$-\frac{b}{2a} = -\frac{-1}{2(-1)} = -\frac{1}{2}.$$
The y-coordinate is
$$-\left(-\frac{1}{2}\right)^2 - \left(-\frac{1}{2}\right) + 12 = -\frac{1}{4} + \frac{1}{2} + 12 = \frac{49}{4}.$$
The vertex is $\left(-\frac{1}{2}, \frac{49}{4}\right)$.

To find the y-intercept we replace x with 0 and compute y:
$$y = -0^2 - 0 + 12 = -0 - 0 + 12 = 12.$$
The y-intercept is $(0, 12)$.

To find the x-intercepts we replace y with 0 and solve for x.
$$0 = -x^2 - x + 12$$
$$0 = x^2 + x - 12 \qquad \text{Multiplying by } -1$$
$$0 = (x + 4)(x - 3)$$
$$x + 4 = 0 \quad or \quad x - 3 = 0$$
$$x = -4 \quad or \qquad x = 3$$
The x-intercepts are $(-4, 0)$ and $(3, 0)$.

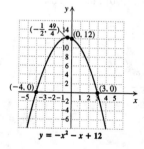

$$y = -x^2 - x + 12$$

24.

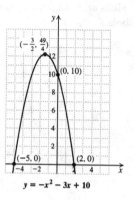

$$y = -x^2 - 3x + 10$$

25. $y = 3x^2 - 6x + 1$

Find the vertex. The x-coordinate is
$$-\frac{b}{2a} = -\frac{-6}{2 \cdot 3} = 1.$$
The y-coordinate is
$$3 \cdot 1^2 - 6 \cdot 1 + 1 = 3 - 6 + 1 = -2.$$
The vertex is $(1, -2)$.

To find the y-intercept we replace x with 0 and compute y:
$$y = 3 \cdot 0^2 - 6 \cdot 0 + 1 = 0 - 0 + 1 = 1.$$
The y-intercept is $(0, 1)$.

To find the x-intercepts we replace y with 0 and solve for x.
$$0 = 3x^2 - 6x + 1$$
$$x = \frac{-b \pm \sqrt{b^2 - 4ac}}{2a}$$
$$x = \frac{-(-6) \pm \sqrt{(-6)^2 - 4 \cdot 3 \cdot 1}}{2 \cdot 3}$$
$$x = \frac{6 \pm \sqrt{36 - 12}}{6} = \frac{6 \pm \sqrt{24}}{6}$$
$$x = \frac{6 \pm 2\sqrt{6}}{6} = \frac{2(3 \pm \sqrt{6})}{2 \cdot 3}$$
$$x = \frac{3 \pm \sqrt{6}}{3}$$

The x-intercepts are $\left(\frac{3 - \sqrt{6}}{3}, 0\right)$ and $\left(\frac{3 + \sqrt{6}}{3}, 0\right)$, or about $(0.184, 0)$ and $(1.816, 0)$.

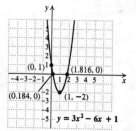

$$y = 3x^2 - 6x + 1$$

26.

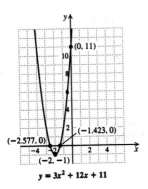

$y = 3x^2 + 12x + 11$

27. $y = x^2 + 2x + 3$

Find the vertex. The x-coordinate is

$$-\frac{b}{2a} = -\frac{2}{2 \cdot 1} = -1.$$

The y-coordinate is

$$y = (-1)^2 + 2(-1) + 3 = 1 - 2 + 3 = 2.$$

The vertex is $(-1, 2)$.

To find the y-intercept we replace x with 0 and compute y:

$$y = 0^2 + 2 \cdot 0 + 3 = 0 + 0 + 3 = 3.$$

The y-intercept is $(0, 3)$.

To find the x-intercepts we replace y with 0 and solve for x.

$$0 = x^2 + 2x + 3$$

$$x = \frac{-b \pm \sqrt{b^2 - 4ac}}{2a}$$

$$x = \frac{-2 \pm \sqrt{2^2 - 4 \cdot 1 \cdot 3}}{2 \cdot 1}$$

$$x = \frac{-2 \pm \sqrt{4 - 12}}{2} = \frac{-2 \pm \sqrt{-8}}{2}$$

Because the radicand, -8, is negative the equation has no real-number solutions. Thus, there are no x-intercepts.

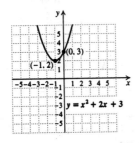

$y = x^2 + 2x + 3$

28.

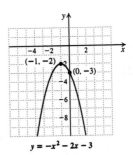

$y = -x^2 - 2x - 3$

29. $y = 3 - 4x - 2x^2$, or $y = -2x^2 - 4x + 3$

Find the vertex. The x-coordinate is

$$-\frac{b}{2a} = -\frac{-4}{2(-2)} = -1.$$

The y-coordinate is

$$y = 3 - 4(-1) - 2(-1)^2 = 3 + 4 - 2 = 5.$$

The vertex is $(-1, 5)$.

To find the y-intercept we replace x with 0 and compute y:

$$y = 3 - 4 \cdot 0 - 2 \cdot 0^2 = 3 - 0 - 0 = 3.$$

The y-intercept is $(0, 3)$.

To find the x-intercepts we replace y with 0 and solve for x.

$$0 = -2x^2 - 4x + 3$$

$$x = \frac{-b \pm \sqrt{b^2 - 4ac}}{2a}$$

$$x = \frac{-(-4) \pm \sqrt{(-4)^2 - 4(-2)(3)}}{2(-2)}$$

$$x = \frac{4 \pm \sqrt{16 + 24}}{-4} = \frac{4 \pm \sqrt{40}}{-4}$$

$$x = \frac{4 \pm 2\sqrt{10}}{-4} = -\frac{2 \pm \sqrt{10}}{2}$$

The x-intercepts are $\left(\dfrac{2 - \sqrt{10}}{2}, 0 \right)$ and

$\left(\dfrac{2 + \sqrt{10}}{2}, 0 \right)$, or about $(0.581, 0)$ and $(-2.581, 0)$.

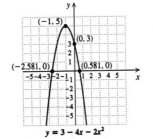

$y = 3 - 4x - 2x^2$

30.

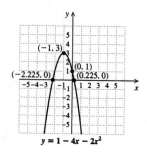

$y = 1 - 4x - 2x^2$

b)

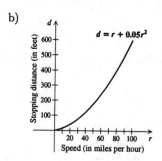

31.

32.

44.

45.

33. $3a^2 - 5a = 3(-1)^2 - 5(-1)$
$= 3 \cdot 1 - 5(-1)$
$= 3 + 5$
$= 8$

34. 22

35. $5x^3 - 2x = 5(-1)^3 - 2(-1)$
$= 5(-1) - 2(-1)$
$= -5 + 2$
$= -3$

36. 68

37. $-9.8x^5 + 3.2x = -9.8(1)^5 + 3.2(1)$
$= -9.8(1) + 3.2(1)$
$= -9.8 + 3.2$
$= -6.6$

38. 0

39.

40.

41. See the answer section in the text.

42. a) 2 sec after launch and 4 sec after launch

    b) 3 sec after launch

    c) 6 sec after launch.

43. a) For $r = 25$, $d = 25 + 0.05(25)^2 = 56.25$ ft

    For $r = 40$, $d = 40 + 0.05(40)^2 = 120$ ft

    For $r = 55$, $d = 55 + 0.05(55)^2 = 206.25$ ft

    For $r = 65$, $d = 65 + 0.05(65)^2 = 276.25$ ft

    For $r = 75$, $d = 75 + 0.05(75)^2 = 356.25$ ft

    For $r = 100$, $d = 100 + 0.05(100)^2 = 600$ ft

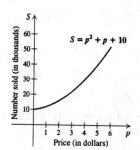

47. $D = (p - 6)^2$

p	D
0	36
1	25
2	16
3	9
4	4
5	1
6	0

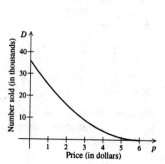

48. \$2; 16,000 units

49. Graph $y = x^2 - 5$ and find the first coordinate of the right-hand x-intercept. We find that $\sqrt{5} \approx 2.2361$.

Exercise Set 9.7

1. Yes; each member of the domain is matched to only one member of the range.

2. Yes

3. Yes; each member of the domain is matched to only one member of the range.

4. No

5. No; a member of the domain is matched to more than one member of the range. In fact, each member of the domain is matched to 3 members of the range.

6. Yes

7. Yes; each member of the domain is matched to only one member of the range.

8. Yes

9. $f(x) = x + 5$
$f(4) = 4 + 5 = 9$
$f(7) = 7 + 5 = 12$
$f(-2) = -2 + 5 = 3$

10. $-5;\ 0;\ 7$

11. $h(p) = 3p$
$h(-7) = 3(-7) = -21$
$h(5) = 3 \cdot 5 = 15$
$h(10) = 3 \cdot 10 = 30$

12. $-24;\ 2;\ -80$

13. $g(s) = 3s + 4$
$g(1) = 3 \cdot 1 + 4 = 3 + 4 = 7$
$g(-5) = 3(-5) + 4 = -15 + 4 = -11$
$g(6.7) = 3(6.7) + 4 = 20.1 + 4 = 24.1$

14. $19;\ 19;\ 19$

15. $F(x) = 2x^2 - 3x$
$F(0) = 2 \cdot 0^2 - 3 \cdot 0 = 0 - 0 = 0$
$F(-1) = 2(-1)^2 - 3(-1) = 2 + 3 = 5$
$F(2) = 2 \cdot 2^2 - 3 \cdot 2 = 8 - 6 = 2$

16. $0;\ 16;\ 21$

17. $f(t) = (t + 1)^2$
$f(-5) = (-5 + 1)^2 = (-4)^2 = 16$
$f(0) = (0 + 1)^2 = 1^2 = 1$
$f\left(-\dfrac{9}{4}\right) = \left(-\dfrac{9}{4} + 1\right)^2 = \left(-\dfrac{5}{4}\right)^2 = \dfrac{25}{16}$

18. $16;\ 8464;\ 10{,}201$

19. $g(t) = t^3 + 3$
$g(1) = 1^3 + 3 = 1 + 3 = 4$
$g(-5) = (-5)^3 + 3 = -125 + 3 = -122$
$g(0) = 0^3 + 3 = 0 + 3 = 3$

20. $-3;\ -2;\ 78$

21. $F(x) = 2.75x + 71.48$
a) $F(32) = 2.75(32) + 71.48$
$= 88 + 71.48$
$= 159.48$ cm
b) $F(30) = 2.75(30) + 71.48$
$= 82.5 + 71.48$
$= 153.98$ cm

22. a) 157.34 cm
b) 171.79 cm

23. $T(5) = 10 \cdot 5 + 20 = 50 + 20 = 70°$ C
$T(20) = 10 \cdot 20 + 20 = 200 + 20 = 220°$ C
$T(1000) = 10 \cdot 1000 + 20 = 10{,}000 + 20 = 10{,}020°$ C

24. $1.\overline{60}$ atm; $1.\overline{90}$ atm; $4.\overline{03}$ atm

25. $W(d) = 0.112d$
$W(16) = 0.112(16) = 1.792$ cm
$W(25) = 0.112(25) = 2.8$ cm
$W(100) = 0.112(100) = 11.2$ cm

26. $16.\overline{6}°;\ 25°;\ -5°$

27. Graph $f(x) = 3x - 2$
Make a list of function values in a table.
When $x = -1$, $f(-1) = 3(-1) - 2 = -3 - 2 = -5$.
When $x = 0$, $f(0) = 3 \cdot 0 - 2 = 0 - 2 = -2$.
When $x = 2$, $f(2) = 3 \cdot 2 - 2 = 6 - 2 = 4$.

x	$f(x)$
-1	-5
0	-2
2	4

Plot these points and connect them.

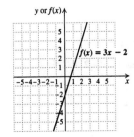

28.

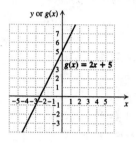

29. Graph $g(x) = -2x + 1$

Make a list of function values in a table.

When $x = -1$, $g(-1) = -2(-1) + 1 = 2 + 1 = 3$.

When $x = 0$, $g(0) = -2 \cdot 0 + 1 = 0 + 1 = 1$.

When $x = 3$, $g(3) = -2 \cdot 3 + 1 = -6 + 1 = -5$.

x	$g(x)$
-1	3
0	1
3	-5

Plot these points and connect them.

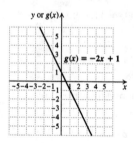

30.

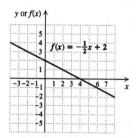

31. Graph $f(x) = \dfrac{1}{2}x + 1$.

Make a list of function values in a table.

When $x = -2$, $f(-2) = \dfrac{1}{2}(-2) + 1 = -1 + 1 = 0$.

When $x = 0$, $f(0) = \dfrac{1}{2} \cdot 0 + 1 = 0 + 1 = 1$.

When $x = 4$, $f(4) = \dfrac{1}{2} \cdot 4 + 1 = 2 + 1 = 3$.

x	$f(x)$
-2	0
0	1
4	3

Plot these points and connect them.

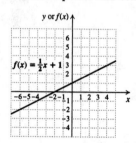

32.

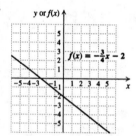

33. Graph $g(x) = 2|x|$.

Make a list of function values in a table.

When $x = -3$, $g(-3) = 2|-3| = 2 \cdot 3 = 6$.

When $x = -1$, $g(-1) = 2|-1| = 2 \cdot 1 = 2$.

When $x = 0$, $g(0) = 2|0| = 2 \cdot 0 = 0$.

When $x = 1$, $g(1) = 2|1| = 2 \cdot 1 = 2$.

When $x = 3$, $g(3) = 2|3| = 2 \cdot 3 = 6$.

x	$g(x)$
-3	6
-1	2
0	0
1	2
3	6

Plot these points and connect them.

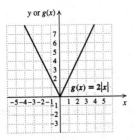

34.

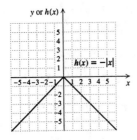

35. Graph $g(x) = x^2$.

Recall from Section 10.5 that the graph is a parabola. Make a list of function values in a table.

When $x = -2$, $g(-2) = (-2)^2 = 4$.

When $x = -1$, $g(-1) = (-1)^2 = 1$.

When $x = 0$, $g(0) = 0^2 = 0$.

When $x = 1$, $g(1) = 1^2 = 1$.

When $x = 2$, $g(2) = 2^2 = 4$.

x	$g(x)$
-2	4
-1	1
0	0
1	1
2	4

Plot these points and connect them.

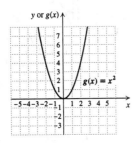

36.

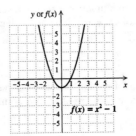

37. Graph $f(x) = x^2 - x - 2$.

Recall from Section 10.5 that the graph is a parabola. Make a list of function values in a table.

When $x = -1$, $f(-1) = (-1)^2 - (-1) - 2 = 1 + 1 - 2 = 0$.

When $x = 0$, $f(0) = 0^2 - 0 - 2 = -2$.

When $x = 1$, $f(1) = 1^2 - 1 - 2 = 1 - 1 - 2 = -2$.

When $x = 2$, $f(2) = 2^2 - 2 - 2 = 4 - 2 - 2 = 0$.

x	$f(x)$
-1	0
0	-2
1	-2
2	0

Plot these points and connect them.

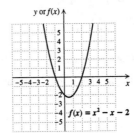

38.

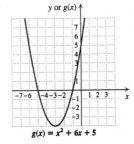

39. The graph is that of a function because no vertical line can cross the graph at more than one point.

40. No

41. The graph is not that of a function because a vertical line, say $x = -2$, crosses the graph at more than one point.

42. No

43. The graph is that of a function because no vertical line can cross the graph at more than one point.

44. Yes

45.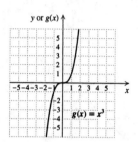

46.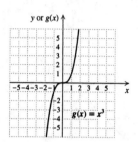

47. The first equation is in slope-intercept form:
$$y = \frac{3}{4}x - 7, \; m = \frac{3}{4}$$
We write the second equation in slope-intercept form.
$$3x + 4y = 7$$
$$4y = -3x + 7$$
$$y = -\frac{3}{4}x + \frac{7}{4}, \; m = -\frac{3}{4}$$
Since the slopes are different, the equations do not represent parallel lines.

48. Yes

49. We write the equations in slope-intercept form.
$$2x = 3y \qquad\qquad 4x = 6y - 1$$
$$\frac{2}{3}x = y \qquad\qquad 4x + 1 = 6y$$
$$\qquad\qquad \frac{2}{3}x + \frac{1}{6} = y$$
Since the slopes are the same $\left(m = \frac{2}{3}\right)$ and the y-intercepts are different, the equations represent parallel lines.

50. No solution

51. $x - 3y = 2$,   (1)
$\quad 3x - 9y = 6$ (2)
Solve Equation (1) for x.
$$x - 3y = 2 \qquad (1)$$
$$x = 3y + 2 \;\text{ Adding } 3y$$
Substitute $3y + 2$ for x in Equation (2) and solve for y.
$$3x - 9y = 6 \quad (2)$$
$$3(3y + 2) - 9y = 6$$
$$9y + 6 - 9y = 6$$
$$6 = 6$$
We get an equation that is true for all values of y, so there are an infinite number of solutions.

52. $\left(\dfrac{9}{11}, -\dfrac{5}{11}\right)$

53.

54.

55. Graph $g(x) = x^3$.

Make a list of function values in a table. Then plot the points and connect them.

x	$g(x)$
-2	-8
-1	-1
0	0
1	1
2	8

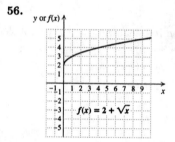

56.

57. Graph $g(x) = |x| + x$.

Make a list of function values in a table. Then plot the points and connect them.

x	$f(x)$
-3	0
-2	0
-1	0
0	0
1	2
2	4
3	6

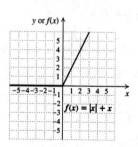

58.

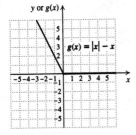

59. Answers may vary.

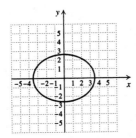

60. $f(x) = \dfrac{15}{4}x - \dfrac{13}{4}$

61. $g(x) = ax^2 + bx + c$

$$-4 = a \cdot 0^2 + b \cdot 0 + c, \text{ or } -4 = c \qquad (1)$$
$$0 = a(-2)^2 + b(-2) + c, \text{ or } 0 = 4a - 2b + c \quad (2)$$
$$0 = a \cdot 2^2 + b \cdot 2 + c, \text{ or } 0 = 4a + 2b + c \qquad (3)$$

Substitute -4 for c in Equations (2) and (3).

$$0 = 4a - 2b - 4, \ \ or \ \ 4 = 4a - 2b \quad (5)$$
$$0 = 4a + 2b - 4, \ \ or \ \ 4 = 4a + 2b \quad (6)$$

Add Equations (5) and (6).

$$8 = 8a$$
$$1 = a$$

Substitute 1 for a in Equation (6).

$$4 = 4 \cdot 1 + 2b$$
$$0 = 2b$$
$$0 = b$$

We have $a = 1$, $b = 0$, $c = -4$, so $g(x) = x^2 - 4$.

62. $\{5, 8, 11, 14\}$

63. $g(t) = t^2 - 5$

The domain is the set $\{-3, -2, -1, 0, 1\}$.

$$g(-3) = (-3)^2 - 5 = 9 - 5 = 4$$
$$g(-2) = (-2)^2 - 5 = 4 - 5 = -1$$
$$g(-1) = (-1)^2 - 5 = 1 - 5 = -4$$
$$g(0) = 0^2 - 5 = 0 - 5 = -5$$
$$g(1) = 1^2 - 5 = 1 - 5 = -4$$

The range is the set $\{-5, -4, -1, 4\}$.

64. $\{0, 2\}$

65. $f(m) = m^3 + 1$

The domain is the set $\{-2, -1, 0, 1, 2\}$.

$$f(-2) = (-2)^3 + 1 = -8 + 1 = -7$$
$$f(-1) = (-1)^3 + 1 = -1 + 1 = 0$$
$$f(0) = 0^3 + 1 = 0 + 1 = 1$$
$$f(1) = 1^3 + 1 = 1 + 1 = 2$$
$$f(2) = 2^3 + 1 = 8 + 1 = 9$$

The range is the set $\{-7, 0, 1, 2, 9\}$.

66. ◿◣

Answers for Exercises in the Appendixes

Exercise Set A

1. $\{3, 4, 5, 6, 7\}$ **2.** $\{82, 83, 84, 85, 86, 87, 88, 89\}$

3. $\{41, 43, 45, 47, 49\}$ **4.** $\{15, 20, 25, 30, 35\}$

5. $\{-3, 3\}$ **6.** $\{0.008\}$ **7.** False **8.** True **9.** True

10. True **11.** True **12.** True **13.** True **14.** True

15. True **16.** False **17.** False **18.** True

19. $\{c, d, e\}$ **20.** $\{u, i\}$ **21.** $\{1, 10\}$ **22.** $\{0, 1\}$

23. $\emptyset$ **24.** $\emptyset$ **25.** $\{a, e, i, o, u, q, c, k\}$

26. $\{a, b, c, d, e, f, g\}$ **27.** $\{0, 1, 2, 5, 7, 10\}$

28. $\{1, 2, 5, 10, 0, 7\}$

29. $\{a, e, i, o, u, m, n, f, g, h\}$

30. $\{1, 2, 5, 10, a, b\}$ **31.** Set-builder notation allows us to name a very large set compactly. **32.** Roster notation allows us to see all the members of a set. It is also useful when a set cannot be named using a general rule or statement. **33.** The set of integers **34.** $\emptyset$ **35.** The set of real numbers **36.** The set of positive even integers **37.** $\emptyset$ **38.** The set of integers **39.** (a) A; (b) A; (c) A; (d) $\emptyset$ **40.** (a) Yes; (b) no; (c) no; (d) yes; (e) yes; (f) no **41.** True

18. $(a + b)(a^2 - ab + b^2)$ **19.** $\left(x + \dfrac{1}{2}\right)\left(x^2 - \dfrac{1}{2}x + \dfrac{1}{4}\right)$

20. $\left(y + \dfrac{1}{3}\right)\left(y^2 - \dfrac{1}{3}y + \dfrac{1}{9}\right)$ **21.** $2(y - 4)(y^2 + 4y + 16)$

22. $3(z - 1)(z^2 + z + 1)$ **23.** $3(2a + 1)(4a^2 - 2a + 1)$

24. $2(3x + 1)(9x^2 - 3x + 1)$ **25.** $r(s - 4)(s^2 + 4s + 16)$

26. $a(b + 5)(b^2 - 5b + 25)$

27. $5(x + 2z)(x^2 - 2xz + 4z^2)$

28. $2(y - 3z)(y^2 + 3yz + 9z^2)$

29. $(x + 0.1)(x^2 - 0.1x + 0.01)$

30. $(y + 0.5)(y^2 - 0.5y + 0.25)$ **31.** Observe that the product $(a - b)(a^2 + b^2) \neq a^3 - b^3$. **32.** The number c is not a perfect cube. Otherwise, $x^3 + c$ could be factored as $(x + \sqrt[3]{c})(x^2 - \sqrt[3]{c}x + (\sqrt[3]{c})^2)$.

33. $(5c^2 + 2d^2)(25c^4 - 10c^2d^2 + 4d^4)$

34. $8(2x^2 + t^2)(4x^4 - 2x^2t^2 + t^4)$

35. $3(x^a - 2y^b)(x^{2a} + 2x^a y^b + 4y^{2b})$

36. $\left(\dfrac{2}{3}x - \dfrac{1}{4}y\right)\left(\dfrac{4}{9}x^2 + + \dfrac{1}{6}xy + \dfrac{1}{16}y^2\right)$

37. $\dfrac{1}{3}\left(\dfrac{1}{2}xy + z\right)\left(\dfrac{1}{4}x^2y^2 - \dfrac{1}{2}xyz + z^2\right)$

38. $\dfrac{1}{2}\left(\dfrac{1}{2}x^a + y^{2a}z^{3b}\right)\left(\dfrac{1}{4}x^{2a} - \dfrac{1}{2}x^a y^{2a}z^{3b} + y^{4a}z^{6b}\right)$

Exercise Set B

1. $(t + 2)(t^2 - 2t + 4)$ **2.** $(p + 3)(p^2 - 3p + 9)$

3. $(a - 4)(a^2 + 4a + 16)$ **4.** $(w - 1)(w^2 + w + 1)$

5. $(z + 5)(z^2 - 5z + 25)$ **6.** $(x + 1)(x^2 - x + 1)$

7. $(2a - 1)(4a^2 + 2a + 1)$ **8.** $(3x - 1)(9x^2 + 3x + 1)$

9. $(y - 3)(y^2 + 3y + 9)$ **10.** $(p - 2)(p^2 + 2p + 4)$

11. $(4 + 5x)(16 - 20x + 25x^2)$

12. $(2 + 3b)(4 - 6b + 9b^2)$ **13.** $(5p - 1)(25p^2 + 5p + 1)$

14. $(4w - 1)(16w^2 + 4w + 1)$

15. $(3m + 4)(9m^2 - 12m + 16)$

16. $(2t + 3)(4t^2 - 6t + 9)$ **17.** $(p - q)(p^2 + pq + q^2)$